Prediction of Long Term Corrosion Behaviour
in Nuclear Waste Systems

EARLIER VOLUMES IN THIS SERIES

European Federation of Corrosion
Publications

NUMBER 36

Prediction of Long Term Corrosion Behaviour in Nuclear Waste Systems

Proceedings of an International Workshop, Cadarache, France, 2002

Edited by

D. Féron AND D. D. Macdonald

*Published for the European Federation of Corrosion
by Maney Publishing on behalf of The Institute of Materials,
Minerals and Mining*

MANEY

publishing

Book Number B0794
First published in 2003 by Maney Publishing
1 Carlton House Terrace, London SW1Y 5DB

on behalf of
The Institute of Materials, Minerals and Mining

ISBN 1-902653-87-4

Typeset in the UK by
Spires Design

Printed and bound in the UK by
The Charlesworth Group, Huddersfield

Contents

Part 4 — R & D Approaches and Results   347

European Federation of Corrosion Publications
Series Introduction

The EFC, incorporated in Belgium, was founded in 1955 with the purpose of promoting European co-operation in the fields of research into corrosion and corrosion prevention.

Membership is based upon participation by corrosion societies and committees in technical Working Parties. Member societies appoint delegates to Working Parties, whose membership is expanded by personal corresponding membership.

The activities of the Working Parties cover corrosion topics associated with inhibition, education, reinforcement in concrete, microbial effects, hot gases and combustion products, environment sensitive fracture, marine environments, surface science, physico–chemical methods of measurement, the nuclear industry, computer based information systems, the oil and gas industry, the petrochemical industry, coatings, automotive engineering and cathodic protection. Working Parties on other topics are established as required.

The Working Parties function in various ways, e.g. by preparing reports, organising symposia, conducting intensive courses and producing instructional material, including films. The activities of the Working Parties are co-ordinated, through a Science and Technology Advisory Committee, by the Scientific Secretary.

The administration of the EFC is handled by three Secretariats: DECHEMA e. V. in Germany, the Société de Chimie Industrielle in France, and The Institute of Materials, Minerals and Mining in the United Kingdom. These three Secretariats meet at the Board of Administrators of the EFC. There is an annual General Assembly at which delegates from all member societies meet to determine and approve EFC policy. News of EFC activities, forthcoming conferences, courses etc. is published in a range of accredited corrosion and certain other journals throughout Europe. More detailed descriptions of activities are given in a Newsletter prepared by the Scientific Secretary.

The output of the EFC takes various forms. Papers on particular topics, for example, reviews or results of experimental work, may be published in scientific and technical journals in one or more countries in Europe. Conference proceedings are often published by the organisation responsible for the conference.

In 1987 the, then, Institute of Metals was appointed as the official EFC publisher. Although the arrangement is non-exclusive and other routes for publication are still available, it is expected that the Working Parties of the EFC will use The Institute of Materials for publication of reports, proceedings etc. wherever possible.

The name of The Institute of Metals was changed to The Institute of Materials in 1992 and to The Institute of Materials, Minerals and Mining (IOM[3]) in 2002 following its merger with the Institution of Mining and Metallurgy.

The EFC Series is now published by Maney Publishing on behalf of The Institute of Materials, Minerals and Mining.

A. D. Mercer
EFC Series Editor, IOM[3], London, UK

EFC Secretariats are located at:

Dr B A Rickinson
European Federation of Corrosion, The Institute of Materials, Minerals and Mining, 1 Carlton House Terrace, London, SW1Y 5DB, UK

Mr P Berge
Fédération Européene de la Corrosion, Société de Chimie Industrielle, 28 rue Saint-Dominique, F-75007 Paris, FRANCE

Professor Dr G Kreysa
Europäische Föderation Korrosion, DECHEMA e. V., Theodor-Heuss-Allee 25, D-60486, Frankfurt, GERMANY

Preface

The planned duration of isolation of high level nuclear waste from the ecosphere is of the order of 10 000 years, which is further into the future than recorded human history is in the past. The reliable prediction of container degradation rate and engineered barrier integrity over this extended period represents one of the greatest scientific and technical challenges that the world has had to face. Clearly, the reliability and viability of integrity predictions, which cannot be easily tested, are of paramount importance in assuring the public that safe disposal can be achieved.

The international workshop entitled "Prediction of Long Term Corrosion Behaviour in Nuclear Waste Systems" (EFC event 256) was held at Cadarache (France), on 26–29 November 2001. Its purpose was to bring together scientists and engineers from various countries that are developing High Level Nuclear Waste (HLNW) disposal technologies, with the goal of promoting scientific and technical exchange concerning the long term behaviour of metallic materials. In particular, the workshop sought to compare the scientific and experimental approaches that are being developed in various organisations worldwide for predicting long term corrosion phenomena, as related to nuclear waste management, including corrosion strategies for interim storage and geological disposal. The systems what are of primary interest include waste canisters, near-field engineered barriers for geological disposal (including the nature, evolution, and influence of corrosion products), and storage infrastructures. The workshop addressed these issues in four sections, comprising:

- R & D corrosion programmes;

- Laboratory and *in situ* testing;

- Historical and archaeological analogues; and

- Fundamental issues: models and prediction

— all of which are reported in these proceedings.

The editors would like to thank the authors, who presented papers of outstanding scientific content and who responded enthusiastically to the discussions and questions raised during the workshop, and the reviewer committee, which was composed of:

T. Ahn (USA)	M. Urquidi-Macdonald (USA)
D.S. Dunn (USA)	T. Saario (Finland)
D.Feron (France)	G. Santarini (France)
J.M. Gras (France)	P. Van Isseghem (Belgium)
D.D. Macdonald (USA)	L. Werne (Sweden)
	C. Wood (UK)

The editors hope that this book will be useful to scientists and engineers in the development of appropriate technologies for HLNW isolation and will also be valuable to operating nuclear waste authorities and regulators, who have a vital interest in monitoring the progress being made toward possible solutions to the HLNW disposal issue.

DAMIEN FÉRON AND DIGBY D. MACDONALD
Editors of this Volume

Foreword

This 36th book of the EFC series contains the proceedings of the International Workshop on "Prediction of Long Term Corrosion Behaviour in Nuclear Waste Systems" that took place in Cadarache, France in November 2001.

The EFC Working Party on Nuclear Corrosion, which sponsored this Workshop, had already considered in 1991 that the long term integrity of Radioactive Waste storage was a crucial problem in the development of nuclear energy for electricity production. At that time, it was realised that public opinion on Nuclear Energy, was concerned not only with the risk of a nuclear accident but also with the safety of long term storage of high level nuclear waste. The nuclear community and the public demanded that any technical barrier for preventing long term radionuclides from entering the biosphere should be effective and guaranteed for 10 000 years or more.

The integrity of the waste containers, particularly their resistance to possible internal and external corrosion, is the most important aspect of this requirement. The European R&D projects devoted to this corrosion problem had been relatively modest, for example, as reported in No.7 in the EFC Series in 1992. These projects were conducted mainly in Switzerland, Germany and Sweden. The corrosion tests on the candidate materials, steels, copper, special stainless steels and nickel or titanium alloys, were run for a maximum of a few thousand hours with no real effort to develop models to foresee the conditions up to 10 000 years exposure, or to extrapolate the corrosion behaviour for such a long period. (The definition of an infinite time, for corrosion tests, has been given by an American Professor as being the time necessary to achieve a PhD, in a laboratory!)

For the last ten years, the importance of programmes on long term corrosion of nuclear waste canisters has become of major significance to the international community.

The Cadarache Workshop included reports on these programmes from eight countries. These approached the long term corrosion evaluation of the containers, not only by corrosion tests, but also by the development of the necessary deterministic and empirical models for extrapolating the behaviour of the selected materials. The difficulties involved in developing such models of corrosion behaviour and in demonstrating their validity has been underlined many times.

An unusual, but remarkable session, was devoted to analysis of the very long term corrosion of historical and archaeological artefacts. Among these, the Delhi pillar — although not built to help corrosionists — does represent 1600 years of exposure of a steel to the atmosphere, and makes it possible to build corrosion models.

The initiative to bring together at this time scientists and engineers from the international community to report and discuss their approaches to the problem of long term corrosion behaviour in nuclear waste systems, their results and strategy for the future is most appropriate.

Damien Féron and Digby Macdonald are to be congratulated on taking this initiative and on having conducted such a highly successful meeting.

PHILIPPE BERGE
Vice president of the CEFRACOR
Former Chairman of the Working Party on Nuclear Corrosion
Past President of the European Federation of Corrosion

Introduction: Long Term Prediction

G. BERANGER

Professor, Member of National Academy of Technologies of France
Université de Technologie de Compiègne, France
ROBERVAL Laboratory, CNRS 6066

First of all, I would like to thank Prof. D.D. Macdonald, and my colleague and friend D. Féron for inviting me to introduce this workshop by giving some main ideas that I consider to be of importance in the scientific and technological problem that we have to solve for the future, i.e. the storage of nuclear waste.

From a general point of view, when we have to decide the right choice for the good use of a material for a given application we have to take into account several parameters of importance as shown in Fig. 1. The life time range is already very large if we consider that only some minutes are required for rocket engines and the thermal shield of spaceships but that several hundred thousand hours are necessary for steam engines and, in the future, for high temperature nuclear reactors (Table 1).

In the case of deep storage of nuclear waste the time required is completely different from these since we have to ensure the integrity of the storage system for several thousand years.

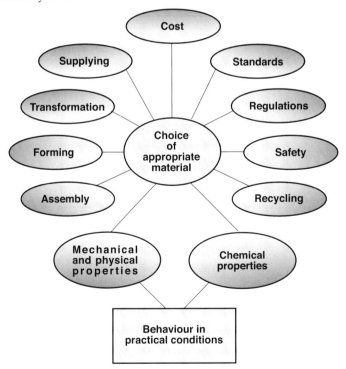

Fig. 1 *Choice of appropriate materials.*

Table 1. *Life time of materials following the applications and the operating temperature (from J. H. Davidson, private communication)*

Typical life time	Applications	Upper temperature or temperature range (°C)
Some minutes	• Rocket engines • thermal shield of spaceship	2400 1650
10-20 h	• Coatings of space shuttle	1750
100–200 h	• Ingot mould for steelmaking	1350–1500
1000 h	• Filament of electric lamp	2500
2000–3000 h	• Gas car engine (automotive)	600–800
10 000–20 000 h	•Diesel engines (trucks) • Electrical resistances for industrial furnaces • Engines and turbojets for civil aircrafts	800–900 1300–1500 1050–1100
100 000 h	• Gas turbines • Petrochemical reactors • Reformer furnaces	1000 950–1000
500 000 h	• High temperature nuclear reactors • Steam turbines	850–900 650

The vitrified nuclear waste should be placed in a metallic canister. Several metals and alloys are candidates for this purpose; among them can be mentioned:

• Iron and carbon steels;

• Stainless steels;

• Copper and copper alloys;

• Nickel alloys;

• Titanium alloys.

Some are passivating metals, i.e. they develop a thin protective film, 2 or 3 nanometres thick — the so-called passive film. This film prevents general corrosion of the metal in a large range of chemical conditions of the environment. But in some specific conditions, localised corrosion, such as the phenomenon of pitting, occurs. Consequently, it is absolutely necessary to determine these chemical conditions and their stability with time, to understand the behaviour of a given material.

In other words, we have to define a system of corrosion constituted by:

MATERIAL / SURFACE / MEDIUM

If one of these three components is changed, the corrosion resistance changes. Whereas some materials have been well known for a long time (copper, carbon steels or cast iron), some others are quite new. These new alloys are mainly stainless steels, titanium alloys and nickel based alloys. For example, stainless steels were first

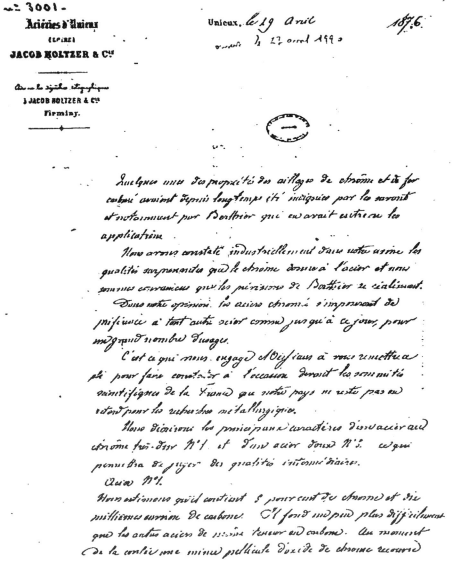

Fig. 2 (Above and opposite): M. Brustlein's sealed envelope deposited in April 1876 at the French Academy of Sciences, where the beneficial role of chromium in steels is pointed out.

used only at the end of the 19th century and then more extensively from the beginning of the 20th century. From an historical point of view, we may mention that the role of chromium added to iron was described in terms of corrosion resistance and of mechanical properties (hardness) by M. Brustlein in a sealed envelope deposited in April 1876 at the French Academy of Sciences. Following the author's request this envelope was opened in April 1990, i.e. 114 years later! (Fig. 2). However, it is necessary to point out that this beneficial role of chromium had been already observed by M. Berthier in 1821; this discovery had not been developed because the steel prepared by co-reduction of iron oxide and chromium oxide by carbon was too carburised and too brittle to be mechanically worked. From the observations reported by M. Berthier and M. Brustlein, a great deal of research and development was conducted, particularly during the first decade of the 20th century. Thus many grades of stainless steel have been designed, giving numerous standards. Later, as a result of progress in analytical and electrochemical methods, much work has been carried out on the passive film grown on the surface of stainless steel which plays such a significant protective role against uniform corrosion.

We now have a good understanding of passive films but not enough to predict their role over a very long period such as that required for the nuclear waste disposal storage. As a matter of fact, the passive film can modify as the environmental medium changes over a long period. A modelling approach is therefore necessary since the stainless steels, like titanium alloys and nickel base alloys are young materials, that is, experience with them is limited to 50 to 180 years compared with the long term of 1000 or 10 000 years required for the high level nuclear waste (HLNW) storage.

During the exposure of a metallic material in a given environment, a dissolution phenomenon takes place. Some ions coming from the metal move to the medium, and change this, even in the case of a passivated material for which only weak dissolution occurs. The long term growth and change in composition of the passive layer is another question of importance: up to now the evolution of the passive layer has only been described for industrial life times (some 10 to 50 years). Will the properties of this very thin layer (a few nanometres) be constant over large periods of time, some thousands of years? As a continuous dissolution occurs during the passivity, could it be possible to have a slow segregation of some species which could modify the passive properties of the oxide layers ?

Over a long time, the corrosion system can vary, but it is very difficult, and may be hazardous, to predict how long one can use data obtained from some tests performed in a laboratory at the human scale for classical durations. Even so, only a few works taking into account the requirement of a long time have been conducted so far; the 45 years study of underground corrosion conducted by NBS including the effect of soils of varying pH on different metals can be mentioned, extensive data were reported (M. Romanof Circular No. 579, Washington, D.C., USA, 1957). Probably, experiments to obtain a good prediction will be of too long a duration, and therefore it will be necessary to go further inside the corrosion mechanisms in order to made some predictions for the very long term (1000–10 000 years).

For high level nuclear waste (HLNW), the main features for resolving the problem are concerned with:

• Geological disposal;

• Deep storage in clay;

• Waste metallic canister;

• Backfill mixture (clay-gypsum) or concrete;

• Long term behaviour;

• Data needed for modelling and for predicting; and

• Choice of appropriate solution among several metallic candidates.

Nevertheless, in order to reach the purpose considered here, we have to conduct more works on archaeological analogues, particularly to have the best knowledge of the limits of their use. These limits relate to the actual surrounding medium, to the material, and to the precise initial surface location before the storage in the ground. In carrying out these investigations, it is absolutely necessary that when these artefacts (iron) are extracted from the ground, a whole block of matter including the ground and the metallic piece is removed. Experiments then have to be performed in the laboratory using analytical methods and metallographic techniques to answer the previous questions. It is clear that we have to extend research works as archaeological analogues.

As pointed out by B. Soerensen and D. Gregory [1], the use and interest of archaeological analogues are limited by the knowledge of the components of the corrosion system, i.e.:

(i) the material (composition, structure, texture, residual stresses.....): old materials are mainly carbon steels or copper alloys. It could be more difficult for other metallic materials;

(ii) the medium (nature of phases, chemical species, water, pH, oxygen content, hydrological conditions, bacteria...);

(iii) the corrosion products (thickness, their nature, passivity,...); and

(iv) the form of corrosion.

Thus, if we consider the determination of long term corrosion rates already reported, in particular, by A. Accary and B. Haijtink [2] and by B. Miller and N. Chapman [3], we find the absence of some of these data. Moreover, a large discrepancy is observed in the determinations made by the former. This discrepancy can be explained by a difference between the different media and perhaps between the materials and demonstrates very clearly the importance of the system (already previously considered):

MATERIAL/SURFACE/MEDIUM

This is of great importance not only in the case of the topics of this workshop but also in other cases because the corrosion resistance of a given material is not an intrinsic property. In the same field, the expected life time has to be taken into account for the choice of adequate material for a given use in well defined conditions. The importance of life time is well illustrated by Table 1, giving some values for classical technological applications.

For this important technological problem dealing with the long time future it is essential to have very accurate data for modelling and predicting the behaviour of materials engaged for the deep storage of high level nuclear waste (HLNW). It is the purpose of this Workshop to bring together articles which have a significant contribution to make, and which take into account all the parameters of importance so that appropriate decisions can be taken.

References

1. B. Soerensen and D. Gregory, *Proc. Metal '98 Conf.* (W.Mourey and L.Robiola, eds), 1998, 94–99.
2. A. Accary and B. Haijtink, *Proc. Journées de Paleometallurgie*, UTC, Compiègne, 1983, 323 and private communication.
3. B. Miller and N. Chapman, *Radwaste Magazine*, 1995, 32–42.

Part 1

R&D Programme Session

1

U.S. Nuclear Regulatory Commission Corrosion Programme for High-Level Nuclear Waste

T. M. AHN

U.S. Nuclear Regulatory Commission Washington, DC, U.S.A.

ABSTRACT

The corrosion programme of the U.S. Nuclear Regulatory Commission (NRC) for the evaluation of the engineered barriers to be emplaced for the geological disposal of high-level radioactive waste (HLW) is reviewed. The risk-informed, performance-based regulation established by NRC for the potential licensing of the proposed repository at Yucca Mountain, Nevada is introduced and resolution methods for HLW technical issues are discussed. Examples of the approach adopted for the resolution of issues related to the degradation of engineered barriers as a result of corrosion are presented including uniform corrosion, localised corrosion, and stress corrosion cracking of HLW container materials. The use of analogues to provide additional lines of evidence to the performance evaluations is addressed. Finally, overall NRC activities in waste package corrosion studies are summarised.

1. Introduction

This paper reviews how the U.S. Nuclear Regulatory Commission (NRC) informs the U.S. Department of Energy (DOE) of its perspective on the long-term corrosion performance of the waste package (waste package) and the drip shield (DS) as the principal engineered barriers to be emplaced for the geological disposal of high-level radioactive waste (HLW) at the proposed Yucca Mountain repository in the USA. The role of the waste package is to confine radioactive waste within a designed boundary while the DS is intended to prevent the dripping of groundwater flowing along rock fractures on the waste package surface. Currently, the DOE's Yucca Mountain Project has selected Alloy 22 (58Ni–22Cr–13Mo–3W–4Fe) as the material for the outer container of the waste package and Titanium grade 7 (Ti–0.15Pd) as DS material. Degradation of these engineered barriers as a result of corrosion processes is one of the key subissues of the NRC's Container Life and Source Term (CLST) Key Technical Issue (KTI) to be evaluated in the NRC licensing of the proposed repository. Other KTIs currently considered by NRC include issues related to the geology, hydrology, and climatology of the proposed site as well as potential disruptive events such as volcanic eruptions and earthquakes. The CLST KTI consists of six subissues related to waste package corrosion, waste package mechanical failure, spent nuclear fuel (SNF) degradation, HLW glass degradation, in-package criticality, and alternative engineered barriers such as DS [1]. This paper addresses the risk-informed, performance-based regulation established by NRC for the potential licensing of the proposed repository, issue resolution methods, examples of the approach adopted for issue resolution including uniform (general) corrosion, localised corrosion, and stress corrosion cracking of HLW container materials. It also addresses the use of multiple lines of evidence in long-term assessment of corrosion behaviour. Finally, the paper summarises the activities of the NRC / Center for Nuclear Waste Regulatory Analyses (CNWRA) waste package corrosion studies.

2. Risk-informed, Performance-based Regulation

The U.S. Nuclear Regulatory Commission (NRC) uses a risk-informed, performance-based approach in pre-licensing reviews and potential licensing of the proposed Yucca Mountain repository. As usual, risk is defined as the product of the probability of occurrence multiplied by the consequence. In this approach, insights, engineering analysis and judgement are used to:

- focus attention on the most important technical sub-issues (e.g. waste package corrosion, and in particular the possibility of localised corrosion);

- establish objective criteria for evaluating performance. Currently, the specific regulation for the proposed Yucca Mountain repository, published in Title 10 of the Code of Federal Regulations, Part 63 (10 CFR Part 63), requires that DOE must demonstrate that, for 10 000 years following disposal: the reasonably maximally exposed individual receives no more than an annual dose of 15 mrem from releases from the undisturbed Yucca Mountain disposal system; the reasonably maximally exposed individual receives no more than an annual dose of 15 mrem as a result of a human intrusion; and for undisturbed performance, releases of radionuclides from waste in the Yucca Mountain disposal system into the accessible environment will not cause the level of radioactivity in the representative volume of ground water to exceed the limits specified in the rule [2];

- provide flexibility on how to meet established performance criteria (i.e. no prescriptive methodologies); and

- focus on overall performance assessment (PA) as primary basis for regulatory decision-making (i.e. no performance measures of subsystems or components of subsystems such as the engineered barrier subsystem or the waste package).

A risk-informed, performance-based approach is used to emphasise the importance of the safety case in the repository performance and achieve reasonable assurance based on the current knowledge. This approach is required because it is impossible to predict all the scientific aspects of the long-term behaviour of the natural and engineered barriers with negligible uncertainty. In this approach, issue resolution is a key component in determining the consequence of various uncertainties associated with mechanisms, models, and parameter values on the overall performance evaluation. The purpose of issue resolution is to assure that sufficient information is available on an issue to enable NRC to conduct a review of a proposed license application. Consistent with the provisions of the Nuclear Waste Policy Act of 1982 and the Amendments Act of 1987 [3], and following a 1992 agreement with DOE, pre-licensing consultation between DOE and NRC is taking place through public technical exchanges and could lead to staff-level resolution prior to the license application.

3. Issue Resolution Methods

The issues associated with the corrosion performance of container and DS may be resolved by

(i)   probability or consequence screening;

(ii) deterministic analysis, or

(iii) PA code calculations.

The first method is to screen these issues using estimates for the probability of occurrence. If such probability is less than 10^{-8}/year, no further analyses or testing are required. For example, hydrogen embrittlement of Alloy 22 is very unlikely under the oxidising conditions expected in the proposed Yucca Mountain.

Additionally, if these issues lead to a very low consequence in terms of the dose limit of 10 CFR Part 63, they will be also screened out. The second method is to conduct experiments and collect data, analyse the acquired data or qualify existing data, and input the data in appropriate models to evaluate the consequences of a given process over the long period of performance. An example is to determine the values and distribution of uniform corrosion rates of Alloy 22 in the expected range of emplacement drift environments. The third method is to conduct a complete system PA. The NRC's Total-system Performance Assessment (TPA) code is an integrated code that includes the effects of engineered barriers and natural barriers in controlling radionuclide releases. In this system analysis, the importance of each process in the total repository system can be determined. For example, crevice corrosion may be a plausible mechanism for Alloy 22 failure. However, within the range of expected environmental conditions (e.g. chloride concentration, temperature, pH), crevice corrosion may not be initiated according to the results of calculations performed with the TPA code and hence, taking into account uncertainties in the estimations, radionuclide release is maintained within the compliance limit.

4. Examples of the Approach Adopted for Issue Resolution

Integrity of the waste package is one of the most important attributes of the engineered barrier system and DOE has indicated that long waste package lifetimes are important for meeting the overall performance requirements for the proposed repository. Alloy 22, presently the material of choice for the waste package outer container is considered to be very resistant to various modes of aqueous corrosion in oxidising environments containing aggressive anions over broad ranges of pH and temperature. Despite its demonstrated corrosion resistance, as a result of the extremely long performance period, corrosion processes are considered to be the most important degradation mechanisms for the WPs. Many possible degradation processes have been identified by DOE [4]. Among them, passive uniform corrosion, localised corrosion and stress corrosion cracking are of particular interest. These degradation processes are central to several issues associated with the obtaining of long term waste package and overall repository performance. Currently these particular issues are closed pending the provision of further information by DOE, i.e. before any licence application — as agreed in discussions in the Container Life and Source Term Key Technical Issue exchange. Differences in the approaches used by DOE and NRC to develop abstractions for these degradation mechanisms are discussed below along with the path to resolving these issues.

4.1. Example 1. Uniform Corrosion of Alloy 22

In the absence of localised corrosion, stress corrosion cracking or mechanical failure, the passive uniform corrosion rate of Alloy 22 may determine the lifetime of the WPs in the proposed repository. The selection of Alloy 22 is based on the resistance of this alloy for the various aqueous corrosion modes. Alloy 22 and other similar Ni–Cr–Mo alloys have been used in a wide variety of applications that require high resistance to localised corrosion. As a result, these alloys have been extensively used in chemical process industries, power plants, oil and gas production, and offshore applications where other materials such as stainless steels may undergo rapid degradation in the form of localised corrosion.

The excellent corrosion resistance of Alloy 22 can be attributed to a protective passive film that restricts the dissolution of the alloy components. For stainless steels and Ni–Cr–Mo alloys, the protective film consists of an inner oxide film that is rich in chromium oxide and an outer hydroxide layer formed by deposition. The stability of the passive film is enhanced by additions of Cr, Mo and W to the alloy. In conditions where the protective oxide film is stable, the corrosion rate of passive metals and alloys is not strongly dependent on temperature, environmental composition, and redox potential [5]. Uncertainties exist with regard to the long term passive corrosion rate, variations in the corrosion rate as a result of changes with time of the environment at the waste package surface and the long term composition and stability of the passive oxide film. Changes to the passive corrosion rate are important in the abstraction of waste package performance.

The corrosion rates used in the DOE Total System Performance Assessment (DOE TSPA) are abstracted from tests conducted at the Lawrence Livermore National Laboratory in the long term corrosion test facility (LTCTF) where numerous test specimens have been exposed to aqueous solutions based on modifications of J-13 well water [6,7]. The J-13 well water is a near neutral (pH ~ 7) representative solution at the Yucca Mountain repository, containing (5 ~ 8) Cl^-, (2 ~ 3) F^-, (5 ~ 9) NO_3, (120 ~ 140) HCO_3 mg L^{-1} and other cations such as Ca and Si [8]. As the dripping groundwater is likely to condense on waste package surfaces, modifications of J-13 well water are also used: simulated J-13 concentrated water, simulated saturated water, and basic saturated water [9]. Corrosion rates of specimens exposed in the

LTCTF were calculated by measuring the mass loss of the specimens [10] after exposures of 6 months, 1 year, and 2 years of exposure. When all exposure times and exposure conditions are considered, the corrosion rates for Alloy 22 range from 60 to 731 nm/year. The highest corrosion rates were observed in the 6 month exposure specimens. Corrosion rates above 200 nm/year were observed for only a few specimens. The average corrosion rates of the Alloy 22 specimens tends to decrease with time from 50 nm/year after 6 months to 30 nm/year after 1 year to 10 nm/year after 2 years based on mass loss measurements. Mass gain was observed on 25 percent of the Alloy 22 mass loss specimens as a result of the deposition of silicates (assumed to be amorphous SiO_2) on the surface of the specimens [4,6].

The abstracted corrosion rate used in the DOE TSPA is based on the 2-year exposure data from the LTCTF. Data from 2 year tests in the LTCTF are described by a normal distribution (mean: 8.3 nm/year, s.d: 13 nm/year). The normal distribution was used to sample corrosion rates and compute waste package failure times. Rates from the negative tail of the distribution were eliminated after the sampling. The positive corrosion rates were increased by a term uniformly distributed between 0 and 63 nm/year to account for the observed formation of silica deposits on samples in the LTCTF. Consistent with suggestions by DOE, corrosion rates were also assumed to be modulated by a factor uniformly distributed between 1 and 2, to account for microbially induced corrosion. Furthermore, it was assumed that a small percentage (3%) of the sampled rates were also affected by a second modulating factor, assumed to be uniformly distributed between 1 and 2.5, to account for thermal ageing. Thermal ageing is assumed to affect only the waste package weld area. The resulting corrosion rates were used to compute failure times, considering that rates are constant in time. Figure 1 shows a comparison of results obtained with the described approach to results by DOE reported in TSPA-SR [11]. Figure 1 includes data derived from three cases and is compared with the DOE data in Fig. 3.4-20 of Ref.[11]. The cumulative distribution function (CDF) is plotted as a function of time. These three cases differ in the way corrosion from the inside to the outside of the waste package is treated. It is assumed that once a corrosion front penetrates the waste package thickness, then electrolyte is available in the interior of the waste package and corrosion from the inside out can occur. In Case 1, it is assumed that 'inside-out' rates are statistically independent from the 'outside-in' rates. In Case 2, it is assumed that the 'inside-out' rates are identical to the 'outside-in' rates. In Case 3, it is considered that 'inside-out' corrosion does not occur.

Although the results of tests conducted in the LTCTF indicate no effect of temperature, DOE has recently evaluated Alloy 22 corrosion rate data obtained from the University of Virginia [12]. The corrosion rates were potentiostatically measured in lithium chloride solutions with 10:1 and 100:1 chloride to sulfate ion ratios (pH 2.8 to 7.8). The activation energy for passive dissolution was determined to be in 36 kJ mol^{-1} using data obtained at temperatures in the range of 80, 85, and 95°C. Using the calculated ±1 standard deviation over the 80 to 95°C temperature range, the activation energy could be as low as 11 kJ mol^{-1}, which would yield calculated passive dissolution rates that are not as strongly dependent on temperature. More recently, DOE presented 17.1 kJ mol^{-1} obtained from tests in a simulated repository solution [13]. Using an activation energy of 36 kJ mol^{-1}, distributions of corrosion rates at 25, 60, and 125°C were subsequently calculated. The 50th percentile corrosion

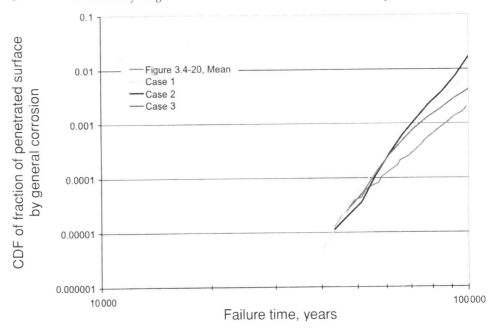

Fig. 1 *Fraction of penetrated surface by general corrosion predicted by the Department of Energy total System Performance Assessment model.*

rates were calculated to be 9.5×10^{-6} mm/year at 25°C, 4.5×10^{-5} mm/year at 60°C, and 3.8×10^{-4} mm/year at 125°C.

Tests conducted at The Center for Nuclear Waste Regulatory Analyses (CNWRA), a federally funded research and development center (FFRDC) for NRC are summarised in Fig. 2 which shows the results of passive current density measurements obtained after 48 h under potentiostatic conditions [14]. At potentials of 200 mV (SCE) or less, the passive current was in the range of 6×10^{-9} to 3×10^{-8} A cm^{-2} corresponding to corrosion rates of 6×10^{-5} to 3×10^{-4} mm/year. The passive current densities were relatively independent of the chloride concentration and solution pH in the range of 2.7 to 8.0. In tests conducted at potentials above 400 m V(SCE), the onset of transpassive dissolution was observed. At pH 0.7 the passive current densities increased to approximately 7×10^{-8} A cm^{-2} at 200 mV (SCE). The increase in the anodic current density observed at pH 0.7 at applied potentials of 200–600 mV (SCE) may be the result of enhanced oxide film dissolution in this highly acidic solution. Temperature, on the other hand, had a significant effect on the passive corrosion rate in dilute chloride solutions. The passive corrosion rate increased by a factor of 10 when the temperature of the 0.028M chloride solution (pH 8.0) was increased from 20 to 95°C (corresponding to an activation energy of 27.5 kJ mol^{-1}). However, practically no temperature effect was noted in 4M chloride. While the main effect of fabrication processes was a decrease in the localised corrosion resistance, CNWRA tests show that improper welding and heat treatment may cause detrimental intergranular attacks, leading to substantial increase in current densities.

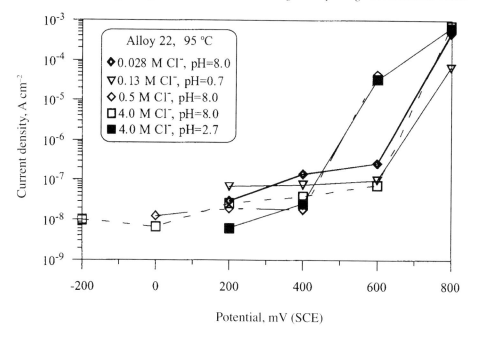

Fig. 2 *Anodic current density of Alloy 22 measured under potentiostatic conditions for a period of 48 h.*

Corrosion rates from the TPA Code are based on Alloy 22 waste package and is calculated using both a low passive current density of 5.0×10^{-9} A cm^{-2} (5.0×10^{-5} mm/year) and a fast dissolution rate of 5.4×10^{-9} A cm^{-2} (5.4×10^{-4} mm/year). Figure 3 shows the output of the TPA Code for the case where the waste package failure time is determined by the rate of passive dissolution. The number of WPs failed versus time according to the TPA Code (Version 4.1j) [15] shows that the earliest failure occurs at 38 000 years. The percentiles and mean were computed from 500 Monte Carlo realisation data.

The independent results obtained by DOE and NRC/CNWRA both indicate that the waste package lifetime in the absence of localised corrosion is longer than the 10 000 year performance period for the proposed repository. Nevertheless, several uncertainties associated with the extrapolation of passive corrosion rates need to be resolved. Abstracted models for uniform corrosion assume that the passive dissolution rate is constant and does not vary with time or exposure conditions. However, recent reviews conducted by the CNRWA suggest that preferential dissolution of alloying elements under passive conditions may lead to passive film instability and increase the overall uniform corrosion rate [14]. The DOE's path forward to reduce the uncertainties associated with the abstraction of long term passive corrosion rate of the waste package outer container will focus on the characterisation of the long term passive behaviour and changes in the passive film chemistry. NRC/CNWRA are independently evaluating these uncertainties to obtain an insight for potential licensing of the proposed Yucca Mountain repository.

4.1.2. Example 2. Localised Corrosion of Alloy 22

Evaluation of localised corrosion susceptibility of the Alloy 22 waste package outer container is necessary because of the possibility of rapid penetration of the Alloy 22 waste package outer barrier in a short period. The DOE selection of candidate container materials has evolved toward more corrosion resistant Ni–Cr–Mo alloys in order to avoid localised corrosion failures of the waste package after emplacement. Alloy 22 in the mill annealed conditions has been determined to be very resistant to pitting corrosion but is susceptible to crevice corrosion in chloride-containing environments. Several uncertainties exist in the abstraction of the localised corrosion susceptibility of waste package outer container. The localised corrosion resistance of Alloy 22 is dependent on both temperature and the composition of the environment. Aggressive species such as chloride and reduced sulfur species promote localised corrosion of Alloy 22 whereas anions such as nitrate and sulfate may inhibit localised corrosion. Minor impurities in the environment such as Pb, Hg, and As may also promote localised corrosion. In addition to the composition of the environment, the effects of fabrication processes such as cold work and welding may alter the localised corrosion resistance of Alloy 22.

The DOE assessment susceptibility of Alloy 22 as well as other Ni-base candidate container alloys to localised corrosion was performed using a comparison of the corrosion potential (E_{corr}) with a DOE's critical potential for localised corrosion ($E_{critical}$) at temperatures of 30, 60, and 90°C. These potentials were measured in short term tests conducted in aerated solutions based on the composition of well water obtained form the proposed repository site. The evaluation of Alloy 22 was conducted using only the base alloy without welds or thermal processing to simulate waste package fabrication effects. The difference between the E_{corr} and $E_{critical}$ was observed to be greater than 400 mV for all environments and temperatures. The evaluation of the possible effects of radiolysis was conducted by measuring the E_{corr} of the alloy in solutions with the addition of H_2O_2. At 25°C, the E_{corr} increased by 190 to 230 mV with the addition of 72 ppm H_2O_2. Because the increase in the E_{corr} was not sufficient to exceed the $E_{critical}$ value, localised corrosion is not considered in the DOE TSPA base case calculations and the Cumulative Distribution Function (CDF) for waste package failure is determined only by the passive corrosion rate (Fig. 1).

The abstraction of localised corrosion susceptibility for the Alloy 22 waste package outer container in the TPA Code uses a comparison of the E_{corr} and the repassivation potential for crevice corrosion (E_{rcrev}). The E_{corr} value is calculated based on the kinetics of anodic and cathodic reactions. If E_{corr} exceeds E_{rcrev} active crevice sites are assumed to initiate and grow — as based on the results of experimental investigations [16,17]. If E_{corr} falls below E_{rcrev}, active crevice sites are assumed to cease growing and the material passivates, corroding uniformly at a very low rate through a passive film. Because E_{rcrev} for mill annealed Alloy 22 is very high and the critical chloride concentration necessary for localised corrosion is near the saturation for NaCl, crevice corrosion of the Alloy 22 waste package outer containers is not predicted by the TPA Code. As a result the waste package lifetime is again dictated only by the passive corrosion rate as shown in Fig. 3.

Although both the TPA Code and the DOE TSPA code do not show localised corrosion, the uncertainties in the data for the localised corrosion susceptibility of the Alloy 22 waste package outer containers must be addressed. The data used by

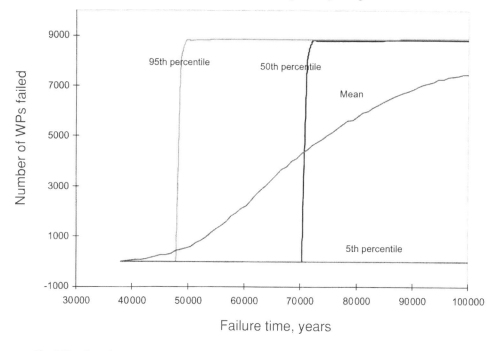

Fig. 3 *Number of waste package failed from uniform corrosion calculated using the NRC/CNRWA TPA code.*

DOE are based on critical potentials that are not consistent with the initiation of localised corrosion. Systematic determination of the effects of aggressive and possible inhibitive anions such as nitrate are not included in either the DOE or the NRC/CNWRA data. In addition, the effects of waste package fabrication may significantly reduce the localised corrosion resistance of Alloy 22 [5,14]. The assessment of these uncertainties is currently being addressed in ongoing investigations conducted by DOE. NRC/CNWRA are independently evaluating these uncertainties to obtain insight for potential licensing of the proposed Yucca Mountain repository.

4.1.3. Example 3. Stress Corrosion Cracking of Alloy 22

Stress corrosion cracking is one of the potential failure modes of an Alloy 22 outer container. The DOE modelling of stress corrosion cracking of Alloy 22 considers a narrow range of expected waste package environments and is limited to the closure lid weld stresses. Two stress corrosion cracking models, the threshold stress intensity model and the slip dissolution/film rupture model, are being used [4]. In the first model, stress corrosion cracking susceptibility of Alloy 22 is evaluated using model parameters obtained from Lawrence Livermore National Laboratory data; whereas, in the second case, experimental data obtained at General Electric (GE) Corporation Research and Development Center for Alloy 22, combined with data reported for stainless steel in boiling water reactor environments, are used. Evaluation of these two alternative models reveals that while a K_{ISCC} value of 33 MPa m$^{1/2}$, determined by Roy *et al.* [18], is adopted in the threshold model, the slip dissolution/film rupture model predicts crack propagation at K_I values less than the experimentally determined value of K_{ISCC}.

Very low rates of crack propagation were measured at GE for Alloy 22 under cyclic loading conditions but no crack growth was detected under constant load conditions at 30 MPa m$^{1/2}$ at 110°C in air saturated alkaline solution (pH = 13.4) produced by evaporation of water containing the anions present in J-13 well water with the pH adjusted with NaOH. Crack propagation rates ranging from 2.1 × 10^{-11} to 7.6 × 10^{-12} ms^{-1} were measured using compact tension specimen at a maximum K_I = 30 MPa m$^{1/2}$ under cyclic unloading conditions with R = 0.5–0.7, and frequency of 0.001 to 0.003 Hz. Under a constant load the crack growth stopped. It is unclear whether the laboratory cyclic condition is real in the repository and the same as the repository seismic excitation. If it is a simple acceleration method, samples with intentional welding defects should be used.

Independently, CNWRA conducted tests using double cantilever beam (DCB) specimens of Alloy 22 in 0.9M Cl$^-$ (5% NaCl, pH = 2.7) at 90°C and in 14.0M Cl$^-$ (40% MgCl$_2$) at 110°C. No cracks were detected after 386-day tests using an initial K_I = 32.7 MPa m$^{1/2}$. Further tests are in progress. In addition, CNWRA investigations to date show no stress corrosion cracking for type 316L stainless steel using DCB, modified wedge-opening-loaded (WOL) and compact tension (CT) specimens in hot concentrated chloride solutions at potentials below the repassivation potential for crevice corrosion, whereas crack propagation was observed above the critical potential.

Although the DOE and NRC/CNWRA test results are in general agreement, indicating that Alloy 22 seems to be extremely resistant to stress corrosion cracking in relevant environments, uncertainties arise mainly from the accurate assessment of stresses to be present. As stress corrosion cracking is limited to the surface area defined by the closure-lid welds, the DOE approach to mitigate or eliminate the possibility of crack growth is to reduce the residual stresses associated with welding. One method proposed for the inner closure lid involves the use of laser peening to introduce compressive stresses on the surface using multiple passes of a laser beam [4]. The other method for the outer closure lid consists of localised annealing of the weld region using induction heating. Nonetheless, residual stresses from waste package fabrication or applied stresses resulting from seismic events need to be better assessed by DOE because they may be sufficient to cause stress corrosion cracking of the outer container. DOE has also agreed to address issues related to the potential detrimental effect of post-weld treatments because variations in annealing parameters may lead to microstructural alterations (i.e. precipitation of topologically close-packed phases) that may reduce the stress corrosion cracking resistance. NRC/CNWRA are independently evaluating these uncertainties to obtain insight for potential licensing of the proposed Yucca Mountain repository.

5. Multiple Lines of Evidence

While the total performance Assessment (TPA) Code is used by NRC to probe the DOE safety case and identify the factors important to performance, it is recognised that the confidence in the PA calculations can be increased by auxiliary analyses derived from multiple lines of evidence. The multiple lines of evidence approach utilises independent process-level models, sensitivity analyses, and observations from laboratory experiments and field service, in addition to simple analysis and calculations, to provide support to the results of TPA Code calculations. Studies of natural and archeological analogues, used in support of PAs of engineered barriers [19], are a good example of such supplementary analyses.

The desire to use natural and archeological analogues stems from two characteristics of this type of evidence:

(i)   the duration of exposure of these systems is much longer than laboratory studies and approaches or exceeds that expected for a high-level nuclear waste repository,

(ii) these systems are exposed to natural environments rather than controlled laboratory environments, and

(iii) they can be persuasive to the public because of the outward similarities in the length of exposure and the typically buried nature of these analogues to radioactive WPs.

Unfortunately, utilising these analogues for predicting the performance of waste package component is fraught with several difficulties, especially from a corrosion point of view. First, an accurate knowledge of the material condition (chemical composition, microstructure, thermomechanical history, geometry, etc.) and the contacting environment (anionic and cationic concentrations, redox potential, temperature, fluid flow, etc.) is required. The effects of environmental and metallurgical factors on long-term corrosion behaviour have been recently demonstrated for the famed iron pillar of Delhi and related iron objects [20]. It has been shown that the atmospheric corrosion rate of the iron pillar may have been low not only because of the propitious climatic conditions at Delhi but also because of the high phosphorus content of the wrought iron that has enabled the formation of a protective hydrated ferrous hydrogen phosphate film. Such a protective film is not typically seen in modern steels. Unfortunately, knowledge of either the environmental or materials conditions or both is usually missing in analogue studies. The predictive uncertainty arising from a lack of information on the initial material conditions is illustrated by the observation that the calculated uniform corrosion rate of iron artifacts ranges from 10^{-2} μm per year to over 10^{3} μm per year [19]. Secondly, even if the environmental and material parameters can be reasonably estimated, the corrosion modes of the analogues can be very different from those of the waste package materials. Even in the case of localised corrosion, the corrosion morphology of an iron object can be different from that of a stainless steel. Lastly, the Ni–Cr–Mo alloys, currently considered by DOE for containers, have been in existence only since the 1930s, first in cast form and then starting in the 1960s in wrought form. Stainless steels have been in existence since about 1912. Therefore, less than a century of industrial experience has been attained with materials similar to the current container materials. Additionally, industrial experience is often not accompanied by adequate measurement of parameters important for modelling the corrosion of these materials.

The focus of our analogue studies is to gain confidence in the models used for long-term prediction. We have shown that the performance of stainless steels and nickel-base alloys in marine applications is consistent with the repassivation potential approach [16]. The localised corrosion of iron meteorites in Antarctica and iron archeological objects in museums has been attributed to the presence of high chloride levels in the crevices and oxidising species outside the crevice, the most important of which is oxygen itself [21,22]. While our present assessment of these archeological objects is not complete, the localised corrosion observed on these objects may shed important light on the model used in PA. This may validate in part modern corrosion theories in assessing the long-term corrosion behaviour. This approach would apply to assessing the long-term passive behaviour, although materials and environments of waste package and analogues, while similar in many respects, are not identical.

Natural analogues of Ni–Fe systems exist in the form of various meteoritic iron–nickel alloys. The well known alloys of this kind, josephenite (also known as awaruite, Ni_2Fe to Ni_3Fe) and tetrataenite (NiFe), have been found intact in some cases where the environment is relatively dry. However, when exposed to aerated, wet atmosphere, severe localised corrosion and disintegration have been noted [21,22]. The nickel–iron system generally exhibits poor passivity and low resistance to localised corrosion. For example, alloys of compositions similar to tetrataenite, known commercially as Invars or NiLO alloys, have been used as controlled thermal expansion alloys in temperature sensors for undegound tests [23], where they have shown significant corrosion and failure in groundwater environments. Natural analogues of Fe–Cr and Ni–Cr systems are rare. However, one such analogue may be the mineral, ferchromide, which has been assigned an empirical formula [24], $(Fe_{2.62}Ni_{0.38})_3(Cr_{0.76}Mn_{0.06}Ti_{0.02}Al_{0.01})_{0.85}$. This intermetallic mineral has been observed in the Karelia and Kola Peninsula, but in micron-sized particles associated with laurite–erlichmanite minerals. The small size of the intermetallic renders the use of this for natural analogue studies difficult.

6. NRC/CNWRA Waste Package Corrosion Programme

While DOE as the prospective licence applicant is implementing the Yucca Mountain project, NRC is focusing on key areas of the repository performance to gain some technical insight for the licensing review. To assist NRC in the pre-licensing activities and the licensing review, CNWRA, a FFRDC, was established in 1987 by NRC at the Southwest Research Institute in San Antonio, Texas. The corrosion programme at CNWRA focuses on the development of some technical insight for the licensing review of the DOE's corrosion assessment of waste package, DS and waste form including cladding. Experiments and models developed at CNWRA provide independent verification of the DOE assessment, input to the TPA Code, and some technical insight for screening features, events, and processes important to the safety and performance of the repository. The programme includes studies on uniform corrosion, passive film stability, pitting and crevice corrosion, stress corrosion cracking, hydrogen embrittlement, microbially influenced corrosion, thermal stability, welding and fabrication processes. CNWRA has established the repassivation criteria for localised corrosion and continues to collect data to support understanding of the assumptions and models uncertainties in the data provided by DOE to support their safety case. In conjunction with CNWRA, the NRC staff are modelling corrosion processes, abstracting these process-level models for code improvement, and conducting sensitivity studies focused on corrosion processes using the TPA Code. In addition, CNWRA is evaluating the manufacturing processes such as welding, and is conducting an extended review of natural, archeological and industrial studies.

7. Summary and Conclusions

the U.S. National Regulatory commission uses risk-informed, performance-based approaches in pre-licensing reviews and potential licensing of the proposed Department of Energy repository at Yucca Mountain. the approach focuses attention on most important activities such as waste package uniform corrosion of the waste package in the Total System Performance Assessment as the primary basis for regulatory decision-making. Technical issues associated with the most important activities are resolved by probability or consequence screening, experiments and process-level modelling, and Total System Performance Assessment Code exercises. Three example cases of the issue resolution are given. Although localised corrosion and stress corrosion cracking of Alloy 22 at the proposed Yucca Mountain site may only occur under very extreme conditions, uncertainties still exist regarding assessments of the environmental conditions and the longer-term corrosion behaviour. Determinations of uniform corrosion rates indicate that Alloy 22 outer container could last well beyond the regulatory period of 10 000 years but reliable assessments still require long-term data and improved modelling. To increase confidence in predicting the Alloy 22 corrosion over a geological time, the use of analogues is discussed. Despite limitations in getting correct information on the historical environment or the nature of analogues, some insights could be gained. Pitting behaviour in iron meteorites appears to be predicted from the current corrosion models. Finally this paper summarises activities associated with waste package corrosion of CNWRA, a federally funded Research and Development Center for the National Regulatory Commission.

8. Acknowledgements

The author acknowledges the contribution of Kien Chang, Richard Codell, David Dancer, David Esh, Charles Greene, and Alvin Henry from NRC and Sean Brossia, Gustavo Cragnolino, Darrell Dunn, Vijay Jain, Oliver Moghissi, Sitakanta Mohanty, Yi-Ming Pan, Osvaldo Pensado, Narasi Sridhar, and Lietai Yang from CNWRA. Views and opinions expressed in this paper reflect those of the author and contributors do not necessarily reflect the views or regulatory position of the U.S. NRC.

References

1. U.S. Nuclear Regulatory Commission, Issue Resolution Status Report, Key Technical Issue: Container Life and Source Term, Revision 3, January 2001.

2. U.S. Nuclear Regulatory Commission, 10 CFR Part 63, Disposal of High-Level Radioactive Wastes in a Proposed Geological Repository at Yucca Mountain, Nevada — Final Rule, Federal Register, Washington, D.C., U.S. Nuclear Regulatory Commission 66 (213), 55732–55816, 2001.

3. Nuclear Waste Policy Act of 1982. Public Law 97-425. 222165, Nuclear Waste Policy Amendments Act of 1987, Public Law 100-203. 223717.

4. CRWMS M&O, Waste package Degradation Process Model Report, TDR-WDS-MD-000002, Revision 00, ICN 01, Las Vegas, NV: Civilian Radioactive Waste Management System, Management and Operating Contractor, 2000.

5. G. A. Cragnolino, D. S. Dunn, C. S. Brossia, V. Jain and K. Chan, Assessment of Performance Issues Related to Alternate EBS Materials and Design Options, CNWRA 99-003, San Antonio, TX: Center for Nuclear Waste Regulatory Analyses, 1999.

6. CRWMS M&O, General Corrosion and Localised Corrosion of Waste package Outer Barrier,. ANL-EBS-MD-000003, Revision 00, Las Vegas NV, Office of Civilian Radioactive Waste Management System, Management and Operating Contractor, 2000.

7. R. D. McCright, Engineered Materials Characterisation, Corrosion Data and Modeling — Update for the Viability Assessment, UCRL-ID-119564, Vol. 3, Revision 1.1, Lawrence Livermore National Laboratory, Livermore, CA, 1998.

8. U.S. Geological Survey, Hydrochemical Data Base for the Death Valley Region, California and Nevada, Open-File Survey 94-305, 1995.

9. CRWMS M&O, Environment on the Surface of the Drip Shield and Waste Package Outer Barrier, ANL-EBS-MD-000001, Revision 00B, Las Vegas NV, Office of Civilian Radioactive Waste Management System, Management and Operating Contractor, 2000.

10. American Society for Testing and Materials, Standard Practice for Preparing, Cleaning, and Evaluating Corrosion Test Specimens: G1-90, Annual Book of Standards, Vol. 03.02, West Conshohocken, Pennsylvania, 1997.

11. CRWMS M&O, Total System Performance Assessment (TSPA) Model for the Site Recommendation MDL-WIS-PA-000002 Revision 00, Las Vegas, NV Office of Civilian Radioactive Waste Management System, Management and Operating Contractor, 2000.

12. G. E. Gdowski, Waste package Corrosion Process Components, Presentation to the Nuclear Waste Technical Review Board, Las Vegas, NV, 20–21 June, 2001.

13. M. Peters, Yucca Mountain Scientific Update, presented to Nuclear Waste Technical Review Board, by Los Alamos National Laboratory, Bechtel SAIC in Pahrump, Nevada, January 29–30, 2002.

14. C. S. Brossia, L. Browning, D. S. Dunn, O. C. Moghissi, O. Pensado and L. Yang, Effect of Environment on the Corrosion of Waste Package and Drip Shield Materials, CNWRA 2001–003, San Antonio, TX: Center for Nuclear Waste Regulatory Analyses, 2001.

15. S. Mohanty and T. J. McCartin, coordinators, Total-system Performance Assessment (TPA) Version 3.2 Code: Module Description and user's Guide (Draft), Version 4.1 (to be published), Center for Nuclear Waste Regulatory Analyses, San Antonio, TX, 1998.

16. D. S. Dunn, G. A. Cragnolino, and N. Sridhar, An Electrochemical Approach to Predicting Long-term Localised Corrosion of Corrosion Resistant High-level Waste Container Materials, *Corrosion*, 2000, **56**, 90–104.

17. N. Sridhar, G. A. Cragnolino and D. S. Dunn, Experimental Investigations of Failure Processes of High-Level Radioactive Waste Container Materials, CNWRA 95-010, San Antonio, TX: Center for Nuclear Waste Regulatory Analyses, 1995.

18. A. K. Roy, D. L. Fleming and B. Y. Lum, Stress Corrosion Cracking of Fe–Ni–Cr–Mo, Ni–Cr–Mo, and Ti Alloys in 90°C Acidic Brine, *Corrosion '98*, Paper No. 157, NACE International, Houston, TX, 1998.

19. A. B. Johnson and B. Francis, Durability of Metals from Archeological Objects, Metal Meteorites, and Native Metals, PNL-3198, UC-70. Richland, WA: Pacific Northwest National Laboratories, 1980.

20. R. Balasubramaniam, On the corrosion resistance of the Delhi iron pillar, *Corros. Sci.*, 2000, **42**, 2103–2129.

21. V. F. Buchwald and R. S. Clarke, Corrosion of Fe–Ni alloys by Cl–containing Akaganeite (beta–FeOOH): The Antarctic Meteorite Case, *Am. Mineral.*, 1989, **74**, 656–667.

22. S. Turgoose, Structure, Composition and Deterioration of Unearthed Iron Objects. Current Problems in the Conservation of Metal Antiquities, Tokyo National Research Institute of Cultural Properties, Japan, pp. 35–53, 1989.

23. W. C. Patrick, Spent Fuel Test — Climax: Evaluation of the Technical Feasibility of Geologic Storage of Spent Nuclear Fuel in Granite — Final Report, UCRL-53702, Lawrence Livermore National Laboratory, Livermore CA, 1986.

24. A. Yu. Barkov, A. I. Lednev, N. N. Trofimov and M. M. Lavrov, Minerals of the Laurite–Erlichmanite Series from Chromite Horizons of layered intrusions in the Karelia-Kola region, Doklad. Akad. Nauk SSSR, 319, pp. 962–965, Translated by Scripta Technica, 1993.

2

Containers and Overpacks for High-Level Radioactive Waste in Deep Geological Disposal Conditions: French Corrosion Programme

D. CRUSSET, F. PLAS and G. SANTARINI*

ANDRA, 1/7 rue Jean Monnet, 92298 Châtenay-Malabry Cedex, France
*CEA-Saclay, DPCS–1CCME, Bât 458, 91191 Gif-sur-Yvette Cedex, France

ABSTRACT

Within the framework of the act of French law dated 31 December, 1991, ANDRA (National Radioactive Waste Management Agency) is responsible for conducting the feasibility study on disposal of reversible and irreversible high-level or long-life radioactive waste in deep geological formations. Consequently, ANDRA is carrying out research on corrosion of the metallic materials envisaged for the possible construction of overpacks for vitrified waste packages or containers for spent nuclear fuel. Low-alloy or unalloyed steels and the passive alloys (Fe–Ni–Cr–Mo) constitute the two families of materials studied and ANDRA has set up a research programme in partnership with other research organisations. The 'broad outlines' of the programme, which includes experimental and modelling operations, are presented.

1. Introduction

The alternatives for the disposal of high-level and long-life radioactive waste in deep geological repositories in France are being studied within the framework of the law of 30 December, 1991, which orients the different research operations in three directions to be pursued jointly for 15 years (1991–2006):

- Research on processes for the separation and transmutation of long-life radioactive elements present in the waste.

- Investigation of the possibilities of reversible or irreversible disposal in deep geological repositories, in particular by the construction of underground research laboratories.

- The study of methods for the packaging and long-term surface storage of waste.

Research on geological repository procedures was entrusted to ANDRA, the French National Radioactive Waste Management Agency which, in 2006, will have to answer the following question:

'Is it possible to design a reversible or irreversible repository, which is safe in the very long term and compatible with the types and volume of waste produced?'

The answer is governed by three factors:

1.   A qualified geological site.

2.   A repository design adapted to the site, industrially feasible in terms of safety and economically viable.

3.   Proof that a repository in the site complies with the very long-term safety and environmental protection objectives.

   The search for potential geological sites where underground research, or so-called 'on-site' laboratories will have a qualification role, was initiated in France in 1994 with reconnaissance surveys. Three sites were finally proposed, two in sedimentary clay formations and one in crystalline rock. On 9 December, 1998, the French government authorised ANDRA to construct an underground research laboratory at a depth of 500 m in a 150-million-year-old clay formation in Bure, straddling the Meuse and Haute-Marne departments (Eastern France), a very slightly permeable shale (claystone) layer in a clay formation (Fig. 1).
   The status of the research programme corresponds to the sinking of shafts for the Underground Research Laboratory, according to the following schedule (major steps) outlined in Table 1.
   Co-operative methodological studies are also in progress in support of other underground facilities, Mol (Belgium) and Mont Terri (Switzerland) for the clay site, the Äspö HRL (Sweden), the Grimsel Test Site (Switzerland) and the Canadian URL for crystalline rock.
   In 1998–1999, a set of preliminary design concepts associated with the claystone layer was studied to help address the issues raised by the feasibility analysis of a potential repository with respect to safety objectives. These concepts concern a

Table 1. Schedule for the Underground Research Laboratory

November 1999	Geophysical campaign (3D seismic)
Starting in January 2000	Borehole drilling for environmental follow-up
April 2000	Borehole drilling for shaft-sinking
June 2000	Beginning of the construction of surface installations
October 2000	Beginning of shaft-sinking
End of 2000 to end of 2002	Observations and experiments relative to shaft-sinking
2003 to end of 2006	Drift experiments

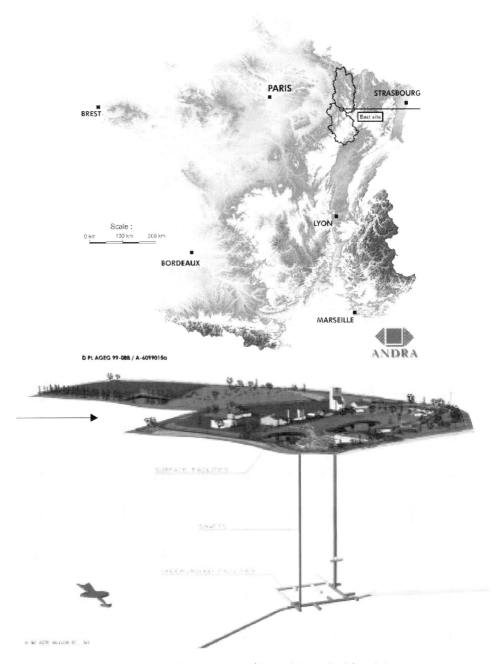

Fig. 1 *French Underground Laboratory: Meuse/Haute-Marne site (clay site).*

preliminary inventory of waste, including transuranic non heat-emitting waste, high-level vitrified waste and potential disposal of spent fuel. Provision was made for a high level of flexibility, in particular regarding the reversibility of the step by step disposal process.

Within the framework of the law and in co-operation with its partners, ANDRA is conducting research programmes in the different scientific areas to propose and assess the different disposal concepts that satisfy the safety objectives for a possible repository from operation to post-shutdown.

An underground research laboratory is under construction on the eastern side of the Paris basin essentially to determine the physicochemical properties of an argillaceous geological site (Fig. 1).

The radioactive waste containers constitute an element in the safety of a repository: they guarantee several functions, one of which is the containment of radionuclides.

The main cause of the degradation envisaged for these containers is corrosion. ANDRA has therefore implemented research programmes with its partners on the corrosion of metallic materials to define those that can be used and to assess their resistance to corrosion. These studies will also determine the size of the containers and assess the impact of their degradation products on the other elements in the repository.

2. Containers and Overpacks for High-Level Radioactive Waste

2.1. Functions of Containers and Overpacks

Research is currently being conducted to determine the containment capacities of the different packages of waste produced or envisaged. Among these, the packages of vitrified waste and spent fuel contain short or medium-life radionuclides (^{137}Cs for example) that play an important role in radiological or thermal impacts on the environment outside the package. Radioactive decay for certain radionuclides in vitrified waste and thermal power of waste packages are shown in Figs 2 and 3 respectively.

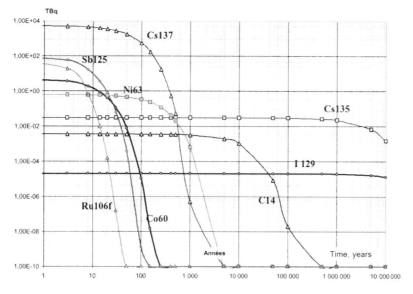

Fig. 2 *Radioactive decay for certain radionuclides in vitrified waste.*

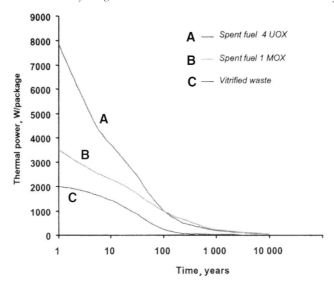

Fig. 3 *Thermal power of waste packages.*

The current plan is for total containment of the radionuclides contained in these waste products for a given length of time, in particular, during the period when the activity of the short and medium-life radionuclides is intense and the thermal power high.

The planned period of containment is approximately one thousand years for current vitrified waste and several thousand years (up to 10 000) for spent fuel.

Containment is guaranteed by an overpack added to the primary stainless steel canister filled with the package of vitrified waste and a container is added for spent fuel. Right from the start of design operations, ANDRA has preferred metallic materials rather than other materials such as concrete or ceramics for overpacks and containers. Examples of containment are shown in Fig. 4.

Apart from its main function of containment, the container or overpack must also allow the handling and installation of packages in the disposal vaults and their removal if necessary, particularly in the case of reversibility, under the required conditions of safety.

Furthermore, the container and the overpack can contribute to limiting the physicochemical effects of the package on its external environment (engineered barrier, geological environment), in particular, the spread of thermal flow and external gamma radiation. (Engineered barrier: materials used to build waste disposal structures. This does not include materials for overpack in the French concept design. This corresponds to 'the second barrier'; the first one is the waste package, including container and overpack and the third is the geological media.)

2.2. Containers and Overpacks in Repository Designs

A set of preliminary designs was selected in 1999 to help address the issues raised by the feasibility analysis of a potential repository. These preliminary designs concern a single level located around the middle of the flat Callovo-Oxfordian clay formation,

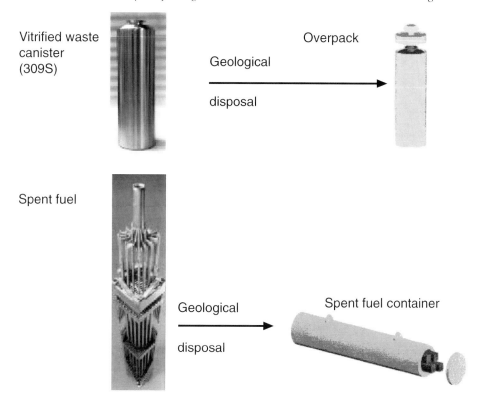

Fig. 4 *Containers and overpacks for high-level waste.*

to allow a significant thickness of clay to retain the radionuclides on either side of the disposal structure.

The architecture of the potential repository distinguishes between disposal zones dedicated respectively to B waste, C waste, UOX and MOX spent fuels. Each zone is itself divided into modules, to segregate the waste types and to guarantee the flexibility of the repository in order to cope with possible variations in inventory or waste management procedures. The segregation and physical distance also aim to simplify the THMC (Thermo-Hydro-Mechanical and Chemical) processes between waste types, including those concerned by reversibility.

Figure 5 presents an example of the preliminary design concept for vitrified waste. Tables 2 and 3 present the main characteristics of the ANDRA's preliminary concepts for vitrified waste and spent fuel.

3. Metallic Materials Envisaged for Containers or Overpacks

Four families of metallic materials are currently studied or envisaged:

- Low-alloy or unalloyed steels.
- The Fe–Ni–Cr–Mo alloys (stainless steels and nickel alloys).
- Titanium and copper alloys.

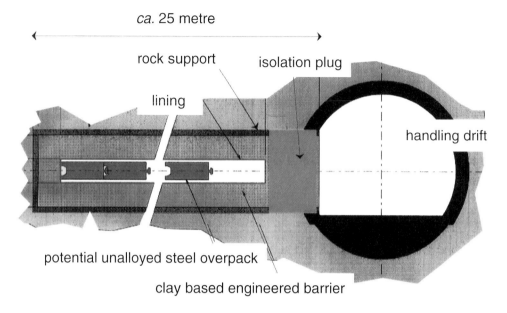

Fig. 5 *Example of preliminary disposal designs, as support to R&D programs.*

Table 2. *ANDRA Preliminary designs for vitrified waste*

General data	Details of the three options
• Three options for disposal cells • Overpack Two options: carbon steel or Ni Alloy, with a lifetime of one thousand years • One canister per overpack • Two options for thermal loading on packages (e.g.: buffer or Callovo-Oxfordian): (a) ≤ 100°C, (b) ≤ 150°C	**1. Plugged Tunnel: expanding clay base for both tunnel buffer and plug** • 8 packages per tunnel (package type C1): dia. 2.5 m; total length 23 m • Steel inner liner for installation of packages • Steel and concrete outer liner **2. Plugged Tunnel: expanding clay base for the tunnel plug but not for the tunnel buffer** **3. Plugged Borehole: expanding clay base for the plug but not for the buffer** • 8 packages per borehole: total length 23 m • Steel outer liner for installation of packages

Only the first two families were adopted by ANDRA for studies in the framework of the preliminary designs. Each family has very pronounced specific characteristics.

Table 3. ANDRA Preliminary designs for Spent Fuel

General data	Details of the two options
• Two options for disposal cells • Two types of package: 4 UOX assemblies per canister and 1 MOX assembly per canister • Two options for canisters: carbon steel or Ni Alloy, with a life time of up to several thousand years • Two options for thermal loading on packages (e.g.: buffer or Callovo-Oxfordian): (a) ≤ 100°C, (b) ≤ 150°C	1. **A vault,** combining floor-installation facilities and **a concrete-based engineered barrier (buffer)** surrounding the packages (the backfill on the upper part is not defined) 2. **A tunnel design similar to that for vitrified waste packages with an expanding clay-based buffer and plug**

3.1. Low-alloy or Unalloyed Steels

On the whole, their characteristics are 'balanced'.

• They have proven technological capacities for implementation in the manufacture of objects (metallurgy, weldability, etc.).

• They possess good physicochemical properties in terms of mechanical resistance and resistance to corrosion. As far as corrosion is concerned, proof of the performance required is based on a set of results, which show that:
 – Generalised corrosion is the corrosion form that is dominant in the medium and long-term.

 – Localised corrosion, by pitting or crevice corrosion, may be observed on these materials, particularly during the periods when oxygen is still present. Experimentally, the rate of localised corrosion (pitting, etc.) observed during short periods decreases more rapidly than that of generalised corrosion with the result that its relative importance is reduced.

 – The grade of low-alloy or unalloyed steels is not a determining parameter for resistance to corrosion.

 – Low-alloy or unalloyed steels are tolerant with respect to water chemistry; this means that they require a lesser degree of accuracy for chemical environment conditions and in the composition of the metal, its structural state and its surface state.

 – The experimental data and modelling give generalised corrosion rates of approximately several hundred nanometres to several micrometres per year under deep geological formation disposal conditions (aqueous reducing medium, temperature from 25 to 150°C).

– A large quantity of experimental data is available for assessing specific corrosion risks such as stress corrosion or hydrogen embrittlement.

– The existence of archaeological parallels with iron dating from more than 2000 years ago gives a time code for the durability of steels on a time scale that is significant in relation to that of a repository.

– The planned thickness (multi-centimetre and decimetre) greatly limits outside radiation (gamma) and consequently, radiolysis of the water in contact with the packages.

· The costs of procuring the basic material and creating the object (container / overpack) are limited in comparison with other types of metallic material.

3.2. The Fe–Ni–Cr–Mo Alloys (Austenitic Stainless Steels and Nickel Alloys)

These are classified in the 'corrosion resistant' or 'passive materials' category of metallic materials: their corrosion results in the creation of a barrier layer of nanometric thickness, the 'passive' layer, which is constantly renewed and protects the 'healthy' metal. Classically, this is why the average rates for generalised corrosion in water (maintenance of passivity) are between 0.01 and 0.1 µm per year. However, note that the environmental conditions of a geological repository ('oxidising' environment then reducing medium with the presence of chlorides) differ from most of the fields in which these alloys are used, for example in 'oxidising' environments, aggressive chemical environments, etc.

The principal characteristics of these alloys with respect to their use in a repository are as follows:

• They are sensitive to localised corrosion and especially to corrosion by pitting or crevice corrosion if the passive layer protecting the metal is destroyed in localised areas.

• The appearance and spread of localised corrosion is highly dependent on chemical environment conditions (in particular, the redox potential and the chloride content).

• Behaviour of these metals is sometimes specific to the disposal context. Thus, there may be:

– Changes in the redox potential (presence of dissolved oxygen during the initial disposal phases followed by an oxygen-deficient environment); risks of localised corrosion are essentially linked to the presence of dissolved oxygen in the water. In a 'reducing' environment (absence of dissolved oxygen) most of these materials are 'kinetically stable.' Localised corrosion cannot start or spread.

– Variations in the thickness of this type of material for the containers. This may have a significant effect of external γ radiation, causing possible radiolysis of the water in contact with the containers. The effects of water radiolysis (creation of oxidising species) are almost unknown at the disposal temperatures in question (data gathered from experiments with nuclear reactors were obtained at much higher temperatures).

- Their mechanical treatment is a determining factor as far as corrosion is concerned. The different forming and welding operations may locally modify the chemical composition and the structural and surface states of the alloy. These parameters are significant for the corrosion resistance of these materials.

- Certain types of corrosion (hydrogen embrittlement, stress corrosion, etc.) must be considered for these materials but can be minimised by choosing the precise grade of material.

- There are no natural parallels that provide a significant time code.

3.3. Titanium and Copper Alloys

These materials constitute 'extreme' solutions with respect to the functions currently required of containers/overpacks, particularly because of their high cost in comparison with other metallic materials. Their use is envisaged only if the low-alloy or unalloyed steels or the Fe–Ni–Cr–Mo alloys are not satisfactory or for containment requirements that are considerably stricter than those that are currently envisaged.

Titanium alloys are sensitive to hydrogen embrittlement and crevice corrosion but, for the latter, they are nevertheless more resistant than most of the Fe–Ni–Cr–Mo alloys.

Copper possesses a very wide thermodynamic stability domain in water (absence of corrosion), particularly in a reducing environment. This special characteristic is dependent upon the absence of complex-forming agents (chloride, sulfur or ammonia species in particular):

– In an oxidising environment, copper corrodes at low generalised corrosion rates (less than one micrometre per year).

– The risk of localised corrosion by pitting or crevice corrosion cannot be excluded. The mechanical resistance of copper is low in comparison with that of steel: the design of a copper container must be associated with an internal structure in steel, which guarantees the mechanical strength function (for example, the Swedish insert for spent fuel containers).

4. Environment of Containers and Overpacks in a Repository

The main cause of degradation foreseen for metal containers in a deep geological repository is corrosion. It is largely dependent on the physicochemical characteristics of the environment surrounding the containers, which may change during the course of time. For example, note the following changes in environmental characteristics:

• 'Atmospheric' environment followed by a waterlogged environment (Fig. 6).

• Chemical: initial presence of oxygen followed by lack of oxygen; addition of chlorides originating from the geological environment.

• Physical: the temperature around the containers is close to 100°C for tens to hundreds of years depending on the type of waste and then decreases to the temperature of the geological environment (Fig. 7).

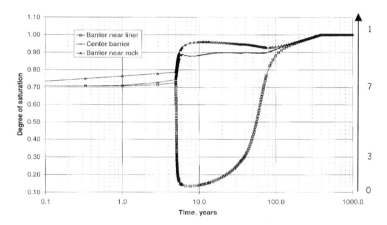

Fig. 6 Degree of water saturation and percentage of RH in the engineered barrier.

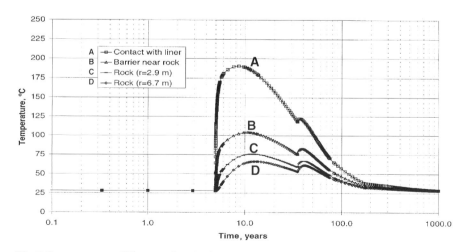

Fig. 7 Temperature at different points in the engineered barrier for vitrified waste packages.

- Radiological: the β/γ radiation, particularly that associated with short or medium-life radionuclides, may favour the creation of oxidising species via radiolysis of the water outside the container if the thickness of the container is not sufficient to attenuate its effect.

5. Study Programme concerning Corrosion of Containers and Overpacks

5.1. Organisation and Methodology for the Establishment of the 2000–2005 Study Programme

At the beginning of 2000, ANDRA formed a group of corrosion specialists, employed by the corrosion laboratories of the CEA, EDF and FRAMATOME and experts from outside these organisations (CEFRACOR and independent experts) with the support of departments of the CNRS (National Scientific Research Center), schools and universities.

The first task of this 'Corrosion Group,' based on an indication of requirements provided by ANDRA and on the basis of current knowledge, consisted of identifying the themes of the studies to be conducted in the field of corrosion. Particular attention was paid to the examination of completeness, the concern being to avoid forgetting the problems to be dealt with and discovering them later. To accomplish this, a systematic approach was followed, using two-way tables (environment/types of corrosion) filled in jointly. It rapidly became apparent that the problem with low-alloy or unalloyed steels was sufficiently different from that of the passive alloys for them to be treated according to two relatively different approaches. The members of the 'Corrosion Group' set up the two sub-programmes:

- By identifying the points to be dealt with by crossing over the corrosion risks (and associated problems) and the environment of the containers/overpacks.

- By justifying the necessity of the point to be dealt with during the crossover study.

- By determining the priorities for these points.

This analysis made it possible to propose 'priority' actions to be carried out starting in the period 2000–2001 and other 'additional' studies that can be conducted during the period 2002–2005.

The full 2000–2005 'corrosion' programme was divided into a 'low-alloy or unalloyed steels' programme and a 'passive alloys' programme. However, to have a full range of solutions available, copper and titanium will be the subject of a scientific watch, based on data drawn from the relevant literature.

5.2. The 'Low-alloy or Unalloyed Steels' Programme

Three 'priority' themes were identified for the period 2000–2001.

5.2.1. The kinetics of general corrosion: experimental determination and modelling.
General corrosion was identified as the principal corrosion risk for unalloyed steel. The corrosion rates obtained experimentally converge on relatively low values (in the region of a few microns per year, sometimes less than one micron per year) and the analyses show the predominant role played by the layers of corrosion products.

The reliability of the behaviour forecasts for the materials over periods of thousands of years and the provability of these forecasts will largely depend on the degree of knowledge acquired on the subject of corrosion mechanisms. Concerning the conditions for storage of waste, these mechanisms are not sufficiently understood and the behaviour forecasts remain, for the most part, empirical.

This part of the programme aims to improve this situation but it does not aspire to the development of complete mechanistic models on an atomic scale. The latter would very probably be outside the reach of any study with a realistic time scale. The approach will be mechanistic but the scale will remain macroscopic.

The experiments and models must enable determination, from among the numerous stages of the global corrosion phenomenon, of those that will be the most relevant to take into consideration for prediction calculations.

Three approaches will be considered for modelling purposes:

- Determination of the inventory of the oxygen and other species (other than water) that are potentially oxidising: knowledge of the total quantity of oxygen present in the engineered barrier at the time of closure of the repository enables the calculation of a maximum thickness of steel likely to be consumed by oxygen, assuming that all the oxygen reacts with the metal.

- Determination of the rates of corrosion by water in the extreme cases of 'pure kinetics': these rates must be assessed in a few borderline cases for which a single kinetic stage controls the rate of the global corrosion process. This approach limits the scope of possible predictions and thus contributes to making the whole modelling approach reliable.

- Determination of corrosion rates in a few situations where several stages are involved in the kinetic control of the global process. This part of the study will be conducted as an iterative process of trial and error that must converge on a model leading to predictions that are increasingly in accordance with experimental data. It will be preceded by a clearly specified phase of identification of the elementary stages and simplifying hypotheses, which is intended to simplify the ensuing iterative approach.

The purpose of the experiments is:

- To provide a support for modelling: identification of the mechanisms, acquisition of the data necessary for modelling, experimental checks and validations of the modelling procedure, etc.).

– To supply 'overall' data: determination of the rates and kinetics of corrosion according to the various physico–chemical paramaters associated with the environment (temperature, pH, redox potential, etc.) and the material (silicon content, etc.).

5.2.2. Risks of localised corrosion by pitting or crevice corrosion.
One of the main arguments for the choice of low-alloy or unalloyed steels is that, unlike 'passive alloys' materials, which are subject to a localised corrosion that is very difficult to model, the corrosion behaviour of these (moderately) 'consumable' materials is less difficult to predict (but their general corrosion rate is greater).

Even with a low alloy content, the steel envisaged for the waste containers or overpacks is nevertheless likely, as shown by the experiments, to be spontaneously covered by a protective film of solid corrosion products on contact with clay. This layer is similar to a passive film and is responsible for a noticeable reduction in the rate of general corrosion over the course of time.

It is therefore to be feared that once covered by this film, the alloy may, like the materials more usually considered as passive (such as stainless steels for example) be sensitive to various forms of localised corrosion such as corrosion by pitting or crevice corrosion.

Two additional actions are proposed in the study programme:

– Experimental work aiming to assess the risks of localised corrosion on low-alloy or unalloyed steels in contact with clay.

– Modelling of the spread of corrosion by crevice corrosion in low-alloy or unalloyed steels. A particular effort will be made to determine the conditions for the slowing down and stopping of deep cracks.

5.2.3. Corrosion products and hydrogen production.
Particular attention is paid to the corrosion products (solids, solutions, gases) that may interact chemically (iron/clay interaction) or physically (mechanical constraints generated by solid or gaseous products) with the surrounding materials (materials in the engineered barrier or in the geological site).

This part of the programme aims to assess the possibility of interaction between the corrosion products of the container and the clay. It will concentrate particularly on specifying the risk of degradation of the waterproofing properties of the clay or concrete and consequently, the increase in the corrosion rate of the container and/or the appearance of localised corrosion phenomena.

5.2.4. Other directions of study.
The other directions of study at the outcome of the hierarchic approach will also be developed, in particular, those concerning stress corrosion cracking, bacterial corrosion, hydrogen embrittlement and galvanic corrosion.

5.3. The 'Passive Alloys' Programme

For passive alloys (essentially the Fe–Ni–Cr–Mo alloys), corrosion by pitting and / or by crevice corrosion is the major risk.

Taking into account initiation times that are sometimes very long, an approach based on exposure tests, even long ones on the scale of the experiment (several years for example), cannot supply a guarantee of long-term resistance and, at best, enables elimination of the least resistant materials.

An electrochemical approach based on an assessment of critical potentials and supported by knowledge and, if possible a model of the processes involved, is more adapted to the problem and is the only approach capable of guaranteeing the 'robustness' and longevity of the criteria adopted.

To avoid all localised corrosion the corrosion potential of passive alloys must always remain lower than the critical initiation potential of the most penalising type of corrosion, which is crevice corrosion. Since the critical initiation potential for crevice corrosion is practically the same as its repassivation potential and that the latter is more conservative and quicker to measure, the practical criterion will be based on the repassivation potential. The materials that will be considered acceptable will be those passive alloys the repassivation potential of which remains higher than the corrosion potential at any given time in the life of the container. This is also the approach developed in Japan or by the CNWRA in the case of the Yucca Mountain repository and in Belgium by the SCK-CEN.

This methodology also has the advantage of removing the surface state problems that are known to affect resistance to the initiation of localised corrosion. Even if localised corrosion were to begin as a result of the bad surface state of the container, the fact that its corrosion potential is lower than the repassivation potential of the base metal guarantees that the spread of corrosion will cease as soon as the faulty metal has been consumed.

The 'passive alloys' programme deals with this problem via an approach that is both experimental and theoretical:

–   Experimental determination, for certain alloys that may be envisaged and under variable physicochemical conditions (pH, temperature, presence of radiation), of their repassivation potential.

–   Modelling of the chemistry in the cracks and of the propagation of the corrosion damage.

5.4. Research Teams Participating in this Programme and their Study Themes

The study approach is based on experiments conducted in 'surface' laboratories and in the underground laboratory on the contact of the argillaceous material in the geological site with industrial equipment and archaeological objects. For the period 2000–2001, the laboratories contributing to the development of this programme are:

- Corrosion of carbon steel in argillaceous mixtures: corrosion measurement and modelling: CEA-Saclay/SCCME.

- Corrosion of low-alloy or unalloyed steels in the presence of compacted clay: CEA-Cadarache/SAMRA.

- Limitation of corrosion by the flow of water in the geological environment: UPC (Universitat politecnica de Catalunya)

- Corrosion of carbon steel in an unsaturated clay material: EDF/R&D.

- Action of silicon on the corrosion resistance of low-alloy or unalloyed steels in an argillaceous environment: ENSCP.

- Corrosion of ferrous archaeological objects: UTC/CNRS (University of Compiègne)

- Influence of the structure and composition of ferrous archaeological objects on their corrosion behaviour in soils: CEA/CNRS/UTC.

- Behaviour of passive alloys under disposal conditions: determination of pitting potentials: CEA-Saclay/SCCME.

- Characterisation of passive alloys with respect to crevice corrosion: FRAMATOME with the 'Ecole des Mines' in St-Etienne and CORREX.

- Influence of the dose rate on the corrosion potential of materials under disposal conditions: preliminary tests: CEA-Saclay/SCCME.

- Modelling of localised corrosion on passive alloys: University of Bourgogne (Burgundy).

4. Conclusions

- Corrosion is the principal degradation mechanism envisaged for high-level waste containers or overpacks.

- A research programme was set up at the beginning of 2000 to provide support for the choice of materials envisaged for the construction of the containers or overpacks. It was also set up to assess and quantify the different risks of corrosion to which the materials may be subject in a deep geological repository.

- For low-alloy or unalloyed steels, the research programme was set up with the objective of consolidating knowledge of the mechanisms and kinetics of corrosion. The work identified should provide:

– Modelling tools that are less empirical.

– Greater reliability in estimating corrosion rates in order to reduce the margins of uncertainty (with the associated consequences on the impact of the corrosion products on the environment surrounding the packages).

– Confirmation that the risk of localised corrosion (and the other forms of corrosion) is not a determining factor.

• For the passive alloys, the research programme aims to determine the 'viability' of certain grades of alloys (with a high nickel and chromium content) in the context of a deep geological repository. In other words, its aim is to show that for certain alloys, corrosion by crevice corrosion cannot spread in the environmental conditions of the repository.

• Since copper and titanium alloys are not among the first choices for preliminary designs, the study of their corrosion is in the form of a watch based on the work carried out by the Swedish and Canadians respectively. However, as far as titanium is concerned, assessment of corrosion by crevice corrosion raises the same problems as for Fe–Ni–Cr–Mo alloys.

• All these studies were engaged on bases of questions stemming from preliminary concepts. Orientations and programmes are susceptible to evolve with the evolution of studied concepts.

5. Acknowledgements

The authors wish to thank all the members of the ANDRA 'Corrosion Group' who participated in setting up the research programme on the corrosion of high-level waste containers and overpacks. B. Besnainou (CEA), P. Combrade (FRAMATOME), A. Desestret (corrosion consultant), D. Féron (CEA), J.M. Gras (EDF) and G. Pinard-Legry (CEFRACOR).

3

In situ Testing of Waste Forms and Container Materials: Contribution to the Identification of their Long Term Behaviour

P. VAN ISEGHEM, B. KURSTEN, E. VALCKE, H. SERRA, J. FAYS
and A. SNEYERS

Waste & Disposal, SCK•CEN, Boeretang 200, B-2400 Mol, Belgium

ABSTRACT

This paper reviews *in situ* projects that have been carried out in the underground laboratory in Boom Clay in Mol in Belgium. The projects involved *in situ* interaction between candidate container materials, nuclear waste glasses or cements and the Boom Clay. The *in situ* tests on container materials showed a strong corrosion of C-steel, so that a new container (overpack) material is being considered actually in Belgium, namely stainless steel. The interaction between waste glass or cements and Boom Clay results in reaction layers of a few hundred μm thickness. These layers were successfully characterised in detail, and the reaction processes based on these analyses interpreted. The outputs of the *in situ* tests on the different materials are compared, and possible interferences discussed. The added value of these *in situ* tests as part of the global objective of evaluating the long term behaviour of the materials in a disposal rock, in conjunction with laboratory tests and modelling is discussed.

1. Introduction

SCK•CEN (Studiecentrum voor Kemenergie/Centre d'etude de l'Energie Nucléaire) has been conducting R&D programmes for more than 20 years to evaluate the suitability of a clay formation to dispose long-living and high-level conditioned radioactive waste. It has constructed an underground laboratory in Boom Clay underneath its site at about 220 m below the surface, and initiated a number of *in situ* projects on various issues such as corrosion, migration, mechanics, and engineering. These *in situ* projects were carried out in complement of laboratory test programmes and modelling studies. The overall objectives are to demonstrate that Boom Clay is a suitable host for geological disposal, and that a disposal concept can be proposed that provides safety on the long term. Note: long term means up to a few thousands of years in the case of the container material, and to hundreds of thousands of years in the case of the waste forms. These periods are set by the acceptance criteria for geological disposal and by the performance assessment studies of the disposal.

The waste forms that received the largest attention are vitrified high-level waste, bituminised intermediate level waste, cemented waste, and more recently spent fuel. For container materials we focussed mainly on C-steel and stainless steel.

This paper will discuss the concepts of the *in situ* tests carried out in direct contact with Boom clay, that is the candidate host rock in Belgium, and thus forms the far field. We do not discuss *in situ* tests that were performed in clay derived atmospheres [1,2], because they are not that relevant for normal situations during disposal. We will discuss some of the most relevant experimental data obtained on candidate container materials, vitrified waste samples, and cements or cemented waste, how the *in situ* tests on the different materials (metal, glass, cement) can have a mutual benefit, and also how we can achieve an understanding of the long term behaviour of the materials considered, and how *in situ* tests can contribute.

2. Description of the Work

2.1. Concept of the *In Situ* Tests

Over the years we developed different 'generations' of *in situ* tests. They have all consisted of fixing the samples (container materials or waste forms) to the outside of a support tube, heated from the inside, and exposing the test tube loaded with the samples to Boom Clay for a certain duration. Afterwards the tube is retrieved by overcoring, and the corrosion of the samples is investigated. The concepts of the various *in situ* tests discussed in this paper are shown in Figs 1–4.

Figures 1 and 2 show the set-ups of the tests providing a direct interaction between the sample and Boom Clay at a predetermined temperature, but without considering any other parameters. We used ring-shaped metallic container or cement rings and coupon shaped waste glass (or glass–ceramic, ceramic, cement or bitumen) samples. These tests aimed to investigate the effect of the composition of the container material or waste form on its corrosion resistance, at different temperatures and durations. We operated tests at natural Boom Clay temperatures of 16°C, 90°C and 170°C, for durations of 1 to 7.5 years. The test tubes loaded with container and essentially waste glass samples were amongst the first tests performed in the underground laboratory — they started around 1986. The test tubes conceived for the cement samples were operated between 1998–2000. Detailed information on these various *in situ* tests can be found in [1–3].

The additional effect of a γ-radiation field was investigated in the CERBERUS (**C**ontrol **E**xperiment with **R**adiation of the **Be**lgian **R**epository for **U**nderground **S**torage) *in situ* test (Fig. 3). This test was conceived to study the effects on the near field in an argillaceous (Boom Clay) environment of a COGEMA (Compagnie Générale des Matières Nucléaires) HLW glass canister shortly after disposal (after 50 years cooling) [4]. ^{60}Co sources and a heating system generated the conditions in which the Boom Clay was studied. We took the opportunity to emplace some waste glass and container samples in the vicinity of the ^{60}Co sources, so that the samples interacted with the natural Boom clay in the presence of a γ-radiation field. The interaction took 5 years at 80–85°C.

The last *in situ* test is called CORALUS (**Cor**rosion of **A**ctive G**l**ass in **U**nderground **S**torage Condition) (Fig 4). This test adds to the CERBERUS test α-activity in the glass samples (about 0.85 wt% of ^{237}NpO$_2$, $^{238-240}$PuO$_2$ or ^{241}Am$_2$O$_3$), the presence of different reacting environments (two candidate backfill materials based on bentonite clay and Boom clay), and the monitoring of the gas generated close to the radiation

Fig. 1 *View of a type I corrosion test tube during installation in the underground research facility (right side: section with metallic samples; left side: section with waste form samples).*

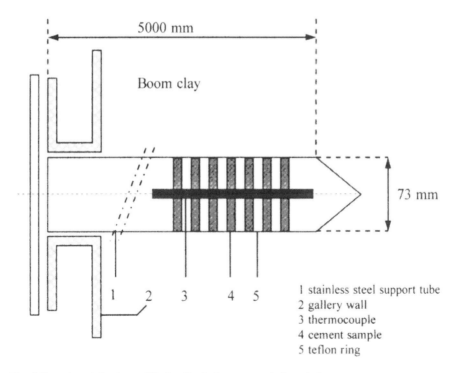

Fig. 2 *Experimental set-up of the* in situ *tests on cemented waste forms.*

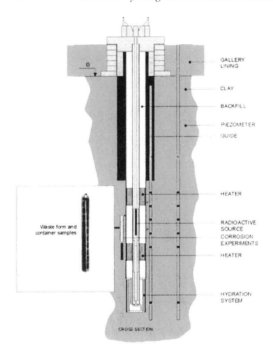

Fig. 3 *Design of the CERBERUS test.*

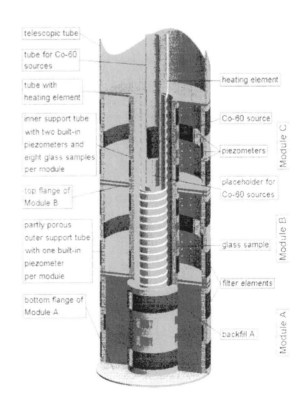

Fig. 4 *Design of a CORALUS* in situ *test tube.*

sources. Some container samples were added as well. The study of different reacting media is possible through the modular concept of CORALUS — one test tube actually consists of three modules containing a specific reacting medium. Four modular test tubes will be installed in the Boom Clay for durations of between two and ten years. The γ-doses to which the samples will be submitted are roughly the same as during the CERBERUS test. Full details of this test can be found in [5]. The first CORALUS test tube is in operation since 1999.

We should comment that in most *in situ* tests discussed here Boom Clay was the interacting or corroding medium, although in the real repository the interacting medium over long periods of time may be a specific kind of backfill. When designing the first *in situ* test concept (Fig. 1) Boom Clay was supposed to contact the waste package 'relatively' soon after disposal. When desinging CERBERUS, the disposal concept was different, but Boom Clay was a candidate backfill material. In the CORALUS test we account for the variety in candidate backfill materials that are presently considered. But in general it can be argued that Boom Clay will be one of the more aggressive media that will interact with the various materials considered here.

2.2. Rationale

The *in situ* tests described above are an integral part of a global approach, together with laboratory parametric experiments and modelling. One of the objectives of our *in situ* tests is to demonstrate the feasibility of manipulations with the engineered barrier materials (container, waste form) in conditions simulating the underground disposal site. *In situ* tests also provide the best means to investigate the pH/Eh and chemical composition of the interstitial clay water [6] — this information being essential to conceive laboratory tests with natural Boom Clay. Another objective of the *in situ* tests is to validate the results from the R&D actions carried out in surface laboratories. Experiments on real systems (real, fully active waste forms or full size packages) is another validation of the surface laboratory studies. The surface laboratory studies focus on identifying specific parameters as input into modelling studies, on performing accelerated tests to obtain information on long term processes, and on integrated effects and processes (e.g. the combined effect of interaction between a waste form or container material and clay, with sorption/migration effects through the clay. Modelling is done to simulate the interaction processes (e.g. to estimate processes in the long term), or to make predictions of the long term behaviour.

Studies on natural analogues are another way to obtain information on the long term behaviour of waste forms or container materials in certain environments.

Several international reviews have commented on the usefulness of *in situ* testing. An international workshop devoted to *in situ* tests on waste forms and container materials in various candidate rocks in 1992 concluded that *in situ* testing is worthwhile and contributes to the characterisation of the performance of the waste package constituents [7]. Two review papers on underground laboratories emphasise besides the technical interests of underground testing (see above) that underground testing is important in view of confidence building for proposed disposal concepts, and for communication to the public [8], and that the output of the *in situ* tests can be various depending on the scale [9].

2.3. *In Situ* Tests on Container Materials

The *in situ* corrosion tests considered a number of candidate container materials: C-steel as corrosion allowance material, and various stainless steels, Ni-alloys and Ti-alloys as corrosion resistant materials. These tests provided a lot of important results concerning the final choice of the overpack material as follows:

(i) the corrosion allowance material, carbon steel, did exhibit severe signs of localised corrosion after interaction with Boom Clay: hemispherical pits up to 240 μm deep were detected after 2 years at 90°C [1]. The rate of pit growth and the surface morphology were similar for 16°C and 90°C, but different for 170°C (the pits looked eroded, showing a succession of wider pits with abraded edges). This fast growth of pits (a conservative reasoning would suggest that an overpack of 50 mm would be needed to withstand the corrosive underground disposal environment during the aerobic period — estimated to last for some 65 years) together with the large amounts of hydrogen gas generated during the subsequent anaerobic period contributed to the decision of the Belgian waste management authorities to abandon carbon steel as reference overpack material.

(ii) the weld material of carbon steel was more susceptible to pitting corrosion than the parent material. The localised attack was even more pronounced in the presence of a radiation field: thickness reductions of about 2000 μm in the heat-affected zone were observed (after direct exposure to Boom Clay for 5 years at 80°C in the CERBERUS test) [10].

(iii) the corrosion-resistant materials (stainless steel, Ni- and Ti-alloys) did not suffer from corrosion. The welds also remained unaffected by a direct contact with clay between 16°C and 170°C. These materials were not affected by the presence of a γ-radiation field.

The specific added value of the *in situ* corrosion tests on container materials discussed above has been that a large number of candidate materials could be tested within the same test. We were also convinced that it is important to know and understand the evolution in the chemistry of the near field (see the change from aerobic to anaerobic phase). This information is very important in view to design new *in situ* tests on container materials.

2.4. *In Situ* Tests on Vitrified Radioactive Waste

Within the *in situ* test programme we investigated various candidate nuclear waste glass compositions. Some 20 different compositions were loaded onto the first test tubes late 1980s [1,2], whereas we restricted to the main glasses of importance to the Belgian programme (the French R7T7 glass and the German PAMELA SM glasses)

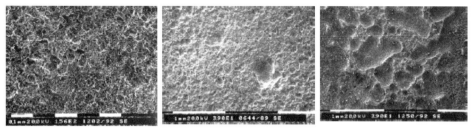

Fig. 5 *Electron micrographs of carbon steel after* in situ *corrosion in Boom Clay for 5 years at 16°C (left), 2 years at 90°C (middle) or 5 years at 170°C (right).*

later on [4,11]. Most of the data on glass dissolution were obtained from the tests at 90°C (about 2, 3 and 7.5 years interaction), which are most relevant for the initial thermal phase of the geological disposal. We briefly refer to some more recent data obtained in the CERBERUS *in situ* test, that added γ radiation sources to the test. The results of this integrated *in situ* test are fully reported in [11,12].

The glass dissolution was analysed by measuring the total mass loss and by extensive surface analysis using scanning electron microscopy, electron microprobe and secondary ion mass spectroscopy. These techniques were used to characterise the outside of the reacted glass, and the composition of the reaction layer as a function of depth. It is worth mentioning that in these *in situ* tests we were successful in separating the contacting Boom clay from the corroded glass samples, without substantially damaging the glass samples and the covering reaction layer.

The most important findings on glass dissolution following the CERBERUS test are [11]:

(i)   Depletion depths in the glass samples between 40 and 400 μm (depending on the glass composition) were found — as calculated from the mass loss.

(ii) Reaction layers between 1 and 450 μm thick formed on top of the glass and consisted of a precipitation layer, a reaction 'gel' layer, and diffusion zones (see Figs 6 and 7).

(iii) Secondary phases formed on top of the reaction layer, the composition of which depended on the glass composition (see Fig. 6).

In Figs 6 and 7 we show some analyses on glass SON68, which is the reference glass used to vitrify the reprocessed spent fuel in the R7T7 plant of COGEMA.

The SIMS profiling analyses in combination with the mass loss data showed that the dissolution mainly proceeded in a selective — substitutional way [13]. In general the 'mobile' alkali elements are almost eliminated in the reaction layer, Si (the matrix element of the glasses) is depleted in some way, other constituents are almost preserved in the layer, while others (H, K, Mg,) show an influx from the clay into the glass surface. Note, This conclusion does not imply that one can conclude on the different dissolution processes that have occurred from the beginning until the end of the test, or that one could use this information to extrapolate to longer terms. The

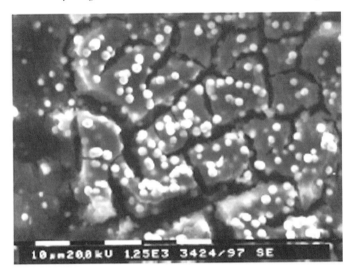

Fig. 6 SEM picture of the outer surface of glass SON68 after five years interaction at 80–85 °C with Boom Clay in the CERBERUS in situ *test. We observe a Zn, S phase (white spots) on top of the reaction layer can be seen.*

predominant selective processes suggest that the dissolution rate decreases with time anyway.

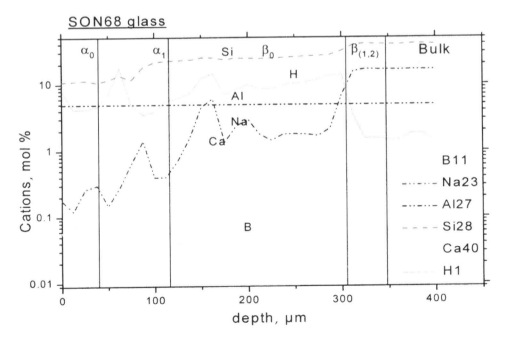

Fig. 7 SIMS profile for some of the main glass constituents recorded by step-scan for glass SON68, after five years interaction at 80–85 °C with Boom Clay in the CERBERUS test.

We also conducted an extensive laboratory R&D programme aiming at identifying the parameters (e.g. the effect of temperature — *in situ* at 16°C we measure a dissolution rate of < 0.3 μm/year) associated with the different dissolution processes [14]. We also proceed through geochemical modelling to improve the basic understanding and to evaluate the role of secondary phases, and analytical modelling to calculate the long term behaviour. Only this way it will be possible to obtain a comprehensive prediction of the long term dissolution behaviour of the waste glasses.

The *in situ* tests on HLW glasses have been particularly interesting to generate data on the long term behaviour [15]. Using standard leach tests in pure water in static condition the total glass mass loss remains limited to ~1 g m^{-2}, whereas glass mass losses as high as ~1000 g m^{-2} were obtained after *in situ* interaction with Boom clay for 7.5 years at 90°C. Such very high reaction progress will lead to very strong accumulation of leached elements in the solution, and to processes that might be impossible to observe otherwise. We mention, the saturation effects in solution, reaction layer formation, precipitation effects on the glass surface, secondary phase formation.

The specific added value of the *in situ* corrosion tests on nuclear waste glass samples reported above is that they have enabled to investigate a large number of glass compositions in the same condition. They also generated data for glass dissolution taking place with a very large reaction progress, which could be thoroughly analysed and interpreted.

2.5. *In Situ* tests on Cement and Cemented Waste

We investigated different cement compositions, including pure OPC (Ordinary Portland Cement), OPC mixed with additives, or cement formulations mixed with specific waste fluxes. All samples were subjected to the same *in situ* test procedure as shown in Fig. 2, and were analysed by electron microprobe analysis, infrared microscopy and X-ray powder diffraction. Details on the sample composition and analytical results can be found in [3].

The various types of cement samples display a quite similar behaviour in the different test conditions. Figure 8 shows an electron micrograph of OPC after exposure to Boom Clay during 12 months at a temperature between 50 and 65°C. An altered zone of a few hundreds of μm forms in both cement and clay. We can distinguish the following sub-layers [16]:

(i)   A Ca and/or Si depleted zone, resulting from leaching from the outer cement zone (zone c in Fig. 8). Other elements (Al, Mg, Fe, K, Na, S) are not visibly altered in the cement;

(ii) A Ca-rich zone in the Boom Clay close to or at the cement–clay contact, formed by precipitation of calcite in the shear zones (disturbed clay) (zone b in Fig. 8);

(iii) A (hydrated) Mg–Al–Si gel precipitation layer in the interface cement–clay (zone i in Fig. 8);

(iv) A calcite (CaCO$_3$) precipitation layer within the cement (zone c in Fig. 8).

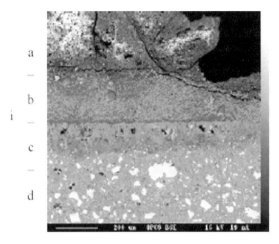

Fig. 8 Backscattered electron image of the Ordinary Portland Cement sample after 12 months contact nominally at 85°C with the Boom Clay. (a: unaltered Boom Clay; b: altered Boom Clay; c: altered cement; d: unaltered cement; i: interface clay–cement).

We conclude that there is a considerable but selective mass transport between cement and clay, together with carbonation reactions [16]. Probably redox reactions will occur as well, for instance involving the oxidation of (part of) the pyrite Fe_2S present in clay. However we consider that these observations are probably not representative for the long term interaction between cement and repository clay, which will be diffusion controlled. It is more likely that convection processes are responsible for a large part of the interaction between cement and clay during our *in situ* test. Clay water would have filled the gap between the test tube and Boom Clay before the clay converged. This water consequently saturates with Portlandite from the cement, and percolates through the fractures or sheared zones in the contacting Boom Clay. This water becomes carbonated by CO_2 supply from the clay, and calcite precipitates — see zones b and d in Fig. 8.

The question is how the interaction between cement and repository clay will proceed in the long term. It is believed that the highly alkaline solution in equilibrium with cement will affect clay in some way, e.g. affecting the chemistry of the interstitial clay water and the sorption characteristics of the clay. The effects of a high alkaline solution ('alkaline plume') on Boom Clay is studied within the ECOCLAY project of the European Commission [17]. Geochemical modelling studies have been conducted to have first estimates of the effect of the alkaline plume [18]. We believe that in order to have a reliable estimation of the long term interaction between cement and Boom clay, one has to process through the following steps:

(i) Experimentally, through interaction experiments between cements and Boom Clay, generating only diffusion processes as mass transfer from cement to clay. The tests must run over sufficiently long times (e.g. 5 to 10 years), and they should be interpreted in terms of a thorough characterisation of the secondary amorphous and crystalline minerals formed at the interface. These secondary phases will affect indeed the mass transport from cement to clay;

(ii) By geochemical modelling; the PHREEQC code allows the cement–clay interaction system to be reproduced, by using input parameters such as the diffusion constant (e.g. of Ca) and the mineral phases in equilibrium with the interstitial solution — all parameters must be determined experimentally. Some geochemical modelling calculations are reported in [19]. They considered either convection or diffusion transport processes. We performed some preliminary calculations in the clay–cement water system, although eventually the two sub-systems cement/cement water and Boom Clay/interstitial clay water will have to be considered.

The specific added value of the *in situ* tests carried out so far on cement has been mainly that we have obtained evidence that various processes such as convection and diffusion control the interaction between the cementitious materials and Boom Clay. The test therefore was very valuable for the design of future tests to study the interaction cement–Boom Clay.

3. Discussion

The main conclusions from the *in situ* tests on the various materials (container, glass, cement) have been discussed in the previous section. More detailed critical assessments of these *in situ* tests are given elsewhere [15, 19]. The conclusions are quite straightforward, in that the *in situ* tests did provide an important added value, and that they did not result in conclusions that were contradicting the laboratory test results.

The question that arises is whether the *in situ* tests on the different metallic container materials, waste glasses, and cements can benefit from each other? One might be very doubtful on this, since for instance metal corrosion is known to be strongly dependent on the electrical potentials existing, for example, between the metal and the solution. In case of glasses and cements other processes or phenomena are predominant (amongst others: solid state diffusion, hydrolysis, chemistry of the solution, sorption on solids). Even the processes are quite different between glasses and cements. But we can identify some statements of common importance for all materials studied.

1. From the projects on glass and cement it is concluded that thorough surface and cross-section analysis of the corroded samples is possible, and that this provides important and reliable information on the reaction mechanisms. Besides the 'common' techniques such as SEM-EDS, EPMA-X-ray and XRD, a high resolution profiling analysis with SIMS proved to be a very powerful tool. It is conceivable that reaction layers are formed on top of corrosion resistant container materials as well, with potential effects on the corrosion rate.

2. The importance of monitoring the environmental conditions was shown particularly in the *in situ* tests on container materials, for example the occurrence of an initial aerated phase (before the clay is setting onto the test tube). The change with time of the chemistry of the near field contacting the various waste forms has been acknowledged as very important, and it should be understood. This chemistry will control for instance the speciation of the

radionuclides leached into the near field, the secondary phases formed upon interaction between the waste form and the near field, etc.

4. Conclusions

A common conclusion from the *in situ* projects on each of the various materials is that they can only have sense as part of an integral approach to study their long term behaviour in disposal conditions, even if the extent of interpretation of the projects was different.

The contribution of the *in situ* tests in predicting the long term behaviour of the materials in disposal conditions is twofold:

(i) It is clear that, at least in the conditions of disposal in Boom Clay, we can only have predictions on the long term behaviour of the different materials studied through the combined information from the various experimental (laboratory, *in situ*) and modelling approaches. None of them alone allows to make extrapolations over the long term. Moreover it is essential that one achieves an understanding of the processes that control the corrosion or dissolution of the container material or waste form in the long term. The design of the different experimental programmes (including the *in situ* tests) therefore must be done with a maximal concern to provide long term data.

(ii) By performing the first kind of *in situ* tests as reported in this work — i.e. evaluating the interaction between a material (container, waste form) and Boom Clay has provided the realisation that other phenomena or processes had a strong effect on the *in situ* test results. Attention can be drawn to the following:

(1) the sequence of aerated/deaerated condition upon contact container material and Boom Clay; thus, more emphasis had to be given to understanding the evolution of the chemistry close to the waste package, and the design of new *in situ* tests has to deal with this parameter;

(2) the *in situ* tests on cement showed that convection and diffusion processes occur, probably sequentially; new test concepts must separate these processes.

4. Acknowledgements

We acknowledge the financial support from the European Commission and NIRAS/ONDRAF for the *in situ* test projects on container and waste glass, and for the European Commission for the *in situ* test project on cement.

References

1. B. Kursten, B. Cornélis, S. Labat and P. Van Iseghem, 'Completion of the corrosion programme in Boom clay — *in situ* experiments', EUR 17105, 1997. Published by the European Commission.

2. B. Kursten and P. Van Iseghem, 'Geological disposal of conditioned high-level and long lived radioactive waste. *In situ* corrosion experiments', Final report for 1995–1996, R-3247, 1998.

3. F. Adenot, *et al.*, 'Barrier Performance of cements and concretes in nuclear waste repositories', EUR 19780, 2001. Published by the European Commission.

4. L. Noynaert, Editor, 'Heat and radiation effects on the near field of a HLW or spent fuel repository in a clay formation (CERBERUS project)', EUR 19125, 2000. Published by the European Commission.

5. E. Valcke, *et al.*, 'An integrated *in situ* corrosion test on α-active HLW glass', EUR 19795 (2001). Published by the European Commission.

6. H. Moors, personal communication.

7. T. McMenamin, Editor, '*In situ* testing of radioactive waste forms and engineered barriers' — Conclusions, EUR 15629 (1993). Published by the European Commission.

8. C. McCombie and W. Kickmaier, 'Underground research laboratories: their roles in demonstrating repository concepts and communicating with the public', Euradwaste 1999 'Radioactive waste management strategies and issues', EUR 19143, 274–281 (C. Davies, ed.), 2000. Published by the European Commission.

9. W. Brewitz, T. Rothfuchs, F. Huertas and B. Neerdael, '*In-situ* testing of underground barrier systems in view of PA-relevant parameters', Euradwaste 1999 'Radioactive waste management strategies and issues', EUR 19143, 282–296, C. Davies Ed, 2000. Published by the European Commission.

10. B. Kursten and P. Van Iseghem, 'Effect of radiation on the corrosion behaviour of candidate container materials for the underground disposal of high-level radioactive waste in Boom Clay — *in situ* experiments'. *SPECTRUM '98*, Published by the American Nuclear Society, 1998, 805–812.

11. P. Van Iseghem, E. Valcke and A. Lodding, 'In situ testing of the chemical durability of vitrified high-level waste in a Boom Clay formation in Belgium: discussion of recent data and concept of a new test', *J. Nucl. Mater.*, 2001, **298**, (1,2), 86–94.

12. L. Noynaert, *et al.*, 'The CERBERUS project. A demonstration test to study the near field effects of a HLW canister in an argillaceous formation', Final report 1986–1998, R-3293, 1998. Published by SCK•CEN.

13. A. Lodding and P. Van Iseghem, 'In depth distributions of elements in leached layers on two HLW waste glasses after burial in clay: step-scan by SIMS', *J. Nucl. Mater.*, 2001, **298** (1,2), 197–202.

14. K. Lemmens, *et al.*, 'Experimental and modelling studies to formulate a source term of nuclear waste glass in representative geological disposal conditions', Final report for 1996–1999. To be published by SCK•CEN.

15. P. Van Iseghem, '*In-situ* corrosion tests on HLW glass as part of a larger approach', GLASS-Scientific Research for High Performance Containment, CEA Valrhô Summer Session Proceedings (1998), 535–544. Published by CEA.

16. A. Sneyers, M. Paul, M. Tyrer, F. P. Glasser, J. Fays and P. Van Iseghem, '*In situ* interaction between cement and clay: implications for geological disposal', in *Scientific Basis for Nuclear Waste Management XXIV*, Mat. Res. Soc. Symp. Proc. Vol. 663 (K. Hart and G. Lumpkin, eds). 2001, pp.123–129.

17. F. Huertas *et al.*, 'Effects of cement on clay barrier performance — ECOCLAY project'. Final Report, EUR 19609 (2000). Published by the European Commission.

18. D. Savage, 'Review of the potential effects of alkaline plume migration from a cementitious repository for radioactive waste', Environmental Agency (Bristol, UK), NAGRA (Switzerland), PNC (Japan). Report DS-97 (1997).

19. D. Read, F. P. Glasser, C. Ayora, M. T. Guardiola and A. Sneyers, 'Mineralogical and microstructural changes accompanying the interaction of Boom clay with ordinary Portland cement', *Adv. Cem. Res.*, 2001, **13**, (4).

20. B. Kursten, F. Druyts and P. Van Iseghem, 'Methodology in Corrosion Testing of Container Materials for Radioactive Waste Disposal in a Geological Clay Repository', *8th Int. Conf. on Radioactive Waste Management and Environmental Remediation (ICEM'01)*, 30 September– 4 October, 2001, Bruges (Belgium), Proceedings published by ASME on CD-Rom.

4
COCON: Corrosion Research Programme for Long Term Interim Storage Conditions

C. DESGRANGES, F. MAZAUDIER, D. GAUVAIN, A. TERLAIN, D. FÉRON
and G. SANTARINI

Service de Corrosion et du Comportement des Matériaux dans leur Environnement,
CEA-Saclay/Bâtiment 458, 91 191 Gif sur Yvette, Cedex, France

ABSTRACT

Two main corrosion phenomena are encountered in long term interim storage conditions: dry oxidation by the air when the temperature of high level nuclear wastes containers is high enough (roughly higher than 100°C) and corrosion phenomena as those encountered in outdoor atmospheric corrosion when the temperature of the container wall is low enough and so condensation is possible on the container walls. Results obtained with dry oxidation in air lead to predict small damages (less than 1 μm on steels over 100 years at 100°C) and no drastic changes with pollutants. For atmospheric corrosion, the first developments of a pragmatic method that gives assessments of the indoor atmospheric corrosivities are reported.

1. Introduction

The corrosion behaviour of metallic materials is generally well predicted for periods of a few tens of years, e.g. thirty to fifty years. For longer periods of time, of one to a few centuries, predictions are less common and probably more uncertain: laboratory and model studies are needed to validate corrosion rates and mechanisms over these long periods of time to have more confidence in the corrosion behaviour. As long term interim storage of nuclear wastes could last from one to several centuries, a specific research and development programme is being conducted on the corrosion behaviour of metallic containers during long term interim storage. This programme is called COCON (Corrosion of CONtainers).

The durability functions attributed to containers during long term interim storage are mainly related to the radionuclide tightness and to the handling or the moving of the containers. To ensure these two functions for periods of over a hundred years storage requires:

- prediction of the corrosion behaviour of metallic materials of container walls over that period, i.e. to describe and to determine the corrosion mechanisms and phenomena, to provide corrosion rates and to confirm the extrapolation methods used for prediction.

- a strategy of corrosion monitoring, as the environment of the interim storage (air) may be open to the outside fluctuations.

The COCON research and development programme includes also the two main conditions expected to be present in interim storage:

- For highly radioactive elements, like nuclear glasses or spent fuel elements, the temperature of the container walls will be high (100–300°C) at the beginning of the interim storage, high enough to avoid any condensation on these walls. During that period, the corrosion phenomena will be due only to dry oxidation by the cooling air.

- After the decrease of the radioactivity of these high radioactive elements, or during the initial period of the storage of the radioactive wastes with low thermal power, the temperature of the container walls will be nearly the same as the temperature of the cooling air. The thermal inertia of the containers, the fluctuations of air temperature and humidity will probably lead to the condensation of water on the containers. Wet and dry cycles will be possible and the corrosion phenomena will be as those encountered in atmospheric corrosion.

These two main corrosion phenomena (dry oxidation by air and indoor atmospheric corrosion) are investigated in the R&D COCON programme. First investigations were performed on steel materials which are widely used as container materials. Results of the bibliographic reviews and of the first experimental and modelling works are presented in this paper.

2. Dry Oxidation

Dry oxidation is one possible mode of degradation of the container candidate materials at the moderately elevated temperatures (50–300°C) expected for the external container walls. Models are required to predict the corrosion over long periods but to have a high reliability they must be based not only on experimental short term data but also on a mechanistic approach.

Few data on dry oxidation of unalloyed or low carbon steels are available in the literature. Those that do exist mainly concern the first steps of the oxide film build-up and have been obtained for very short oxidation periods or low oxygen partial pressures. They must therefore be complemented by more relevant data for our study. Two main theories of dry oxidation have been developed with the common approximation that only one of the partial processes involved controls the rate of the whole oxidation process. The Wagner theory [1] for the high temperature oxidation in which the transport of the reactants in the oxide layer is mainly accomplished by thermally activated diffusion via point defects leads to a parabolic law of the oxide layer growth. The Cabrera–Mott theory [2] and others derived for low temperatures or thin oxide films considers that the reactant transport is due to the electric field across the oxide film. Various oxide layer growth rates (parabolic, logarithmic or cubic) are deduced from these theories. Some attempts have been made to develop models taking into account all the partial processes of the oxide growth, but a lot of developments wil be required for such models.

The first objective of the experimental work is to obtain data over longer periods of time than those given in the literature at different temperatures between 100 and 500°C in order to determine whether an extrapolation of these data is possible by using an apparent activation energy. These data are then interpreted using allowable oxidation models and used for long term oxidation prediction. The influence of some pollutants in the atmosphere on the oxidation will also be presented.

2.1. Experimental Procedure

The experimental procedures are detailed in [3,4]. The investigated materials were pure iron and a low alloy steel. The possible chemical compositions are given in Table 1. These materials can be considered as reference materials for candidate steels likely to be offered as container materials. In the extrapolations it was also considered that corrosion would be uniform, without any occurrence of intergranular corrosion. This assumption was verified in all the tests conducted but more investigation would be necessary for reliable long term predictions about the potential risk of intergranular corrosion.

Oxidation tests were carried out in quartz reactors heated by electrical furnaces for 260 to 305-hour periods at temperatures between 100 and 400°C. The atmospheres were flowing air with 2% vol. H_2O (the reference atmosphere), wet air (12% vol. H_2O) or reference atmosphere with a pollutant 10 vppm HCl or 2 vppm SO_2. Oxidation of the specimen was evaluated from mass change measurements, observations with optical and electron microscopes and analyses by means of electron microprobe, XRD (X-ray diffraction), RBS (Rutherford backscattering spectroscopy) or XPS (X-ray photoelectron spectroscopy), depending on the scale thickness. The oxidation kinetics under air with 2% H_2O were also determined by continuously recording the mass gain of a specimen by means of a symmetrical microbalance.

2.2. Results in the Reference Atmosphere

The main results obtained in the reference atmosphere (air with 2% vol. H_2O) are gathered in Table 2. The oxide thickness have been determined by XPS or RBS for oxidation temperatures lower than 300°C and deduced from mass gain measurements at 300°C and higher temperatures. For the thicker scales, SEM observations of specimen fractures were made. No intergranular oxidation was detected and scale

Table 1. Chemical composition of the investigated materials (mass%)

	C	Mn	Si	S	P	Ni	Cr	Mo	Cu	Al	Fe	Sn	N
Iron	0.001	0.05	0.01	0.003	0.004	0.019	0.013	0.001	0.005	0.001	bal.	0.002	0.003
Steel	0.403	0.692	0.214	0.0007	0.09	0.014	0.330	0.030	0.005	0.013	bal.		

Table 2. *Iron and steel oxide thickness measured after oxidation tests*

Temperature	Duration	Oxide thickness	
		Iron	Steel
20°C		5.5 nm	5.0 nm
100°C	260 h	13.5 nm	13.0 nm
180°C	260 h	38.5 nm	55.0 nm
300°C	260 h	1 μm	0.2 μm
400°C	260 h	8.7 μm	3.3 μm

thickness observed were in good agreement with estimations from mass gain measurements for both materials. The oxides which have been identified by X-ray analyses are Fe_2O_3 and Fe_3O_4, at 300°C or higher, in agreement with the data reported in the literature. As the expected oxidation kinetics at temperatures lower than 300°C are very low and the oxides are roughly the same whatever the temperature, the first modelling approach was to work at higher temperatures and then to extrapolate the data to lower temperatures. As most of the literature data on iron oxidation in air or oxygen at temperatures higher than 300°C report that the oxide growth kinetics follow a parabolic law as predicted by the Wagner theory, we have assumed the same type of law for oxide growth kinetics. Therefore, from the following equation:

$$W^2 = K_p\, t$$

W being the mass gain, K_p the parabolic constant and t the time, with $K_p = K_0 \exp(-E_a/RT)$, where K_0 is a constant and E_a an apparent activation energy, we have calculated a K_p value at 300, 400 and 500°C and from an Arrhenius plot we have deduced the K_0 and E_a values. Figure 1 shows such a plot. The apparent activation energy deduced from these data is between 90 and 130 kJ mol^{-1} which is in agreement with the 100–150 kJ mol^{-1} values from the literature for iron oxidation in dry air or oxygen.

From these K_p data and assuming that the oxidation mechanism and kinetics remain the same during all the considered period, we can expect a 1 μm thick oxide layer on iron after 100 years at 150°C in the reference atmosphere. At 300°C, which is the maximum expected temperature, the calculated oxide thickness is 40 μm.

The limitation of the confidence in this extrapolation comes from the absence of experimental kinetic law measured over a long-time experiment.

We have therefore determined the oxide thickness variations with the time at 300°C for iron and steel by performing 300 to 700 h microbalance experiments. The results are shown in Fig. 2.

The oxide growth kinetics at 300°C on iron is better fitted by a parabolic law if only the first 300 h oxidation are considered. This is the most reported type of law

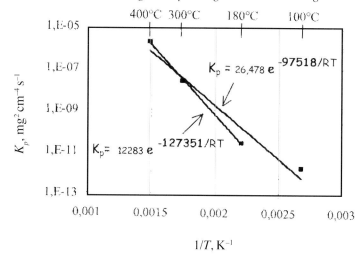

Fig. 1 *Parabolic constant deduced from oxidation of iron for 260 h in the reference atmosphere.*

most often reported in the literature. However, if we take into account all the data, the results are better fitted by a logarithmic law. This result has never been observed at this temperature for such long oxidation times and oxide thicknesses and needs to be confirmed.

As far as the steel is concerned, the oxide thickness variations with time at 300 and 250°C are well fitted by a logarithmic law. An extrapolation over a 100-year period leads to predict the formation of a 1 μm thick oxide layer at 300°C on steel by using logarithmic laws. It is less than what is predicted by Larose and Rapp [5] but they use a parabolic law.

All these calculations predict that the damage due to dry oxidation should be limited and acceptable for the external container walls. However, all the extrapolations

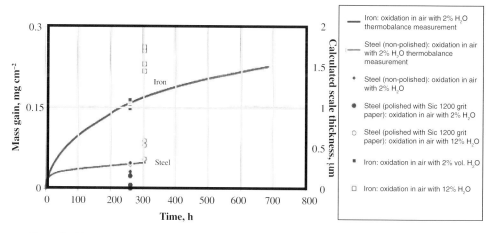

Fig. 2 *Oxidation kinetics in the reference atmosphere at 300°C of iron (upper curve) and of unalloyed steel (lower curve).*

are based on the same assumption that no detachment of the oxide layer arises and that oxide growth mechanism — which is not yet well defined in the temperature and the environmental conditions under consideration — remain unchanged during the whole interim oxidation storage period. A model, taking into account several elementary processes involved in the oxide growth process must be developed in order to increase the reliability of the extrapolations of the short term data to long periods.

2.3. Results with Pollutants

The effect of pollutants, HCl and SO_2, and of wet air have been investigated at 100 and 300°C. Similar results have been obtained for the two materials. The X-ray analyses of the specimens oxidised at 300°C in the polluted atmospheres do not show any constituent other than Fe_2O_3 and Fe_3O_4 in the oxide layers. XPS analyses show that the specimens oxidised in the wet atmosphere have a higher hydroxyl content near the gas/oxide interface than those oxidised in the reference atmosphere. SO_x ($x = 3$ or 4) has been detected in the first superficial 1.5 nm of the oxide layer of the specimens oxidised in the presence of an SO_2-containing atmosphere. Specimens oxidised in the HCl-containing atmosphere exhibited an oxide layer inside which chlorides were detected. Figure 3 shows the measured oxide thickness on the iron and steel specimens oxidised in the different atmospheres that were investigated.

Figure 4 also shows the mass gain variations with time of iron during oxidation at 300°C in the wet atmosphere and, for comparison, in the reference atmosphere. In these conditions, the measured oxide layer increase in the wet atmosphere with time is best fitted by a parabolic law. These results show a small thickness increase of the oxides on iron and steel formed during 300 h oxidation tests as the water content in the atmosphere increases from 2 to 12 % vol. at 300 and 400°C, and no thickness increase at 100°C. Furthermore, since oxidation kinetics measured in the wet atmosphere exhibit the same shape as those measured in the reference atmosphere, in the conditions of interim storage, we can expect that the impact of an increase of vapour content in the air is inconsequential for the dry oxidation stage.

As far as the HCl pollutant is concerned, its presence in the atmosphere leads to

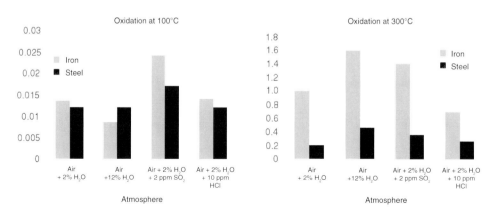

Fig. 3 *Mass gains of iron and steel specimens oxidised at 100 °C and 300 °C for 260 h in various atmospheres.*

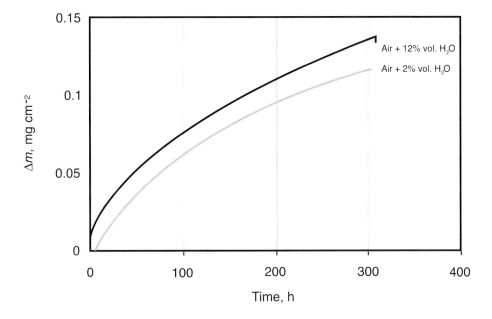

Fig. 4 *Oxidation kinetics at 300°C of iron in the wet atmosphere (black line) and in the reference atmosphere (grey line).*

an increase in oxide thickness and the presence of chloride in the oxide layer. The corrosion mechanism in this intermediate temperature range is probably different from that at high temperature ($>500°C$) because of the high volatility of the chlorides. Longer duration tests and determination of the oxidation kinetics are planned in further work.

The results show that, in the test conditions, the presence of SO_2 in the atmosphere does not increase the growth oxide kinetics at 100 and 300°C and modifies the iron and steel oxide layers only superficially at 100°C and more deeply at 300°C. This pollutant seems less detrimental than HCl.

These first results on dry oxidation of iron and an unalloyed steel allow some models to be proposed for predicting the damage by dry oxidation for long periods in a reference atmosphere. Their reliability must be increased by developing more sophisticated models taking into account more numerous partial processes that contribute to the whole oxidation process. Nevertheless, all our results, in agreement with those from the literature but obtained for short oxidation periods, agree in predicting little damage if there is no change in the oxidation mechanism. The first data obtained with the pollutants do not predict a drastic change in the oxidation rates but they must be confirmed.

3. Indoor Atmospheric Corrosion

Numerous data are available on outdoor atmospheric corrosion of iron or low alloy steels, including some from quite long-term testing periods (i.e. over 10 years). Results from a number of atmospheric corrosion tests show that the behaviour of low alloy steels or iron can be characterised by two parameters K and n in a linear bilogarithmic law $P = K.tn$ (metal penetration P and time t) as far as this is valid for outdoor conditions [6]. These laws are helpful in extrapolating results of corrosion test relating to the long term, i.e. 20–30 years, but not for our need which is the very long term behaviour of a few hundred years.

On the other hand there is a lack of data concerning indoor atmospheric corrosion. The stored containers can experience this particular kind of atmospheric corrosion during cyclic wet and dry conditions, depending on meteorological conditions and numerous other parameters. Nevertheless, these interim storage indoor conditions are not so far from those of the classical atmospheric exposure except that, of course, no rain is expected inside the disposal. So, the first requirement was to determine how to apply outdoor data to the indoor conditions of an interim storage.

3.1. Atmospheric Corrosivity according to Standards

According to the ISO/DIS 9223 standard [7], outdoor atmospheric corrosivity is the capacity of a defined atmospheric environment (in terms of time of wetness and pollutant contents, see later) to cause corrosion damage on metallic materials.

The corrosivity of the indoor atmospheres has been assessed using the classification tabulated for outdoor conditions in ISO/DIS 9223 and 9224 standards [7,8], i.e. using one of the main corrosion parameters, the so called time of wetness τ_i (days per year). A widely adopted standardised definition of the time of wetness is the number of days τ_i with relative humidity of air equal to or higher than 80% at positive temperatures. It is the time during which an aqueous film (i.e. about 10 adsorbed monolayers of water at room temperature [9]) is supposed to exist in these conditions of relative humidity and temperature on a metallic surface and allow electrochemical processes of atmospheric corrosion to occur. However, such a definition should not be taken too literally, because airborne gases and salt particles can retain humidity on the surface, decreasing the deliquescence point and thus the critical relative humidity value thereby leading to a sharp increase of the atmospheric corrosion rate.

The five categories of corrosivity C_i are connected with the corrosion rate for the first year of exposure, the average corrosion rate for the first ten years of exposure and the steady state corrosion rate of well-defined metal panels (i.e. carbon steel, zinc, copper and aluminium). The correspondence between the corrosivity categories and the atmospheric corrosion rates of carbon steel is reported in Table 3.

The atmospheric corrosivity depends not only on the time of wetness but also on the content of the main pollutants involved in atmospheric corrosion processes, i.e. the sulfur dioxide SO_2 (industrial environment) and chloride ions (e.g. marine environment). That is the reason why the atmospheric corrosivity categories C_i involves in consideration three atmospheric parameters:

τ_i :   the time of wetness;
P_i :   the sulfur-containing compounds content, represented by the sulfur dioxide SO_2 level (mg m^{-2} d^{-1});
S_i :   the airborne salinity, represented by the deposition rate (mg m^{-2} d^{-1}) of chloride ion.

Table 3. Corrosion rates (µm/year) of carbon steel for the different corrosivity categories [7]

Corrosivity category		Corrosion rate (µm/year)		
		First year	$t < 10$ years	$t > 10$ years
C1	very weak	0.1–1.3	< 0.5	< 0.1
C2	weak	1.3–25	0.5–5	0.1–1.5
C3	mid	25–51	5–12	1.5–6
C4	high	51–83	12–30	6–20
C5	very high	> 83	30–100	20–90

Different categories of these three parameters are defined, depending on their values. Thus, a corrosivity category C_i is connected with a chosen combination of the value of these three parameters and a well-defined metal [7,8].

3.2. Time of Wetness on Container Wall

Following our intention to work on a textbook case, we defined the dimensions of a waste container, whose size and shape are quite realistic, to be of 1.5 m dia. and 2 m height. The container is supposed to be a thin metallic canister filled with concrete with no thermal power that is assumed to control the thermal behaviour of the container which is disposed on the storage floor. The floor temperature T_{floor} of the storage is fixed arbitrarily to 13°C.

As soon as the temperature at the interface T_i falls below the dew-point, i.e. the saturation temperature of the ambient air T_{dew}, condensation will begin on the container walls.

Principles of the thermal modelling and methodology used for the calculations of the time of wetness on container walls, can be found elsewhere [10]. As input data for the modelling, we used one-year climatic parameters to determine the thermal behaviour of the container. These data have been given by the French Meteorological Office for a quite humid place with low annual temperature range in the north of France ('oceanic climate'), and, by contrast, for a warm Mediterranean country with wide annual temperature range in the south of France ('Mediterranean climate').

As the interim storage of the nuclear waste canister takes place inside an airy disposal, global heat transfer coefficients h at container walls result from natural convection (no ventilation) or from forced convection if any ventilation occurs: these assumptions on the ventilation regimes mean that different global heat transfer coefficients h ($h = 0.1$; 1 and 10 W m^{-2} K^{-1}). The calculated times of wetness for different heat transfer coefficients are reported in Table 4.

The total time of wetness of the well-defined container stored in the northern place (oceanic climate) is far more significant than it is in the southern location

(Mediterranean climate) and is about twice as much (Table 4). For the northern and humid location, the time of wetness increases with the heat transfer coefficient value, i.e. the ventilation efficiency, whereas it decreases in the southern site location.

3.3. Corrosivity Assessment

For any one metal, the atmospheric corrosivity categories involve a combination of the three parameters previously mentioned. However, it is possible to assess the corrosivity for a given couple of pollutant content S_i/P_i and for one particular time of wetness (see section 3.2).

Two opposite cases, i.e. two pairs of pollutant contents S_i/P_i, have been considered for the assessment of C_i: a free pollutant site (S_0, P_0), i.e. a rural location, and a heavy polluted site (S_3, P_3), i.e. a marine-industrial location. The results of the numerical simulations are reported in Table 5 for three values of h.

Even if the values of the time of wetness associated to the lateral and the upper face are different, equivalent corrosivity categories have been obtained for both faces (Table 5). Comparing Oceanic and Mediterranean climates for weak atmospheric pollutant level and similar convection, the atmosphere is generally less corrosive in the south. As a matter of fact, results show the huge influence of ventilation on the calculated time of wetness (Table 4) and then on corrosivity (Table 5).

As a general rule, results show the importance of simulations for prediction, as the long-term container behaviour seems closely linked, among other things, to the repository design, and cannot thus be derived from direct extrapolations of classical atmospheric corrosion data.

3.4. Wet–Dry Cycles

The wet–dry cycles are another important corrosion parameter for these 'corrosion allowance' materials as their number and frequency are correlated to the corrosion rate and the rust layer transformation. Earlier investigations have shown that during a wet–dry cycle, the atmospheric corrosion of iron or low alloy steel can be divided into three stages. As initially proposed by Evans [11] and then studied by Stratmann *et al.* [12,13], during the first stage (wetting), the anodic dissolution of iron is balanced

Table 4. *Time of wetness (days/year) for the two site locations, on both exposed walls and for different values of the heat transfer coefficient in the air (h)*

	Oceanic climate		Mediterranean climate	
h (W m^{-2} K^{-1})	Lateral wall	Superior wall	Lateral wall	Superior wall
0.1	214	223	125	113
1.0	231	241	89	80
10	255	260	95	96

Table 5. *Corrosivity classes for extreme values of Cl⁻ and SO₂ contents for the two site locations*

	Extreme pollutant parameters couples S_i/P_i			
	Oceanic climate		**Mediterranean climate**	
h (W m^{-2} K^{-1})	$S_0/S_1 - P_0/P_1$	$S_3 - P_3$	$S_0/S_1 - P_0/P_1$	$S_3 - P_3$
0.1	C_3	C_5	C_3	C_5
1.0	C_4	C_5	C_2	C_5
10	C_4	C_5	C_2	C_5

by the reduction of the ferric species (namely the oxyhydroxide γ-FeOOH) within the rust layer. According to Stratmann [13] and Cox [14], a quite acidic pH (from 3.5 to 6.0) is required for this cathodic reaction.

In our description, we assume that the metal damage during this first stage is proportional to the rust layer specific area, since reduction occurs at the nanopore surfaces. As the rust layer become thicker, the internal surface area of reducible rust layer is assumed to decrease and, similarly, the metal damage.

After the reducible γ-FeOOH is used up, the second stage (wet surface) begins: the oxygen reduction starts and takes place at the pore surfaces. It is assumed, in our work, that this step is a diffusion limited process.

During the end of the drying, the third stage of the cycle, the reduced layer of γ-FeOOH and the other ferrous species are reoxidised by the oxygen and the electrolyte film is used up, stopping the corrosion process completely.

This description, which is not discussed further here, agrees with the fact that the average corrosion rate decreases with time as is obviously shown by the well-known bilogarithmic laws. This first part of our work has made it clear that the next step is to model the evolution of the rust layer porosity whereas future calculations considering the previously mentioned well-defined container have to concentrate on the assessment of the main wet-dry cycle features regarding the physico–chemical processes described before. For this mechanistic approach the number of wet–dry cycles has been calculated as well as their average duration (Table 6).

Regarding the storage location, the simulations illustrate the influence of the environmental parameters: except for an hypothetical thermally isolated waste container (i.e. $h = 0.1$ W m^{-2} K^{-1}), the number of wet–dry cycles in the north of France is nearly twice that in the south. We can thus make the assumption that the number of actual interruptions of electrochemical atmospheric processes (i.e. an electrolyte-free container wall) is more important in the south. The average duration of the wet phase of wet-dry cycles is nearly the same for both locations, i.e. about 20 h.

This pragmatic method gives assessments of the indoor atmospheric corrosivities according to the ISO/DIS 9223 standards. The calculation of indoor atmospheric corrosivities is of practical interest for the selection of materials and for the prediction of the corrosion behaviour of classical engineering structures. Nevertheless, this

Table 6. 1997 — number and average duration (h) of wet–dry cycle for the two locations on both exposed walls

h (W m^{-2} K^{-1})	Oceanic climate		Mediterranean climate	
	Lateral wall	Superior wall	Lateral wall	Superior wall
0.1	28/125	41/60	49/37.4	55/18.8
1.0	66/31.5	76/16	35/18.5	35/17.6
10	64/7	64/6	35/2.9	41/3.3

approach has a low discriminating power and does not permit reliable long-term predictions of the corrosion behaviour of a waste canister. However, as far as atmospheric corrosion of corrosion allowance materials like carbon or weathering steels are concerned, simulations give all the fundamental characteristics for deriving corrosion rates for future developments that should involve all parameters including wet–dry cycles.

4. Conclusions

The two main corrosion phenomena encountered in long-term interim storages, dry oxidation by the air and atmospheric corrosion due to condensation on the container walls have been investigated.

For dry oxidation, few data are available in the literature for the expected range of temperatures (below 300°C). The first experiments confirm the few available data of the literature, but over a longer period of exposure time, and validate an extrapolation method based on the apparent activation energy of oxidation. The results obtained with non-polluted air lead to the expectation of uniform corrosion and low corrosion rate on iron and steel materials (less than 1 μm is expected on these alloys exposed 100 years at 100°C in air). The effects of pollutants (chloride and sulfur compounds) do not bring about large increases of these corrosion rates. These first results have to be confirmed while modelling of the oxidation mechanisms will secure the extrapolation. The possible occurrence of kinetic transitions such as breakaway corrosion phenomena will have to be specially investigated.

A number of data are available concerning atmospheric corrosion, even during long periods of exposure (over 10 years), but nearly exclusively for outdoors exposure conditions. The use of these data with the ISO standards using a thermohydraulic modelling of the indoor exposure conditions which leads to the classification of an indoor atmosphere as function not only of the atmospheric conditions (Oceanic and Mediterranean climates have been considered), but also as the function of the thermal inertia of the containers. To confirm the corrosion rates deduced from this practical and pragmatic method, the next developments of the indoor atmospheric corrosion modelling will include the use of semi-empirical well-known bilogarithmic laws, and further the physical description of the corrosion phenomena and the effects of wet–dry cycles.

The last objective of the COCON programme will be the strategy for corrosion monitoring. This will be based on the results obtained during dry oxidation and atmospheric corrosion tests; the analyses of the corrosion oxide layers seem to be a promising strategy (thickness, structural and chemical composition). The purpose is to link the relevant oxide layer parameters to the evolution of the types and rates of corrosion.

References

1. C. Wagner, *Z. Physik Chem.*, 1936, **B21**, 447.
2. N. Cabrera and N. F. Mott, *Rep. Prog. Phys.*, 1948/1949, **12**, 163.
3. A. Terlain, D. Féron, A. Galtayries, F. Malengreau, J.-M. Siffre and P. Marcus, Oxidation at low or moderate temperatures of metallic materials, *EUROCORR '99*, Aachen, September, 1999.
4. A. Terlain, C. Desgranges, D. Gauvain, D. Féron, A. Galtayries and P. Marcus, Oxidation of materials for nuclear waste containers under long term disposal, *Corrosion 2001*, Paper 0119, NACE International, Houston, TX, USA, 2001.
5. S. Larose and B. Rapp, Review of low-temperature oxidation of carbon steels and low alloyed steels for use as high level radioactive waste package materials. Nuclear Regulatory Commission, Contract NRC-02-93-005, Report CNWRA 97-003.
6. M. Pourbaix, 'The linear bilogarithmic law for atmospheric corrosion' in *Atmospheric Corrosion* (W. H. Ailor, ed.). John Wiley & Sons Inc., 1982, p.107.
7. ISO 9223:1992, Corrosion of metals and alloys — Corrosivity of atmospheres.
8. ISO 9224:1992, Corrosion of metals and alloys — Corrosivity of atmospheres. Guiding values for the corrosivity categories.
9. C. Leygraf and T. E. Graedel, *Atmospheric Corrosion*. Electrochemical Society Series, John Wiley & Sons, Inc., New York, USA, 2000, p.96.
10. M. Baklouti, F. Mazaudier, D. Féron and N. Midoux, Atmospheric corrosion in interim storage, *EUROCORR 2000*, London, UK, 10–14 Sept, 2000. Publ. The Institute of Materials, London, UK, 2000.
11. U. R. Evans, *Corros. Sci.*, 1972, **12**, 227.
12. M. Stratmann, K. Bohnenkamp and H. J. Engell, *Corros. Sci.*, 1983, **23**, 969.
13. M. Stratmann and K. Bohnenkamp, *Corros. Sci.*, 1987, **27**, 905.
14. A. Cox and S. B. Lyon, *Corros. Sci.*, 1994, **36**, 1177–1192.

5

Approaches Chosen to Predict the Effect of Different Forms of Corrosion on Copper in Disposal Conditions of Spent Fuel

M. BOJINOV, J. HINTTALA*, P. KINNUNEN, T. LAITINEN, C. LILJA[†],
K. MÄKELÄ and T. SAARIO

VTT Industrial Systems, Finland
*Radiation and Nuclear Safety Authority, Finland
†Swedish Nuclear Power Inspectorate, Sweden

ABSTRACT

This paper presents deterministic approaches chosen to predict the effect of stress corrosion cracking, creep, pitting corrosion and general corrosion on the lifetime of copper in disposal conditions of spent fuel. The interest in copper in Finland and Sweden is because of the plans to use it as a corrosion shield for the cast iron canister containing high activity nuclear waste. Our approach is based on assessing whether the occurrence of different types of corrosion can be excluded in the disposal conditions. If this is not the case, we shall recommend the maximum corrosion penetration to be predicted on the basis of experimental results and to assess whether this is below an acceptable level. We have applied this approach to show that the occurrence of stress corrosion cracking in the presence of nitrite ions in the disposal conditions can very likely be excluded. In addition, we have shown that assessing the occurrence of localised corrosion due to bicarbonate ions and the occurrence of general corrosion in the disposal conditions requires further experimental work.

1. Introduction

In the present Finnish and Swedish design spent nuclear fuel is packed in a canister made of spheroidal graphite cast iron. The cast iron canister has an outer shield made of copper. The copper shield is responsible for the corrosion protection of the canister. The design thickness of the wall of the copper shield has been decreased to about 50 mm, and this sets a limit for the maximum corrosion allowance of copper in the disposal conditions. The design life, a minimum 100 000+ years, sets challenges for the prediction of the longevity of the corrosion resistance of the copper shield. To assess the corrosion resistance of the copper shield, three major corrosion types (phenomena) are considered in this paper, namely stress corrosion cracking, pitting corrosion and general corrosion. Our proposed approach is first to assess whether the occurrence of any of these in the disposal conditions can be excluded (exclusion principle). If this is not the case, the assessment of the severity of the corrosion will be assessed. In other words, we propose to estimate the maximum damage caused by the specific corrosion type. If the estimated maximum corrosion damage is not below acceptable limits, the estimation has to be repeated based on a more mechanistic understanding and on more quantitative experimental data. We discuss in this paper the application of this approach to the three different forms of corrosion and the status of knowledge on the critical issues related to them.

2. Disposal Conditions

2.1. Pressure

The pressure in the final disposal vault consists of two components. The first part is the hydrostatic pressure of water. This depends on the depth at which the vault is built. If the vault is at a depth of 700 m, the hydrostatic pressure increases up to 7 MPa. The bentonite clay surrounding the copper canister will swell when it is wetted. This swelling causes an additional pressure, which is estimated to be 7 MPa. Thus the total pressure estimated to prevail in the vault at the copper canister surface is then a maximum of 14 MPa.

2.2. Temperature

The canisters reach their maximum temperature of about 90°C in the repository within 20 years after the disposal, after which the temperature will slowly decrease towards the temperature of the surrounding bedrock.

2.3. Redox Conditions

During the transient phase immediately after the closure of the repository the conditions will be first oxidising, after which reducing conditions will gradually be re-established. The undisturbed natural redox conditions of the groundwater at depth of all the Finnish investigation sites including Olkiluoto and Hästholmen sites have been estimated to result in redox potentials in the range –0.2 V to –0.3 V(SHE), (at ambient temperature) depending on the pH (e.g. Anttila *et al.* [1]). Because of the uncertainties related to the measurement of the redox potential this estimate is based not only on the measured values but also on theoretical studies and reaction path calculations with EQ3/EQ6 software. The calculation also takes into account a number of parameters affecting the redox potential, such as the concentration of sulfide ions, iron, dissolved hydrogen and methane. Also included in the estimate is the effect of fracture mineralogy, e.g. the presence of pyrite.

In the nearfield of the spent-fuel canister, bentonite is expected to bring about changes to the chemistry of the groundwater equilibrated with it. The chemical changes brought about depend to a great deal on the type of groundwater, fresh, brackish, saline and brine. For saline groundwaters, such as those occurring at depth in Olkiluoto and Hästholmen, the main observed changes have been related to pH buffering, ion-exchange (possibly also including NH_4^+), bicarbonate ion concentration (calcite dissolution), sulfate ion concentration (due to gypsum dissolution), and increase in ionic strength [2]. Bentonite is also expected to affect the redox potential (e.g. oxygen consumption by pyrite).

2.4. Chemistry

At some sites chloride ions have been found in the groundwater up to a concentration of 40 000 mg L^{-1} close to the planned depth of the repository. In the analysis of the groundwater composition in Finland and Sweden small amounts of ammonium ions (NH_4^+) have been found. The maximum concentrations have been about 3 mg L^{-1} at the Hästholmen site and 1.1 mg L^{-1} at the Olkiluoto site [1,3]. Ammonium ions have been mainly found in the upper part of the bedrock at the depth of 100–200 m, in the brackish groundwater affected water of the ancient Litorina Sea. Ammonium ions, as well as nitrite and nitrate ions, are known to make copper prone to stress corrosion cracking. The nitrite and nitrate ion concentrations in some single samples from the Olkiluoto site have been 0.01 mg L^{-1} NO_2^- and 0.2–0.3 mg L^{-1} NO_3^-. The corresponding concentrations at Hästholmen site have been 0.01 mg L^{-1} NO_2^- and 0.03–0.1 mg L^{-1} NO_3^- also in some single samples. Nitrite and nitrate ion concentrations are mostly below detection limits (0.01 mg L^{-1}), as would be expected in deep reducing groundwater and the reported analytical results are associated with uncertainties due to analytical problems. Carbonate ions, which are known to make copper prone at least to pitting corrosion, have been found at Olkiluoto site in a maximum concentration of 400 mg L^{-1} and at Hästholmen site in a maximum concentration of 250 mg L^{-1} in the brackish groundwater in the upper part of the bedrock. In the saline water at the depth below about 500–600 m the corresponding concentrations have been found to be 20–50 mg L^{-1}. Acetogenic bacteria have been observed in the groundwater of the investigation sites, these may form acetate ions [4]. Acetate ions may be consumed as energy sources in methanogenesis or by iron reducing or sulfate reducing bacteria [5].

3. Stress corrosion Cracking and Creep

In view of the required longevity of the corrosion resistance of the copper shield, the occurrence of stress corrosion cracking should be completely avoided. This is because once initiated, the cracking is likely to cause severe damage. Accordingly, the application of the exclusion principle when assessing the occurrence of stress corrosion cracking seems especially justified and attractive. On the other hand sufficient knowledge is not yet available to assess the effect of the multiaxial stress state and the effect of environment on creep. It is thus not yet possible to decide exactly what kind of approach should be chosen to assess the risk related to creep phenomena.

Effective stress levels are likely to be important from the viewpoint of both stress corrosion cracking and creep. The expected momentary maximum stress levels from e.g. canister handling would be normally of the order of 15 to 35 MPa [6], but forced straining beyond yield could occur locally due to external pressure loads in the repository. The canister shell base material is in the hot–formed condition and the electron beam (EB) welds are in the cast condition. The yield strength of this kind of copper is about 50 MPa at room temperature and the ultimate strength more than 200 MPa.

As mentioned above, the canisters reach their maximum temperature of about

90°C in the repository within 20 years after the disposal. Now we presume that simultaneously when reaching the maximum temperature the canister overpack is deformed due to creep or plastic deformation under the external pressure load. As a result of these deformations the whole radial 1 mm gap between the cylindrical overpack and insert will be closed. Due to the possibly very slowly increasing external pressure load the copper overpack will be deformed until a full contact is reached on all surfaces between the overpack and the iron insert. According to earlier calculations, when the 1 mm radial gap between the overpack and the insert is forced to close the actual maximum local strain in the copper overpack will be about 2%. This 2% strain corresponds to a stress level of *ca.* 80 MPa in hot-formed copper [6].

The local tensile stress component in the corner area of the copper shell thus exceeds the copper yield strength (50 MPa) on the surface. However, the stress distribution over the wall thickness is such that most of the wall thickness is in a compression state. In addition, these kinds of residual stresses are displacement controlled; in other words, the residual stresses are relaxed, if cracking or creeping occurs.

Raiko *et al.* [6] have suggested that the postulated cracking cannot penetrate the whole wall, because the crack growth is stopped when the tension stress is relaxed in the area and the crack front reaches the compressive stress area. In spite of possible local crack initiation and growth, the remaining wall thickness (more than half of the nominal wall) will be thick enough for the corrosion allowance. The possible combined effect of creep and stress corrosion cracking cannot be properly dealt with at the moment because of the lack of relevant creep data. Ongoing experimental work has indicated that a multiaxial stress state accelerates creep in copper markedly (unpublished work). The repository environment is expected further to accelerate creep. Therefore, experimental data on creep of copper under multiaxial stress state and in closely simulated repository conditions is needed to assess fully the effect of creep on the lifetime of the copper shield.

We recommend as the next step — according to the approach we have adopted — assessing whether the occurrence of stress corrosion cracking can be excluded in the chemical conditions of the waste repository environment. The chemical constituents known to be able to induce stress corrosion cracking in copper [7] and to exist in the Scandinavian groundwater [1,3] are nitrite, ammonium and acetate ions. In the two recent investigations of stress corrosion susceptibility of copper in simulated ground water conditions in the presence of ammonium ions [8,9] no sign of stress corrosion cracking has been found. In the case of acetate ions, more experimental data are needed in order to exclude the possibility of stress corrosion cracking [10]. The case of nitrite ions is discussed in more detail below.

3.1. Example of the Exclusion Principle; Nitrite Ions

The effect of potential on the reduction of area at fracture in slow strain rate tests in solutions containing nitrite ions can be demonstrated as shown in Fig. 1 [11].

The smaller the reduction in surface area, the less ductile the material behaviour and the larger the effect of stress corrosion cracking. The reduction in area at fracture at –0.1 V(SCE) (+0.12 V (SHE)) is comparable to that in air (59%), which indicates that below this potential copper is not susceptible to stress corrosion cracking. Stress corrosion cracking (SCC) of copper may on the other hand occur at higher potentials where Cu(II) can be formed. Benjamin *et al.* [11] have shown that at room temperature the minimum amount of $NaNO_2$ needed for SCC to occur is higher than 0.001M and that the lower the concentration the higher is the potential above which SCC may occur. They have also shown that in 0.6M $NaNO_2$ increasing the temperature to 80°C tends to diminish the susceptibility of copper to stress corrosion cracking.

The potential above which SCC may occur has been found to be relatively insensitive to variation in pH in the range 5.5–10.3 [12,13]. Aaltonen *et al.* [14] have shown that at 80°C the potential above which SCC may occur is roughly the same as at room temperature. Salmond and Atrens [15] have found that in 1M $NaNO_2$ at room temperature the threshold stress for copper is 120 MPa. This stress is higher than the maximum stress predicted to take place in the copper canister [6].

Based on the results from the above literature references, nitrite ions do not cause SCC in copper at concentrations below 0.001M, which is equal to 69 mg L^{-1}, or at potentials lower than +0.1 V(SHE) in slightly alkaline solutions. The reported concentration of nitrite ions in some groundwater samples at the Hästholmen and Olkiluoto sites have been 0.01 mg L^{-1} (although with large uncertainties). This is roughly four decades lower than the concentration of 69 mg L^{-1}. Additionally, the potential which has to be exceeded for SCC of copper to occur in nitrite ion containing environment, +0.1 V(SHE), is high (a potential where Cu(II) can be formed is required), in comparison with the redox potentials (and thereby also corrosion potential of copper) measured and predicted to prevail in the ground water of the disposal vault environment. It thus seems clear that the occurrence of SCC of copper due to the

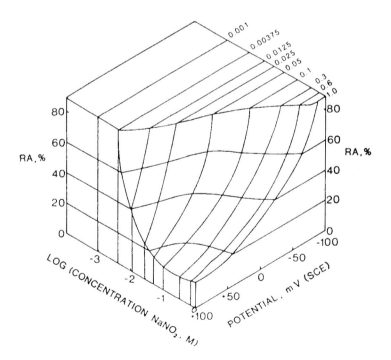

Fig. 1 *Effects of sodium nitrite concentration and potential on reduction of surface area at fracture, RA, of OFHC (oxygen free high conductivity) and PDO (phosphorus deoxidised) copper [11].*

effect of potentials and nitrite ions can be excluded in the conditions prevailing at the Hästholmen and Olkiluoto sites.

4. Localised Corrosion

Localised corrosion may cause local penetration of the copper canister. We propose first to consider, whether the occurrence of localised corrosion can be excluded within the potential, pH and concentration limits of the disposal conditions. If this is not the case, we shall recommend that the maximum penetration depth should be predicted and to assess whether it is below an acceptable level.

In the safety analysis of the disposal of spent nuclear fuel it has been assumed that the penetration rate will be very low and that it will decrease with time [16]. The possible penetration rates have been estimated based on the measurements obtained from items that have been buried underground for long periods of time. From these items the pitting factor P_f has been determined as the ratio of the maximum pit depth and the average general penetration depth, determined from weight loss ($P_f = P_{max}/P_{av}$). The pitting factor has been found to be at a maximum of 25, but this has been considered as a very conservative value. The pitting factor has also been estimated for archaeological artefacts, e.g. copper coins, but in these cases the average penetration rate is impossible to define because the original dimensions of the items are not known, and so the pitting factor is only a guess. It has to be remembered that it is not reasonable to use the concept of a pitting factor when pitting or general corrosion is only slight [18].

As pitting of copper has almost exclusively been reported to occur in the region where Cu(II) can be formed, i.e. under oxygenated conditions, longer term exposure tests under oxygenated conditions would be needed to assess the probability of occurrence reliably. A possible mechanism for pitting of copper in low potential conditions containing sulfide ions has been recently proposed [18]. It is however still unclear whether the mechanism is operable in repository conditions.

An example of an experimental approach chosen to assess whether the occurrence of localised corrosion in the presence of bicarbonate ions can be excluded is given below. The studies reported in the literature for different environments [19,20] indicate that HCO_3^- ions increase the solubility of copper in the stability region of Cu(II), possibly by complexing the divalent copper ions. Thus they can be concluded to render the oxide film formed on copper susceptible to local damage and to localised corrosion at positive potentials. To clarify the influence of bicarbonate ions further, an example of the contact electrical resistance data measured recently [21] for copper in water with 100 mg L^{-1} bicarbonate ions is shown in Fig. 2.

After oxidising copper in the potential region where Cu(I) is stable (0 V (SHE)) and switching to open circuit, the electrical resistance of the film first increases with decreasing potential and then remains at a constant and high level for several hours indicating that the stability of the Cu(I) oxide film is not affected by bicarbonate ions in the solution. After oxidising at a potential at which also Cu(II) oxide can be formed (0.4 V (SHE)) and switching to open circuit, the electrical resistance again increases first with decreasing potential as well. However, in the presence of HCO_3^- ions the R–t curve following oxidation at 0.4 V shows a clear minimum when the potential

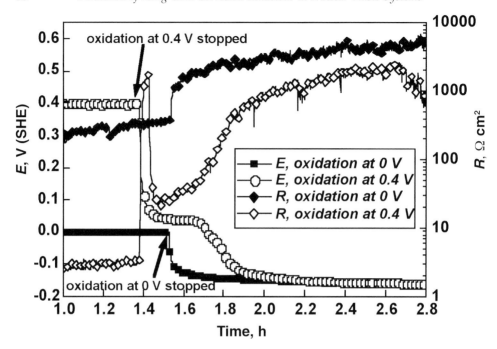

Fig. 2 *Open circuit potential decay curves and the corresponding resistance vs time curves for Cu oxidised for 1 h at 0 V and 0.4 V (SHE) in 0.1M Na$_2$B$_4$O$_7$ + 100 mg L^{-1} HCO$_3^-$ at 80°C and 2 MPa Ar+3% H$_2$.*

has decayed to 0.0–0.1 V. This kind of minimum has been found to be much weaker in the absence of bicarbonate ions. This points to a more defective structure of the film in the Cu(II) potential region in the presence than in the absence of HCO$_3^-$ ions. Thus the results gained in simulated repository conditions strongly indicate that copper is more susceptible to metastable pitting and related localised corrosion phenomena in the presence of bicarbonate ions at positive potentials where divalent copper Cu(II) is stable than at lower potentials.

The concentration of bicarbonate ions in the borehole analysis in Finland has been found to be between 25 mg and 400 mg L^{-1} [1,3]. As discussed above, the stability of the oxide film can be questioned at high potentials in the presence of 100 mg L^{-1} HCO$_3^-$ ions. Thus, the repository conditions may lie in the concentration window for susceptibility to localised corrosion. Another factor to be taken into account when considering the possibility for excluding the occurrence of localised corrosion is the potential. The potential where Cu(II) may start to form at the expected slightly alkaline pH of 7.5 to 8.5 is +0.07 to +0.14 V(SHE). These potentials are about 0.3 to 0.4 V higher than the measured natural redox potential for anoxic environment in the undisturbed condition of the bedrock. Thus, for this condition there seems to be no overlap of the potentials with the potential window for susceptibility to localised corrosion. Higher positive potentials may however be relevant in repository conditions in the case of the oxygen-containing period immediately after the closure of the repository or during flushing of the repository with fresh oxygenated water e.g. during an ice age. Such a period with oxygenated conditions can be expected, depending on the source, to last from a couple of weeks to 30 years after which all the oxygen is consumed and anoxic conditions resume. Accordingly, a possible hazard caused by bicarbonate ions to the stability of copper oxide films cannot be totally excluded during such a period. Therefore, the next step in assessing the probability of occurrence of localised corrosion would be to check first if such high positive potentials are achievable in repository conditions, and second to verify with longer term experiments at the highest achievable potential if a real risk exists for pitting by bicarbonate ions in simulated oxygenated repository conditions.

5. General Corrosion

The occurrence of general corrosion of copper cannot be excluded in repository conditions. It has for instance been recently predicted [22] on the basis of thermodynamic calculations that copper dissolves actively in highly saline environments even in anoxic conditions. However, thermodynamic calculations do not give any estimate for the actual corrosion rate. Accordingly, one has to arrive at either experimental or theoretical results showing that the rate of general corrosion is acceptably low to prove the suitability of copper as a shield material. Taking the corrosion allowance of the copper canister as 50 mm and the demanded life time of 100 000 years, the allowed maximum corrosion rate is 5×10^{-4} mm/year.

Recent experimental studies [23,24] in simulated repository conditions (80°C, 14 MPa) have reached the conclusion that active dissolution of copper does take place both in highly saline (5.4% Cl⁻) and saline (1.4% Cl⁻) simulated ground water. Figure 3 shows an example of the data supporting this conclusion. The electrical

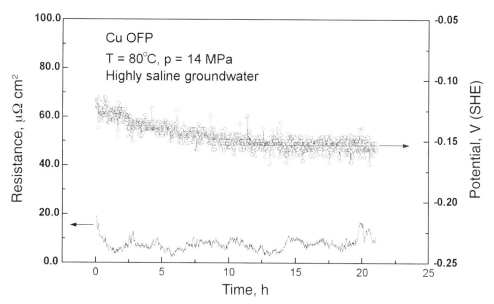

Fig. 3 *Electrical resistance of Cu OFP fresh surfaces exposed to the highly saline groundwater at open circuit potential* (E_{corr}).

resistance of the surface of copper in the highly saline simulated ground water has been found to be negligible (less than 20 $\mu\Omega cm^2$, typical for an adsorption layer) during the whole 20 hour monitoring period, indicating that no passive surface film is present.

The corrosion rate measured in mass loss coupon tests of relatively short duration (7 days) has been found to be 2×10^{-2} mm/year. This corrosion rate would indicate a lifetime of roughly 2500 years. This is clearly lower than the required 100 000+ years. However, such an estimate rests on the assumption that the transport rate of the reactants (e.g. oxygen containing species) towards the surface and that of reaction products (cuprous ion complexes) away from the surface are fast enough to allow the corrosion process to proceed. However, it is likely that the bentonite layer next to the copper surface will act as a diffusion/migration barrier slowing down the transport. To draw more definite conclusions, this needs to be experimentally verified and supported by theoretical modelling of transport in bentonite.

6. Conclusions

An approach for dealing with the three forms of corrosion that may pose a risk for the stability of the copper canister in the repository conditions, namely stress corrosion cracking, pitting corrosion and general corrosion has been described. The approach is based on an attempt to find out whether the occurrence of any corrosion type can be excluded in the disposal conditions. If this is not the case, the next step is to evaluate whether the maximum extent of corrosion is below an acceptable limit. In case of potential risks, more detailed mechanistic and quantitative corrosion studies are recommended. This leads to a slightly different research strategy for each form of corrosion.

Stress corrosion cracking has been considered as a mechanism that cannot be allowed to occur, and therefore the exclusion principle has been chosen for this type of corrosion. The exclusion principle, involving concentration, potential and stress level as variables has successfully been applied to exclude the possibility of stress corrosion cracking caused by nitrite ions. In the case of other known species that may cause stress corrosion cracking in pure copper, i.e. ammonium ions and acetate ions, the situation is different, and more experimental data are needed in order to be able to exclude the risk caused by them.

Pitting corrosion of copper has been reported to occur almost exclusively at high positive potentials where Cu(II) can be formed and may dominate in the passive film on copper. These potentials are thought to be not achievable in the repository conditions. However, a solid experimental proof for this is not yet available. In addition, the recently proposed mechanism of localised corrosion of copper in anoxic conditions in the presence of high amounts of sulfides needs to be further investigated. If this mechanism were proven to be valid, the next step would be to use the exclusion principle with regard to the concentration of the sulfide species causing the phenomena.

General corrosion has been indicated to take place at an unacceptably high rate in highly saline environments that are representative of some possible ground water environments. These findings call for an experimental and theoretical verification of the transport rate of oxygen to the reacting surface and of the corrosion product away from the surface under realistic repository conditions.

References

1. P. Anttila, *et al.*, Final disposal of spent nuclear fuel in Finnish bedrock — Hästholmen site report. POSIVA 99-08.

2. Vuorinen, U. & Snellman, M. 1998. Finnish reference waters for solubility, sorption and diffusion studies. Helsinki. Finland. Posiva Working Report 98-61, 39 pp.

3. P. Anttila, *et al.*, Final disposal of spent nuclear fuel in Finnish bedrock — Olkiluoto site report. POSIVA 99-10.

4. S. Haveman, K. Pedersen and P. Ruotsalainen, Geomicrobiological investigations of groundwaters from Olkiluoto, Hästholmen, Kivetty, and Romuvaara, Finland. Helsinki, 1998. Posiva Oy. 40 p. POSIVA 98-09. ISSN 1239-3096.

5. K. Pedersen, Investigations of subterranean micro-organisms and their importance for performance assessment of radioactive waste disposal. Results and conclusions achieved during the period 1995 to 1997. SKB Technical report 97-22, 1997.

6. H. Raiko and J.-P. Salo, Design report of the disposal canister for twelve fuel assemblies. POSIVA-99-18, Posiva Oy, Helsinki, 1999, 64 p.

7. B. M. Ikeda and F. King, State of knowledge on stress corrosion cracking of copper for used-fuel containers in a deep geologic repository. AECL Report No. 06819-REP-01200-10058-R00. 2001, 50 p.

8. F. King, C. D. Litke and B. M. Ikeda, The effects of oxidant supply and chloride ions on the stress corrosion cracking of copper. AECL Report No. 06819-REP-01200-10013-R00, 1999, 60 p.

9. E. Arilahti, M. Bojinov, K. Mäkelä, T. Laitinen and T. Saario, Stress corrosion cracking investigation of copper in groundwater with ammonium ions. Posiva Working Report 2000-46.

10. T. Saario, T. Laitinen, K. Mäkelä and M. Bojinov, Literature survey on stress corrosion cracking of Cu in presence of nitrites, ammonia, carbonates and acetates. Posiva Working Report 99-57, October 1999, 17 pp.

11. L. A. Benjamin, D. Hardie and R. N. Parkins, Stress corrosion resistance of pure coppers in ground waters and sodium nitrite solutions. *Brit. Corros. J.*, 1988, **23**, (2), 89–95.

12. T. B. Cassagne, J. Kruger and E. N. Pugh, Role of the oxide film in the transgranular stress corrosion cracking of copper, in Environmentally assisted cracking: Science and engineering, ASTM STP 1049 (W.B. Lisagor, T.W. Crooker and B.N. Leis, eds.), ASTM, Philadelphia, 1990, pp. 59–75.

13. J. Yu and R. N. Parkins, Stress corrosion crack propagation in α-brass and copper exposed to sodium nitrite solutions, *Corros. Sci.*, 1987, **27**, (2), 159–182.

14. P. Aaltonen, H. Hänninen, H. Illi and M. Kemppainen, On the mechanism of environment sensitive cracking of pure OFHC-copper. In Predictive capabilities in environmentally assisted cracking - PVP-Vol. 99 (R. Rungta, ed.). ASME, USA, 1987, pp. 329–340.

15. J. Salmond and A. Atrens, SCC of copper using the linearly increasing stress test, *Scripta Met. et Material.*, 1992, **26**, 1447–1450.

16. T. Vieno, Safety analysis of disposal of spent nuclear fuel. VTT Publications 177. Technical Research Centre of Finland, Espoo 1994. 6+256 p. + app. 3 p.

17. ASM Handbook, 1992a. Vol. 13, Corrosion. ASM International, 1992, pp. 232.

18. H.-P. Hermansson and S. Eriksson, Corrosion of the copper canister in the repository environment. SKI Report 99:52, Swedish Nuclear Power Inspectorate, Stockholm, 1999. 112 p. + app. 27 p.

19. S. B. Ribotta, M. E. Folquer and J. R. Vilche, Influence of bicarbonate ions on the stability of prepassive layers formed on copper in carbonate–bicarbonate buffers, *Corrosion*, 1995, **51**, (9), 682–688.

20. M. Sanchez-Perez, M. Barrera, S. Gonzalez, R. M. Souto, R. C. Salvarezza and A. J. Arvia, Electrochemical behaviour of copper in aqueous moderate alkaline media, containing sodium carbonate and bicarbonate, and sodium perchlorate, *Electrochim. Acta*, 1990, **35**, (9), 1337–1343.

21. P. Sirkiä, T. Saario, K. Mäkelä, T. Laitinen and M. Bojinov, Electric and electrochemical properties of surface films formed on copper in the presence of bicarbonate ions, STUK-YTO-TR 157 (1999).

22. B. Beverskog and I. Puigdomenech, Pourbaix diagrams for the system copper–chlorine at 5–100°C. SKI Report 98:19, 1998. Swedish Nuclear Power Inspectorate, Stockholm. 18 p. + app. 16 p.

23. T. Laitinen, K. Mäkelä, T. Saario and M. Bojinov, Susceptibility of copper to general and pitting corrosion in highly saline groundwater, SKI Report 01:2. Swedish Nuclear Power Inspectorate, Stockholm, 2001. 24 p.

24. T. Saario, K. Mäkelä, T. Laitinen and M. Bojinov, Susceptibility of copper to general and pitting corrosion in saline groundwater, STUK-YTO-TR 176, February 2001.

Part 2

Modelling Session

6

The Holy Grail: Deterministic Prediction of Corrosion Damage Thousands of Years into the Future

D. D. MACDONALD

Center for Electrochemical Science and Technology, Pennsylvania State University, 201 Steidle Building, University Park, PA, USA

ABSTRACT

This paper summarises work being carried out by the author, under Department of Energy (DoE) sponsorship, to develop general corrosion models for Alloy C-22 HLNW canisters in a Yucca Mountain type repository. Two models have been developed to date: (1) the General Corrosion Model (GCM), which yields the corrosion potential and corrosion rate at a specified state point (i.e. a point on the evolutionary path), and (2) the Accumulated Damage Model (ADM), which integrates the corrosion rate calculated for closely spaced state points along the evolutionary path to yield the corrosion loss over the history of the repository.

The GCM is based upon the Point Defect Model (PDM) for the steady state growth of passive films on metal surfaces and upon the general Butler–Volmer equation for the kinetics of redox reactions (hydrogen evolution and oxygen reduction), both of which are combined within the framework of the Mixed Potential Model (MPM) to yield the desired corrosion potential and corrosion rate. Importantly, the solution of the constitutive equations contained within the MPM is constrained by the conservation of charge, thereby imparting a high degree of determinism to the predictions. Application of the ADM to evolutionary paths defined by the repository temperature, and assuming that the canisters are always in contact with saturated brine under ambient oxygen pressure conditions, yields a corrosion loss of about 1.7 mm over the 10 000-year service life of the repository. It is estimated that complete loss of the 2cm Alloy C-22 wall would require an exposure time of about 200 000 years and that the corrosion loss after 1 000 000 years would be slightly less than 9 cm.

1. Introduction

The safe disposal of High Level Nuclear Waste (HLNW) represents one of the greatest technical challenges ever to face humankind. The principal difficulty is that the service life horizon is twice as far into the future (10 000 years) as recorded human history is in the past, and no direct experience exists with the degradation of engineered materials over that time period. Indeed, some of the materials that are proposed for use in HLNW repositories did not exist thirty years ago (e.g. Alloy C-22: Cr22, Mo13.5, W3, Fe4, C 0.015, Si0.08 (at.%), bal. Ni). A search of the literature reveals that no material that is currently contemplated for HLNW service has been investigated in a single corrosion study for more than a few thousand hours, or for more than a few one thousandths of one percent of the intended service life. Not only is the corrosion rate unknown over even a very small fraction of the service life, we also cannot be sure that the mechanism of corrosion will remain the same as the conditions within the repository evolve.

It is evident from the above that purely empirical methods cannot meet the challenge of assuring the public of safe disposal of nuclear waste within the engineered barrier concept. In response to that conclusion, the author and his colleagues, under DOE/NERI/IDT sponsorship, have developed deterministic models for predicting the accumulation of general and localised corrosion damage to Alloy C-22 in simulated HLNW repository environments. The ultimate goal is to develop deterministic models that capture the mechanistic essence of the damaging processes, that require minimal calibration, and that can be used to predict the evolution of corrosion damage over the requisite time to within the required engineering accuracy. This paper focuses on the deterministic prediction of damage due to general corrosion and possibly also due to transpassive dissolution. Damage due to localised corrosion, in the form of pitting, is discussed in an accompanying paper in this Volume (pp.103-117) [1].

2. Determinism in the Evolution of Science

'Science' may be defined as the transition from empiricism to determinism, with the former implying that everything that we ultimately can know must have been experienced. On the other hand, determinism (as used in the scientific sense) is the philosophy that the future may be predicted from the past, provided that:

(i)   the phenomena can be described in terms of viable physical mechanisms;

(ii) the solutions to the constitutive equations (i.e. those equations that predict the response of the system to given inputs) are constrained by natural laws that are invariant in space and time (conservation of mass, charge, etc); and

(iii) the path to the future state is continuous and can be specified.

The philosophy of determinism does not mean that we must predict the behaviour of every particle in the system, since much of the system may be described in terms of average properties (density, temperature, etc). Indeed, this is precisely the approximation that is used in planning the trajectory of a spacecraft to a far distant planet; the constitutive equations (equations of motion and the law of gravitation) are solved subject to the constraints imposed by the conservation of mass, energy (or mass-energy), and momentum, assuming that the planets, which exert gravitational forces on the craft, can be regarded as bodies of uniform, average properties. Clearly, this course is preferable (and less expensive) to the empirical method of firing rockets at random to find which two bracket the target, and then repeating the process within the bracket to find which spacecraft actually reaches the target.

One of the greatest challenges in devising a deterministic model is to decide what phenomena or processes should be included in describing the global system. While no dogmatic rules exist to guide the theorist on this matter, a rule-of-thumb used by the author is that all processes that have a perceptible impact on the dependent variable, when the independent variables are varied over prescribed ranges, should be included, but that begs a definition of the term 'perceptible'. The most general definition of 'perceptible' is in terms of the desired precision or accuracy of the prediction; if the impact is greater than the desired accuracy, when an independent variable is changed over a prescribed range, then the response is said to be 'perceptible' and the process or phenomenon should be included in the model. As an example, consider the case referred to above of a spacecraft flying to a far distant planet. Provided that the craft was sufficiently far from earth, a volcanic eruption, while catastrophic to an observer on earth, would have an imperceptible impact on the trajectory and hence it would not need to be included in the model. However, if the spacecraft was sufficiently close to the earth, and the volcanic eruption caused a significant redistribution of mass, the impact on the trajectory may well be perceptible and hence the eruption would have to be included in the physical description of the system. Finally, it is worth recalling Einstein's admonishment that processes and phenomena must not be introduced into a model in an ad hoc fashion, simply to make the model function correctly, and that ultimately all models must be based on (and justified by) observation. The process by which this is ensured is one of prediction followed by testing, commonly known as the 'scientific method'.

It is important to note that a 'model' is simply our perception of reality, and should not be mistaken for reality itself. Because we can never perceive reality absolutely (we make observations only through imperfect senses and interpret the results through an imperfect intellect), there is no such thing as 'the correct model', notwithstanding the occasional claims to the contrary. All models are ultimately deficient in some respect, as revealed by experiment, and hence eventually must be replaced when they are no longer able to predict correctly within the accuracy of the observation. However, there is an aspect to this issue that is commonly ignored, but which can have a profound impact on the evolution of models and that is that there must exist 'confluence' between theory and experiment. Simply put, it is necessary to insure that experiments that are designed to test a theory, and theories that are developed to account for observations, must be inter-compatible with regards to the underlying assumptions and postulates. A classic example in electrochemistry and corrosion science of where the 'confluence' condition is violated is the use of potentiodynamic techniques for measuring polarisation data, which are then used to evaluate steady state models for the passive state, without first demonstrating that the voltage scan rate was sufficiently low that the experimental measurements were made under acceptable, quasi steady state conditions. In this case, the lack of agreement between theory and experiment cannot be taken as evidence that the theory or model is wrong, but rather that the theory and experiment are not in confluence, and hence the issue of 'right or wrong' is moot. It is the author's opinion that many models have been prematurely condemned on the basis of data that were measured under conditions where confluence did not exist between theory and experiment. Contrariwise, other models have survived even though data from experiments that are in good confluence with the theory are at odds with the predictions of the model!

The condition that the solution to the constitutive equations be constrained by the natural laws and that the laws must be invariant in space and time is obvious, but it contains many subtle undertones. Thus, it is important to note that the empirical-to-deterministic evolution, which we refer to as 'science', is compiled in the form of the natural laws that are *universally true*. These laws are the summation of a wide range of empirical observation and theoretical deduction, and contain information that extends well beyond the system of particular interest. The universal applicability of the 'natural' laws has played a pivotal role in the evolution of science. For example, it was Lorentz's realisation that the (natural) laws of motion must be valid in all inertial frames of reference, simultaneously, that led him to formulate the famous transformations that bear his name (e.g. $m = m_0 / \sqrt{1 - v^2 / c^2}$, where m_0 is the 'rest' (observer's) mass, m is the mass in a frame of reference moving at a uniform velocity v with respect to the 'rest' frame, and c is the velocity of light), which ultimately led to Einstein's formulation of the Theory of Relativity. Clearly, the spatial and, in particular, the temporal invariance of the natural laws is a critical property for the prediction of physico–chemical phenomena extending ten thousand years or more into the future.

Another question that commonly arises in exploring the role of determinism in science concerns which of the natural laws must be invoked in describing the behaviour of a system. The short answer to this question is simple; all of the laws must be obeyed. However, only certain laws are operatively important, depending on how the system and observer are defined. Consider once more the spacecraft, and imagine that a component in the environment control system that is in contact with an aqueous phase is undergoing stress corrosion cracking. Now, stress corrosion cracking is a mechano–electrochemical phenomenon, so that the relevant (operatively important) laws are those that constrain mechanical and electrochemical phenomena. These laws include the conservation of mass / energy, momentum, charge, and mass / charge (Faraday's law). Assuming that the observer is in the spacecraft and recognising that the natural laws are the same for all frames of reference, the laws as applied to the spacecraft, as a whole, need not be considered, because the inertial frame of reference is common to the spacecraft and the observer. Accordingly, the conservation laws, as applied to the center of mass of the system, are of no relevance, but only those applied to the system within the local inertial frame of reference need be considered. Some of these laws are implicit, while others must be explicitly invoked. For example, the conservation of mass is implicit, if the system is 'closed' (in the thermodynamic sense) and balanced relationships (e.g. reactions) describe the various processes that occur in the system (e.g. metal dissolution). Likewise, invocation of the Griffith criteria of fracture satisfies the conservation of energy, while the use of Faraday's law ensures the conservation of mass / charge. On the other hand, some laws must be invoked explicitly, because of the nature of the problem. For example, in the case of stress corrosion cracking, it has been established experimentally that a positive current flows from the crack through the solution to the external surfaces where it is consumed by some cathodic reaction. Except in a few cases, the role played by the external environment has been totally ignored in modelling SCC, with the result that the models fail to conserve charge and hence cannot satisfy the requirements of determinism. Close examination of these models generally reveals that they are largely, if not entirely, empirical in nature, albeit in a sophisticated manner.

It is the author's postulate that we must use the same deterministic philosophy as that outlined above in predicting the evolution of corrosion damage to HLNW canisters. In light of the fact that a repository has yet to be built and operated, our 'experience' is minimal, if not non-existent, and hence there does not exist a basis for the empirical prediction of damage. Furthermore, repositories are unique systems, so that even if we had operated repositories for the last few hundreds or thousands of years, insufficient data would be available to formulate statistically significant predictions over the intended 10 000-year operating life. These factors alone are devastating for the prospects of empirical methods for predicting corrosion damage, in this instance.

3. Deterministic Prediction of Damage

Recent work by the author and his colleagues has focused on the development of models for the deterministic prediction of general and localised corrosion damage. The resulting algorithms are generally based upon the Point Defect Model (PDM) for the growth and breakdown of passive films on metal surfaces and on the Coupled Environment models for predicting corrosion cavity growth rate, culminating in Damage Function Analysis [2,3]. our recent work on HLNW disposal has focused on Alloy C-22 and has endeavoured, where possible, to employ repository conditions (temperature, pH, p_{O_2}, etc) predicted in other programmes for the Yucca Mountain site. For the sake of brevity, the mathematical bases of the models will not be described here; the fundamental aspects of both the PDM and DFA are available elsewhere [2,3]. The present paper concentrates on the prediction of damage due to general corrosion only; the prediction of localised corrosion damage is described in this Volume, pp.103-117 [1].

3.1. General Corrosion

As noted above, general corrosion is described in terms of the steady-state form of the Point Defect Model (PDM) [2]. Briefly, the PDM describes the barrier layer of the passive film as a defective oxide, with the principal defect, in the case of the Cr_2O_3 oxide on C-22, being the cation interstitial (Cr_i^{3+}). The concentration and flux of this defect, and of other defects (e.g. oxygen vacancies, $V_O^{\bullet\bullet}$), in the barrier layer is determined by the kinetics of generation and annihilation of defects at the metal/film and film/solution interfaces and by the electric field strength (transport of the defects is primarily via migration, not diffusion). In any event, the PDM provides explicit expressions for the thickness of the barrier oxide layer on the alloy surface and for the steady-state anodic current density. When combined with the generalised Butler–Volmer equations for describing the principal cathodic processes that occur during the corrosion of metals in aqueous systems (the evolution of hydrogen and the reduction of oxygen), within the framework of the Mixed Potential Model (MPM) [4–6], the PDM permits deterministic prediction of the accumulation of general corrosion damage. In order to distinguish this model from its (significantly different) predecessors (e.g., the MPM), it is now referred to as the General Corrosion Model (GCM).

The primary output of the GCM is the corrosion rate and the corrosion potential. other output parameters include the pH and [NaCl] of the thin electrolyte (NaCl) film that is assumed to cover the metal surface, the transpassive dissolution potential, and the potential of zero passive film thickness. In calculating the transpassive dissolution potential, it is assumed that dissolution becomes spontaneous once the potential is more positive than the equilibrium potential for the Cr_2O_3/CrO_4^{2-} reaction, with the chromate ion activity being set by the chemical equilibrium $Cr_2O_3/CrO_4^{2-}/O_2$. This is tantamount to assuming that transpassive dissolution coincides with oxygen evolution. The accumulated damage is calculated by first using the general corrosion model (GCM) [4–6] to calculate the corrosion rate at closely spaced state points along the corrosion evolutionary path (CEP). In the present case, the CEP is defined by the evolution of the repository temperature, which in turn is used to calculate the saturation NaCl concentration, $[O_2]$, and the pH for the thin electrolyte film that is assumed to cover the canister surface over the history of the repository. The corrosion rate is then integrated along the CEP to yield the accumulated general corrosion damage.

The corrosion potential (ECP) calculations shown in Fig. 1 employed kinetic data for the evolution of hydrogen and the reduction of oxygen from our previous work on Type 304 SS in high temperature aqueous systems [4], because at the time of the calculations no comparable data for C-22 were available (these are now being measured, see elsewhere in this Workshop). On the other hand, the kinetic parameters for the anodic reactions were adjusted to reproduce the measured passive current density for C-22 in specific solutions [7,8]. While the solutions from which the data were taken were not saturated NaCl, it is believed that the anodic kinetic data used in the calculations are reasonably accurate. In spite of this uncertainty, the calculated ECP is found to describe the characteristic sigmoid shape with increasing log (p_{O_2}). Also, the ECP is predicted to increase slightly with temperature, which is the opposite trend to that for stainless steel [9], but which is consistent with data for C-22 in sulfuric acid [10].

Predicted polarisation curves for C-22 at 90°C, as a function of pH, are shown in Fig. 2. These curves were constructed using the PDM to describe the partial anodic reaction (anodic oxidation of C-22 to form a Cr_2O_3 barrier oxide layer) and the Butler–Volmer equation to describe the partial cathodic reactions of oxygen reduction and hydrogen evolution, all within the framework of the Mixed Potential Model, assuming the Wagner–Traud hypothesis. The cathodic branch displays a single wave due to the reduction of oxygen, although a second wave due to hydrogen evolution can be made to appear at more negative potentials by suitably adjusting the exchange current density for this reaction. Furthermore, the cathodic branch displays a well-defined Tafel region, culminating in the limiting current for the transport of oxygen to the alloy surface. In calculating the latter quantity, it is assumed that the ideal Nernst diffusion layer thickness is the same as the thickness of the electrolyte film. Note that the corrosion potential displays the expected negative shift with increasing pH. Finally, at more positive potentials than the corrosion potential, the current is dominated by the passive current for the oxidation of the substrate. The passive

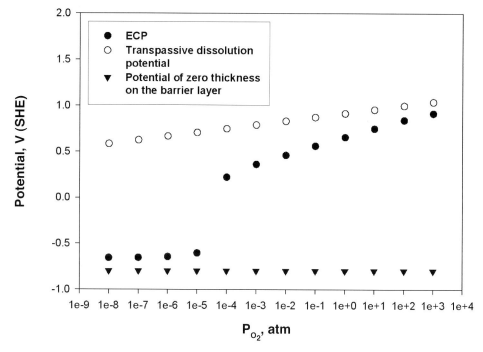

Fig. 1 *Dependence of the corrosion potential (ECP), transpassive dissolution potential, and the potential of zero thickness of the barrier layer on the partial pressure of oxygen. pH = 3, T = 90 °C, saturated NaCl solution ([NaCl] = 6.59M), electrolyte film thickness = 0.01 cm.*

current is independent of potential, which is expected for a passive film in which the principal defect is the oxygen vacancy or cation interstitial.

The most surprising prediction of the PDM was the existence of transpassive dissolution, because this was not introduced in an *ad hoc* fashion (remember Einstein's admonishment!), but rather results directly from the MPM. Transpassive dissolution arises from the thermodynamic prediction that, at a sufficiently high potential (greater than the equilibrium potential for the Cr_2O_3/CrO_4^{2-} reaction, with the chromate ion activity being set by the chemical equilibrium $Cr_2O_3/CrO_4^{2-}/O_2$), the species being ejected from the passive film, and the product of film dissolution, are in the +6 state (i.e. chromate), rather than in the +3, chromic state. The origin of the sudden increase in the current density has been traced to a concomitant decrease in the thickness of the oxide film, due to enhanced passive film dissolution, as shown in Fig. 3. Transpassive dissolution has been found to occur on C-22 (like most, if not all chromium-containing alloys) at roughly the predicted potential [10]. The importance of transpassive dissolution is that it might represent a mechanism for the depassivation of C-22 in the repository, provided that the potential is displaced by some process to sufficiently positive values. If this did occur, catastrophic corrosion of the canister might be expected. This issue is dealt with in further depth below. It should be noted that the barrier layer is not predicted to disappear under all circumstances. For example, using the same parameters values used in the GCM for predicting the plots shown in Figs 2 and 3, but allowing the oxidation state of the

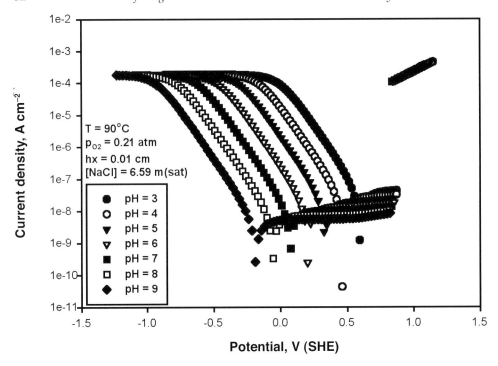

Fig. 2 *Predicted polarisation curves for Alloy C-22 in contact with a thin electrolyte film (saturated NaCl) on the surface as a function of pH.* p_{O_2} = 0.21 atm, electrolyte film thickness hx = 0.01 cm, T = 90°C.

species being ejected from the film to vary from 3 to 6, results in reduction of the steady state barrier layer thickness to progressively lower values, with only that at the highest oxidation state being zero. Concomitantly, the current density in the transpassive state is predicted to increase to progressively higher values along the same series.

An enormous literature exists on the transpassive dissolution of metals and alloys, including chromium-containing alloys. No attempt is made to review the literature here, except to note that the findings with respect to the nature of the oxide film in the transpassive state, compared to that in the passive state, are equivocal. Thus, some authors [11] report, on the basis of ellipsometric studies, that transpassive dissolution is accompanied by an increase in the thickness of the oxide film, while others [12] found that the contact resistance decreases as the potential is displaced into the transpassive region. Noting that passive films generally form as bilayer structures [2], it is unfortunate that Jacobs *et al.* [11] did not analyse their data in terms of a two-layer model, so that the changes in the thickness and complex refractive index of the barrier layer alone were not isolated. The observed increase in the thickness of the 'passive film' could well be explained by the expected increase in the thickness of the porous outer layer resulting from the increase in the interfacial concentration of metal dissolution products under transpassive dissolution conditions. At any rate, the observed increase in the current density upon entering

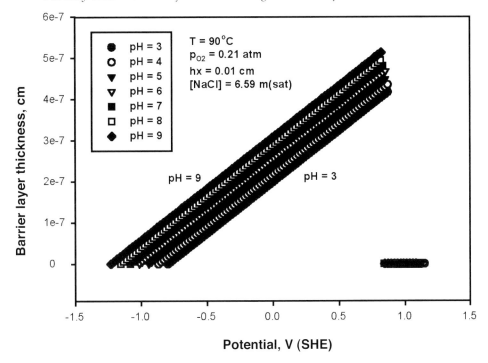

Fig. 3 *Predicted barrier oxide layer thickness for Alloy C-22 in contact with a thin electrolyte film (saturated NaCl) on the surface as a function of pH.* $p_{O_2} = 0.21$ *atm, electrolyte film thickness* hx = 0.01 *cm,* T = 90 °C.

the transpassive state is inconsistent with an increase in the steady state film thickness, if the film is the barrier layer. With regards to the contact resistance observations by Bojinov and *et al.*, [12], the observed decrease in resistance could be due to an increase in the defect (donor) concentration or to a decrease in the thickness of the barrier layer. However, the barrier layer on Fe–Cr–Mo alloys is a degenerate n-type conductor, due to the preponderance of oxygen vacancies ($V_O^{\bullet\bullet}$) and/or cation ((Cr_i^{3+})) interstitials. Studies on austenitic stainless steels [13], zinc [14], and tungsten [15], all of which form *n*-type barrier layers, show that the defect concentrations decrease as the potential is made more positive, a finding that is inconsistent with the suggestion of Bojinov, *et al.* [12]. Thus, the more likely explanation for the onset of transpassive dissolution is that the barrier layer thickness suddenly decreases as the potential becomes more positive than the transpassive dissolution potential.

As noted above, the GCM now exists as a subroutine in a more general code for predicting accumulated general corrosion damage to canisters over the life of the repository. The reader will recall from an earlier section in this paper that the deterministic prediction of damage requires that the evolutionary path of the system be continuous and specified. For the present purposes, we describe the evolutionary path in terms of the temperature, as predicted by researchers at the Lawrence Livermore National Laboratory (LLNL) (Fig. 4) for the Yucca Mountain, Nevada repository over a period of a million years. Three cases were defined:

(1) Base Case,

(2) Low Temperature Operating Mode (LTOM), and

(3) High Temperature Operating Mode (HTOM). The three evolutionary paths are shown in Fig. 4.

The Base Case and the HTOM display similar temperature profiles, in which the temperature rises to a maximum of about 160°C soon after closure of the repository and then decays with the decay in radionuclide activity in the waste. On the other hand, the LTOM, which corresponds to the scenario where closure of the repository is delayed until considerable decay of the most active radionuclides has occurred, displays significantly lower temperatures, with a maximum of a little over 85°C. The corresponding [NaCl], [O_2], and pH are calculated, assuming that the surface is covered with a thin layer (0.01 cm thick) of neutral, saturated sodium chloride solution. These data, in turn, are used to predict the corrosion potential and the corrosion rate (Fig. 4, lower plots) of C-22 over the same evolutionary path.

The reader will note that the corrosion rate correlates closely with the temperature. However, the model predicts only a weak correlation with the ECP and pH. This is because the PDM predicts that the corrosion current density is only weakly dependent on pH and potential, provided that the principal defects in the barrier layer is the oxygen vacancy and/or the cation interstitial, and that the passive current is only weakly dependent upon the activity of H^+ (an experimental finding).

The accumulated damage (corrosion loss) is calculated by simply integrating the corrosion rate (Fig. 4) over time to yield the plots shown in Fig. 5. The integrated damage is surprisingly monotonic with time and the model predicts that all three proposed operating modes will result in essentially identical damage. However, if we focus on the design life of 10 000 years, the similarity in the general corrosion damage profiles shown in Fig. 5 is illusory, because of the logarithmic abscissa. Replotting the data on a linear abscissa (Fig. 6) shows that there is a significant difference between the LTOM and the other two proposed modes as one might expect from the predicted general corrosion rate. At any rate, the model predicts that a little less than 9 cm of Alloy C-22 will be lost over the one million year period for which the temperature has been calculated. Importantly, the corrosion loss that is estimated for the 10 000-year life of the repository is modest (a little more than 1.8 mm for the Base Case and the HTOM, and slightly less for the LTOM), compared with the wall thickness of the canister of 2 cm. Importantly, the model predicts that the C-22 wall will not perforate due to general corrosion for a period of about 200 000 years, thereby indicating a generous corrosion allowance in the design. In the author's opinion, localised corrosion, as discussed later (this Volume, pp.103–117) [1], is a more likely mode of failure, particularly if the delayed repassivation constant for pits is small.

Of course, we hasten to stress that the calculations displayed above are very preliminary in nature, in that we do not yet possess a complete set of model parameter values for Alloy C-22 in the appropriate environments. Furthermore, little information

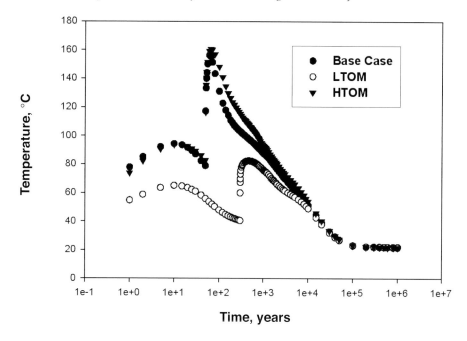

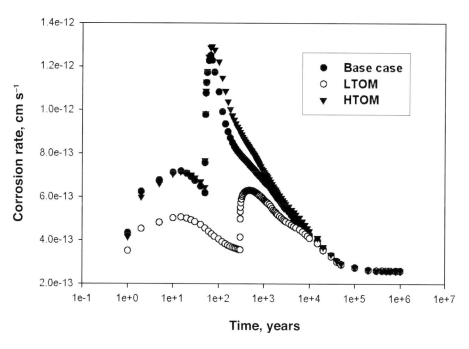

Fig. 4 *Evolutionary paths with respect to temperature and corrosion rate for Alloy C-22 in contact with saturated NaCl for the Yucca Mountain HLNW repository for the three scenarios defined by the LLNL.*

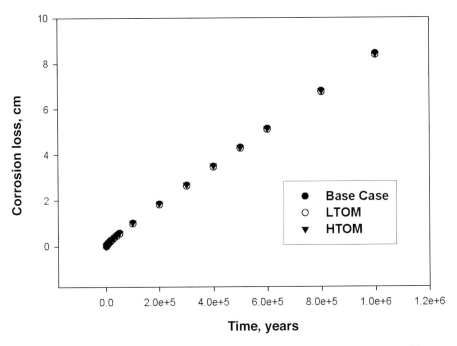

Fig. 5 *Evolutionary paths with respect to corrosion loss for Alloy C-22 in contact with saturated NaCl for the Yucca Mountain HLNW repository for the three scenarios defined by the LLNL.*

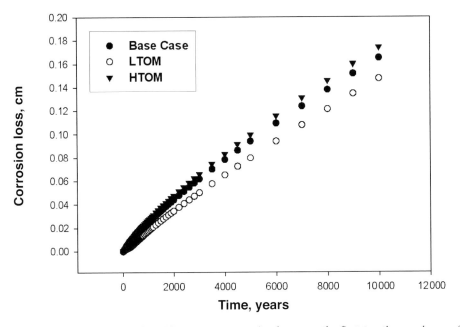

Fig. 6 *Evolutionary paths with respect to corrosion loss over the first ten thousand years for Alloy C-22 in contact with saturated NaCl for the Yucca Mountain HLNW repository for the three scenarios defined by the LLNL.*

is currently available on the chemical evolution of the repository environment, particularly with respect to the chemical composition of the electrolyte layer that is postulated to exist on the canister surface.

3.2. Transpassive Dissolution

We now return to the issue of the transpassive dissolution phenomenon, because this process represents a possible mechanism for the catastrophic dissolution of the canister and subsequent release of the radioactive waste to the geosphere. Considerable discussion has taken place regarding the possibility of transpassive dissolution, because of the well-recognised threat to canister integrity. This issue is addressed in Fig. 7, in which is plotted the ECP and the transpassive dissolution potential over the million year period for which the temperature was estimated for the three hypothetical operating modes.

For transpassive dissolution to occur, the corrosion potential (the ECP) must be more positive than the transpassive dissolution potential (E_{trans}). The prediction of this work is that the condition that ECP > E_{trans} will never occur during the million year repository period. Indeed, the greatest potential for transpassive dissolution, as measured by the difference between ECP and E_{trans} (the smaller the difference, the greater the potential), would seem to exist shortly after closure of the repository, corresponding to the sharp rise in temperature, when inspection would still be feasible. However, the ECP and E_{trans} values are predicted to never become closer than about 400 mV; this difference would seem to be insurmountable in terms of current corrosion theory (it would require the presence at high concentrations of a much more powerful oxidising agent than oxygen). Finally, it is important to note that transpassive dissolution is more akin to a phase change phenomenon (a thermodynamic condition) than it is to pitting corrosion, which is a progressive kinetic process. Accordingly, and given the time and space invariance of the laws of thermodynamics, which describe the conversion of Cr(III) to Cr(VI), it is concluded that transpassive dissolution of Alloy C-22 poses little risk to canister integrity.

4. Summary and Conclusions

This paper outlines a deterministic methodology for predicting general corrosion rate and the accumulation of general corrosion damage on Alloy C-22 HLNW canisters over the evolution of a Yucca Mountain-type repository. While the work is not yet complete, a number of significant advances and findings have been made, as follows:

- Two deterministic models for predicting the general corrosion behaviour of Alloy C-22 canisters in the Yucca Mountain HLNW repository have been developed. Thus, the general corrosion model (GCM) calculates the corrosion potential, the transpassive dissolution potential, and the corrosion current density for the alloy in contact with a thin electrolyte film containing a specified concentration of NaCl (including the saturated, default system), which in turn

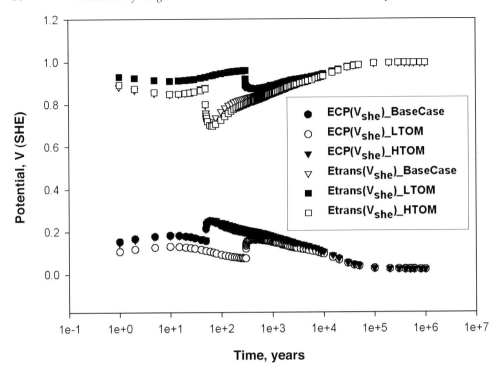

Fig. 7 Predicted evolutionary paths with respect to corrosion potential and transpassive dissolution potential over one million years for Alloy C-22 in contact with saturated NaCl in the Yucca Mountain HLNW repository for the three operating modes defined by the LLNL.

is in contact with the ambient atmosphere. The second model, the accumulated damage model (ADM), then employs the CGM at closely spaced state points to calculate the corrosion loss over a specified evolutionary path for the repository. The models are deterministic, in that the predictions are based upon specific corrosion mechanisms, as described by the Point Defect Model (PDM) for the passive current density and the Butler–Volmer equation for the partial cathodic processes that occur at the surface, that the predictions are constrained by the appropriate natural laws (the conservation of charge and Faraday's law), and that the route to the future is described in terms of a continuous evolutionary path. The models have not been fully optimised, because of the lack of data for key model parameters

- The GCM and ADM account for the existence of the transpassive state in terms of known electrochemical phenomena [oxidation of Cr(III) to Cr(VI)]. In fact, the PDM predicts the existence of the transpassive state without the need to adjust the theory in any manner whatsoever.

- The ADM (Accumulated Damage Model) predicts that the corrosion loss to the Alloy C-22 canister after 10 000 years in contact with saturated brine over the specified evolutionary path, as developed by the Lawrence Livermore National Laboratory will be of the order of 1.8 mm, compared with a wall thickness of 2 cm. Consumption of the wall thickness is predicted to take about 200 000 years, whereas the general corrosion loss after a million years exposure in the repository is estimated to be slightly less that 9 cm.

- The ADM predicts that at no time during the evolution of the repository will the corrosion potential become more positive than the transpassive dissolution potential. Indeed, the prediction of the models is that the ECP will remain more negative than the transpassive dissolution potential by at least 400 mV at all times. Accordingly, the threat of transpassive dissolution to the integrity of the canister is considered to be highly unlikely.

5. Acknowledgements

The authors gratefully acknowledge the support of this work by the US Department of Energy through Innovative Design Technology's Subcontract A20257JG1S. Many valuable discussions with Dr. Joon Lee of Sandia National Laboratory are also gratefully acknowledged.

References

1. D. D. Macdonald, G. R. Engelhardt, P. Jayaweera, N. Priyantha and A. Davydov, 'The Deterministic Prediction of Localized Corrosion Damage to Alloy C-22 HLNW Canisters'. This Volume, pp.103–117.

2. D. D. Macdonald, 'Passivity: The Key to our Metals-Based Civilization', *Pure Appl. Chem.*, **71**, 951–986 (1999), and references cited therein.

3. G. R. Engelhardt and D. D. Macdonald, 'Deterministic Prediction of Pit Depth Distribution', *Corrosion*, 1998, **54**, 469–479.

4. D. D. Macdonald, 'Viability of Hydrogen Water Chemistry for Protecting In-Vessel Components of Boiling Water Reactors', *Corrosion*, 1992, **48**, 194–205.

5. D. D. Macdonald, I. Balachov and G. Engelhardt, 'Deterministic Prediction of Localized Corrosion Damage in Power Plant Coolant Circuits', *Power Plant Chemistry*, 1999, **1**, 9.

6. D. D. Macdonald, C. Liu, M. Urquidi-Macdonald, G. Stickford, B. Hindin and A. K. Agrawal, 'Prediction and Measurement of Pitting Damage Functions for Condensing Heat Exchangers', 1994, *Corrosion*, **50**, 761–780.

7. D. S. Dunn, C. S. Brossia and O. Pensado, 'Stress Corrosion Cracking, Passive, and Localized Corrosion of Alloy C-22 High Lever Radioactive Waste Containers', *Corrosion 2001*, Paper No. 00206, NACE International, Houston, TX, USA (2001).

8. A. C. Lloyd, D. W. Shoesmith, N. S. McIntyre and J. J. Noel, 'Investigating the Localized Corrosion Properties of Ni–Cr–Mo Alloys for Their Use in Nuclear Waste Disposal Systems', *Stainless Steel World*, 2001, **13**, 29–33.

9. G. Cragnolino and D. D. Macdonald, 'Intergranular Stress Corrosion Cracking of Austenitic Stainless Steel at Temperatures Below 100°C — A Review', *Corrosion*, 1982, **38**, 406–424.

10. G. Bellanger and J. J. Rameau, 'Behaviour of Hastelloy C-22 Steel in Sulphate Solutions at pH 3 and Low Temperatures', *J. Mat. Sci.*, 1996, **31**, 2097–2108.

11. L. C. Jacobs, H. P. De Vogel, K. Hemmes, M. M. Wind and J. H. W. De Wit, 'Potential Modulated Ellipsometric Measurements on an Fe–17Cr Alloy in Sulphuric Acid', *Corros. Sci.*, 1995, **37**, 1211–1233.

12. M. Bojinov, G. Fabricus, T. Laitinen and T. Saario, 'Transpassivity Mechanism of Iron–Chromium–Molybdenum Alloys Studied by AC Impedance, DC Resistance and RRDE Measurements', *Electrochim. Acta*, 1999, **44**, 4331–4343.

13. E. Sikora, C. B. Breslin, J. Sikora and D. D. Macdonald, 'Semiconductive Properties of Passive Films on Stainless Steels', *Proc. Electrochem. Soc.*, 1995, **95-1**, 344–354.

14. K. Ismail, E. Sikora and D. D. Macdonald, 'Characterization of the Passive State on Zinc', *J. Electrochem. Soc.*, 1998, **145**, 3141–3149.

15. E. Sikora, D. D. Macdonald and J. Sikora, 'The Point Defect Model vs. the High Field Model for Describing the Growth of Passive Films', in *Proc. 7th Int. Symp. Oxide Films, Mets. Alloys* (K.R. Herbert and G. E. Thompson, eds). PEC, **94–25**, 139–151 (1994).

7

Semi-Empirical Model for Carbon Steel Corrosion in Long Term Geological Nuclear Waste Disposal

F. FOCT and J.-M. GRAS

EDF R&D, Département Etude des Matériaux, Site des Renardières, 77818 Moret sur Loing Cedex, France

ABSTRACT

In France and other countries, carbon and low alloy steels have been proposed as suitable materials for nuclear waste containers for long term geological disposal since, for such types of steels, general and localised corrosion can be fairly well predicted in geological environments (mainly argillaceous and granitic conditions) during the initial oxic and the following anoxic stages.

This paper presents a model developed for the long term estimation of general and localised corrosion of carbon steel in argillaceous and granitic environments. In the case of localised corrosion, the model assumes that pitting and crevice corrosion propagation rates are similar. The estimations are based on numerous data coming from various experimental programmes conducted by the following laboratories: UKAEA (United Kingdom); NAGRA (Switzerland); SCK-CEN (Belgium); JNC (Japan) and ANDRA-CEA-EDF (France). From these data, the corrosion rates measured over long periods (from six months to several years) and derived from mass loss measurements have been selected to construct the proposed models.

For general corrosion, the model takes into account an activation energy deduced from the experimental results (Arrhenius law) and proposes three equations for the corrosion rate: one for the oxic conditions, one for the early stage of the anoxic conditions and one for the long term anoxic corrosion.

Concerning localised corrosion, a semi-empirical model, based on the evolution of the pitting factor (ratio between the maximum pit depth and the average general corrosion depth) as a function of the general corrosion depth, is proposed. This model is compared to other approaches where the maximum pit depth is directly calculated as a function of time, temperature and oxic or anoxic conditions.

Finally, the presented semi-empirical models for long term corrosion estimation are applied to the case of nuclear waste storage. The results obtained by the different methods are then discussed and compared with other predictions.

1. Introduction

In France, carbon or low alloy steels are the reference materials for high level nuclear waste containers used for long term geological disposal. Compared to passive materials, one of the main advantages of these materials is the good confidence that can be placed on the prediction of their corrosion behaviour over a long period. General and localised (pitting or crevice) corrosion are the two types of degradation affecting the lifespan of the container. A semi-empirical model is proposed for long term corrosion evaluation. It is based on numerous data from the literature and gives kinetic laws for both types of corrosion in oxic and anoxic conditions. This paper firstly presents the way the model has been constructed for general and localised corrosion; the corrosion depth is then quantified for different cases typical of nuclear waste storage conditions.

2. General Corrosion

This study is based on numerous data obtained by different laboratories which conduct experimental programmes to study the possibility of nuclear waste disposal in granitic/argillaceous or argillaceous geological conditions with non or low alloyed steels containers. The principal sources of data are the following:

- UKAEA (Atomic Energy Authority, UK);

- NAGRA (Nationale Genossenschaft für die Lagerung Radioaktiver Abfälle from Switzerland);

- SCK-CEN (Studiecentrum voor Kernenergie _ Centre D'Etude de l'Energie Nucléaire in Belgium);

- JNC (Japan Nuclear Cycle Development Institute); and

- ANDRA-CEA-EDF (Agence Nationale pour la Gestion des Déchets Radioactifs — Commissariat á d'Energie Atomique — Electricité de France).

This work takes into account only data resulting from experience of at least several months. Most of the testing conditions were in compacted clay (MX80 or Fo–Ca types of clay), mixed water and clay or local clay for *in situ* experiences (Mol). General corrosion rates have been estimated by metal mass loss techniques.

2.1. Influence of the Oxidising/Reducing Conditions

In oxidising conditions (Fig. 1a), the results obtained by UKAEA [1] show that the corrosion rate is roughly constant over 5 years (between 22 and 28 μm/year at 90°C).

In anoxic conditions, the corrosion rate decreases as the exposure time increases and is much lower than in aerated conditions: it decreases to below 5 μm/year after 5 years at 90°C (Fig. 1b) [2,3]. The corrosion depth (X) could then be described by a power law as a function of time ($X = k\, t^n$). Nevertheless, for the model proposed in paragraph 2.5, we propose to estimate the long term corrosion by a constant rate as in oxic conditions. This simplified method (based on a mean corrosion rate) tends to overestimate the corrosion depth in the case where the oxide layer would be protective and if the experiment does not last long enough.

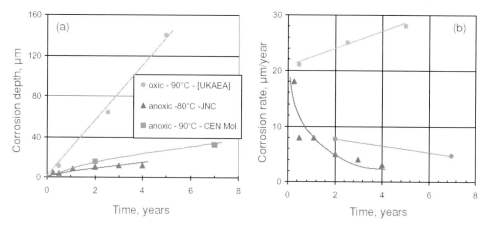

Fig. 1 *Evolution of the corrosion depth (a) and corrosion rate (b) of carbon steels as a function of time in oxic conditions and anoxic conditions [1–3].*

2.2. Influence of the Water Chemistry

The major influence of the redox potential of the environment — which is mainly fixed by the aeration conditions — on the corrosion of carbon steels has been presented in the previous paragraph (2.1). The effects of the composition of the water are discussed in this paragraph.

The Japanese Nuclear Corporation (JNC) [4] has measured the corrosion rate of a carbon steel in compacted bentonite over 3 months with various anions concentrations : $Cl^- \leq 50$, $F^- \leq 0.1$, $SO_4^{2-} \leq 3$ and $CO_3^{2-} \leq 10$ gL^{-1}). Results (Table 1) show that the corrosion rate — between 6 and 11 µm/year — of this steel is not significantly affected by the chemical composition of the water. This lack of effect of the anionic composition of the water has also been observed at 100 and 250°C in deaerated conditions [5].

Hence, in the model developed in paragraph 2.5, the anionic composition of the water is not taken into account, since it does not have a major influence on the corrosion rate of carbon steel.

2.3. Influence of the Type of Carbon Steel

The influence of minor alloying elements and of fabrication/elaboration conditions has been studied through the results obtained by UKAEA [1] on various carbon steels (CS) and cast iron (which has a higher carbon content than CS). The steels were forged, and some had a low carbon content. Moreover some samples were also electron beam welded. After 5 years of exposure to oxidising or reducing waters, the mass loss measurements showed no significant differences between these materials (Fig. 2). This general trend, of no effect of material type on carbon steel corrosion behaviour, is confirmed by the field experience of the American National Bureau of Standards which found no evidence of any effect of steel type of various unalloyed steels placed in 128 different soils over periods of up to 45 years [6]. Therefore, in the model proposed in paragraph 2.5, we have chosen to neglect any influence of the fundamental trend.

Table 1*. Influence of the anionic composition of the water on corrosion rate of a carbon steel [4]*

Material: carbon steel (C 0.10, S 0.003, P 0.14, Si 0.22, Mn 1.01%) Corrosion tests in compacted bentonite — non deaerated initial conditions Temperature: 80°C — Test duration: 3 months					
Conditions: compacted bentonite	Cl⁻(ppm)	F⁻ (ppm)	SO_4^{2-} (ppm)	CO_3^{2-} (ppm)	CR (μm/year)
+ distilled water	–	–	–	–	6
+ sea water	19 990	1.4	2 798	–	11
	100	–	–	–	7
	10 000	–	–	–	8
	50 000	–	–	–	11
+ distilled water +	–	10	–	–	8
	–	100	–	–	7
	–	–	30	–	8
	–	–	3 000	–	10
	–	–	–	100	9
	–	–	–	10 000	8

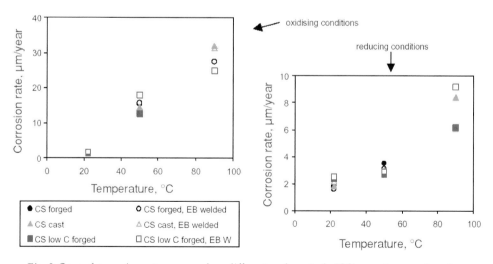

Fig. 2 *General corrosion rate measured on different carbon steels (CS) over 5 years at various temperatures and under oxic/anoxic conditions [1].*

2.4. Influence of Temperature

Most of the data obtained in granitic or argillaceous conditions, after more than 6 months of exposure, are shown in Fig. 3 as an Arrhenius plot. Archaeological analogues (iron specimens buried in various soils) have also been plotted in this Figure. The activation energy of the corrosion in aerated conditions (11.1 kJ mol^{-1}) has been determined in a previous work [7]. It is represented by the solid line in Fig. 3. In deaerated conditions, the activation energy has been calculated with the data from CEN Mol (Belgium) obtained at four different temperatures in similar environmental conditions (lower dotted line on Fig. 3). Its value (11 kJ mol^{-1}) is very close to the results obtained by PNC for deaerated conditions (10.5 kJ mol^{-1}) [4] and to the value measured in aerated conditions.

2.5. Kinetic Law for General Corrosion

The previous paragraphs have provided evidence that the general corrosion kinetics for carbon steel in neutral or slightly alkaline media, depend only on the aeration/deaeration conditions and on temperature. Therefore three experimental laws are proposed for the corrosion rate as a function of these parameters.

The first deals with oxic conditions where we have seen that the corrosion kinetics are linear. Taking into account the previously determined activation energy, the corrosion rate (*CR*) can be written as follows:

$$CR \ (mm / year) = 1.042 e^{-\frac{1340}{T}} \qquad (1)$$

This equation is represented as the uppermost solid line in Fig. 3.

In anoxic conditions, the corrosion kinetics are probably described by a power law. Nevertheless, we have chosen a simplified but slightly overestimating method

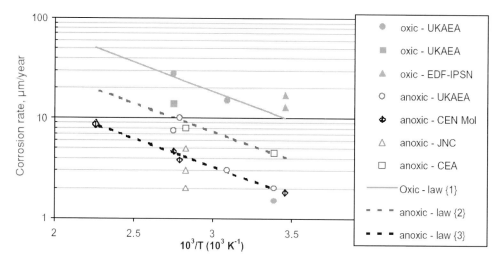

Fig. 3 *Influence of temperature on the corrosion rate of carbon steel in contact with granitic or argillaceous environments for more than 6 months.*

which consists of evaluating the corrosion rate with two linear laws. The first (eqn (2)) corresponds to a short term estimation of the corrosion (between 6 months and 10 years). It is plotted as the upper dashed line in Fig. 3, which is the upper limit for corrosion rate values measured in deaerated environments (the dots near this line were obtained after less than one year's exposure of the specimens). The second (eqn (3)) gives the long term prediction of the carbon steel corrosion rate (after more than 10 years) and is plotted as the lower dashed line in Fig. 3. These equations use the activation energy measured in the previous paragraph.

$$\text{Short term evaluation (6 months to 10 years): } CR \ (mm/year) = 0.364e^{-\frac{1300}{T}} \quad (2)$$

$$\text{Long term evaluation (more than 10 years): } CR \ (mm/year) = 0.162e^{-\frac{1300}{T}} \quad (3)$$

3. Localised Corrosion

3.1. Kinetic Laws for Localised Corrosion

Kinetic laws for propagation of localised corrosion of carbon steel have been put forward in the literature [1,8]. They are mainly based on pitting, rather than crevice, corrosion. In this paper, we assume that the propagation rates of these two type of localised corrosions are equivalent, the only difference being in the initiation mechanism which we have chosen to neglect in our estimations. According to these studies, the propagation of the localised corrosion can be described by a power law $k.t^n$ where k is a function of the temperature and the oxidising/reducing conditions and t is time. Kinetic laws have been proposed by Marsh *et al.* [1] in aerated conditions at 90°C and by Ishikawa *et al.* [8] for both aerated and deaerated conditions at 80°C. They estimate the pit depth as a function of time as follows:

$$\text{In oxic conditions: } d_{pit} = 7.02t^{0.42} \qquad (4) \text{ from [1]}$$

$$d_{pit} = 9.4t^{0.5} \qquad (5) \text{ from [8]}$$

$$\text{In anoxic conditions: } d_{pit} = 0.44t^{0.5} \qquad (6) \text{ from [8]}$$

3.2. Evaluation of the Localised Corrosion through the Pitting Ratio Evolution

Another approach to the evaluation of localised corrosion can be described. This method consists of estimating the maximum depth of the localised corrosion as a function of the general corrosion depth. The JNC [2] has used a similar approach in the case of carbon steels. It is based on a pitting factor criterion (P) which can be defined as the ratio between the maximum pit depth (d_{pit}) and the average general corrosion depth (X) observed at the same time:

$$P = \frac{d_{pit}}{X} \tag{7}$$

The data obtained by Romanoff [6], UKAEA [1,9], CEN Mol [3,10, 11–13], CEA [14], Reynaud [15] and EDF [16], have been used in determining the pitting factor for carbon steel for the various conditions as described below:

- temperature range: ~15–90°C;
- pH: neutral to ~12 (conditions in concrete);
- aerated or deaerated conditions;
- various chemical conditions (argillaceous water, granitic water, sea water, different soils etc.);
- pitting corrosion or crevice corrosion;
- exposure time up to 18 years.

All these data are plotted in Fig. 4 as a function of the average corrosion depth. Whatever the experimental conditions, these results are systematically placed under a line (eqn (8)) which represents the decrease of the pitting factor with increasing general corrosion depth. In other words, the localised corrosion rate decreases faster than the general corrosion rate.

$$P = 4.64X^{-0.67}(X \text{ in mm}) \tag{8}$$

The maximum pit depth (d_{pit}) can be deduced from eqn (9):

$$d_{pit}(\text{mm}) = 4.64X^{0.33} \tag{9}$$

Finally the total corrosion depth (d_{corr}) corresponds to the addition of the

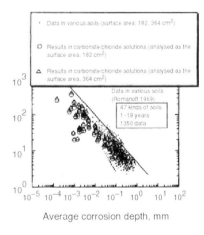

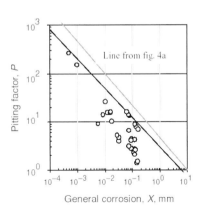

(a) Data compiled by JNC [2]   (b) Other data compiled in this work

Fig. 4 *Evolution of the pitting factor in function of the average corrosion depth. Results obtained for different carbon steels or cast iron in contact with various soils or waters.*

general corrosion (calculated from eqns (1–3)) and of the local corrosion (estimated with eqn (9)):

$$d_{corr}(\text{mm}) = X + 4.64X^{0.33} \tag{10}$$

The evolution of d_{corr} as a function of the general corrosion depth has been plotted in Fig. 5. This Figure illustrates the decreasing importance of localised corrosion compared to general corrosion with time: the localised corrosion is 100 times deeper when the general corrosion is 10 µm and becomes equivalent when the general corrosion reaches 10 mm.

This method for localised corrosion evaluation presents some important advantages. First, it is a unified approach which can be used whatever the thermal and chemical characteristics of the environment since the influence of these parameters is already taken into account through the general corrosion kinetics. Therefore, this method does not exclude the possibility of local corrosion growth in anoxic conditions (i.e. as a result of the presence of other oxidants than oxygen). Secondly, it is an overestimating but realistic approach since the pitting ratio evolution law has been obtained from numerous data.

3.3. Comparison between the Two Approaches of Localised Corrosion

The evolution of pit depth has been represented in Fig. 6 for the different laws presented in the previous paragraphs. The localised corrosion depth evaluated with the pitting factor approach is much lower — about one order of magnitude — than that calculated from the kinetic laws. In the first case, the localised corrosion

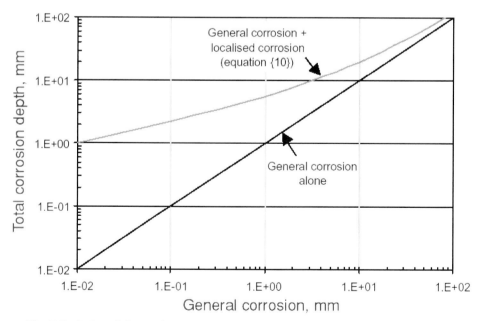

Fig. 5 Evolution of the maximum corrosion depth (general + localised) as a function of the general corrosion depth (eqn (10)).

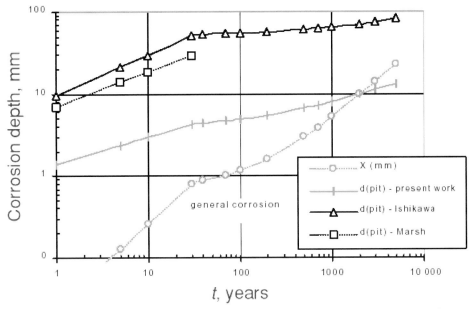

Fig. 6 *Evolution of pit depth over 5000 years (aerated phase during 30 years) at 80°C according to eqns (4) from Marsh (aerated phase only) [1], (5) and (6) from Ishikawa [8] and (10) from this work.*

contributes less than the general corrosion whereas the kinetic laws still have the major influence on localised corrosion.

4. Corrosion Prediction in the Case of Nuclear Waste Storage

In order to test this semi-empirical model (eqns (1–3), (8) and (9)) for the long term corrosion of nuclear waste containers in carbon steel, the influence of different parameters has been evaluated. These environmental conditions have been chosen to be within the range of the realistic nuclear waste storage conditions. They are the following:

- temperature between 15 and 150°C;

- duration of the aerated stage up to 50 years; and

- corrosion depth calculated over 5000 years.

The corrosion depths after 5000 years at 100°C have been evaluated as a function of the duration of the aerated stage (Fig. 7). Although the influence of the aerobic stage is important during the first hundreds of years of storage, it becomes negligible after 5000 years (less than 3%).

Temperature is a much more important factor and multiplies the total corrosion depth by nearly 3 between 15 and 150°C (Fig. 8). Nevertheless, it appears that, after 5000 years, the maximum localised corrosion depth varies little with the temperature.

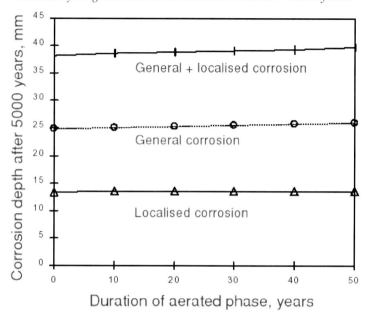

Fig. 7 *Influence of the duration of the aerated stage on the corrosion depth after 5000 years at 100°C.*

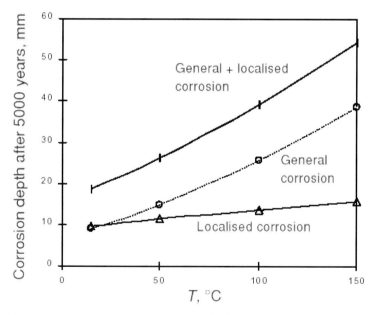

Fig. 8 *Influence of temperature on the corrosion depth after 5000 years (aerated conditions during the first 30 years).*

Most of the effect of temperature is due to the general corrosion evolution.

The evolution of the different types of corrosion are shown in Fig. 9 as a function of time. After 5000 years at 100°C, the corrosion depth estimated with this model

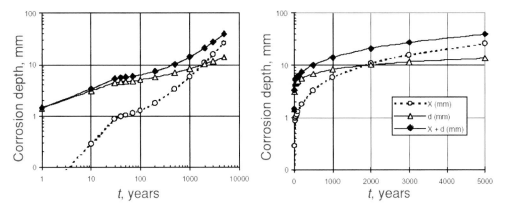

Fig. 9 *Evolution of the corrosion depth over 5000 years at 100 °C with aerated conditions during the first 30 years.*

reaches nearly 40 mm and the localised corrosion contributes less than 50% after 2000 years of storage.

5. Conclusions

A semi-empirical model is proposed for carbon steel corrosion in long term geological nuclear waste disposal. It is based on numerous data obtained by different laboratories in various conditions of temperatures, environments and duration. Its main features are:

- the estimation of the influence of the temperature on general corrosion through an activation energy of 11 kJ mol^{-1} common for both aerated and deaerated conditions;

- the slow down of the general corrosion in deaerated conditions which is taken into account for long term evaluation;

- the estimation of the localised corrosion through the evolution of the pitting factor as a function of the general corrosion depth.

Using this model, the maximum corrosion depth after 1000 years is close to 2 cm. Afterwards, the corrosion depth continues to increase slowly and stays within the order of magnitude of a few centimetres.

References

1. G. P. Marsh, K. J. Taylor, S. M. Sharland and A. J. Diver, Corrosion of carbon steel overpacks for the geological disposal of radioactive waste, Report of the Commission of the European Communities No. EC EUR13671, 1991.
2. JNC, H12: Project to establish the scientific and technical basis for HLW disposal in Japan, Supporting Report 2, April 2000, p. IV–23.

3. B. Kursten and P. Van Iseghem, *CORROSION '99*, Paper No. 473, NACE International, houston, TX, 1999.

4. A. Honda *et al., Mat. Res. Soc. Symp. Proc.*, 1991, **212**, 287–294.

5. R. P. Anantatmula, *Mat. Res. Soc. Symp. Proc.*, 1985, **44**, 273–278.

6. M. Romanoff, Underground corrosion, National Bureau of Standards Circular 579, US Government Printing Office, Washington, 1957.

7. J. M. Gras, Résistance à la corrosion des matériaux de conteneur envisagés pour le stockage profond des déchets nucléaires. 1ère partie: Aciers non ou faiblement alliés, fontes, 1996, EDF report, N° HT-40/96/002/A.

8. H. Ishikawa, A. Honda and N. Sasaki, Long life prediction of carbon steel overpack for geological isolation of high-level radioactive waste, in *Life Prediction of Corrodible Structures*, Published NACE, Houston, TX, 1994.

9. G. P. Marsh and K. J. Taylor, *Corros. Sci.*, 1988, **28**, 289–320.

10. W. Debruyn, J. Dresselaers, Ph. Vermeiren, J. Kelchtermans and H. Tas, Corrosion of container and infrastructure materials under clay repository conditions, Report EC EUR13667, 1991.

11. B. Cornélis and P. Van Iseghem, *In situ* tests of canister and overpack materials in Boom clay: results after five years, Report CEC EUR 15629, 1992.

12. B. Kursten and P. Van Iseghem, Geological disposal conditioned high-level and long lived radioactive waste. *In situ* experiments, Report SCK-CEN R-3247, 1998.

13. L. Noynaert, Heat and radiation effects on the near field of a HLW or spent fuel repository in a clay formation (CERBERUS Project), Rapport EC EUR19125, 2000.

14. F. Papillon *et al.*, Expérience 'Corrosion à 80°C': synthèse et analyse des résultats (June 2000), Report CEA, NT SESD 00-13, June 2000.

15. A. Reynaud, Résistance à la corrosion des fontes moulées. Synthèse bibliographique, Report of the Centre Technique des Industries de la Fonderie (CTIF), Report No. FO.149, 1988.

16. P. Vigne and F. Foct, 2002, EDF report HT-44/02/015/A.

8

The Determimistic Prediction of Localised Corrosion Damage to Alloy C-22 HLNW Canisters

D. D. MACDONALD, G. ENGELHARDT*, P. JAYAWEERA[†], N. PRIYANTHA[†] and A. DAVYDOV[§]

Pennsylvania State University, University Park, PA 16802, USA
*OLI Systems, Morris Town, NJ 07950, USA
[†]SRI International, Menlo Park, CA 94025, USA
[§]Frumkin Institute of Electrochemistry, Russian Academy of Sciences, Moscow, Russia

ABSTRACT

This paper summarises DOE-funded research programmes currently underway by researchers at SRI International, Penn State University, OLI Systems, and the Frumkin Institute of Electrochemistry (Moscow, Russia) that are aimed at exploring the corrosion behaviour of Alloy C-22 as the canister material for the disposal of high-level nuclear waste (HLNW) in Yucca Mountain-type repositories. The ultimate objective of these programmes is to develop deterministic models for predicting the accumulation of damage due to general corrosion and localised corrosion over the specified evolutionary path of the repository. Additionally, the programme seeks to measure important electrochemical parameters and diagnostic functions under conditions (steady-state) that are in good confluence with the theories and models used in the predictions. The present paper deals with the prediction of accumulated localised corrosion damage in the form of pitting; the prediction of general corrosion damage is dealt elsewhere in the Volume (pp.75–90).

1. Introduction

Localised corrosion is generally considered to be a significant threat to the long-term integrity of High Level Nuclear Waste (HLNW) canisters, because of the rapid penetration rates that can result from localised corrosion processes. The localised corrosion processes of greatest concern include pitting corrosion (PC) and stress corrosion cracking (SCC), both of which occur on 'corrosion resistant' alloys in a variety of environments. As reported in this work, Alloy C-22, which is the canister barrier alloy of choice for the Yucca Mountain Repository, suffers from pitting attack in laboratory observation times, but only under extremely aggressive conditions. On the other hand, SCC under constant load apparently has not been detected in this alloy under even severe laboratory conditions. In assessing the possibility of any given form of corrosion occurring in the field from laboratory data, a major difficulty arises with respect to the service life. Thus, the principal difficulty is that the service life horizon is twice as far into the future (10 000 years) as recorded human history is in the past, and no direct experience exists within the degradation of engineering materials over that time period. Indeed, some of the materials that are proposed for use in HLNW repositories did not exist thirty years ago (e.g. Alloy C-22). A search of the literature reveals that no material that is currently contemplated for HLNW service has been investigated in a single corrosion study for more than 60 000 hours, or for more than a few hundredths of one percent of the intended service life. Not only is the corrosion rate unknown over even a very small fraction of the service life, but also we cannot be sure that the mechanism of attack will remain the same even if the external conditions were invariant.

As noted elsewhere [1], it is evident that purely empirical methods cannot meet the challenge of assuring the public of safe disposal within the engineered barrier concept, primarily because of the stark disparity between the time available for laboratory experiments and the design lifetime of the repository. In response to that conclusion, we have instigated, under DOE/NERI sponsorship, a programme to develop deterministic models within the framework of Damage Function Analysis (DFA) [2,4] for predicting the accumulation of general and localised corrosion damage to Alloy C-22 in stimulated HLNW repository environments. The ultimate goal is to develop deterministic models that capture the mechanistic essence of the damaging processes, that require minimal calibration, and that can be used to predict the evolution of corrosion damage over the requisite time to within the required engineering accuracy.

The present paper deals with the possible occurrence of localised corrosion in the form of pitting on Alloy C-22 in contact with concentrated sodium chloride solution, which was selected to stimulate a geochemical brine and because the properties of this system are well understood. The pH and temperature were selected as being within the ranges expected for these two parameters within the repository and because they yield observable results within laboratory times. Alloy C-22 is a nickel base alloy having the nominal composition (at.%) of Cr22, Mo13.5, W3, Fe4, C0.015, Si0.08. Both experimental and theoretical studies of the nucleation and growth of pits are described, with the theoretical work being presented within the framework of Damage Function Analysis (DFA) [2,4]. A principal, but yet largely unaccomplished, goal of the work is to accurately characterise pitting under very aggressive conditions in the laboratory and then to use DFA to extrapolate the accumulation of damage to the (possibly) less aggressive conditions in the field (repository), but which exist over much longer times. The prediction of the accumulation of damage due to general corrosion is discussed elsewhere in this Workshop [1].

2. Pitting Corrosion Alloy C-22

2.1. Does Alloy C-22 Pit?

Given the well-known ability of pitting corrosion to cause premature failure, it is necessary to examine the possibility that this form of corrosion might occur over the ten thousand year design life of the canisters in a Yucca Mountain type of repository. It is important to note, at the outset, that considerable disagreement exists in the scientific community as to whether Alloy C-22 suffers pitting corrosion under any conditions. However, it is important to recognise that, because pitting is a progressive kinetic phenomenon, any claims that the alloy is immune to pitting corrosion may simply reflect the shortness of the experimental observation time. Indeed, examination of pit nucleation theory shows that if pitting can be demonstrated under severe conditions in the laboratory, then it will eventually occur under less severe conditions in the field, given sufficient time [2]. Accordingly, our first task was to explore whether C-22 can be made to pit under severe conditions in the laboratory. The implications for long-term exposure in the field are explored later in this paper.

As Alloy C-22 is resistant to passivity breakdown, localised corrosion processes are generally not observed under mild environmental conditions in the laboratory. However, the breakdown voltage and other electrochemical properties of Alloy C-22 are highly dependent on experimental conditions, including solution pH, temperature, activity of anions in the electrolyte medium, and instrumental variables, such as potential scan rate [1]. The effect of the activity of chloride on potentiodynamic polarisation curves recorded at pH 3 and at 80°C is shown in Fig. 1. While the data shown in this figure are not sufficiently precise to discern the dependence of the breakdown potential on chloride activity, it is evident that the current generated as the result of passivity breakdown is a strong function of [Cl$^-$]. This suggests that chloride ion plays a direct role in the dissolution process, possibly due to the formation of Cr(III)-chloro complexes [3].

Furthermore, in saturated chloride solutions at elevated temperatures, the breakdown process, which is usually recognised as a sudden increase in current, is overlapped with the oxygen evolution. Based on potentiodynamic polarisation experiments conducted at scan rates of 0.5 mV s^{-1} or lower, the average breakdown voltage in saturated NaCl solutions at 80°C, maintained at pH 3 with HCl using a 'pH-stat', was estimated to be about 948 mV vs the Standard Hydrogen Electrode (SHE). Repeated application of cyclic polarisation

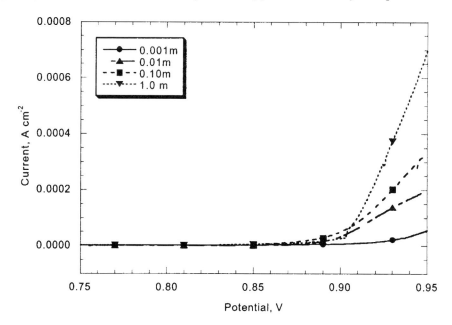

Fig. 1 *Selected linear polarisation scans for unpolished Alloy C-22 in chloride solutions (pH 3, 80°C, N$_2$ deaerated, scan rate 0.5 mV s^{-1}. Concentration of chloride: 0.001M, 0.011M ,0.101M and 1.001M).*

programmes has some effects, such as variations in the zero current potential, on the current–potential characteristics of the specimen. These changes are clearly visible when the data are presented in logarithmic current scale as shown in Fig. 2. After long-term cycling of the potential, Tafel behaviour was significantly altered, presumably as a result of changes in the passive film on the surface.

Alloy C-22 under various conditions, including aeration and deaeration (N_2) sparging, and at different chloride concentrations, showed very low rates of general corrosion, much less than 2 μm/year, indicating its high stability. Nevertheless, long-time exposure of C-22 at potentials higher than the breakdown voltage, during either cyclic polarisation and/or potentiostatic polarisation, made the C-22 surface susceptible to pitting damage, especially at elevated temperatures.

Application of potentials greater than the breakdown voltage (948 mV vs SHE) in pH 3 HCl saturated with NaCl at 80°C under N_2 deaeration was sufficient to initiate accelerated corrosion of C-22, which was visibly signalled by the appearance of a yellowish coloration in the solution. Continuation of this process turned the solution to pale green. However, the corrosion current gradually decreased, indicating that the surface repassivates with time. This suggests that pitting corrosion is more predominant than general corrosion in Alloy C-22, in this particular environment.

Figure 3 shows pits produced on a polished (600 grit SiC + 0.05 Al_2O_3) Alloy C-22 specimen at 948 mV (SHE) in saturated NaCl solution (≈ 6.2M) at pH = 3, and at 80°C, for a period of 10 days. The pits were a few microns in size and some were combined to form elongated fissures of more than 10 μm in length. In comparison, application of 998 mV (SHE) for the same period of time under identical conditions produced a larger number of small pits. These pits were very shallow, and the dimensions were similar to the pits formed at the lower applied potential. This observation supports the idea that, although pits are easily initiated under severe conditions, they are easily repassivated, possibly because of their open morphology. Pits of larger dimensions are formed if a potential of greater than 1 V(SHE) is applied for a long time, or if the potential is scanned beyond 1 V(SHE) at low scan rates. Potentials of this magnitude were required to induce pitting within the experimental time.

Figure 4 shows a scanning electron micrograph of a polished Alloy C-22 surface after 10 days exposure to such corrosive conditions. In this case, it is evident that passivity breakdown has occurred at the periphery of an emergent second phase particle, eventually resulting in significant ditching. Also apparent is significant secondary breakdown.

2.2. Is Pitting a Significant Threat?

Demonstration that pitting occurs on Alloy C-22 under severe laboratory conditions is, by itself, possibly not convincing that pitting corrosion will be a threat to canister integrity over that proposed life of the repository. This issue can only be addressed by resorting to deterministic models that are constrained by natural laws that are invariant in space and time and by specifying a continuous path into the future. The basic models for predicting the evolution of localised corrosion damage over repository lifetimes have been developed in the form of Damage Function Analysis (DFA) [2,4]. Briefly, DFA combines the PDM (Point Defect Model) for passivity breakdown with a 'coupled environment' model for cavity (pit or crack) growth and

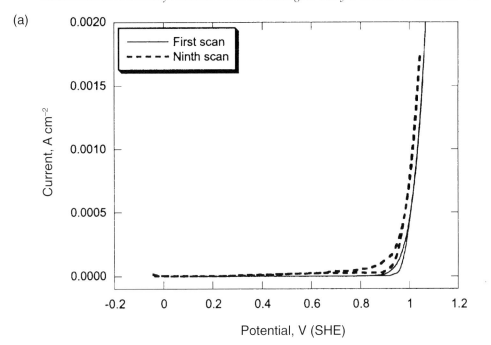

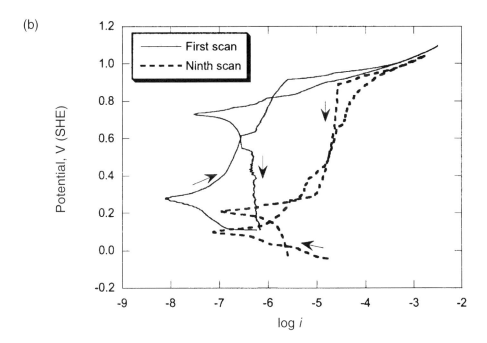

Fig. 2 *Polarisation curves for Alloy C-22 in a saturated NaCl solution (pH 3, 80 °C, N₂ deaerated, scan rate 0.5 mV s⁻¹). (a) Linear polarisation (i–E) curves and (b) cyclic polarisation [E–log(i)] plots.*

Fig. 3 *Scanning electron micrograpps of a Hastelloy C-22 electrode surface after exposure to a bias voltage of 948 mV (SHE) in saturated NaCl over a period of ten days (80°C, pH 3, N_2 deaeration).*

models for prompt and delayed repassivation. While DFA has yet to be modified to predict the accumulation of localised corrosion damage along an evolutionary path (i.e. over the future 'history' of the repository), the way for doing so in the near future is clear. Instead, in the present analysis, we describe the 'constant conditions' case, in which the repository conditions (temperature, pH, ECP (Electrochemical Corrosion Potential), [Cl⁻], etc.) are assumed to be invariant with time.

Damage Function analysis (DFA) expresses damage in terms of the damage function (DF), which is the histogram of the event (pit) frequency vs the increment in event depth. Predicted damage functions (DFs) for the pitting of C-22 in 'repository

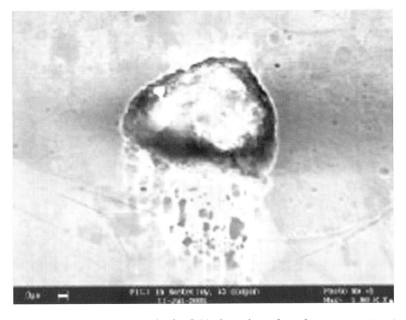

Fig. 4 *Scanning electron micrograph of a C-22 electrode surface after exposure to potentials greater than 1 V in saturated NaCl solution over a period of 10 days (80°C, pH 3, N_2 deaeration).*

brine' after exposure for 1000, 5000 and 10 000 years are shown in Fig. 5, as a function of the delayed repassivation constant (γ) [4]. The delayed repassivation constant, γ, describes the rate of repassivation of stable pits, assuming first order kinetics. Among the processes that are postulated to be responsible for delayed repassivation are the inability of the external surfaces to supply the resources in the form of oxygen reduction to maintain the required separation between the local anode in the pit and the local cathode on the adjacent external surface and hence to sustain the pit ('death by old age'), competition between neighbouring pits for the external surfaces ('death by competition'), and inadvertent events, such as drying of the surface ('death by misadventure') [2]. The PDM estimated the mean passivity breakdown potential as being 0.76 V(SHE), which is in good agreement with experiment (see Fig. 2). The first set of damage functions was developed for the rate of delayed repassivation, as measured by the first order rate constant, γ, being zero. In this limit, no stable pit that nucleates on the surface ever dies, so that the distributions inherent in the damage functions result from the progressive nature of nucleation and growth, such that the deepest pits are those that nucleated first. Pits of shallower depth nucleated at a later time. The parameter values chosen for the calculation result in the deepest pit extending to about 2.0 cm after 10 000 years, although the values were not chosen to give that result. The other two sets of DFs shown in Fig. 5 were calculated for $\gamma > 0$, which implies that all stable pits eventually die (repassivate) at a rate that is measured by the value of the rate constant, γ. Upon increasing γ to 10^{-3}/year, the shape of the DF changes dramatically and the maximum penetration is significantly reduced. The shape that is calculated is essentially that for an ensemble of dead pits, although it is clear that many of the pits are predicted to penetrate to a considerable depth before death occurs (to about 1.5 cm for the longest time). On the other hand, when γ is made even still larger (10^{-1}/year), all pits that nucleate at any time repassivate before they extend in depth beyond the first increment. Clearly, the value of the rate constant for delayed repassivation has a profound impact on the development of pitting damage on metals in general and on a canister surface in particular. Indeed, controlling the value of γ may be a viable way of limiting the amount of pitting damage that accumulates on the surface, and this possibility is currently being explored in our programme.

The corrosion potential is also predicted (and known) to have a profound impact on the rate of accumulation of pitting corrosion damage on a metal surface. Thus, both the nucleation rate and the rate of growth of pits on a surface increase significantly with increasing potential, as found experimentally and as predicted by the PDM [2,4]. As an illustration of the impact of the ECP, we show in Fig. 6 plots of the number of pits nucleated as a function of time for ECP values ranging from 0.35 V(SHE) to 0.60 V(SHE). Note that at the highest potential, essentially all of the pits nucleate at the shortest time, corresponding to 'instantaneous nucleation'. On the other hand, at a voltage of 0.35 V(SHE) only about 25% of the pits that could have nucleated have done so after 10 000 years. This latter case is a prime example of pitting being a progressive nucleation/growth/death process, in which new damage nucleates as existing pits grow and die.

There are other factors that must be considered in further development of DFA that could have an important impact on the predicted damage. For example, it has been found, at least in the case of the pitting of nickel in chloride-containing solutions,

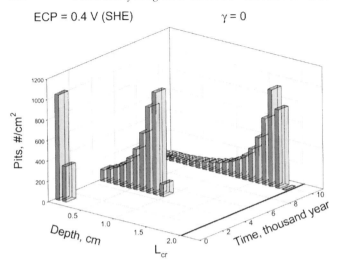

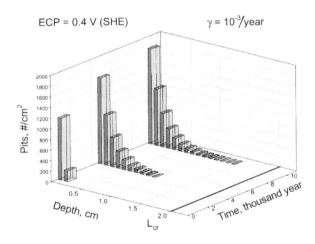

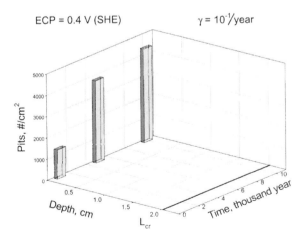

Fig. 5 *Predicted damage functions for the pitting of Alloy C-22 at 1000, 5000 and 10 000 years after exposure to saturated brine ($a_{Cl^-} = 6.2$) at pH = 7, and T = 100 °C and as a function of the delayed repassivation constant, γ. The ECP was assumed to be 0.40 V(SHE) and the mean passivity breakdown potential was taken as 0.7607 V(SHE). L_{Cr} = critical pit depth = wall thickness.*

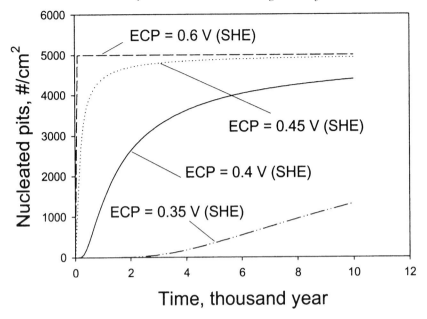

Fig. 6 *Number of nucleated pits vs time as a function of potential for Alloy C-22 in NaCl brine ($a_{Cl} = 12$) at 100°C. The parameter values assumed in the calculation are the same as those employed in Fig. 5.*

that each pit has an associated hemisphere of influence (HOI) on the external surface, within which oxygen reduction occurs [5]. However, it was also found that no new pits nucleated within an existing (active) pit's HOI; this being attributed to the fact that the area under the HOI is cathodically protected by the pit itself. Accordingly, the area upon which new pits can nucleate becomes a complex function of time that is expected to have a substantial impact upon the shape of the damage function, although it may not significantly affect the maximum pit depth.

Another way of expressing the tendency toward failure is to plot the failure probability against the wall thickness. The failure probability is calculated as being unity if the maximum pit depth exceeds the specified wall thickness at the specified observation time, or being equal to zero if the maximum pit depth does not exceed the wall thickness. Typical failure probability plots calculated using the Point Defect Model are shown in Fig. 7 for the parameter values specified in Fig. 5. The plots in Fig. 7 show that for an observation time of 10 000 years, the wall thickness at which failure is predicted to occur is of the order of 1.9 cm; which is greater than the 1.6 mm value predicted for general corrosion by a factor of more than ten [1].

Thus, at first sight, the model would seem to predict that pitting corrosion does pose a substantially greater risk for failure of the canister than does general corrosion. However, this conclusion must be tempered by the fact that the damage functions are not calculated over the evolutionary path of the repository and by the fact that we do not know the value of γ for the canister in the repository environment. Indeed, the value of γ, which is the first order rate constant for the repassivation (death) of stable pits, is possibly the most important parameter in

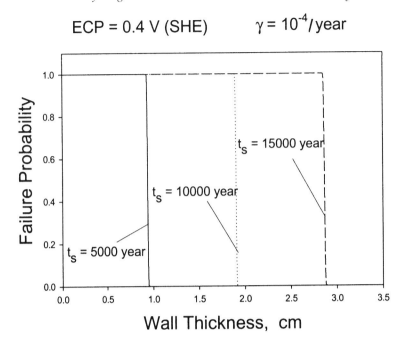

Fig. 7 *Failure probability vs wall thickness as a function of observation time for C-22 in contact with a thin electrolyte film on the canister surface. NaCl brine ($a_{Cl} = 6.2$) at 100°C. The parameter values assumed in the calculation are the same as those employed in Fig. 5.*

the calculation of the damage function and hence the failure time. At this point in time, we do not have a sound theoretical basis for calculating γ, but methods are now under development.

2.3. What Do We Need to Know?

It is argued in this paper and elsewhere in this Volume [1] that the only viable philosophy for predicting the accumulation of corrosion damage to HLNW canisters is determinism, as constrained by time- and space-invariant natural laws. This is so, because of the infeasibility of performing continuous experiments over even a small fraction of the required repository lifetime. Once this view is accepted, then the nature of the experimental work that must be done becomes clear; the work must aim at providing high quality information on model parameter values, and at providing specific data against which the models can be tested, even if these data correspond only to the short term. In performing the experiments, it is vital that the experiments and theory are in confluence; that is that the experiments are carried out under the conditions that are specified by the theory and vice versa. Thus, if the model that will be used in the predictions assumes steady state conditions, then the experiments must be carried out under demonstrably steady state conditions. Part of our DOE/ NERI programme aims at generating high quality data for model parameters. The current work has focused on measuring kinetic data for the hydrogen evolution and oxygen reduction reactions, and for alloy passive dissolution, on Alloy C-22 in 'relevant' environments, including NaCl brines under conditions that are typical of the repository during its lifetime. Much of this work is being carried out at the Frumkin Institute of Electrochemistry (FIE) in Moscow, Russia, and a brief synopsis of their work is given below.

The experiments being carried out at the FIE have concentrated on establishing conditions that reasonably approximate the steady state, particularly when determining cathodic and anodic polarisation curves. Careful analysis of the experimental options suggests that steady state current density vs potential curves for oxygen reduction should be measured in the direction of increasingly positive potential values, and that the potential range should not include the potential range for cathodic hydrogen evolution. When cathodic hydrogen evolution is studied, it is necessary to avoid the potential range of significant hydrogen penetration into the metal, which leads to irreversible changes in the state of specimen surface layer. Figure 8 shows cathodic polarisation curves for the reduction of oxygen on Alloy C-22 at various temperatures in 5M NaCl solution of pH = 8 saturated with oxygen. At 20°C, under the adopted experimental conditions (solution stirring with a magnetic stirrer), the limiting current for oxygen reduction according to the reaction $O_2 + 4H^+ + 4e^- \rightarrow 2H_2O$ is $i_1 = 4.3 \times 10^{-4}$ A cm^{-2} the transfer coefficient is $\alpha = 0.34$, and the exchange current density is calculated as $i_0 = 3.46 \times 10^{-10}$ A cm^{-2}. At 40°C, $i_1 = 6.7 \times 10^{-4}$ A cm^{-2}, and $i_0 = 1.8 \times 10^{-8}$ A cm^{-2}. At higher temperatures, a new wave arises in the polarisation curve, corresponding to oxygen reduction to hydrogen peroxide: $O_2 + 2H^+ + 2e^- \rightarrow H_2O_2$. Then (at more negative potentials), there follows a polarisation response corresponding to the reaction: $H_2O_2 + 2H^+ + 2e^- \rightarrow 2H_2O$.

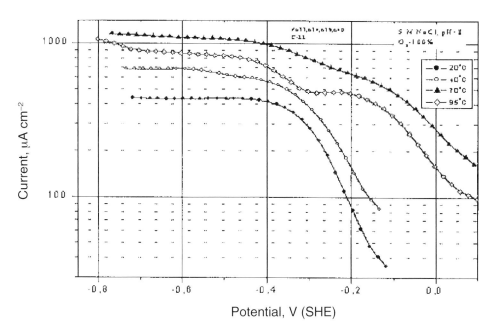

Fig. 8 Oxygen reduction polarisation curves on Alloy C-22 in 5M NaCl, pH = 8 at 20, 40, 70, and 95°C.

(Note that hydrogen peroxide is produced during the cathodic reduction of oxygen.) The kinetic parameters for the cathodic reaction then change substantially. It should be noted that the curve for 95°C lies in the lower current density range than that for 75°C, which may be associated with the decrease in oxygen solubility at high temperatures.

Figure 9 shows typical polarisation curves for the cathodic evolution of hydrogen on Alloy C-22 in the same solution, but in this case the solution is saturated with hydrogen, so as to establish a definite open circuit potential (corresponding approximately to the hydrogen electrode equilibrium potential). From the Tafel portions of these curves, α (the transfer coefficient) and i_0 (the exchange current density) can be obtained. At 20°C, $\alpha = 0.29$ and $i_0 = 4.36 \times 10^{-6}$ A cm^{-2}; at 40°C, $\alpha = 0.31$ and $i_0 = 11.94 \times 10^{-6}$ A cm^{-2}; at 70°C $\alpha = 0.32$ and $i_0 = 35.3 \times 10^{-6}$ A cm^{-2}; and at 95°C $\alpha = 0.37$ and $i_0 = 82.93 \times 10^{-6}$ A cm^{-2}.

Clearly, both α and i_0 increase with increasing temperature. When α was determined, single-electron transfer was considered to be the limiting stage in all cases.

In order to demonstrate the difficulty in measuring steady state parameters for an alloy that is as corrosion resistant as is C-22, we show in Fig. 10 polarisation curves measured potentiodynamically at a voltage scan rate of 0.0001 Vs^{-1}. The polarisation curve displays a potential-independent current in the passive range, but on the reverse sweep the current does not retrace the behaviour on the forward scan. This hysteresis is typical of systems that are not at steady state. Furthermore, the data shown in Fig. 10 display a transpassive dissolution phenomenon at 0.7–0.8 V(SHE), which compares favourably with that calculated from the Point Defect Model [2], allowing for the differences in conditions. It is important to note that the behaviour at high potentials is indicative of the process that dominates the electrochemistry of the interface. Thus, in Fig. 10 we see that, in the reverse scan, a decrease in the potential leads to a decrease

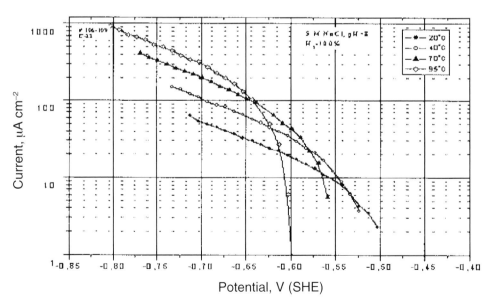

Fig. 9 *Polarisation curves for the hydrogen evolution reaction on Alloy C-22 in 5M NaCl solution, pH = 8, as a function of temperature.*

in the current (the onset of the reverse scan is started). This is typical of passive behaviour. On the other hand, at 95°C (Fig. 11), we observe a quite different behaviour that is typical of pitting; in spite of the change of scan direction, the current continues to increase. This to be taken as strong, additional evidence that Alloy C-22 is susceptible to pitting corrosion under the most severe conditions (a high NaCl concentration, low pH, and high temperature) and approximately in the same potential range as transpassive dissolution (0.8 V(SHE)).

3. Summary and Conclusions

This paper summarises Department of Energy (DOE)-funded research programme currently being carried out by researchers at SRI International, Penn State University, OLI Systems, and the Frumkin Institute of Electrochemistry (Moscow, Russia) that are aimed at exploring the corrosion behaviour of Alloy C-22 for the disposal of high-level nuclear waste (HLNW) in Yucca Mountain-type repositories. The ultimate objective of these programmes is to develop deterministic models for predicting the accumulation of damage due to general corrosion and localised corrosion over the specified evolutionary path of the repository. Additionally, the programme seeks to ensure important electrochemical parameters and diagnostic functions under steady state conditions that are in good confluence with the theories and models used in the predictions. While the work is not yet complete, a number of significant advances and important findings have been made, as follows:

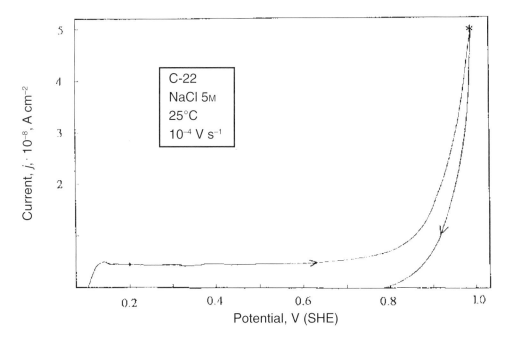

Fig. 10 *Typical polarisation behaviour of a passive system, in this case Alloy C-22 in 5M NaCl at 25 T. Note the transpassive behaviour at potentials above 0.7–0.8 V(SHE).*

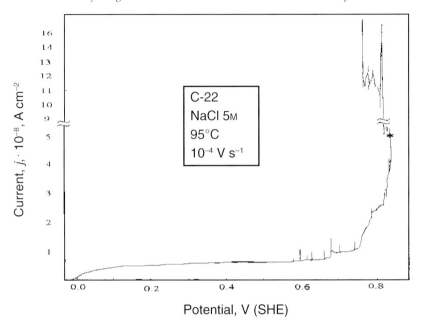

Fig. 11 *Typical polarisation curve for Alloy C-22 under conditions of pitting attack (5M NaCl, 95 °C). Note that the current on the reverse scan continues to increase as the potential is swept in the negative direction. At a sufficiently low potential, the localised corrosion sites will passivate resulting in the 'repassivation potential'.*

- Alloy C-22 is found to undergo pitting corrosion in very aggressive, low pH, high [Cl⁻], high temperature environments, with the breakdown potential being near the transpassive dissolution potential. However, because of the distributed nature of the breakdown potential, pits are expected to nucleate at potentials that are significantly more negative than the mean, so that pitting is expected under considerably less aggressive conditions, provided that the observation time is sufficiently long. The pits, as observed in short term experiments, are open and of low aspect ratio (depth/diameter).

- Damage Function Analysis (DFA), which is based upon the Point Defect Model (PDM) for passivity breakdown, upon deterministic models for cavity growth, and upon the concept of delayed repassivation, has been used to predict damage functions for pitting corrosion in simulated repository environments for times up to 10 000 years. The maximum depth of penetration is predicted to be of the order of 1.9 cm, which is of the same order as the wall thickness of the canister. However, the maximum penetration depth is a sensitive function of the rate constant for the delayed repassivation of active pits, with the maximum depth decreasing as the rate constant increases. No data are currently available for the delayed repassivation rate constant, but methods have been devised in the present program for measuring this important parameter.

- Experimental data are being measured for a variety of model parameters, including the kinetic parameters (exchange current densities and Tafel constants) for the reduction of oxygen and the evolution of hydrogen, and for the oxidation of the substrate, on Alloy C-22 in sodium chloride solution as a function of temperature and pH. The greatest challenge has been to achieve quasi steady state conditions, so as to conform to the constraints of the models. Achieving steady state conditions is particularly difficult with respect to the anodic oxidation current, because of the extraordinarily long time that the transients last. Nevertheless, a reasonably complete set of parameter values is being assembled, which will be of value not only in the present work, but also in studies being carried out elsewhere on the most challenging problem of HLNW disposal.

4. Acknowledgements

The authors gratefully acknowledge the support of this work by the US Department of Energy under the Nuclear Energy Research Initiative, Project 99-217.

References

1. D. D. Macdonald, 'The Holy Grail: Deterministic Prediction of Corrosion Damage Thousands of Years into the Future', this Volume, pp.75–90.
2. D. D. Macdonald, 'Passivity: The Key to Our Metals-Based Civilization', *Pure Appl. Chem.,* 1999, **71**, 951–986, and references cited therein.
3. F. A. Cotton and G. Wilkinson, *Advanced Inorganic Chemistry*, 5th Edn. Wiley Interscience, NY, 1988.
4. G. R. Engelhardt and D. D. Macdonald, 'Deterministic Prediction of Pit Depth Distribution', *Corrosion*, 1998, **54**, 469–479.
5. D. D. Macdonald, M. Urquidi-Macdonald, S. J. Lenhart and C. English, 'Minor Alloying Elements in the Pitting Behaviour of Metals and Alloys', Final Report from SRI International to DOE/BES, Contract DE- FG03-84ER 45164, 1987.

9

Numerical Simulations of Simple Processes Associated with Corrosion, Diffusion and Formation of a Passive Layer

F. ROUX, J. STAFIEJ*, A. CHAUSSÉ and J. P. BADIALI†

Laboratoire Analyse et Environnement, UMR 8587, Université d'Evry, Rue du Père Jarlan, 91025 Evry Cedex, France
*Institute of Physical Chemistry of Polish Academy of Sciences, ul Kasprzaka, 44/52 01- 224 Warsaw, Poland
†Laboratoire d'Electrochimie Chimie Analytique, UMR 7575, boite 39, Université P. et M. Curie, 4 Place Jussieu, 75230 Paris Cedex 05, France

ABSTRACT

A combination of simple processes that may be assumed to be present in a large class of corrosion phenomena taking place in natural environment conditions are described. A heterogeneous chemical reaction coupled with a diffusion reaction process and a precipitation phenomenon is considered, and a simple model for describing a non-uniform corrosion probability introduced. Although extremely simple processes are retained the numerical simulations lead to a large number of types of behaviour. In particular, it is possible to investigate the transition from general to localised corrosion. The model may predict the existence of pseudo-cyclic evolution in the structure of the corrosion front.

1. Introduction

It is well known that passive metals are protected from corrosion in many aqueous environments by the formation of oxide layers. The structure of these layers controls their protective properties. Many attempts have been made in the past to determine the layer composition but most studies have been limited by the difficulties either in measuring small amounts of materials in wet environments by *in situ* analysis or because of the instability of materials when they are extracted from wet environments for *ex situ* analysis. Moreover an extra difficulty arises with the formation of corrosion transient products that transform after a time into products that are thermodynamically more stable. In what follows we will investigate simultaneously metal corrosion and passive layer formation.

The corrosion processes can be investigated by experiments at constant potential or under open-circuit conditions, which mimic the natural circumstances of the corrosion process. Hereafter we focus on this second case because we intend to describe the behaviour of materials in their natural environments. Corrosion of metals results from a large number of processes [1]. Basically we have to deal with some chemical and/or electrochemical reactions at the metal/electrolyte interface coupled with mass transport phenomena. This may lead to the formation of a passive layer at the interface via precipitation reactions; this layer greatly reduces the rate of corrosion but it can also lead to pitting corrosion if it is susceptible to localised breakdown (see for instance [2]). Partial dissolution of this layer into solution can also occur via precipitation–dissolution equilibria.

Numerical simulations represent a very powerful theoretical tool for investigating complex processes such as those appearing in natural corrosion. From simulations, we may analyse separately various processes that appear simultaneously in real systems [3]. In this way we can use very simple models to start to understand how the competition between the elementary processes may determine the morphology of a large class of metal/solution interfaces. Moreover, the simulations might be useful to give some predictions on the long time behaviour of real interfaces. Accordingly our main goal is not to explain the behaviour of a particular system but rather to analyse some basic processes that might be useful in a general understanding of corrosion phenomena. At the same time it was hoped that numerical simulations might be useful in analysing common assumptions that are introduced in simple models such as, for example, those used in the analysis of impedance measurements. Another goal of this paper is to generalise a model proposed recently [4] by introducing a non-uniform corrosion probability; from this extension of the model we plan to investigate the transition from general corrosion to localised corrosion. As we shall see, a small number of processes is sufficient to induce a large class of behaviours.

This paper is organised as follows. In Section 2, we describe the physical model. In Section 3, the physical model is put in a tractable form that we can use in numerical simulations; in particular the model of non-uniform corrosion probability is introduced. The results are presented in Section 4. We first recall briefly the results already obtained for the case of a uniform corrosion probability and then we mainly focus on the case of a non-uniform corrosion probability. Some conclusions are given in Section 5.

2. A Simple Physical Model

In order to understand the main features of the corrosion processes, we select some basic phenomena that may appear in a large class of metal/solution interfaces. We assume that many 'details' are irrelevant for our goal although these 'details' can determine some important aspects in a specific experiment. Our approach corresponds to a description at a mesoscopic (10–100 µm) scale. We assume that all the neglected processes are implicitly taken into account by introducing a random aspect in the processes involved in the model [4].

We consider a class of corrosion processes in which the corrosion is initiated by a given chemical species, s, existing in the ionic solution. In contact with these species and the ionic solution, several phenomena can take place at a microscopic level; the net result of these processes is metal corrosion, i.e. in our mesoscopic description, the detachment of a piece of metal containing a given number of metal atoms. One part of these metal atoms, M, immediately reacts with a set, S, of s species located in their neighbourhood; the result is the formation of a given amount, L, of a new compound, which precipitates on the corroded site. At this coarse-grained level, we represent this process by the formal reaction, $M + S \rightarrow L$.

Another part of the metal atoms involved in the corrosion, denoted D, are inserted in the layer. The species M and D have the same chemical nature but M is fixed while D is moving. A D species moves according to a self-diffusion process until it reaches a position where it can react with the species S forming a new L; L precipitates on the layer that has already formed. We can summarise these processes by writing the formal chemical reactions:

$$S + (\phi +1) M \rightarrow L + (\phi) D \text{ and } D + S \rightarrow L \qquad (1)$$

where $(\phi +1)M$ represents the total amount of metal involved at a corrosion site. Our basic processes are then a heterogeneous chemical reaction coupled with a reaction diffusion process. The parameter ϕ takes into account the fact that the species L has a volume larger than the one of M.

This is an extremely crude model compared to what happens in real systems. Firstly, we are only concerned with one global chemical reaction while with real materials several chemical and electrochemical reactions are involved. Secondly, we assume that the limiting transport phenomenon is determined only by the D species diffusion. In particular, the S species diffusion towards the metal surface is assumed to be very fast. Thirdly, all the electrical processes are neglected; no migration or explicit electrochemical reactions are taken into account. Finally, our assumption is that all the L species precipitate on the surface and then no partial dissolution of the layer is considered. Nevertheless, as we shall see, the simple model corresponding to (1) is already rich enough to describe a lot of processes.

In order to give a quantitative description of the processes involved in (1), we have to introduce some unavoidable ingredients. First we have to describe how the process starts. This means that we have to know the initial corrosion rate and the diffusion coefficient initial value. Since two dynamic processes are in competition, we have to set their relative time scale. In order to describe the corrosion, we have to decide whether it takes place uniformly on the metal surface or not. In addition we have to decide how the species D move and where it moves, for example, through the layer in formation or into the ionic solution.

3. Model Used in Numerical Simulations

To save computational time, the simulations are performed on a two dimensional square lattice as frequently the case in simulations of growth on a surface [5]. Note that some experiments performed in quasi-two dimension electrochemical cells that now exist represent good candidates to test the models (see for instance [6]). The lattice spacing, a, represents the minimum scale to get some information, i.e. any process associated with a shorter length is assumed to be irrelevant in our description. On the infinite lattice, we consider a slice of N_c columns; its 'area' is $A = N_c a$ or N_c but since we take as unit of length $a = 1$; N_c will measure the 'area' of our initial metal/solution interface. The number of sites in each column is unlimited. Each site is referred to two coordinates (i,j); i refers to the column position in the slice and j indicates the row position in this column.

Outside this slice we have periodic boundary conditions. On this lattice we have to define the rules on which we put, transform and move the various species.

3.1. Definition of Sites

First we assume that S and L have the same volume and that they may occupy one lattice site. We assume that a site can be occupied by $(\phi + 1)$ entities of M species. The entities S, L and M cannot coexist on the same site. We define a corrosion site as a metal site in contact with S or L sites; a growth site is an S site first nearest neighbour of sites M or L. Only growth and corrosion sites can be transformed during the processes. By these choices the passive layer will exhibit a given mechanical rigidity. We assume that a fresh D species can stay located on the corrosion site where it has been created or be injected into the layer during its formation. Here we assume that the motion of a D species takes place on the part of the lattice occupied by the L sites previously created. Only one D species can coexist with a site occupied by L. Since we mimic a motion through a solid phase, we may expect that it is the determining mass transport phenomena. A D species may also jump to a growth site where it is immediately transformed into an L site that is integrated into the layer.

3.2. Diffusion Process

We mimic the diffusion by considering a random walk on the part of the lattice occupied by the L species. In order to take into account the steric interaction between walkers, we assume that a site can be occupied by only one D species. If a walker tries to move towards a site already occupied by a D species, its motion is not realised. In this way the diffusion motion depends on the walkers' concentration and we introduce a coupling between corrosion rate and transport phenomena. For instance, in the part of the layer rich in D species, the effective diffusion coefficient will be smaller than near the growth front where the D concentration is low [4].

By definition, during the unit of simulation time, we try to move all the D species so that the diffusion rate is defined. In parallel, we choose a corrosion site number N_{corr} and we analyse whether they can be corroded or not during t_{corr}, that is a multiple of the unit of simulation time. As shown in [4], this choice sets the ratio between the characteristic times for diffusion and the corrosion rate when we start the simulation. Then, increasing t_{corr} leads to a decrease of the corrosion rate with regard to the diffusion rate. However, increasing N_{corr} leads to the inverse effect.

3.3. Model for Corrosion Probability

A corrosion site C referred to (i, j) is chosen at random. The probability of transforming this site into an L species depends on two elements. First, we have to decide if the corrosion may take place considering the occupation states of its nearest neighbours; this first point is connected with the value of ϕ. Second, if the corrosion is possible, we have to decide with which probability it can be effectively realised; this leads to the need to specify whether the corrosion probability is uniform.

The environment of C must be such that a number ϕ of D species can be inserted into the layer. If $\phi = 1$, the corrosion can take place with certainty since the D species can stay at the position of the corrosion site if there is no free site for it in the neighbourhood of (i, j). Thus, in the case $\phi = 1$, the corrosion rate remains constant and equal to the initial corrosion rate during all the simulation time. If ϕ is larger than one, we have to see if amongst the nearest neighbours of the site (i, j), there are, at least $(\phi - 1)$ free sites for the D species. If this is not the case, there is no corrosion and the averaged corrosion rate becomes smaller than the initial one. By this simple procedure, we introduce a feedback effect of the layer formation on the corrosion rate. In terms of standard electrochemistry, this mimics the existence of a concentration under-potential. This aspect will be developed in a subsequent paper [7].

When the corrosion can take place, we have to determine the probability with which it can be effectively realised. For a number of reasons, a non-uniform corrosion probability is expected in real systems. For instance, the aggressive species, s, may not be uniformly distributed near the metal surface. In particular, in the case of pit corrosion, it is frequently assumed that H^+ concentration depends on the position in the pit [8]. Very often, in the theoretical description of localised corrosion, it is assumed that the pits have a cylindrical form and the metal dissolution takes place only at the bottom of the cavity [9]. In addition to these effects, we may also consider that the metal oxidation can be conditioned by the local electric potential which may depend on the surface topology; this corresponds to the so-called tip effect. We may also assume that the roughness of the surface may induce a distribution of mechanical constraints that may enhance the corrosion locally. These examples suggest that the existence of non-uniform corrosion probability must be considered as the general phenomena.

In our model we assume that the probability of corrosion of the site (i, j) is conditioned by the topology of the layer in its vicinity. The local environment of (i, j) is defined as the set of sites located in a square box of size L_f centred on the site (i, j). To characterise the corrosion front topology near the site (i, j), we only retain the positions h_{max} and h_{min}, which represent the two extreme values of the rows where we can find a metal site in the columns located in the region $i \pm L_f/2$. After that, for the corrosion site (i, j), we compare the value of j to h_{max} and h_{min} using the ratio $(h_{max} - j)/(h_{max} - h_{min})$. If this ratio is smaller than a given number λ $(0 < \lambda < 1)$, we set the corrosion probability equal to ε; otherwise the corrosion probability will be $(1 - \varepsilon)$. The general corrosion corresponds to $\varepsilon = 0.5$. In this case, whatever the value of λ, the corrosion probability is $1/2$. If $\varepsilon < 0.5$, we favour the corrosion of the sites for which $(h_{max} - j)/(h_{max} - h_{min})$ is larger than λ.

We have chosen this model of non-uniform corrosion probability as a simple example; its main interest is to introduce the smallest number of parameters and from it, we can go continuously from general to localised corrosion. Of course, if we had a more precise view about the process origin we can introduce a more sophisticated model.

3.4. Quantitative Properties Associated with a Simulation

A simulation requires a choice of a given area for our cell, N_c, and the number N of steps in the simulation, i.e. how many times we will proceed to a diffusion process for the D species. The choices concerning N_c and N are only determined by the restrictions concerning the computation time. A simulation will be defined by a choice of the model parameters defined above: N_{corr}, L_f, λ and ε.

After a given number of simulation steps N, the snapshots give a picture of the layer as we shall see below. From these snapshots, we can extract some averaged quantity such as the front mean position ($h_c(N)$, $h_g(N)$) and their thickness ($\sigma_c(N)$, $\sigma_g(N)$) where the indices c and g refer to the corrosion and the growth fronts respectively. For the corrosion front, for instance, $h_c(N)$ and $\sigma_c(N)$ are defined according to:

$$h_c(N) = (1/N_c)\Sigma h(i) \quad \text{and} \quad \sigma_c(N) = \left[(1/N_c)\Sigma \left[h(i) - h_c(N) \right]^2 \right]^{1/2} \tag{2}$$

where i runs from 1 to N_c and $h(i)$ is the smallest height at which we can find an L species in the column i. For a given value of N, we may also investigate the distribution profile for the walkers across the layer.

4. Results

From now on we use mainly the model introduced above to investigate the transition from general to a localised corrosion. In order to have the smallest number of parameters, we only consider the case $\phi = 1$ for which the corrosion rate is constant during the simulation.

4.1. Uniform Corrosion Probability

A series of results have been already discussed in [4]. In order to compare these results with those obtained below we briefly summarise them. From these simulations, we essentially investigate the role of the diffusion on the layer morphology at a given corrosion rate.

The mean position of the corrosion front, $h_c(N)$, increases linearly with N according to $h_c(N) = p\,(N/N_c)$ where p is the corrosion probability. Here N is also a measure of time. The random choice of the corrosion sites leads to a given thickness $\sigma_c(N)$ for the corrosion front; $\sigma_c(N)$ verifies the scaling law predicted by the Eden model [5]. In the regime $h_c(N) \ll N_c$, we have $\sigma_c(N) \approx [h_c(N)]^{1/3}$.

The growth front exhibits a more original structure. The existence of a finite diffusion coefficient leads to a constant value for $\sigma_g(N)$ ($\sigma_g(N) \approx 1$). This value is small in comparison with $\sigma_c(N)$, at least when N is large enough. The mean position of the growth front $h_g(N)$ is no more a linear function of N. The overall layer thickness, $h_g(N) - h_c(N)$, increases approximately as the square root of N, at least for the cases that we have investigated. The D species distribution inside the layer decreases linearly from the corrosion front to the growth front at the beginning of the process or when the D species concentration is low enough. However if the corrosion rate dominates the diffusion process (i.e. for large values of N_{corr}), the walker's distribution does not exhibit a uniform gradient through the layer. Near the corrosion front we may form a D species layer with a very low effective diffusion coefficient.

4.2. Non-Uniform Corrosion Probability

To use the model of non-uniform corrosion probability described above, we have to take some values for the three parameters (λ, ε, L_f). In what follows L_f will be maintained fixed to $L_f = N_c/2$. For $\varepsilon = 0.5$, we have general corrosion whatever the value of λ, the results are identical to those published in [1] with $p = {}^1/2$. Hereafter, we must first analyse the results obtained for $\varepsilon = 0.3$ and several values of λ.

We start with $\lambda = 0.5$. In this case, after choosing a corrosion site (i,j) and getting the values of h_{max} and h_{min} corresponding to its neighbourhood, we have to see whether or not j is smaller than $(h_{max} - h_{min})/2$; then the probability of corrosion will be respectively 0.3 and 0.7. In Fig. 1, we can see the evolution of the snapshots for N varying from $N = 10^4$ to 10^5; they do not show any localised corrosion. The results are close to those obtained with $\varepsilon = 0.5$. As expected, the main change is an increase of the corrosion front thickness $\sigma_c(N)$; moreover, since the corrosion probability is not uniform, $\sigma_c(N)$ no longer verifies the scaling laws of the Eden model. As a consequence of the D species diffusion, we still have $\sigma_g(N) \approx 1$ whatever the value of N. The mean position of the corrosion front $h_c(N)$ is a linear function of N that we may approximate by the relation $h_c(N) = p_{eff} (N/N_c)$ in which the random choice of the sites is neglected and p_{eff} is the effective corrosion probability defined by $p_{eff} = \varepsilon\lambda + (1 - \varepsilon)(1 - \lambda)$; here we have $p_{eff} = 0.5$.

If we increase λ from 0.5 to 0.7 we favour the sites located near the corrosion front bottom. The corresponding snapshots given in Fig. 2, clearly show that the corrosion front cannot be considered as uniform whatever the value of N.

For $N = 25 \times 10^3$ and $N = 60 \times 10^3$, we can distinguish the existence of two domains corresponding to two different mean positions for the corrosion front. After some simulation steps, these domains interact and we may observe their coalescing, the snapshot obtained with $N = 38 \times 10^3$ corresponds to a situation after a coalescing. As a consequence of the coalescing, for $N = 44 \times 10^3$, we observe a smoothing of the fronts. After that, the domain formation process may restart, as we can see on the snapshot corresponding to $N = 60 \times 10^3$. Of course, when a new non-uniform corrosion restarts, the mean position of the new domains is random. This time evolution is also illustrated on Fig. 3 where we can see that $\sigma_c(N)$ has an oscillatory behaviour.

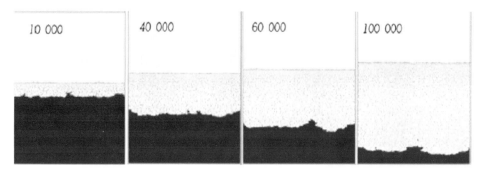

Fig. 1 *Snapshots obtained with* $N_c = 200$, $\lambda = 0.5$, $\varepsilon = 0.3$ *and* $\phi = 1$ *for N running from* 10^4 *to* 10^5. *The black part corresponds to the bulk metal, the white part to the bulk domain for the s species. In the layer that is being formed the white squares correspond to the moving D species.*

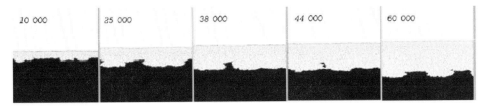

Fig. 2 *Snapshots corresponding to λ = 0.7, ε = 0.3 and φ = 1. Same conditions as in Fig. 1.*

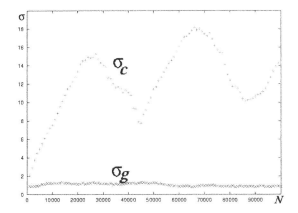

Fig. 3 *Thickness $σ_g$(N) and $σ_c$(N) of the growth and corrosion front as a function of N in the case λ = 0.7, ε = 0.3 and φ = 1. + corresponds to $σ_c$(N) and x to $σ_g$(N).*

These results are magnified if we take λ = 0.9 as shown in Fig. 4.

Thus, the simple model that we have introduced may lead to a great diversity of behaviour including some quasi-periodic regimes. At least partly, the existence of these regimes results from the fact that the evolution in the corrosion front structure

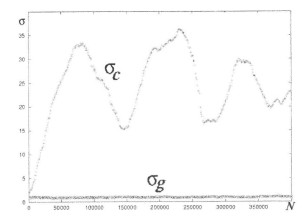

Fig. 4 *Thickness $σ_g$(N) and $σ_c$(N) of the growth and corrosion front as a function of N in the case λ = 0.9, ε = 0.3 and φ = 1. Same notations as in Fig. 3.*

depends not only on ε and λ but also, implicitly, on the ratio of the number of corrosion sites for which we have $(h_{max} - j)/(h_{max} - h_{min}) < \lambda$ to the total number of corrosion sites in the front. This ratio is not constant when we increase the value of N. Due to this, for a given value of N, the mean corrosion probability averaged on all the reactive sites can be a complicated function of N and p_{eff} as defined above may be a poor estimation of it.

Although the thickness of the corrosion front exhibits a new behaviour compared to the case of uniform corrosion probability we see in Fig. 5 that the mean front positions $h_g(N)$ and $h_c(N)$ do not reveal any new behaviour. Moreover, we have still a relation such as $h_c(N) = p(N/N_c)$ if we accept that p may deviate by about 20% from p_{eff} defined above. The thickness of the growth front is not quantitatively modified since we have $\sigma_g(N) \approx 1$. The D species distribution for different N, given in Fig. 6, is qualitatively similar to the case of uniform corrosion probability.

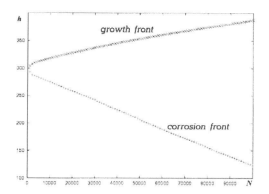

Fig. 5 *Mean position h$_g$(N) and h$_c$(N) of the growth and corrosion fronts as a function of N in the case λ = 0.9, ε = 0.3 and φ = 1. Same notations as in Fig. 3.*

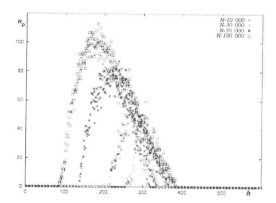

Fig. 6 *Number of D species in each row and for several value of N in the case λ = 0.9, ε = 0.3 and φ = 1.*

In order to enhance the formation of domains, we have considered the case for $\lambda = 0.9$ and $\varepsilon = 0.1$. The corresponding snapshots are given in Fig. 7.

They clearly show the formation of localised corrosion. In addition to general corrosion we see the formation of holes that are roughly of cylindrical forms. For $N = 10^5$ we can observe their coalescing. This is illustrated by the existence of extrema in the variation of $\sigma_c(N)$ with N that we can see in Fig. 8. However, $\sigma_c(N)$ that is averaged on the N_c columns forming the elementary cell in our simulation gives a smoothed representation of the snapshot content. More generally the quantities defined in (2) now give a poor representation of the corrosion front structure. This can be also illustrated by the fact that the corroded volume mean thickness, $h_c(0) - h_c(N)$, on one hand and $\sigma_c(N)$ on the other hand have the same order of magnitude. Now $\sigma_c(N)$ can be ten times larger than the results corresponding to the uniform corrosion. The snapshots in Fig. 7 also show that the D species are mainly localised inside the holes and, to a lesser extent, on the top of them. Due to this, the walkers reach the growth front preferentially in the regions near the holes as we can see from the snapshots.

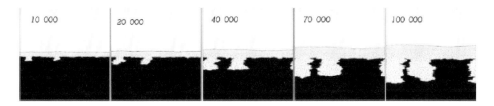

Fig. 7 *Snapshots corresponding to $\lambda = 0.9$, $\varepsilon = 0.1$ and $\phi = 1$. Same conditions as in Fig. 1.*

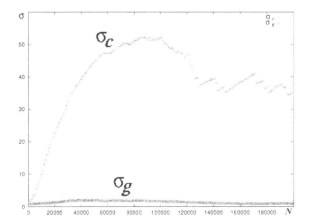

Fig. 8 *Thickness $\sigma_g(N)$ and $\sigma_c(N)$ of the growth and corrosion front as a function of N in the case $\lambda = 0.9$, $\varepsilon = 0.1$ and $\phi = 1$. Same notations as in Fig. 3.*

5. Conclusions

The extension of the model presented in [4] to more realistic situations in which a non-local corrosion process appears yields new important features. We have shown that even a simple model of non-homogeneous corrosion leads to a very rich class of behaviour going from general corrosion to localised corrosion. As expected, the usual definitions of $h_c(N)$ and $\sigma_c(N)$ give a very poor layer description in presence of localised corrosion. All the metal surface is not protected because of the front roughness. Moreover, in some cases we can observe quasi-cyclic evolution of the corrosion front roughness. These results are interesting for environmental protection where the formation of such a layer and the associated corrosion pattern is a very important phenomenon. The preliminary results presented in this work show that the model might be very useful for investigating the long term evolution of localised corrosion. Moreover, in the $\phi = 2$ case, the feedback effect of the layer changes the effective corrosion rate and $h_c(N)$ is no more a linear function of N, this case may offer a rich class of new behaviours. For the future, two kinds of model improvements are in preparation. First, we would like to organise a mapping between this approach and experimental results performed in two dimensional cell as in [6]. Second, we plan to develop the model so that it becomes closer to the usual chemical model.

References

1. F. Mansfeld, ed., *Corrosion Mechanisms*. Marcel Dekker, New York, 1987.
2. G. S. Frankel, *J. Electrochem. Soc.*, 1998, **145**, 2186.
3. M. Lafage, V. Russier and J. P. Badiali, *J. Electroanal. Chem.*, 1998, **450**, 203.
4. A. Taleb, J. Stafiej, A. Chaussé, R. Messina and J. P. Badiali, *J. Electroanal. Chem.*, 2001, **500**, 554.
5. A. L. Barabasi and H. E. Stanley, *Fractal Concepts in Surface Growth*. Cambridge University Press, Cambridge (1995).
6. P. Schilardi, S. Mendez, R. C. Salvarezza and A. J. Arvia, *Langmuir*, 1998, **14**, 4308.
7. A. Taleb, A. Chaussé, M. Dymitrowska, J. Stafiej and J. P. Badiali, Simulations of corrosion and passivation phenomena: diffusion feedback on the corrosion rate, submitted to *J. Phys. Chem. B*.
8. J. R. Galvele, *J. Electrochem. Soc.*, 1976, **123**, 464.
9. G. Engelhardt, M. Urquidi-Macdonald and D. D. Macdonald, *Corros. Sci.*, 1997, **39**, 419.

10

An Integrated Stochastic Model for Long Term Performance of Waste Package for High Level Nuclear Waste Disposal

J. H. LEE , K. G. MON*, D. E. LONGSINE[†] and B. E. BULLARD*

Sandia National Laboratories, Las Vegas, NV, USA
*Duke Engineering & Services, Las Vegas, NV, USA
[†]Duke Engineering & Services, Austin, TX, USA

ABSTRACT

The technical basis for Site Recommendation (SR) of the potential repository for high level nuclear waste at Yucca Mountain, Nevada has been completed. Long term containment of the waste and subsequent slow release of radionuclides from the engineered barrier system (EBS) into the geosphere will rely on a robust waste package (WP) design, among other EBS components as well as the natural barrier system. The SR WP design has two layers: an Alloy 22 outer barrier and a 316NG stainless steel inner shell. A Titanium Grade 7 drip shield (DS) is placed over the WP at the closure of the repository. No backfill is used. The WP outer barrier has two Alloy 22 closure lids: one outer lid and one inner lid.

An integrated stochastic WP degradation simulation model, WAste Package DEGradation (WAPDEG), was developed to analyse long term degradation of WPs in the potential repository. The WP and DS degradation analyses for the total system performance assessment (TSPA) baseline model for the SR have shown that, based on the current corrosion models and assumptions, both the DSs and the WPs do not fail within the regulatory compliance time period (10 000 years). From the perspective of initial WP failure time, the analysis results are encouraging because the upper bounds of the baseline case are likely to represent the worst case combination of key corrosion model parameters that significantly affect long term performance of WPs in the potential repository.

The estimated long life-time of the WPs in the current analysis is attributed mostly to the following two factors that delay the onset of stress corrosion cracking (SCC):

(1) the stress mitigation to substantial depths from the outer surface in the dual closure-lid weld regions; and
(2) the very low general-corrosion rate applied to the closure-lid weld regions to corrode the compressive stress zones.

Uncertainties are associated with the current WP SCC analysis. These are stress mitigation on the closure-lid welds, characterisation of manufacturing flaws applied to SCC, and general corrosion rate applied to the closure-lid weld regions. These uncertainties are expected to be reduced as additional data and analyses are developed.

1. Introduction

The technical basis for Site Recommendation (SR) of the potential repository for high level nuclear waste at Yucca Mountain, Nevada has been completed. Long term containment of the waste and subsequent slow release of radionuclides from the engineered barrier system (EBS) into the geosphere will rely on a robust waste package (WP) design, among other EBS components as well as the natural barrier system. The SR WP design has two layers: an Alloy 22 outer barrier and a Type 316NG stainless steel inner shell. A thickness of 20 mm is assumed for the WP outer barrier in the current analysis. A Titanium Grade 7 drip shield (DS) (15 mm thick) is placed over the WP at the closure of the repository. No backfill is used. The stainless steel inner shell is to provide the structural integrity of the WP, and no barrier performance credit is taken for the inner shell. However, in reality, the inner shell would provide 'some' waste containment performance after breach of the WP outer barrier, and would also serve as a barrier to radionuclide transport after WP failure. This is a conservative modelling approach. The WP outer barrier has two Alloy 22 closure lids. The outer and inner lids are assumed to be 25 mm and 10 mm thick respectively for the current analysis. The lids are welded to the outer barrier after the waste is loaded.

The WAste Package DEGradation (WAPDEG) stochastic simulation model was developed to analyse long term degradation of WPs in the potential repository [1–5] and used in the previous total system performance assessments (TSPA) of the repository [6,7]. The primary role of the WP degradation analysis is to provide information on the lifetime and overall degradation of the WP. For the TSPA-SR analysis [8], the WAPDEG model was further improved by incorporating the latest data and analyses for corrosion degradation of the WP and DS materials. The WAPDEG model analyses the WP degradation considering general corrosion, localised corrosion and stress corrosion cracking (SCC) of the outer barrier. The WAPDEG model also includes effects of microbiologically influenced corrosion (MIC), aging and phase stability, and manufacturing flaws on the WP degradation. The exposure conditions the WAPDEG model uses are temperature and relative humidity (RH) at the WP and DS surfaces as a function of time, and the bounding corrosive chemistry of aqueous solutions contacting the DSs and WPs. This paper discusses the integrated stochastic WP degradation simulation (WAPDEG) model and analysis results for the TSPA-SR baseline case. Details of the model and analyses are found elsewhere [8,9].

2. Conceptual model of the Integrated Waste Package Degradation Model

In the TSPA-SR analysis, the WAPDEG model considers various types of corrosion mechanisms that may occur on a WP and DS as a function of the exposure time and conditions. (For convenience of discussion the DS is considered as an integral part of the WP; and except where it is necessary, no separate discussion is given for the DS). In the TSPA-SR baseline case analysis, the WP outer barrier and DS are included in the WP degradation analysis. As indicated above, no performance credit is taken for the stainless steel inner shell in the WP degradation analyses. A dual closure-lid design has been adopted for the closure-end of WP outer barrier to mitigate the potential for early failure of the closure-lid weld regions by SCC. This design concept is illustrated schematically in Fig. 1. There is a physical 'gap' between the two closure-lids. Thus, any SCC cracks penetrating the outer closure-lid stop at the gap between the closure-lids. The inner closure-lid weld regions are then subject to SCC initiation and subsequent crack growth. This design feature is included in the WP degradation analysis as described below.

A schematic illustrating the conceptual approach of the WAPDEG model is shown in Fig. 2. Effects of spatial and temporal variations in the exposure conditions over the repository region are modelled by explicitly incorporating relevant exposure condition time-histories into the WP degradation analysis. The exposure condition parameters that are considered are RH and temperature at the WP surface, seepage into the emplacement drift, and chemistry of the seepage water. In the WP degradation analysis, the humid-air corrosion condition occurs when there is no dripping water contacting the WP surface and the RH at the WP surface is equal to or greater than the no-drip threshold RH (i.e. the threshold RH in the absence of drips). The threshold RH is defined as the RH at which a stable water film can form and sustained electrochemical reactions occur to support corrosion. The aqueous corrosion condition requires the presence of dripping water and the RH at the WP surface to be equal to or greater than the drip threshold RH (i.e. the threshold RH in the presence of drips). As discussed later, in the current analysis, the no-drip and drip threshold RH distributions are identical (see Section 4.1 for more details).

In the WAPDEG analysis the WP (or DS) surface is discretised into many sub-areas referred to as 'patches.' It is at the patch-level that corrosion degradation processes are modelled (i.e. the relevant corrosion model parameter values and/or

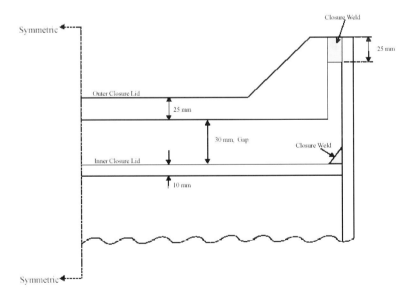

Fig. 1 *Schematic of the dual-closure-lid design for WP outer barrier.*

corrosion rates are sampled and applied to each patch). Figure 2 shows the patch conceptual approach. This approach allows for representation of potentially 'variable' degradation processes on the patches composing a single WP. For example, patches contacted by dripping water (those marked with 's') could be subject to localised corrosion depending on the exposure conditions (especially chemistry of contacting water) that those patches experience. Patches with closure-lid welds (those marked with 'w') could be subject to SCC depending on the stress state and exposure conditions. In addition, the general corrosion rate may vary over the WP surface. This potential variability is represented by populating the general corrosion rate over the patches.

3. Total System Performance Assessment — Site Recommendation (TSPA-SR) Baseline Waste Package and Drip Shield Degradation Model

In general, the approach used in the TSPA-SR baseline analysis in dealing with the uncertainties in each of the process models (and the abstraction models input to the TSPA) is to employ conservative assumptions and bound the models. The conservatism and uncertainties in each of the WP corrosion models were incorporated into the WAPDEG model [8,9]. Figure 3 shows a logic flow diagram for the WAPDEG model for WP (or DS) degradation analysis for TSPA-SR. In the Figure each of the shaded boxes represents a model abstraction or a set of model abstractions. The logic flow repeats for a DS, if included in the analysis, and each barrier of the WP. Exposure conditions that are included in the analysis are temperature and RH at the WP (and DS) surface, in-drift dripping water, and pH of the water contacting WP (and DS). The temperature and RH histories at the WP and DS surfaces are provided by the multi-scale thermal-hydrologic model abstraction [10]. The

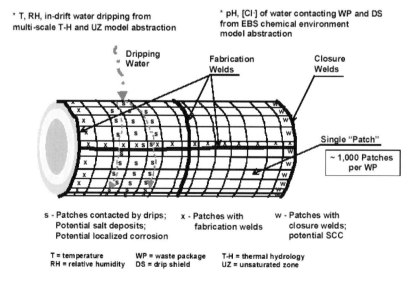

Fig. 2 *Schematic illustrating the conceptual approach of the integrated stochastic WAPDEG model for TSPA-SR.*

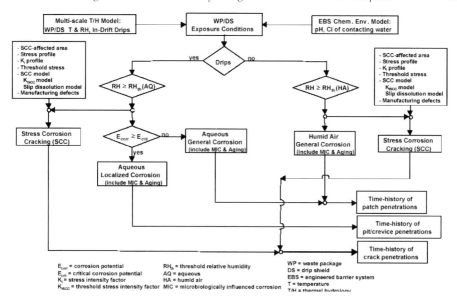

Fig. 3 *Logic flow for the TSPA-SR baseline case model for the integrated WP and drip shield degradation model.*

evolution of pH of solution contacting WP is provided by the EBS chemical environment abstraction [11].

A threshold RH (RH_{th}) is used to initiate corrosion. The threshold RH is based on the deliquescence point of pure $NaNO_3$ salt, which is dependent on temperature. In the analysis, the WP surface RH is compared to the threshold RH for corrosion initiation. If the surface RH becomes greater than or equal to the threshold RH, the WP (or DS) undergoes corrosion. Depending on whether the WP (or DS) is dripped on or not, it could undergo different corrosion degradation modes. In the current analysis, the same threshold RH is used for both the dripping and no-drip conditions. Both the upper and under sides of DS are assumed to be subject to corrosion if the initiation threshold is met, that is, if the RH on the DS is equal to or greater than the threshold RH. This is because both sides are exposed to the conditions in the emplacement drift.

Waste packages that are not dripped on are assumed to undergo humid-air corrosion during the entire time that the threshold RH is exceeded. Under humid-air conditions, the WPs undergo general corrosion and eventually fail by gradual thinning of the container wall. No localised corrosion is assumed to occur in the absence of dripping water. The general corrosion rates of the WP outer barrier (and DS) are very low (see Fig. 5 and related discussions below). For the WP outer barrier and DS, the current model assumes the same general corrosion rate distribution for both humid-air general corrosion and aqueous general corrosion (i.e. regardless of whether the WP is dripped on or not).

The WAPDEG model conservatively assumes that MIC is possible if RH at the surface is greater than the threshold RH for corrosion initiation. The MIC effect is modelled with a corrosion rate multiplier (enhancement factor), which is applied to the general corrosion rate of the barrier (no localised corrosion is assumed to occur for no-drip condition under the expected repository exposure conditions as discussed later in this section and Section 4.3). The enhancement factor is assumed to be uniformly distributed between 1 and 2 [8]. The MIC enhancement factor is applied to the entire WP outer barrier surface whenever the RH is above the threshold RH (i.e. corrosion initiation threshold RH). The DS is assumed not to be subject to MIC under expected repository exposure conditions [8].

The current model assumes that the (Alloy 22) WP outer barrier closure-lid weld regions are subject to aging under expected repository conditions. The effect is modelled with a corrosion rate multiplier (enhancement factor) that is applied to the general corrosion rates sampled for the weld region patches. The ageing enhancement factor is assumed to be uniformly distributed between 1 and 2.5 [8], and is applied only to the closure lid weld regions of the outer barrier [9]. The DS is assumed not to be subject to ageing and phase instability [9]. The MIC and ageing enhancement factors are applied independently, thus the maximum possible enhancement factor applied to the closure-lid weld region is 5.

In the baseline model, SCC is assumed to be possible if the RH at the WP surface is greater than or equal to the threshold RH for corrosion initiation (RH_{th}) (i.e. water chemistry is not considered). As indicated in the logic flow diagram, the inputs to the SCC analysis are:

- the area subject to SCC, such as the closure-lid weld regions as in the current analysis;

- stress and stress intensity factor (K_I) vs depth of the affected area;

- SCC crack propagation model (slip dissolution/film rupture model for the current analysis);

- threshold stress for SCC initiation;

- density and size of incipient cracks in the affected area; and

- manufacturing flaw occurrence and size in the affected area.

Incipient cracks are microscopic cracks on the smooth surface for which the slip dissolution model can be applied [8,9]. For SCC analysis with the slip dissolution model, the following criteria must be met for initiation of SCC on a patch:

(1) the stress intensity factor (K_I) should be positive, and

(2) the stress at the crack tip must be greater than or equal to the threshold stress.

For those patches with a compressive stress zone in their outer surface, it is assumed the compressive stress zone must be removed by general corrosion before SCC can initiate. The delay time depends on the compressive zone thickness and the general corrosion rate sampled for the patch.

The drip shield (DS) is assumed to be fully annealed before it is placed in the emplacement drift. The DS could be subject to rockfall-induced SCC and through-wall cracks. However, as discussed later, these degradation processes are not considered in the baseline case analysis. Likewise, all the fabrication welds in the WP, except the welds of the closure lids, are assumed to be fully annealed and not subject to SCC. Therefore, only the closure-lid weld regions are considered in the SCC analysis.

The manufacturing flaws considered in the current analysis include both the initially surface-breaking and embedded flaws. These manufacturing flaws are assumed to be all surface-breaking. The flaws are assumed to grow at the general corrosion rate sampled for the patch in which the flaws reside. This is also a highly conservative modelling assumption that assumes the same exposure condition that a patch experiences during a given time step is also applicable to the interior of the flaws in the patch. Also, the flaws are assumed to maintain their shape (i.e. constant crack aspect ratio) while growing at the general corrosion rate of the patch. This also a highly conservative modelling assumption. Patches with pre-existing flaws would be subject to SCC earlier than other patches with incipient cracks only.

For a WP (or a DS) that is dripped on, the wetted area (or patches) of the WP is assumed to undergo aqueous corrosion if the RH at the surface is greater than the threshold RH. As indicated above, the same threshold RH is used for both the dripping and no-drip conditions. It is assumed that the entire surface of WP (or DS) is wetted by the drips if it is dripped on. This is a conservative assumption. While the DS is operative (i.e. not failed), the WP underneath the DS is assumed to undergo humid-air corrosion. General corrosion occurs all the time under aqueous corrosion conditions. For the WP outer barrier and DS, the baseline model assumes the same general corrosion rate distribution for both humid-air general corrosion and aqueous general corrosion (i.e. regardless of whether it is dripped on or not). The same MIC and aging enhancement factors for the humid-air general corrosion are also applied to the aqueous general corrosion. Also, the same SCC model and inputs are applied for both humid-air and aqueous conditions. These modelling approaches result from the no-dependence of the process models on drip and no-drip conditions. Therefore this approach renders no difference in the WP and DS degradation analysis results for drip and no-drip conditions. Initiation of localised corrosion (pitting and crevice corrosion) is dependent on the local exposure environment on the wetted patches. In the baseline analysis, it is assumed that localised corrosion of the DS and WP outer barrier can initiate only under dripping conditions. This is because of the necessary presence of aggressive ions (such as chloride) in order to initiate and sustain pit and crevice growth and because the only mechanism for these species to gain ingress to the drift is assumed to be by dripping. Localised corrosion of a patch is assumed to initiate if the corrosion potential (E_{corr}) of the patch is greater than or equal to the 'threshold' critical corrosion potential (E_{crit}) sampled for the patch. After initiation, localised corrosion continues while $E_{corr} \geq E_{crit}$. If E_{corr} becomes less than E_{crit}, or dripping ceases, localised corrosion is assumed to stop. Note that, as mentioned earlier and discussed in detail in Section 4.3, the WP outer barrier and DS are not subject to localised corrosion for the exposure conditions expected in the repository. The potential for MIC and ageing effects on localised corrosion is represented with the same corrosion rate enhancement factors as for general corrosion, that is, the MIC enhancement factor sampled from a uniform distribution between 1 and 2, and the aging enhancement factor sampled from a uniform distribution between 1 and 2.5. The enhancement factors are based on the electrochemical testing results [8]. As stated above, because the exposure conditions on the DS and WP surface are not severe enough, localised corrosion does not initiate for the WP (Alloy 22 outer barrier) and DS [8]. Consequently, the current modelling approach renders no difference in the WP and DS degradation analysis results for drip and no-drip conditions.

After failure of the DS, it is assumed that the dripping water finds the opening(s) in the DS regardless of the opening location in the DS (i.e. top or side of the DS), and the WP underneath the failed DS is assumed to be subject to dripping. Although, in reality, the area of the WP surface wetted by drips underneath the failed DS would depend on the drip rate and the location, size and number of the penetration openings in the DS, the current model assumes that the entire surface of the WP is wetted by the dripping water flowing through the failed DS. This is a highly conservative modelling assumption. However, this would not affect the WP degradation analyses results due to insensitivity of the corrosion models to dripping conditions.

When a WP fails, the WAPDEG model considers corrosion degradation of the WP from the inside-out corrosion. The inside-out corrosion analysis includes general corrosion and localised corrosion of the WP outer barrier. The inside-out corrosion would cause penetrations by general and localised corrosion in addition to those by the outside-in corrosion only. The inside-out general corrosion is assumed to initiate at the time of the WP failure. Like the outside-in localised corrosion, initiation of the inside-out localised corrosion is based on the corrosion potential and critical corrosion potential, which are a function of the pH of water inside the failed WP. The in-package water chemistry results from degradation of the waste form and other internal structural materials (such as basket materials) [12]. Note that in the TSPA-SR baseline analysis, the WP outer barrier is not subject to inside-out localised corrosion for the estimated in-package water chemistry ranges. The rates and initiation parameters for the inside-out corrosion are sampled independently of the outside-in corrosion.

The TSPA-SR baseline model does not include the radiolysis-enhanced corrosion of WP outer barrier and DS because the materials are not subject to radiolysis enhanced corrosion under the repository conditions [8]. Also the baseline model does not consider potential rockfall-induced stresses, which could lead to SCC of the DS. The process could result in through-wall cracks in the DS, and the crack wall will continue to corrode at very low passive corrosion rates until the tight crack opening is 'plugged' by the corrosion products and mineral precipitates (such as carbonate). Any water transport through this plugged crack area will be mainly by diffusion-dominant transport processes [8]. Thus, the effective water flow rate through the cracks in the DS is expected to be extremely low and should not contribute significantly to the overall radionuclide release rate from the underlying failed WP. Therefore, because the primary role of the DS is to keep water from contacting the WP, SCC of the DS would not compromise the intended function of the DS. In addition, analyses have shown that for the current WP and DS design, both the WP and DS are not subject to failure by hydrogen induced cracking (HIC) under the exposure conditions expected at the repository, thus HIC is not considered in the TSPA-SR analysis [8].

As stated above, the WAPDEG analysis considers corrosion degradation of WPs (and DSs) for three types of penetration modes: crack penetration by SCC, pit and crevice penetration by localised corrosion, and patch penetration by general corrosion. The analysis provides, as output, the cumulative probability of WP failure by one of the three penetration modes as a function of time, and the number of penetrations for each of the penetration modes as a function of time. Note that in the TSPA-SR baseline analysis, the WP and DS are not subject to penetration by localised corrosion. The WP failure time and penetration number profiles are used as input to other TSPA analyses such as waste form degradation and radionuclide release rate from WPs. Details of the TSPA-SR baseline WP degradation model are given in Refs [8,9].

4. Individual Model Components

This section provides an overview of the individual corrosion models and their parameters that have been incorporated into the WP and DS degradation analysis for the TSPA-SR baseline analysis [8]. In general, the approach used in the TSPA-SR baseline analysis in dealing with the uncertainties in each of the individual model components is to employ conservative assumptions and bound the models. The corrosion modes that are included in the TSPA-SR baseline analysis are:

- Humid-air phase general corrosion of DS.
- Aqueous phase general corrosion of DS.
- Localised corrosion (pitting and crevice corrosion) of DS.
- Humid-air phase general corrosion of WP outer barrier.
- Aqueous phase general corrosion of WP outer barrier
- Localised corrosion (pitting and crevice corrosion) of WP outer barrier.
- SCC of WP outer barrier (closure-lid weld region).

In addition, the following corrosion parameters were developed for the analysis:

- RH threshold for corrosion initiation of DS and WP outer barrier.
- Corrosion potential-based threshold for localised corrosion initiation of DS and WP outer barrier.
- Probability of the occurrence and size of manufacturing flaws in closure-lid welds of WP outer barrier.
- Stress and stress intensity factor vs depth profiles in the closure-lid weld regions of WP outer barrier incorporating post-welding stress mitigation techniques.
- Threshold stress for the initiation (nucleation) of SCC of WP outer barrier, used in conjunction with the slip dissolution/film rupture model.

- Corrosion enhancement factor for ageing of the closure-lid weld regions of WP outer barrier.

- Corrosion enhancement factor for MIC of WP outer barrier.

4.1. General Corrosion Initiation Threshold

In the baseline TSPA-SR analysis, a carbonate-base water (which has characteristics of J-13 well water*) was considered the most likely to contact the DS and WP [8]. Sodium nitrate salt is the most soluble salt formed during the evaporative concentration testing of simulated J-13 well water and has the lowest deliquescence point among salts produced from the testing. The threshold RH for humid-air and aqueous-phase corrosion is based on the deliquescence point of sodium nitrate salt [8], which is temperature-dependent as shown in Fig. 4. Because the hygroscopic salt could form from the salts entrapped in the dusts settled on the WP and DS surfaces, the deliquescence point of sodium nitrate salt is conservatively assumed also for the threshold of corrosion initiation for humid-air condition. With the threshold RH model corrosion can initiate at a RH as low as 50% at 120°C. The threshold RH gradually increases with temperature, for example 65% RH at 90°C to 75% RH at 20°C.

4.2. General Corrosion

General corrosion is relatively uniform thinning of materials without significant localised corrosion. The mode of general corrosion differs depending on the local environment; the repository environment will result in three general corrosion modes (dry oxidation, humid-air corrosion, and aqueous-phase corrosion). Dry oxidation

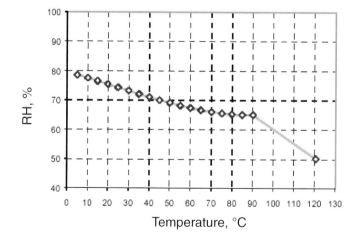

Fig. 4 *Deliquescence point of NaNO₃ salt as a function of temperature [8].*

---

*J-13 well water is a carbonate-base water from the saturated zone of the Yucca Mountain site. the well water is considered to be the same type of water as the perched waters from the unsaturated zone of the site.

occurs at a RH below the threshold for humid-air corrosion. This process results in the formation of an adherent, protective oxide film of uniform thickness. The rates of dry-air corrosion are very low for Alloy 22 and Titanium Grade 7 [8]. Given the extremely small magnitudes and uncertainties of these rates, dry-air corrosion is expected to have no significant impact on WP and DS performance, and so this process is not considered in the TSPA-SR baseline analysis.

As discussed previously, humid-air corrosion is assumed to occur above the threshold RH (provided that there are no impinging drips), and aqueous-phase corrosion is assumed to occur in the presence of dripping water. The general corrosion rate distribution for Alloy 22 used in the TSPA-SR baseline analysis is based upon mass-loss data over a two-year period, and that of Titanium Grade 7 is over a year period. The test programme includes sample coupons exposed to a variety of test media (e.g. water chemistry, temperature) to cover a potential range of corrosion conditions in the repository. The cumulative distribution functions (CDFs) of the general corrosion rate for the WP and DS are shown in Fig. 5. The median (50th percentile) general corrosion rate of Alloy 22 is 0.045 µm/year, and the upper bound (100th percentile) rate is 0.14 µm/year. The median general corrosion rate of Titanium Grade 7 is 0.068 µm/year, and the upper bound rate is 0.39 µm/year.

To date, corrosion data for Alloy 22 and Titanium Grade 7 indicate that the general corrosion rates for humid-air and aqueous-phase corrosion are about the same. The data also show little sensitivity to the water chemistry and temperature ranges considered in the test programme [8]. Because the general corrosion rate is expected to decrease with exposure time, the current general corrosion rate models (shown in Fig. 5) represent reasonably conservative corrosion behaviours of those highly corrosion-resistant materials under the expected repository conditions. Therefore, in the TSPA-SR baseline analysis, it was conservatively assumed that the rate of humid-air corrosion is represented

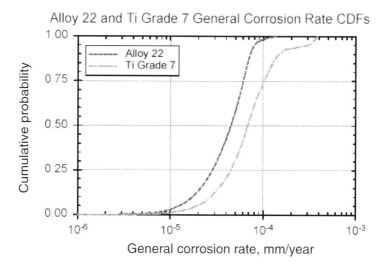

Fig. 5 Cumulative distribution functions (CDFs) of general corrosion rate of Alloy 22 and titanium grade 7 used in the TSPA-SR baseline analysis.

by the same corrosion rate distribution used for aqueous-phase corrosion during the period where humid-air corrosion is operable [8]. It was further assumed that the corrosion rate is constant and does not decrease with time.

The general corrosion rate distributions for Alloy 22 and Titanium Grade 7 include substantial uncertainties. Most of the uncertainties result from insufficient resolution of the mass-loss measurements of the sample coupons due to the extremely low corrosion rates of the materials in the test media. The testing data and information are not sufficient to quantify the uncertainty fully, and the fraction of the corrosion rate distribution that is attributed to spatial variability (and conversely uncertainty) is not known. The conceptual approach in the WP and DS degradation analysis in the TSPA-SR baseline analysis assumes that this fraction is unknown and varies uniformly, ranging from pure uncertainty to pure variability. The Gaussian variance partitioning technique was used to accomplish the separation of the uncertainty and variability from the original distribution [8].

4.3. Localised Corrosion

The generic localised corrosion model for WP materials assumes that localised corrosion occurs if the open circuit corrosion potential (E_{corr}) exceeds the threshold (critical) potential for breakdown of the passive film (E_{crit}). In some cases, the threshold potential is assumed to be the repassivation potential. The repassivation potential is the level at which a damaged passive film repassivates, or heals, thereby protecting the metal. In other cases, the threshold potential is assumed to be the pit or crevice initiation potential.

A correlation was developed to compare corrosion and threshold potentials (i.e. $E_{crit} - E_{corr}$) as a function of exposure conditions (pH, chloride concentration and temperature) and determine whether localised corrosion initiates for the WP (Alloy 22) and DS (Titanium Grade 7) [8]. The correlation is based upon standard cyclic polarisation measurements in a variety of test solution media covering a broad range of temperatures. Because the pH is the dominant parameter, the correlation for the localised corrosion initiation threshold was reduced to as a function of pH only [8]. Figures 6(a) and (b) show the resulting correlation for the median potential difference ($E_{crit} - E_{corr}$) as a function of pH for Titanium Grade 7 and Alloy 22 respectively. Also shown in the Figures are the measured data and the ±3 and ±4 standard-deviations confidence intervals for the alloys. The current potential-based localised-corrosion initiation threshold model is considered to represent the realistic behaviour of Alloy 22 and Titanium Grade 7 under the expected repository conditions. Estimates of uncertainty in the selection of corrosion and threshold potential have been made and are incorporated in WAPDEG analyses. The uncertainty in the corrosion potential due to gamma radiolysis (i.e. a maximum positive shift of about 250 mV for Alloy 22) has been addressed [8]. As shown in Fig. 6, because the threshold potentials on the WP and DS surfaces are not exceeded, both the WP and DS are never subject to localised corrosion for the entire simulation period [8,9].

When exposure conditions are such that the threshold potential is exceeded, localised corrosion rates are applied to simulate penetration of the WP and DS by the rapid-penetrating localised corrosion. The rates of localised corrosion for

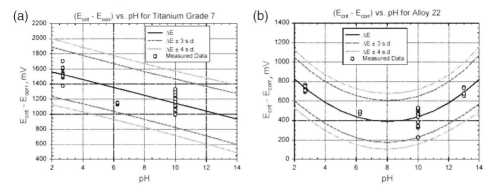

Fig. 6 ($E_{crit} - E_{corr}$) *vs pH for (a) Titanium Grade 7 and (b) Alloy 22 showing the ±3 and ±4 standard deviations confidence intervals and the measured data.*

Alloy 22 and Titanium Grade 7 have been bounded with the range of values found in the published literature. The localised corrosion rates for Alloy 22 are assumed to be log-uniformly distributed between the bounds of 0.0127 mm/year and 1.27 mm/year, and the distribution is assumed to represent uncertainty in the localised corrosion rate [8]. The localised corrosion rates for Titanium Grade 7 are assumed to be uniformly distributed between the bounds of 0.49 mm/year and 1.12 mm/year, and the distribution is assumed to represent uncertainty in the localised corrosion rate [8].

4.4. Stress Corrosion Cracking

Stress corrosion cracking (SCC) is a potential degradation mode that can result in penetration of the WP outer barrier (Alloy 22). Stress corrosion cracking of materials may occur when an appropriate combination of material susceptibility, tensile stress, and aggressive environment is present. SCC is assumed to occur only in the regions around the closure welds of the Alloy 22 outer barrier because the residual stress in this weld may not be relieved completely. Residual stress in all other welds used in the fabrication of the WP outer barrier and Titanium Grade 7 DS are assumed to be fully annealed, and are not subject to SCC [8]. The effect of rockfall on the WP outer barrier is excluded because of protection provided by the DS. Stress corrosion cracking of the DS is possible under the applied stresses resulting from rockfall; however, as discussed previously, it is assumed that SCC is of low consequence to DS performance, so it is not modelled [8].

An effective approach to eliminating the threat of SCC and the resultant through-wall cracking in the closure weld is to implement a post-welding stress mitigation process to remove either residual tensile stresses in the weld region or reduce them below threshold values for SCC initiation and growth. The closure of the WP outer barrier is designed to include two lids (inner and outer) with two separate post-welding stress mitigation processes: laser peening of the inner closure-lid weld and induction annealing stress relief of the outer closure-lid weld. This two-lid design and stress mitigation strategy forms part of the conceptual model for the SCC model. The slip dissolution model is used to calculate SCC crack growth rates.

4.4.1. Slip dissolution model.

The slip dissolution model is used to calculate the crack growth rate, which is expressed as follows.

$$V = \overline{A}(K_1)^{\overline{n}} \tag{1}$$

$$\overline{A} = 7.8 \times 10^{-2} n^{3.6} (4.1 \times 10^{14})^{n} \tag{2}$$

$$\overline{n} = 4n \tag{3}$$

where V is the crack growth rate in mm s^{-1}, and K_I is the stress intensity factor in MPa(m)$^{1/2}$. Parameter n (referred to also as the repassivation slope) is a function of environmental and materials parameters such as solution conductivity, corrosion potential, and alloy composition (i.e. chromium depletion in the grain boundary) [8]. The model employs a threshold stress for crack initiation, that is, SCC is initiated if the threshold stress is exceeded on a smooth surface [8]. The SCC propagation rate is calculated as a function of local environment and stress intensity factor. The time-to-failure is determined by integrating the calculated propagation rate. In the TSPA-SR baseline analysis, the slip dissolution model and its parameters are based primarily on published data. Limited data have been obtained under repository-relevant conditions. The parameters for the slip dissolution model are based on data for stainless steel subjected to the boiling water reactor conditions. Because stainless steels are much more prone to SCC than Alloy 22, these parameter estimates are conservative. Effect of the rate of repassivation at the crack tip is captured by the repassivation-slope parameter n. A characteristic of the slip dissolution (or film rupture) model is that SCC susceptibility decreases with increasing values of the repassivation slope. In the TSPA-SR baseline analysis, the model parameter for repassivation slope is represented by a uniform distribution with an upper bound of 0.84 and a lower bound of 0.75, and the variation of the parameter value is due to uncertainty [8].

Figure 7 shows the model responses for the SCC failure time of the outer lid (25 mm thick) and inner lid (10 mm thick) of the WP outer barrier as a function of stress intensity factor (K_I) for the bounding values of n (0.75 and 0.84). As shown in the Figure, the stress intensity factor is the dominant parameter in the model, and the time to failure by SCC increases exponentially as the stress intensity factor decreases. The failure time by SCC is less than 100 years for the stress intensity factors greater than 20 MPa (m)$^{1/2}$. The failure time increases to well above 1000 years if the stress intensity factor is kept below 6 MPa (m)$^{1/2}$. The analysis demonstrates that, once a SCC crack initiates, it penetrates the closure-lid thickness rapidly. It also demonstrates importance of stress mitigation in the closure-lid welds to avoid early failures of waste packages by SCC.

4.4.2. Threshold stress model for crack initiation.

Although the slip dissolution model assumes that crack growth can initiate at any surface defect/flaw that can generate a stress intensity factor, examination of the relevant SCC literature indicates that there is a threshold stress below which SCC will not initiate (or nucleate) on a 'smooth' surface (i.e. free of surface breaking flaws).

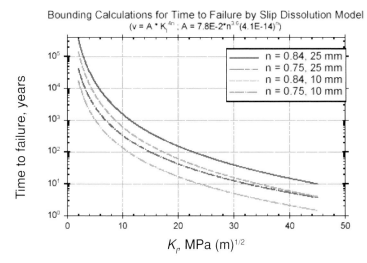

Fig. 7 *Bounding calculations for the slip dissolution model responses for the time to failure of the outer and inner closure lids by SCC as a function of stress intensity factor* K_I *for the bounding values of repassivation slope* n.

The uncertainty in the threshold stress for initiation of stress corrosion cracking in the TSPA-SR baseline analysis is conservatively estimated to be approximately 20 to 30% of the Alloy 22 yield strength [8]. The threshold stress and its uncertainty range are conservative and based on the determination of such thresholds for more susceptible alloy systems (e.g. stainless steels) exposed to very aggressive environments, such as boiling magnesium chloride. Uniform distribution between the bounds indicated above is assumed for the threshold stress uncertainty.

4.4.3. Closure-weld residual stress uncertainty.
The stress profile (stress vs depth) and corresponding stress intensity factor profile in the closure-lid weld region, before and after stress mitigation, are determined from analyses of measured residual stresses in welds as well as finite element modelling. The residual stress analysis shows that the dominant stress in the closure-lid weld region after stress mitigation that could result in a through-wall crack by SCC is the hoop stress. The hoop stress drives radial cracks through the closure lid weld region. The radial stress state, which drives circumferential cracks, does not remain tensile throughout the thickness of the closure lid weld region; therefore, there will be no through-wall cracking caused by circumferential cracks [8]. This analysis indicates that only radial flaws are potential sites for through-wall SCC if it occurs.

In the TSPA-SR baseline analysis, the most conservative (or tensile) stress profile and corresponding stress intensity factor profile were used as the representative 'mean' profile for the entire WPs that are analysed. The uncertainties in the residual stress profile (and corresponding stress intensity factor profile) were estimated from literature data. A separate uncertainty range was developed for each of the following three cases: optimum, realistic, and worst case. A stress uncertainty range of ±5 % of the yield strength was used to represent the optimum case that is achievable through stringent control of such processes as welding, stress mitigation, material variability, and other fabrication steps. The stress uncertainty range of ±10 % of the yield strength is used to represent the realistic case that is achievable through appropriate levels of process controls. The stress uncertainty range of ±30 % of the yield strength is used to represent the worst case that might result from inadequate control of the processes [8]. The TSPA-SR baseline analysis conservatively employs an uncertainty range of the worst case (±30 % of the yield strength). Triangular distributions around the conservatively chosen mean (discussed above) are assumed for the residual stress uncertainty. Figure 8(a) shows the uncertainty range of the hoop stress as a function of depth at the location with the largest tensile stress in the outer closure-lid weld region (25 mm thick). The figure shows that the hoop stress is compressive at the surface (from the stress mitigation using the local induction annealing) and becomes tensile at a depth of 6 mm for the maximum upper bound (+30% of the yield strength) and 10 mm for the minimum lower bound (–30% of the yield strength). The corresponding stress intensity factor profiles as a function of radial crack depth are shown in Fig. 8(b). The stress intensity factor is negative at the surface, consistent with the compressive stress at the surface, and becomes positive at a depth of 3 mm for the maximum upper bound. No SCC crack growth will initiate until this layer is removed.

The uncertainty range of the hoop stress as a function of depth at the location with the largest tensile stress in the inner closure-lid weld region (10 mm thick) is shown in Fig. 9(a). The hoop stress in the inner closure-lid welds is compressive at the surface, transitioning to the tensile state at a depth between 1.5 and about 3 mm, and then back to the compressive state at a depth between 6.8 and 9.8 mm. The corresponding stress intensity factor profiles as a function of the radial crack depth are shown in Fig. 9(b). The stress intensity factor is negative at the surface and becomes positive at a depth of about 2 mm for the maximum upper bound case. No SCC crack growth will initiate until this layer is removed. The uncertainty of the stress intensity factor increases with depth.

4.4.4. Manufacturing flaws in closure weld.

Another important input to the SCC analysis is the characteristics of incipient defects and manufacturing flaws. Pre-existing manufacturing flaws in the closure lid welds

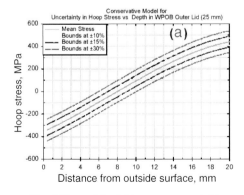
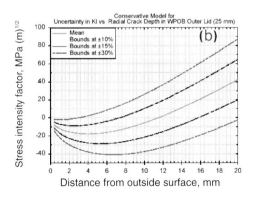

Fig. 8 *Plots of the (a) hoop stress and (b) corresponding stress intensity factor as a function of depth in the Alloy 22 outer closure-lid weld region (25 mm thick) using uncertainty bounds of ±5, 10, and 30% of the yield strength used in the TSPA-SR baseline analysis.*

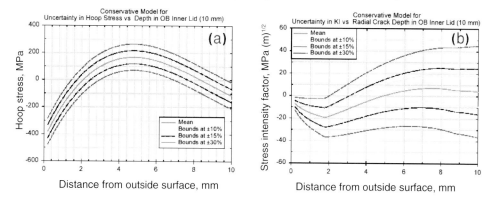

Fig. 9 *Plots of the (a) hoop stress and (b) corresponding stress intensity factor as a function of depth in the Alloy 22 outer closure-lid weld region (10 mm thick) using uncertainty bounds of ±5, 10, and 30% of the yield strength used in the TSPA-SR baseline analysis.*

are the most likely sites for SCC failure. The frequency and size distributions for manufacturing flaws in the closure welds are based on published data for stainless steel piping welds in nuclear power plants. The published data used to develop the manufacturing flaw model are for relevant welding techniques and post-welding inspection methods.

In the TSPA-SR baseline analysis, pre-existing surface-breaking manufacturing flaws and flaws embedded in the outer 25 percent of the weld thickness are considered as potential sites for SCC crack growth [8]. There is uncertainty associated with this modelling assumption because, as general corrosion propagates, some of the existing surface-breaking flaws may disappear, and some of the embedded flaws may become surface-breaking flaws. Use of this assumption is conservative because the WAPDEG model does not allow existing surface-breaking flaws to be removed by general corrosion processes during the simulation. This allows a greater number of flaws available for potential SCC sites. In addition, weld flaws are assumed to be randomly distributed spatially, as represented by a Poisson process [8].

Figure 10(a) shows several probability density functions (PDFs) for manufacturing flaw sizes in the closure-lid welds for various combinations of values of the flaw location parameter b and the scale parameter v. The same PDFs are used for both the outer and inner closure-lid welds. As shown in the figure, the size of most flaws is between 1 and 3 mm. Few manufacturing flaws could be longer than 4 mm. Figure 10(b) shows the cumulative probability for the average number of manufacturing flaws per WP in the welds of the outer (25 mm thick) and inner (10 mm thick) closure-lids of the WP outer barrier. As shown in the Figure, almost 100% of WPs have at least one flaw. At the 50th percentile the outer closure-lid weld has an average of 18 measurable flaws per WP, and the value for the inner closure-lid weld is about 15 flaws per WP. An upper bound value for the outer and inner closure-lid weld is about 40 flaws per WP.

As described above, the residual stress analysis shows that the dominant stress in the closure-lid weld region after stress mitigation is hoop stress, which drives radial cracks. The analysis indicates that only radial flaws are potential sites for through-wall SCC if it occurs. The TSPA-SR baseline analysis assumes that all manufacturing

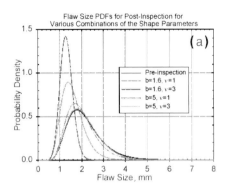
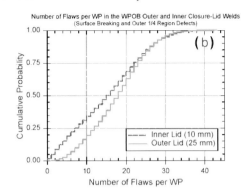

Fig. 10 *Manufacturing flaw model results: (a) Conditional PDFs of manufacturing flaw sizes in the closure-lid welds for various combinations of values for the location and scale parameters (b and ϖ); and (b) CDFs for the average number of manufacturing flaws per WP in the welds of the outer (25 mm thick) and inner (10 mm thick) closure lids of WP outer barrier including surface breaking and embedded flaws.*

flaws are oriented in such a way that they could grow in the radial direction in the presence of hoop stresses. This is a highly conservative assumption. More realistically, most weld flaws, such as lack of fusion and slag inclusions, would be expected to be oriented within a few degrees of the weld centreline [8]. Available published data and limited flaw measurements from the mock-ups for the Viability Assessment (VA) WP design also show that most weld flaws (about 99 percent) tend to be oriented in a circumferential direction [8]. Analyses show it is extremely unlikely that cracks initiating from circumferential flaws grow in the radial direction [8].

4.5. Thermal Ageing

The long term ageing of Alloy 22 at elevated temperatures may cause the precipitation of intermetallic phases, affecting the corrosion resistance of the metal. Experiments to obtain the data needed to develop the model and analysis for the aging process are typically conducted at temperatures higher than those expected in the repository. This is necessary because the precipitation of intermetallic phases at expected repository temperatures would be extremely slow, and would be difficult to observe for reasonable test periods.

The TSPA-SR baseline analysis assumes the closure weld of the WP outer barrier and heat-affected areas near the closure weld are subject to thermal aging, and such effect on corrosion is accounted for with a corrosion rate enhancement factor. The enhancement factor was determined from the ratio of measured non-equilibrium passive current densities for fully aged and unaged samples using the potentiodynamic polarisation technique [8]. The enhancement factor is represented with a range of values to account for uncertainty in the thermal aging process and the effects on corrosion. Based on measured data, the factor is assumed to be a uniform distribution with an upper bound of 2.5 and a lower bound of 1 [8]. The enhancement factor is applied only to the closure weld region of the WP outer barrier and assumed to represent variability in the aging effect on the closure weld region. Thermal ageing of Titanium Grade 7 under repository-relevant thermal conditions is expected to have little impact on the corrosion resistance of this material [8].

4.6. Microbiologically Influenced Corrosion (MIC)

It has been observed that Titanium Grade 7 and Alloy 22 are resistant to MIC. Furthermore, microbial growth in the repository will be limited by the availability of nutrients. In general, the impact of MIC can be accounted for by adjusting E_{corr}, E_{crit}, pH, and the sulfide concentration in the area affected by the microbial activity [8].

The TSPA-SR baseline analysis assumes that when the RH at the surface of the WP and DS is greater than the threshold value for general corrosion initiation, microbial activity becomes significant. The analysis assumes conservatively that when the RH condition is met, microbial activity is at such a level that the biofilm covers the entire surface of all the WPs and DSs. The TSPA-SR baseline analysis assumes that Titanium Grade 7 is not subject to MIC [8].

The effect of MIC of Alloy 22 is modelled with a corrosion rate enhancement factor. The enhancement factor is determined from the ratio of measured corrosion current densities for abiotic and inoculated samples. Based on measurement data, the enhancement factor is assumed to be a uniform distribution with an upper bound of 2 and a lower bound of 1. This enhancement factor is applied to the entire surface of the WP outer barrier and assumed to represent variability in the MIC effects on corrosion [8].

5. Total System Performance Assessment – Site Recommendation (TSPA-SR) Baseline Analysis

The individual models and their parameters that are included in the TSPA-SR baseline analysis are discussed in the previous section, and detailed descriptions are given in Ref. [8]. For most of the degradation models and parameters used, data and analyses are available to quantify their uncertainty and variability. Thus, uncertainty and variability are represented explicitly in the WAPDEG analysis. Variability in the degradation of the WPs is represented by allocating the total variability variance of the individual degradation models and their parameters to WP-to-WP variability among WPs and to patch-to-patch variability within a single WP. The fraction of the total variability variance due to the WP-to-WP variability (and the remains due to the patch-to-patch variability) is treated as an uncertain parameter and sampled randomly for each realisation. For general corrosion, the uncertainty and variability in the general corrosion rate are not quantifiable, and the total variance is considered to represent a mix of uncertainty and variability. The fraction of the total variance due to uncertainty is treated as an uncertain parameter and sampled randomly for each realisation. Detailed discussions on the variance partitioning for the uncertainty and variability representation are found elsewhere [9].

Because, except for the RH threshold for corrosion initiation, temperature and RH do not affect WP and DS degradation, a representative set of temperature and RH histories are used in the current analysis. The representative histories are shown in Fig. 11. No separate analysis is conducted for different waste-type

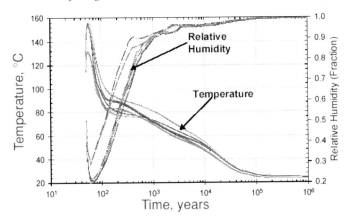

Fig. 11 *A representative set of temperature and RH histories at the WP surface used in the TSPA-SR baseline analysis.*

WPs (i.e. commercial spent nuclear fuel WPs, high level waste WPs, etc.). The corrosion models in the baseline case analysis are not dependent on dripping conditions (i.e. drip vs no-drip), therefore no separate analyses are conducted for different dripping conditions. Details of the approaches and assumptions for the analyses are described in Refs [8,9].

In the TSPA-SR analysis, WP degradation is analysed using multiple realisations of the uncertain degradation parameters used in the WAPDEG model. Each WAPDEG realisation, for a given number of WP and DS pairs, represents a characterisation of the variability of the degradation processes modelled. Accordingly, the WAPDEG analysis outputs are reported as a group of 'curves' that represent the potential range of the output parameters. The baseline case analysis uses 100 realisations of the uncertain degradation parameters (or 100 WAPDEG simulations). The major simulation parameters used in the analysis are summarised below:

- Temperature, RH, and contacting solution pH as a function of time
- 400 WP and DS pairs
- 20 mm thick WP outer barrier (Alloy 22)
- 15 mm thick DS (Titanium Grade 7)
- 1000 patches per WP
- 500 patches per DS.

The WAPDEG analysis results (i.e. WP and DS failure time and number of crack and patch penetrations) are reported as a group of 'degradation profile curves' that represent the potential range of the output parameters. The analysis results are presented for the upper bound (100th percentile), lower bound (0th percentile), mean, and 95th, 75th, 25th and 5th percentiles as a function of time for the following output parameters:

- WP first breach (or failure)

- DS first breach (or failure)

- WP first crack penetration

- WP first patch penetration.

The upper and lower bounds, mean, and 95th, 75th, 25th and 5th percentile curves do not correspond to single realisations. They are summary statistics related to consideration of all 100 realisations. The origin of the upper and lower bounds, mean, and 95th, 75th, 25th and 5th percentile curves for the WP and DS degradation profiles are discussed elsewhere [9]. Note that localised corrosion does not initiate for the WP (Alloy 22 outer barrier) and DS because the exposure conditions on the DS and WP surface are not severe enough to initiate localised corrosion. Therefore no pit or crevice penetration is reported. Also note that the DS is assumed to be not subject to SCC and so there is no SCC failure of DS. Thus, for the DS, the first patch-breach time profile is the same as the failure time profile. Figure 12(a) shows the upper and lower bounds, mean, and 95th, 75th, 25th and 5th percentile confidence intervals of the first breach profile (i.e. failure) of the WPs with time. The upper bound profile, which is the upper extreme of the probable range of the first breach time, indicates that the earliest possible first breach time for a WP is approximately 11 000 years. Note that the estimated earliest possible first breach time has an extremely low probability. It is shown by comparing with the upper bound profile of the first crack breach profiles of WPs with time (discussed below) that the first breach is by SCC. The median estimate (50 percent of WPs failed) of the first breach time of the upper bound profile is approximately 34 000 years. For the 95th percentile profile, which is the reasonably conservative case, the WP failure initiates at approximately 20 000 years. The median estimate of the first breach time of the mean profile is approximately 97 000 years. The time to fail 10% of WPs for the two profiles (95th percentile and mean) is approximately 32 000 and 49 500 years, respectively.

Figure 12(b) shows the first breach profiles of DSs with time. Because the DSs are

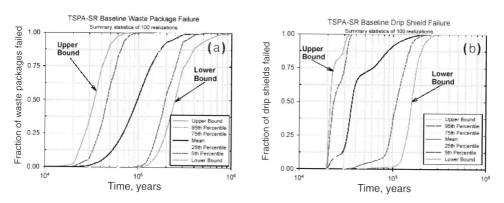

Fig. 12 *The upper and lower bounds, mean, and 95th, 75th, 25th and 5th percentile confidence intervals of the first breach profile of (a) WPs and (b) drip shields with time for the TSPA-SR baseline analysis.*

not subject to SCC and localised corrosion, the first breach profiles shown in the Figure are all by general corrosion only. Both the upper and under sides of DS are exposed to the exposure conditions in the emplacement drift and are subject to corrosion. In addition, the both sides experience the same exposure conditions regardless of whether the DSs are dripped on or not. Thus, in the analysis, the general corrosion rate for the DSs is sampled twice independently, once for the patches on the upper side and the other for the patches on the under side. This results in reduced variability in the degradation profiles and thus rapid failure rate (i.e. many DSs failing over a short time period). This is shown in the upper bound profile, in which the DS failure starts at approximately 20 000 years and 50 percent of the DSs fail within a couple of thousand years after the initial failure. Similar trends are also seen with the 95th, 75th and mean profiles. In terms of the number of patch penetration openings per failed DS with time (not shown here; see [8,9]), the upper bound profile shows that as the DSs fail, a large number of patches are perforated over a relatively short time period (a few thousand years). A similar trend is seen for the 95th percentile profile. However, a greater spread of the failure profile is shown for the other profiles.

Figures 13(a) and (b) show, respectively, the first crack breach and first patch breach profiles of the WPs with time. The first crack breach times of the upper bound and 95th percentile profiles are about 11 000 and 20 000 years respectively (Fig. 13(a)), and the first patch breach times of the upper and 95th percentile profiles are about 36 500 and 41 000 years, respectively (Fig. 13(b)). Comparison of the first crack and patch breach profiles with the first breach profiles in Fig. 12(a) indicates that the initial breach (or failure) of the WPs is by SCC in the Alloy 22 WP outer barrier closure-lid weld region. For the 75th percentile profiles in the Figures, the first crack and patch penetration times are about 30 000 and 50 000 years, respectively.

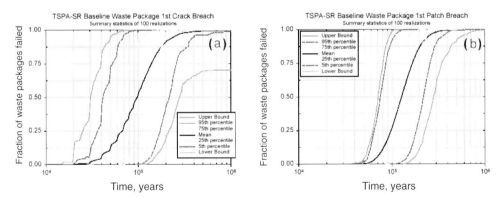

Fig. 13 The upper and lower bounds, mean, and 95th, 75th, 25th and 5th percentile confidence intervals of (a) the first breach profile and (b) the first patch breach profile of WPs with time for the TSPA-SR baseline analysis.

6. Summary and Conclusions

The technical basis for Site Recommendation (SR) of the potential repository for high level nuclear waste at Yucca Mountain, Nevada has been completed. Performance of the Yucca Mountain repository depends on both the natural barrier system (NBS) and the engineered barrier system (EBS) and on their interactions. In particular, long term containment of the waste and subsequent slow release of radionuclides from the EBS into the geosphere will rely on a robust waste package (WP) design, among other EBS components. The WP protected by a drip shield (DS) will be exposed to processes and conditions in the repository environment that will eventually have an impact on its post-closure performance. Some of the important conditions contributing to WP degradation include: humidity and temperature in the emplacement drift, chemistry of the water dripping onto the WP, and corrosion properties of the WP outer barrier. An integrated stochastic simulation model (WAPDEG) was developed to analyse long term degradation of WPs in the potential repository. The primary role of the WP degradation analysis is to provide information on the lifetime and overall degradation of the WP. The WAPDEG model analyses the WP degradation considering general corrosion, localised corrosion, and SCC of the outer barrier. The WAPDEG model also includes effects of MIC, ageing, and manufacturing flaws on the WP degradation. The exposure conditions the WAPDEG model uses are temperature and RH at the WP and DS surfaces as a function of time and the bounding corrosive chemistry of aqueous solutions contacting the DSs and WPs.

The WP and DS degradation analyses for the Total System Performance Assessment — Site Recommendation (TSPA-SR) baseline model have shown that, based on the current corrosion models and assumptions, both the DSs and WPs do not fail within the regulatory compliance time period (10 000 years). From the perspective of initial WP failure time, the analysis results are encouraging because the upper bounds of the baseline case are likely to represent the worst case combination of key corrosion model parameters that significantly affect long term performance of WPs in the potential repository. The upper bound curve has an extremely low probability. The materials selected for the DS (Titanium Grade 7) and the WP outer barrier (Alloy 22) are highly corrosion resistant and, under the repository exposure conditions, are expected to be immune to the degradation processes that, if initiated, could lead to failure in a shorter time period. Those degradation modes are localised corrosion (pitting and crevice corrosion), stress corrosion cracking (SCC), and hydrogen induced cracking (HIC) (applicable to the DS only). Both the DS and WP degrade by general corrosion at very low passive dissolution rates. The current experimental data and detailed process-level analyses, upon which the model abstractions incorporated in the WAPDEG analysis are based, have also indicated that the candidate materials would not be subject to those rapidly penetrating corrosion modes under the expected repository conditions, except for the closure-lid welds of the WP, if SCC were to occur there. To preclude SCC, post-welding stress mitigation will be implemented on a dual-lid WP closure weld design.

The estimated long life-time of the WPs in the current analysis is attributed mostly to the following two factors:

(1) the post-welding stress mitigation to substantial depths from the outer surface in the dual closure-lid weld regions, which delays the onset of SCC until the compressive zone is removed by general corrosion processes; and

(2) the very low general-corrosion rate applied to the closure-lid weld regions to corrode the compressive stress zones, which results in a long delay time before initiating SCC.

Uncertainties are associated with the current SCC analysis, especially post-welding stress mitigation on the closure-lid welds. The uncertainty in the dominant stress (i.e. hoop stress) and associated stress intensity factor profiles used in the current SCC analysis needs to be re-evaluated to better quantify the uncertainties. These uncertainties are expected to be reduced as additional data and analyses are developed. In addition, inclusion of all the embedded manufacturing flaws in the SCC analysis of the WP outer barrier closure-lid weld regions is highly conservative because most embedded flaws do not necessarily lead to radial cracks. Another major uncertainty in the current analysis is the general corrosion rate applied to the closure-lid weld regions. The general corrosion model needs to be refined as additional data and analyses are developed.

7. Acknowledgements

This work was performed and funded under DOE contract DE-AC08-01RW12101 for the Civilian Radioactive Waste Management System (CRWMS M&O) led by the prime contractor (Bechtel SAIC Company, LLC). The authors would like to acknowledge helpful discussions and comments by the following project staff: Joseph Farmer (LLNL), Gerald Gordon (FANP), V. Pasupathi (BSC), Tammy Summers (LLNL), Ahmed Monib (BSC) and Gregory Gdowski (LLNL).

References

1. J. H. Lee, J. E. Atkins and R.W. Andrews, in *Scientific Basis for Nuclear Waste Management XIX* (W.M. Murphy and D.A. Knecht, eds), (Mater. Res. Soc. Proc. **412**, Pittsburgh, PA, 1996), pp. 603–611.

2. J. E. Atkins, J. H. Lee and R. W. Andrews, in *Proc. 7th Int. Conf. on the High-Level Radioactive Waste Management*, Am. Nucl. Soc., LaGrange Park, IL and Am. Soc. of Civil Engrs., New York, NY, pp. 459–461 (1996).

3. J. H. Lee, J. E. Atkins and B. Dunlap, in *Scientific Basis for Nuclear Waste Management XX* (W.J. Gray and I.R. Triay, eds), (Mater. Res. Soc. Proc. **465**, Pittsburgh, PA, 1997), pp. 1075–1082.

4. J. H. Lee, K. G. Mon, D. Longsine and B. E. Bullard, in *Scientific Basis for Nuclear Waste Management XXII* (D.J. Wronkiewicz and J.H. Lee, eds), (Mater. Res. Soc. Proc. **556**, Pittsburgh, PA, 1999) pp. 567–574.

5. J. H. Lee, K. G. Mon, D. E. Longsine and B. E. Bullard, in *Scientific Basis for Nuclear Waste Management XXII* (D.J. Wronkiewicz and J.H. Lee, eds), (Mater. Res. Soc. Proc. **556**, Pittsburgh, PA, 1999) pp. 515–523.

6. CRWMS M&O, Total System Performance Assessment — 1995: An Evaluation of the Potential Yucca Mountain Repository, B00000000-01717-2200-00136 Rev. 01, Las Vegas, NV (1995).

7. CRWMS M&O, Total System Performance Assessment-Viability Assessment (TSPA-VA) Analyses Technical Basis Document: Chapter 5 Waste Package Degradation Modeling and Abstraction, B00000000-01717-4301-00005 REV 01, Las Vegas, NV (1998).

8. CRWMS M&O, Waste Package Degradation Process Model Report, TDR-WIS-MD-000002 REV 00 ICN 02, Las Vegas, NV (2000).

9. CRWMS M&O, WAPDEG Analysis of Waste Package and Drip Shield Degradation, ANL-EBS-PA-000001 REV 00 ICN 01, Las Vegas, NV (2000).

10. CRWMS M&O, Abstraction of NFE Drift Thermodynamic Environment and Percolation Flux, ANL-EBS-HS-000003 REV 00 ICN 02. Las Vegas, NV (2001).

11. CRWMS M&O, In-Drift Precipitates/Salts Analysis, ANL-EBS-MD-000045 REV 00 ICN 02, Las Vegas, NV (2001).

12. BSC (Bechtel SAIC Company), In-Package Chemistry Abstraction, ANL-EBS-MD-000037 REV 01, Las Vegas, NV (2001).

11
Application of the Point Defect Model to Modelling the Corrosion of Iron Based Canisters in Geological Repository

C. BATAILLON

CEA/DEN/DPC/SCCME Laboratoire d'Etude de la Corrosion Aqueuse Bât 458 91191, Gif-sur-Yvette, France

ABSTRACT

A corrosion model for iron and low alloy steels is proposed. This model is essentially based on the Point Defect Model. A modification of the original model has been introduced. To take into consideration the free corrosion case, electron flux has been added. So the mean electric field becomes voltage dependent. As a consequence, no analytical solution for this model is available even in stationary conditions.

1. Introduction

In the project being carried out by the French Agency for Nuclear Waste Management (ANDRA), the long-term behaviour of nuclear waste canisters in geological repository is shown to depend mainly on the

(i)   thermo-hydro-mechanical behaviour of ground soil around the canister,

(ii)   metallic materials used to manufacture canisters, and

(iii)   geochemical features of the ground soil surrounding the canisters.

The disposal of nuclear waste canisters must be considered as a disturbance of the state of the clay fossil soil, not only because (i) the canister is a source of heat but also because (ii) metallic canisters corrode. In clay soil, corrosion products like hydrogen and/or metallic cations induce geochemical transformations of clay which lead to changes in pH and redox potential of the clay surrounding canisters which, in turn, brings about changes in corrosion conditions of canisters. So, the 1000-years behaviour of metallic nuclear waste canisters depends greatly of the coupling between corrosion kinetics and geochemical transformations of clay around canisters.

The purpose of this paper is to propose a corrosion model for iron and low alloy steels. This model is essentially based on the Point Defect Model [1] because Electrochemical Impedance Spectroscopy studies have shown that results obtained on cast iron in different clay-ground water mixtures can be successfully interpreted with this model.

The final target of this work is to link corrosion and geochemical models to describe the long-term behaviour of the canister-surrounding clay system.

2. The Point Defect Model

The Point Defect Model assumes that a barrier layer covers the metal. This layer is a compact oxide which grows at the inner interface. This growth is due to the oxidation of iron at the inner interface and the diffusion of oxygen anion from the outer to the inner interfaces. At the outer interface, oxide dissolves into the solution by an unknown mechanism. This general scheme is shown in Fig. 1. Transport through the barrier layer is due to electronic and non-electronic point defects. Each point defect flux is independent of the others. The point defects are created at one interface (inner or outer) and consumed at the other (outer or inner).

This system (solution/barrier layer/metal) is described by the following equations. First is the flux equation in which it is assumed that a hopping diffusion migration process governs the transport of point defects [2]. Assuming that the potential gradient inside the layer is not too high, the flux of point defect X is given by:

$$\vec{J}_X = -D_X \overrightarrow{grad}\, c_X \; + \; z_X \gamma D_X c_X \overrightarrow{grad}\, \phi \tag{1}$$

where D_X is the diffusion coefficient, c_X the concentration profile inside the layer, z_X the point defect charge, ϕ the potential profile and $\gamma = F/RT$.

The second equation gives the time variation of the concentration profile:

$$\frac{\partial c_X}{\partial t} = -div\, \vec{J}_X = D_X \nabla^2 c_X \; - z_X \gamma D_X \left[c_X \nabla^2 \phi + \; \overrightarrow{grad}\, \phi \cdot \overrightarrow{grad}\, c_X \right] \tag{2}$$

where ∇^2 is the Laplacian. The potential profile ϕ is given by the Poisson equation:

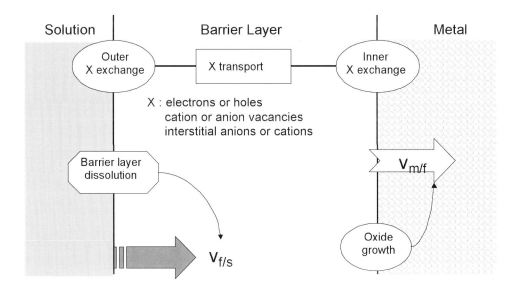

Fig. 1 *General description of the Point Defect Model.*

$$\nabla^2 \phi + \frac{F}{\chi \, \chi_o} \sum_X z_X c_X = 0 \tag{3}$$

where χ is the dielectric constant of the oxide and χ_o the dielectric constant of vacuum. To solve eqns (2) and (3), boundary conditions are needed. To integrate eqn (2), two boundary conditions (one at the inner interface and the other at the outer interface) and one initial condition have to be defined. The initial condition is the initial concentration profile. The inner and outer interface conditions are the kinetics of point defect production and consumption reactions. It is assumed that Butler–Volmer laws give these kinetics.

$$k_c = k_c^o \exp\left(-|z_X| b_X \gamma \Delta\phi_{interface}\right) \quad k_a = k_a^o \exp\left[|z_X|(1 - b_X)\gamma\Delta\phi_{interface}\right] \tag{4}$$

where k's are kinetic constants, b_X the transfer coefficient and $\Delta\phi_{interface}$ the outer or the inner interface voltage.

To integrate the Poisson equation two boundary potentials are needed (i.e. the inner phase potentials ϕ_M and ϕ_S (see Fig. 3)). From an electrochemical point of view it is more convenient to introduce two voltages. One is trivial, that is $V = \phi_M - \phi_s$. The other can be any voltage linked to ϕ_M or ϕ_s. It seems useful to introduce $V_{redox} = \phi_{MR} - \phi_s$ where ϕ_{MR} is the inner phase potential of a noble metal electrode. This voltage V_{redox} will be given by the geochemical model in the NHE reference scale. The voltages V and V_{redox} are not measurable quantities but the difference $E = V - V_{redox}$ is measurable.

The last equation concerns the time variation of the barrier layer thickness L. This layer grows at the inner interface and the fluxes of anionic oxygen vacancy and interstitial anionic oxygen give the kinetics of the growth. This layer dissolves in the solution at the outer interface by the following reaction:

$$\text{Fe}_x\text{O}_y + 2y \ \text{H}^+ \xrightarrow{\quad k_s \quad} x\text{Fe}^{n+}_{sol.} + (xn - 2y)e \ + \ y\text{H}_2\text{O} \tag{5}$$

It is assumed that the barrier layer cannot be built with cations and anions in solution. A general kinetic law gives the kinetics of the dissolution reaction:

$$v_{f/s} = k_s a_{H^+}^p \tag{6}$$

where k_s could be voltage dependent for electrochemical dissolution; p is the order of the dissolution reaction. So the time variation of L is given by:

$$\frac{1}{\Omega_f} \frac{\partial L}{\partial t} = -J_{V_O^{\cdot\cdot}} + J_{O_i^{\prime\prime}} - v_{f/s} \tag{7}$$

where Ω_f is molar volume of oxide.

It is now well known [3] that the oxide layer on iron in neutral solution has a maghemite like structure (γ–Fe_2O_3). Oxygen anions arrange in closed compact cfc lattice; so interstitial anions can be discarded. Only anionic vacancies are considered. The oxide layer on iron is an *n*-type semi-conductor [4–6] with a band gap around 2 eV [7]. So within the temperature range considered ($T<150°C$), holes can be discarded. Only electrons are considered and only reduction of oxidising species at the outer interface is considered. A schematic representation of the model is presented in Fig. 2 in the case where the oxidising species is the proton and assuming that the dissolution of the barrier layer is not an electrochemical process.

3. Model Resolution

3.1. Introduction

The equations (1)–(3) and (7) cannot be solved easily even numerically. There are two main problems. Firstly, the diffusion migration process takes place between moving boundaries. Secondly, eqns (2) and (3) are coupled. It is necessary to know concentration profiles to solve the Poisson equation or to know potential profile to get concentration profiles.

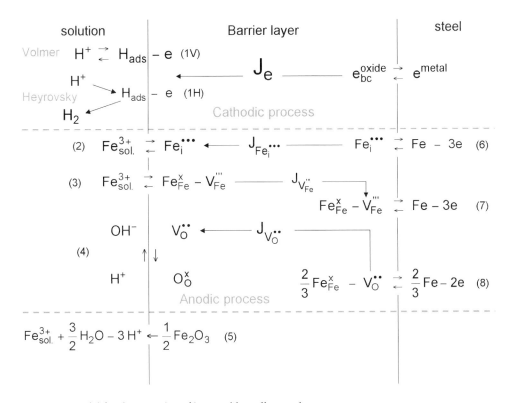

Fig. 2 *Model for the corrosion of iron and low alloy steels.*

The first problem is simplified in using the quasi-stationary approximation, i.e. all partial derivatives vanish with time. Then (7) implies that both interfaces move with the same kinetics. If the spatial coordinate is referred to the outer interface, the diffusion migration process takes place between fixed boundaries. This approximation implies also that the model gives a stationary state of the system if it exits but not the time evolution to this stationary state. This restriction seems not to be a problem to describe the coupling between the corrosion process and the transformation of the clay around canisters because it is believed that the kinetics of clay transformation are much slower than the kinetics of the corrosion process.

The second problem can be reduced by assuming a realistic potential profile (ϕ) and introducing it into the equations profile (2). On an *n*-type semi-conductor, two profiles may be considered depending on the state of semi-conductor. This is shown in Fig. 3. If the semi-conductor is not degenerated the space charge layer plays a major role and an exponential like profile must be considered. If the semi-conductor is degenerated the space charge layer vanishes and a linear profile must be considered. The Point Defect Model is based on linear potential profile (i.e. mean field approximation). This approximation is assumed to be valid for non-oxidising conditions ($V < V_{fb}$, flat band voltage).

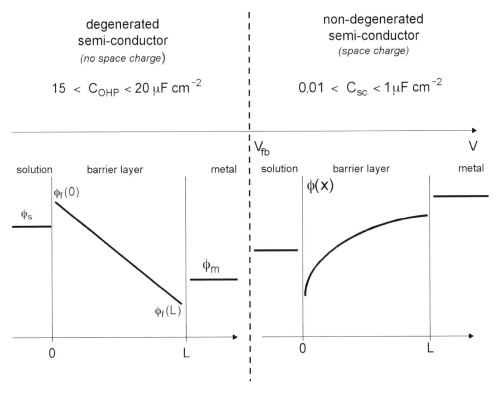

Fig. 3 *Potential profile* ϕ *in the case of degenerated and non-degenerated* n-*type semi-conductor* [8].

Under this assumption, the voltage V is given by $V = \Delta\phi_{m/f} + <\varepsilon>L + \Delta\phi_{f/s}$ where $<\varepsilon>$ is the mean electric field. For a fixed value of V, two more equations are needed to complete the problem. One defines the mean electric field and the other concerns the polarisation of the outer interface.

3.2. Mean Field Equation

In the original Point Defect Model, D. D. MacDonald *et al.* [1] considered that $<\varepsilon>$ is constant. This assumption is based on experimental work on the passive film; that is the steady state film thickness varies linearly with the applied voltage V. From a physical point of view, it is assumed that at high field, the electron-hole generation process could buffer the field by some tunnelling process like the Esaki interband tunnelling [9]. As a result of high field, electrons in the valence band could tunnel to the conduction band at the same energy or reverse (interband exchange).

It is believed that this interband exchange could be valid only at high anodic applied voltage as a result of high field $<\varepsilon>$. At lower voltage, it is believed that the field is not so high. An oxide *n*-type semi-conductor is equivalent to a diode. It will let electrons pass in only one direction (to reduce oxidant). When the diode is conducting, it can be regarded as a metal and Ohm's law can be applied:

$$\vec{J}_e \approx -\gamma D_e < c_e > \vec{\varepsilon} \tag{8}$$

where $<c_e>$ is the mean electrons concentration inside the barrier layer and is given by:

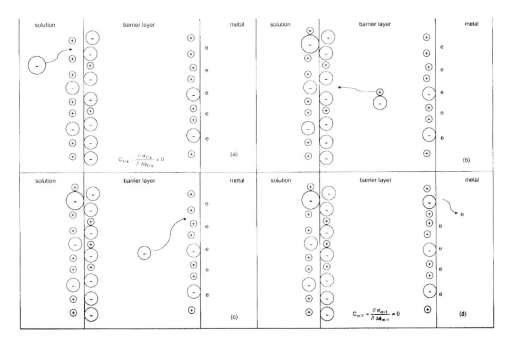

Fig. 4 *Modification process of charge of outer and inner interfaces.*

$$< c_e > = \frac{1}{L} \sum_{X=Fe_i^{...}, V_O^{..}, V_{Fe}^{...}} \int_0^L z_X c_X(x) \mathrm{d}x \tag{9}$$

This equation sets up the whole electroneutrality of the barrier layer. This layer is regarded as metal whose mean electronic conductivity $\gamma D_e < c_e >$ is not constant. As the electrons flux is equal to the kinetics of the cathodic reaction at the outer interface, the following equation stands:

$$j_c\left(\Delta\phi_{f/s}\right) = -<\varepsilon> F\gamma D_e < c_e > \tag{10}$$

This equation defines the mean electric field $<\varepsilon>$ which depends on $\Delta\phi_{f/s}$.

3.3. Polarisability of the Outer Interface

This equation was introduced in the Point Defect Model. It is assumed that the outer voltage $\Delta\phi_{f/s}$ is a part of the whole voltage V:

$$\Delta\phi_{f/s} = \alpha V + \Delta\phi_{f/s}^o \tag{11}$$

which implies a linear relationship between inner and outer voltages:

$$\Delta\phi_{m/f} = \frac{1-\alpha}{\alpha} \Delta\phi_{f/s} - <\varepsilon> L - \frac{\Delta\phi_{f/s}^o}{\alpha} \tag{12}$$

This last relation can be explained qualitatively by the scheme shown in Fig. 4. Let us assume that for any reason, the outer voltage $\Delta\phi_{f/s}$ changes. As the outer interface capacitance $C_{f/s}$ is not zero, this change implies a change in the outer interface charge $\sigma_{f/s}$. Let us also assume that an anion arrives from solution (see Fig. 4a). To preserve electroneutrality of the whole outer interface, a positive point defect has to balance the anion (see Fig. 4b). To preserve the electroneutrality inside the barrier layer, the counter point defect charge (i.e. negative point defect) has to move to the inner interface (see Fig. 4c). To preserve the whole electroneutrality of the inner interface, an electron has to move away (see Fig. 4d). As the inner interface capacitance is not zero the inner interface voltage has to change. These pictures show qualitatively that under mean field approximation and to preserve the whole electroneutrality of the system, a relationship between inner and outer interface voltages must exist (eqn (11)) where α is the polarisability of the outer interface.

The polarisability eqn (11) of the outer interface introduces a second term $\Delta\phi_{f/s}^o$. To outline the role played by this term, let us introduce eqn (11) in anodic (v_a) and cathodic (v_c) Butler–Volmer laws:

$$v_c = k_c^o \left| a_X \; \exp\left(-|z_X| b_X \gamma \Delta\phi_{f/s}^o\right) \right| \exp\left(-|z_X| b_X \gamma \alpha V\right) \tag{13a}$$

$$v_a = k_a^o \left| a_X \exp\left[|z_X|(1-b_X)\gamma\Delta\phi_{f/s}^o\right] \right| \exp\left[|z_X|(1-b_X)\gamma\alpha V\right] \tag{13b}$$

In these equations the term inside the brackets can be considered as an effective point defect activity which is different from the actual point defect activity a_X. This correction introduced in the Butler–Volmer law is equivalent to the correction introduced by Bockris *et al.* [10] to take into consideration the Gouy–Chapman layer in solution. This correction is introduced when a solution does not contain supporting electrolyte. An equivalent state exists in the barrier layer because no unreactive point defects are available. Each kind of point defect has to contribute to the charge of outer and inner interface and simultaneously reacts at these interfaces. Consequently, the term $\Delta\phi_{f/s}^o$ describes interaction between kinetics and interface structure.

3.4. Resolution Processing

The resolution processing of the model is shown in Fig. 5. The activity of Fe^{3+} $\left(a_{Fe^{3+}}\right)$, ionic product of the water (K_w) and pH are input data. $K_w(T)$ is given in the literature [11]. The geochemical model will give pH and cation Fe^{3+} activity. Output data are fluxes (hydrogen and Fe^{3+} fluxes). Depending on the geochemical model, Fe^{3+} cations may or may not react with hydrogen and water to reduce to Fe^{2+}. These cations could also to participate in the formation of new phases (magnetite, haematite, etc.).

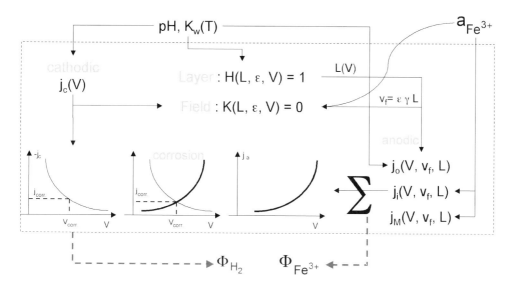

Fig. 5 *Resolution processing for the model.*

The model consists in linked modules. The cathodic module gives the cathodic current density $j_c(V)$. The kernel of the model consists in the 'Layer' and 'Field' modules. Resolution of equations $H(L, \varepsilon, V) = 1$ (14) and $K(L, \varepsilon, V) = 0$ (15) gives $L(V)$ and $v_f(V) = \varepsilon(V).\gamma.L(V)$. These functions are used to calculate point defect current densities (j_o, j_i and j_M). The sum of these current densities is the anodic current density $j_a(V)$.

The root of equation: $j_a(V) + j_c(V) = 0$ is the free corrosion voltage V_{corr}. Then $j_c(V_{corr})$ or $j_a(V_{corr})$ is the free corrosion current density which is proportional to H_2 and Fe^{3+} fluxes. If reductive dissolution of the barrier layer is introduced, a third output flux (Fe^{2+}) will exist. Then the Fe^{3+} flux will be proportional to $j_i + j_M$ and the Fe^{2+} flux will be proportional to j_o.

The function $H(L, \varepsilon, V)$ is defined by:

$$H(L,\varepsilon,V) = \frac{2D_o}{L\ k_s\ K_W}\frac{\ln Y}{Y^2-1}\left[\begin{array}{l}\dfrac{a_{H^+}^{-p}K_8K_We^{2(1-\alpha)\gamma V}}{Y^2} - \dfrac{a_{H^+}^{2-p}Y^2e^{-\alpha_I V}}{K_4}\\[2mm] -\dfrac{a_{H^+}k_sY\ e^{-\alpha(1-b_4)\gamma V}}{k_4} - \dfrac{k_sK_We^{2b_8(1-\alpha)\gamma V}}{k_{-8}\ Y^{2b_8}}\end{array}\right] = 1 \qquad (14)$$

where $Y = e^{2V_f}$ and $v_f = \varepsilon\ \gamma\ L$. K's and k's are respectively the equilibrium constants and kinetic constants of interfacial reactions numbered in Fig. 2. The function $K(L, \varepsilon, V)$ is defined by:

$$K(L,\varepsilon,V) = j_c(V) + FD_e < \varepsilon > \gamma\left[\frac{2}{L}\int_0^L c_o(u)du + \frac{3}{L}\int_0^L c_i(u)du - \frac{3}{L}\int_0^L c_M(u)du\right] = 0 \quad (15)$$

where $c_o(u)$, $c_i(u)$ and $c_M(u)$ are the concentration profiles of oxygen vacancies, interstitial cations and cationic vacancies. These profiles are stationary solutions of eqn (2).

For a fixed temperature, the coupling between corrosion and geochemical models runs in the following way: corrosion model gives H_2 and Fe^{3+} fluxes that depend on pH and Fe^{3+} activity. If H_2 and / or Fe^{3+} fluxes lead to variations in pH or Fe^{3+} activity in the surrounding of canisters then the free corrosion current density, H_2 and Fe^{3+} fluxes change.

4. Conclusions

- A corrosion model for iron and low alloy steels has been proposed. This model is essentially based on the Point Defect Model. A modification of the original model has been introduced. Electrons flux has been added so that the mean electric field is now voltage dependent.

- There is no analytical solution to this model because eqns (14) and (15) cannot be solved analytically in the general case. Then, the resolution must be numerical. So to go forward, it is necessary to evaluate values for the parameters of the model.

- This work is in progress step by step. The next step could be to define a plan for testing the basic assumptions of the model.

List of Symbols

a_X	activity of X in solution
b_X	transfer coefficient
c_X	concentration profile of defect X
D_X	diffusion coefficient
j_a	anodic current density
j_c	cathodic current density
j_i	current density for interstitial cations
j_M	current density for cation vacancies
j_o	current density for oxygen vacancies
k	kinetics constant
K	equilibrium constant
L	thickness of the barrier layer
J_X	flux of defect X
t	time
z_X	charge of defect X
Ω_f	molar volume of the oxide
α	polarisation of outer interface
ε	electric field
ϕ	potential profile
γ	F/RT
χ	dielectric constant

References

1. C. Y. Chao, L. F. Lin and D. D. MacDonald, *J. Electrochem. Soc.*, 1981, **128**, 1187–1194.
2. Y. Adda and J. Philibert, 'La diffusion dans les solides', Vol. 2, Ch. XVI: la diffusion en champ électrique, INSTN (Saclay) 1966, p.893–937.
3. A. J. Davenport, L. J. Oblonsky, M. P. Ryan and M. F. Toney, *J. Electrochem. Soc.*, 2000, **147**, 2162–2173.
4. W. Schultze and M. M. Lohrengel, *Electrochim. Acta*, 2000, **45**, 2499–2513.
5. E. B. Castro and J. R. Vilche, *Electrochim. Acta*, 1993, **38**, 1567–1572.

6. A. M. Sukotin, M. S. Grilikhes and E. V. Lisovaya, *Electrochim. Acta*, 1999, **34**, 109–112.

7. J. S. Kim, E. A. Cho and H. S. Kwon, *Corros. Sci.*, 2001, **43**, 1403–1415.

8. S. R. Morrison, *Electrochemistry at Semiconductor and Oxidized Metal Electrodes*. Plenum Press, New York, 1980, p.107–118.

9. L. Zhang, D. D. MacDonald, E. Sikora and J. Sikora, *J. Electrochem. Soc.*, 1998, **145**, 898–905.

10. J. M. Bockris and A. K. N. Reddy, *Modern Electrochemistry*, Volume 2. Plenum Press, New York, 1970, p.911–916.

11. Data available on http://www.iapws.org/ — also in 'Physical Chemistry of Aqueous Systems', *Proc. 12th Int. Conf. on Properties of Water and Steam* (H.J. White, J.U. Sengers, D.B. Neumann and J.C. Bellows). Begell House, New York, and Wallingford, UK.

12
Transients in the Growth of Passive Films on High Level Nuclear Waste Canisters

M. URQUIDI-MACDONALD and D. D. MACDONALD

Center for Electrochemical Science and Technology, Pennsylvania State University, 201 Steidle Bldg, University Park, PA 16802, USA

ABSTRACT

A new rate law for the growth of anodic passive films on metal surfaces that was recently derived from the Point Defect Model (PDM) is used to predict the transients in current density and film thickness on Alloy C-22 over extended periods of time in an environment (saturated brine) that is postulated to exist in high level nuclear waste repositories. The model recognises both the growth of the barrier oxide layer into the metal via the generation of oxygen vacancies at the metal/film interface and the dissolution of the barrier layer at the film/solution interface, as well as the current carried by cation interstitials within the Cr_2O_3 barrier layer. The derived rate law accounts for the existence of a steady state in film thickness as well as for the transients in thickness and film growth current as the potential is stepped in the positive or negative direction from an initial steady state. The predicted transients in film thickness and growth current density for Alloy C-22 in the prototypical HLNW (High Level Nuclear Waste) environment employed demonstrate that the kinetics of dissolution of the barrier oxide layer at the barrier layer/solution interface control the rate of passive film thinning when the corrosion potential is stepped in the negative direction, whereas the kinetics of oxygen vacancy generation at the metal/film interface control the rate of film thickening when the potential is displaced in the positive direction. While the transients are predicted to persist for considerable time, the times are short compared with the design life of the repository and we conclude that the accumulated damage due to general corrosion is readily predicted by using quasi steady-state models.

1. Introduction

Recently [1], it was demonstrated that the potentiostatic current transients for the growth of passive films on zirconium, titanium, tungsten, and tantalum are inconsistent with the predictions of the high field model (HFM) [2–4]. Thus, the HFM postulates that the electric field strength within the film decreases with the inverse of the film thickness as the film thickens. On the other hand, the current transient data were found to be consistent with the postulate that the electric field strength is insensitive to changes in film thickness, which is the basis of the Point Defect Model (PDM) [5]. The insensitivity of the field strength to thickness is explained by the occurrence of band-to-band tunnelling ('Esaki tunnelling') of charge carriers (electrons and holes) within the film, such that the field strength becomes buffered against any process that tends to increase its value [5]. Thus, because the internal tunnelling current is exponentially dependent on the field, any increase in the potential gradient will cause a massive separation of charge and will result in a counter field that effectively buffers the potential distribution. Discrimination between the two models (HFM and the PDM) was made on the basis of the diagnostic relationship

$$\sqrt{-i' / i_{ss}(i - i_{ss})} = \sigma \ln(i)$$

where i' is the differential di/dt, i is the current density, i_{ss} is the steady state current density (i.e. the current density for $t \rightarrow \infty$), and σ is a constant. The HFM predicts that, for films that are formed primarily by the movement of oxygen vacancies (which is the case for the 'valve' metals), the quantity σ should be greater than 0, whereas the PDM predicts that $\sigma = 0$. The latter relationship was found for the metals employed in the previous study [1].

Subsequent to that study, Macdonald *et al.* [6] derived a new rate law for the growth and thinning of the barrier layer of a passive film as the potential is stepped in the positive or negative direction, respectively from an initial steady state. This new rate law is shown to account for the transients in thickness and current for passive tungsten in acidic phosphate buffer solutions [6]. The purpose of the present communication is to describe predictions of transients in film thickness and film growth current density for Alloy C-22 under conditions that are postulated to exist in High Level Nuclear Waste (HLNW) repositories.

2. Theory

According to the PDM (Ref. 5 and citations therein), the rate of change in thickness of a barrier layer passive film can be written in the form

$$\frac{dL}{dt} = a.e^{-bL} - c \tag{1}$$

where $a = \Omega k_3^0 e^{\alpha_3(1-\alpha)\chi\gamma V} e^{-\alpha_3\chi\beta\gamma pH}$, $b = \alpha_3\chi\varepsilon\gamma$, $c = \Omega k_7 (C_{H^+} / C_{H^+}^0)^n$, ε is the electric field strength, C_{H^+} is the concentration of hydrogen ion at the film/solution interface, $C_{H^+}^0$ is the standard state concentration, such that k_7 (the rate constant for the dissolution of the film, Reaction 7, Fig. 1) has units of mol cm^{-2} s^{-1} (independent of the reaction order, n), χ is the oxidation state of the cation in the barrier layer, Ω is the mol volume of the film per cation, n is the kinetic order of the dissolution reaction at the film/solution interface with respect to H$^+$, α is the polarisability of the film/solution interface, β is the dependence of the potential drop across the film/solution interface on the pH, V is the applied voltage, $\gamma = F/RT$, F is Faraday's constant, R is the universal gas constant, T is the Kelvin temperature, and α_3 and k_3^0 (mol cm^{-2} s^{-1}) are the transfer coefficient and the standard rate constant, respectively, for the generation of oxygen vacancies at the metal/film interface (Reaction 3, Fig. 1).

Metal	Barrier Oxide Layer	Solution

(1) $m + V_M^{\chi'} = M_M + v_m + \chi e'$ (4) $M_M = M^{\delta^+} + V_M^{\chi'} + (\delta - \chi)e^-$

(2) $m = M_i^{\chi^+} + v_m + \chi e'$ (5) $M_i^{\chi^+} = M^{\delta^+} + (\delta - \chi)e^-$

(3) $m = M_M + \dfrac{\chi}{2} V_{\ddot{O}} + \chi e^-$ (6) $V_{\ddot{O}} + H_2O = O_o + 2H^+$

(7) $MO_{\chi/2} + \chi H^+ = M^{\delta^+} + \dfrac{\chi}{2} H_2O + (\delta - \chi)e^-$

Fig. 1 *Interfacial defect generation/annihilation reactions that occur in the growth of anodic barrier oxide films according to the Point Defect Model [1].* m = *metal atom,* $V_M^{\chi'}$ = *cation vacancy on the metal sub-lattice of the barrier layer,* $M_i^{\chi^+}$ = *interstitial cation,* M_M = *metal cation on the metal sub-lattice of the barrier layer,* $V_{\ddot{O}}$ = *oxygen vacancy on the oxygen sub-lattice of the barrier layer,* O_o = *oxygen anion on the oxygen sub-lattice of the barrier layer,* M^{δ^+} = *metal cation in solution.*

For steady state conditions, $dL/dt = 0$, so that

$$L_{ss} = (1/b)\ln(a/c) \qquad (2)$$

where L_{ss} is the steady state thickness of the film. Upon substitution of the constants defined above, it is a straightforward matter to show that eqn (2) is identical to the expression that was previously derived for the steady state thickness of the film (barrier layer) [5]:

$$L_{ss} = \frac{1}{\varepsilon}\left[1 - \alpha - \frac{\alpha\alpha_7}{\alpha_3}\left(\frac{\delta}{\chi} - 1\right)\right]V + \frac{1}{\varepsilon}\left\{\frac{2.303n}{\alpha_3\chi\gamma} - \beta\left[\frac{\alpha_7}{\alpha_3}\left(\frac{\delta}{\chi} - 1\right) + 1\right]\right\}pH + \frac{1}{\alpha_3\chi K}\ln\left(\frac{k_3^0}{k_7^0}\right) \qquad (3)$$

where $K = \varepsilon\gamma$, k_7^0 and α_7 are the standard rate constant and transfer coefficient, respectively, for the film dissolution reaction, and δ is the oxidation state of the cation being ejected from the barrier layer into the external environment or that results from barrier layer dissolution. In deriving this expression, it is assumed that a change may occur in the oxidation state of a cation upon ejection from the film/solution interface or upon dissolution of the film. If no change in oxidation state occurs ($\delta = \chi$), a somewhat simpler expression results in the form:

$$L_{ss} = \left(\frac{1 - \alpha}{\varepsilon}\right)V + \frac{1}{\varepsilon}\left\{\frac{2.303n}{\alpha_3\chi\gamma} - \beta\right\}pH + \frac{1}{\alpha_3\chi K}\ln\left(\frac{k_3^0}{k_7^0}\right) \qquad (4)$$

Note that eqns (3) and (4) predict that for a given pH the steady state film thickness varies linearly with applied voltage, which is an almost ubiquitous finding in the study of the formation of anodic passive films [5]. Furthermore, the equations predict that, for a given voltage, L_{ss} varies linearly with pH, and that for a given voltage and pH, the thickness is controlled by the ratio of the standard rate constants for the generation of oxygen vacancies at the metal/film interface (Reaction (2), Fig. 1) and the dissolution of the film (Reaction (5)).

2.1. Transient in Film Thickness

Equation (1) is readily integrated over the limits $(L_0, 0, V)$ to $(L, t, V+\Delta V)$, where ΔV is the step in voltage at $t = 0$, to yield (for $\delta = \chi$) the transient in film thickness as

$$L(t) = L_0 + \left(\frac{1}{b}\right) \ln\left[\left(\frac{a'}{c}\right)e^{-bL_0}(e^{bct} - 1) + 1\right] - ct \tag{5}$$

where the constants and parameters are as given above, except that c is the dissolution rate at voltage $V+\Delta V$. Note that $a' = k_3^0 e^{a_3(1-\alpha)\gamma(V+\Delta V)}e^{-\alpha_3\beta\gamma pH}$. It is also a straight-forward matter to show that, for $t = 0$, $L = L_0$, and as $t\to\infty$, the term in square brackets $\to (a'/c)e^{-bL_0}e^{bct} \gg 1$, and hence $L(t)$ approaches a steady state that is given by eqn (4), but with a being replaced by a'. The initial thickness, L_0, is given by eqn (4) which was obtained by using eqn (2), which uses a (a function of V) instead of a' (a function of $V + \Delta V$). Examination of eqn (5) shows that the transient in film thickness is a combination of a linear relationship and a logarithmic function.

If the rate of dissolution of the passive film (c in eqn (1)) is sufficiently small that $e^{bct} \approx 1 + bct$ and ct is much smaller than the second term in eqn (5), then the rate law reduces to [5]

$$L(t) = L_0 + \left(\frac{1}{b}\right) \ln\left[1 + a'be^{-bL_0}t\right] \tag{6}$$

where the parameters are as defined above. If $a'be^{-bL_0t} \gg 1$, eqn (6) collapses into the classical logarithmic relationship [4]. Note that for $t = 0$, the film thickness equals the initial thickness, L_0, which is the correct limiting form.

2.2. Transient in Film Growth Current

The transient in film growth current density due to the generation of oxygen vacancies at the metal/film interface and the dissolution of the film at the film/solution interface is readily calculated as

$$i = \chi F k_3 + (\delta - \chi)F k_7 (C_{H^+}/C_{H^+}^0)^n \tag{7}$$

Noting that $k_3 = k_3^0 \exp(a_3 V - b_3 L - c_3 pH)$ and $k_7 = k_7^0 \exp(a_7 V + c_7 pH)$, where $a_3 = \alpha_3(1-\alpha)\chi\gamma$, $b_3 = \alpha_3\chi\gamma\varepsilon$, $c_3 = \alpha_3\chi\gamma\beta$, $a_7 = \alpha_7\alpha(\delta-\chi)\gamma$, and $c_7 = \alpha_7\beta(\delta-\chi)\gamma$, we obtain the transient in the current as

$$i = \chi F k_3^0 e^{a_3 V} e^{-c_3 pH} X + (\delta - \chi) F k_7^0 e^{a_7 V} e^{c_7 pH} (C_{H^+} / C_{H^+}^0)^n \qquad (8)$$

where

$$X = \frac{ce^{-b_3 L_0}}{a' e^{-b_3 L_0} - (a' e^{-b_3 L_0} - c)e^{-bct}} \qquad (9)$$

with a', b, and c being as defined in eqn (1). Note that eqn (8) applies strictly to an oxygen ion conducting film, although it is believed (but not proven) that the theory is also applicable to cation interstitial conducting barrier layers. Finally, it is important to note that eqn (5) is general to all three cases, because only the generation of oxygen vacancies via Reaction (3), Fig. 1, and their annihilation at the film/solution interface (Reaction (6)), coupled with film dissolution (Reaction (7)), lead to a change in dimension of the film.

3. Discussion

Equations (5) and (8) represent completely new rate laws for the growth of a passive film under potentiostatic transient conditions. These laws result from the postulate that the electric field strength is independent of the applied voltage and that the film undergoes dissolution at the film/solution interface. Before analysing the laws in detail, it is instructive to compare the PDM with the classical HFM as developed by Cabrera and Mott (CM) [3]. This is necessary, because there is still a great tendency for researchers to adopt the HF model, in spite of the fact that it has been discredited in numerous studies on a wide variety of passive metals and alloys. The continued use of the HF model is, no doubt, a result of the mathematical simplicity of the differential rate law (see eqn (10) below). Among the devastating shortcomings of the HFM is that it does not predict the existence of a steady state in the barrier layer thickness under potentiostatic conditions, and it cannot account for the thinning of the barrier when the potential is stepped in the negative direction. Both observations are established experimental facts and both are accounted for by eqn (5) [6].

In the HF model, and in variants thereof, such as the Cabrera–Mott model [3], the electric field strength is assumed to decrease as the film thickens, and film growth is assumed to occur only at the film/solution interface, because no movement of oxygen vacancies is recognised. Furthermore, dissolution is assumed not to occur at the film/solution interface and rate control is postulated to reside in cation ejection from the metal into the film. This model gives the film growth rate law, for potentiostatic conditions, as

$$\frac{dL}{dt} = Ae^{B/L} \qquad (10)$$

where A and B can be expressed in terms of fundamental parameters (see Refs. [3] and [4]). Equation (10) does not have an analytical solution, and the one offered by Cabrera and Mott [3] is strictly not valid (see Burstein and Davenport [8] for an excellent discussion of this issue). For our purposes, we need only note that as $t \to \infty$, $L \to \infty$ and $dL/dt \to 0$. Thus, the film thickness is predicted to never reach a steady state. Furthermore, regardless of the magnitude or sign of B, which contains the same voltage dependence as a', the sign of dL/dt never changes, although the magnitude clearly does. Thus, regardless of whether ΔV is positive or negative, and irrespective of the magnitude of the step in voltage, the CM theory predicts that the film will continue to grow, which is contrary to experimental observation [6,7]. Accordingly, the classical HFM, regardless of the form [2–4], must be rejected as being capable of describing the growth of passive films on metal surfaces.

An attempt has been made to correct the deficiencies of the HFM by introducing, in an *ad hoc* fashion, dissolution of the film/solution interface [7]. This modification removed some of the objections to the HFM, in that the model was capable of predicting a steady state in film thickness and was able to account for a change in the sign of dL/dt as the sign of ΔV is changed. However, it was not able to account for the pH dependence of the steady state passive film thickness vs applied voltage and so the modified HFM was eventually rejected [7].

We now return to eqns (1) and (5). Equation (1) clearly predicts a steady state in film thickness, resulting in eqns (3) and (4). Furthermore, as ΔV changes sign, and hence as a' changes value, but never sign, dL/dt may be positive or negative, corresponding to film growth or film thinning, respectively, depending upon the relative values of the two terms on the right side of eqn (1). Note that, because Reaction (3), Fig. 1, is considered to be irreversible in the direction indicated, film thinning occurs exclusively via dissolution at the film/solution interface (Reaction 7, Fig. 1) with the rate being greater than the rate of film growth at the metal/film interface (Reaction 3). This is considered to be a reasonable approximation if the voltage is significantly more positive than the equilibrium potential for the formation of the passive film phase. Furthermore, eqn (4) correctly predicts the linear dependencies of L_{ss} on the applied voltage (V) and pH [5]. We now show below that eqn (5) correctly accounts for the transients in film thickness.

Prior to doing this, it is first necessary to identify the rate control process in film growth according to the PDM. Thus, as noted previously [1,5], rate control resides in the injection of charge into the film at the metal/film interface. With respect to film growth, this occurs via the generation of oxygen vacancies at the metal/film interface by Reaction 3, Fig. 1. The charge is injected into an electric field that is independent of the thickness of the film and the applied voltage. Accordingly, the potential drop across the metal/film interface ($\phi_{m/f}$), which drives the oxygen vacancy formation reaction (Reaction 3, Fig. 1), decreases linearly with increasing film thickness, viz

$$\phi_{m/f} = (1-\alpha)(V + \Delta V) - \varepsilon L - \beta.pH - \phi^0 \tag{11}$$

for $t > 0$ and $L(t) > L_{ss}$, where ϕ^0 is a constant that depends upon the selection of the potential scale. The other parameters are as previously defined. It is the formation of oxygen vacancies at the metal/film interface, in the presence of the decreasing potential drop, $\phi_{m/f}$, that constitutes rate control in the case of the PDM. In contrast, the CM form of the HFM postulates that the rate-determining step in film growth is the injection of cations from the metal into the film in the presence of an electric field that decreases with increasing film thickness. The CM model gives rise to the same rate law as does the original Verwey model [2] (but with small differences in some parameters), which postulates that rate control resides in the transfer of a cation from one lattice position to another within the bulk of the film.

3.1. Steady State Behaviour of Alloy C-22

Calculated steady state thickness and steady state current density plots for Alloy C-22 in contact with saturated brine at 25 and 90°C are shown in Figs 2 and 3, respectively. At voltages below *ca.* 0.8 V(SHE), the barrier layer thickness varies linearly with voltage, as predicted by eqns (3) and (4). However, at more positive potentials, where δ is no longer equal to χ (6 and 3, respectively), oxidative dissolution of chromium species from the Cr_2O_3 barrier layer implies that $L_{ss}(V)$ is described by

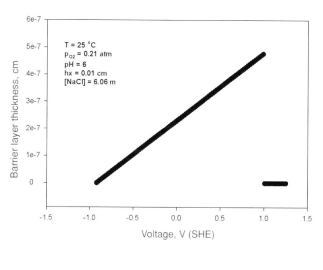

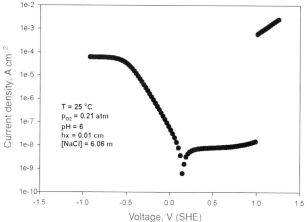

Fig. 2 *Plots of steady state barrier layer film thickness and current density for Alloy C-22 in saturated NaCl solution at 25°C.*

eqn (3) and not by the simpler form, eqn (4). The change in oxidation state results in a sharp reduction in film thickness and hence in a correspondingly sharp increase in the potential drop across the metal/film interface (see eqn (11) above), which in turn causes a sharp increase in the current density (Fig. 3). This sudden increase in the current density marks the transition to the transpassive state. Note that in constructing the current/voltage curve in Fig. 3, we have included the cathodic reduction of oxygen and we have assumed that the Cr_2O_3 barrier layer on C-22 is either an oxygen vacancy conductor or a cation interstitial conductor, or both. This assignment of the majority defect type appears to be consistent with the available experimental data, although the assignment should not be regarded as being unequivocal. The parameter values used in constructing the plots shown in Figs 2 and 3 are as follows:

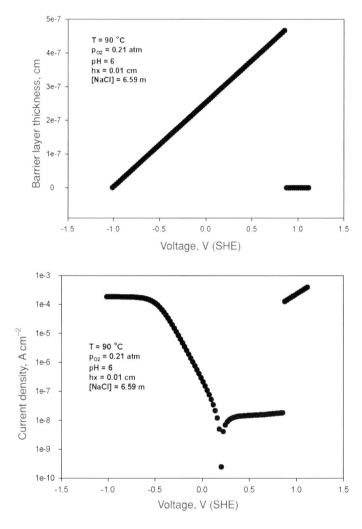

Fig. 3 *Plots of steady state barrier layer film thickness and current density for Alloy C-22 in saturated NaCl solution at 90 °C.*

$\alpha = 0.5$; $\beta = -0.001$ V; $\varepsilon = 2.0 \times 10^6$ V cm^{-1}; $n = 0.1$; $\Omega = 29.80$ cm^3 mol^{-1} (for Cr$_2$O$_3$);
$\chi = \delta = 3$; $\{\alpha_{i'} i = 1–7\} = \{0.1\}$; $\{k_i^{00}, i = 1–7\} = \{1 \times 10^{-30}, 7 \times 10^{-14}, 1.93 \times 10^{-12}, 1 \times 10^{-30}, 1 \times 10^{-18}, 1 \times 10^{-10}, 5 \times 10^{-14}\}$ cm s^{-1}; and $\{E_{a,i}, i = 1–7\} = \{500\}$ cal/mol.

The parameters for the cathodic reactions (hydrogen evolution and oxygen reduction) were assumed to be the same as those for Type 304 stainless steel [9]. All of the parameter values should be regarded as being tentative, because few experimental data are currently available for C-22 in relevant environments.

3.2. Transients in Barrier Layer Thickness and Current

Equations (5) and (8) were used to calculate the transients in film thickness and current for Alloy C-22 upon stepping the applied voltage over the cycle

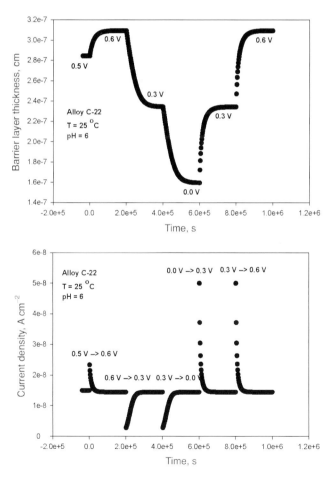

Fig. 4 *Calculated transients in the barrier layer thickness and anodic (passive) current density for Alloy C-22 upon stepping the potential in the negative and positive directions. Alloy C-22, T = 25 °C, pH = 6. Voltages are with respect to the Standard Hydrogen Electrode (SHE).*

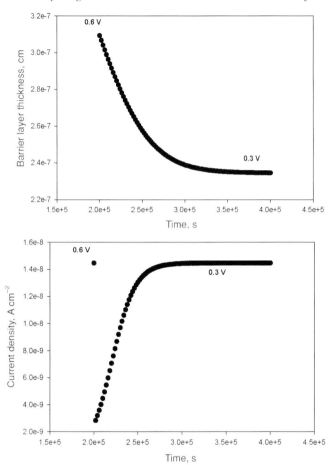

Fig. 5 *Transients in the barrier layer thickness and current density for a step decrease in potential from 0.6 V to 0.3 V(SHE) for Alloy C-22, T = 25 °C, pH = 6.*

0.5 V – 0.6 V – 0.3 V – 0.0 V – 0.3 V – 0.6 V(SHE) in solutions of pH = 6 at 25 and 90°C (Figs 4–7). The voltages were chosen, such that the system is always in the passive state ($\delta = \chi$). No relevant experimental data are available for comparison with the predicted transients, but we note that the theory is quite general and has previously accounted for the transients in both passive film thickness and current density for tungsten in phosphoric acid (pH = 1.5) at 25°C when subjected to positive or negative steps in potential. In that case, passive film thickness data were available from both capacitance and reflectance studies [6]. Accordingly, deviations of the theory from (future) experimental results will probably result from incorrect values for various model parameters.

The initial increase in film thickness on stepping the potential from 0.5 V to 0.6 V(SHE) is due to film growth at the metal/film interface (Reaction (3), Fig. 1). As the film thickens, the voltage drop across the m/f interface decreases and hence the rate of film growth also decreases, as does the growth current, with dL/dt eventually becoming zero. Upon stepping the potential in the negative direction, an initial linear decrease in $L(t)$ is predicted by eqn (1), and hence by eqn (5), by noting that, under these conditions, ΔV is sufficiently negative that the first term in eqn (1) is negligible with respect to the second term, and hence that $dL/dt \approx -c$ (a constant for a given pH). Thus, the film thickness is predicted to decrease linearly with time, as observed in the case of tungsten [6], and as predicted in Fig. 5 for short times.

Eventually, the two terms on the right side of eqn (1) become equal as the film thins (see eqn (11)), principally due to the increase in the potential drop across the metal/film interface. Film thinning, due to dissolution, gives rise to a continual increase in the value of the first term in eqn (1), resulting in the attainment of a steady state whose thickness is described by eqn (4) with V being replaced by $V + \Delta V$ (where ΔV, in this case, is negative). At longer times, $L(t)$ remains constant in the steady state until the voltage is stepped back to another value. For example, in Step 3 (0.3 V → 0.0 V(SHE)), the barrier layer eventually attains a thickness of 1.6 nm, but upon stepping the potential back to 0.3 V $L(t)$ is predicted to increase

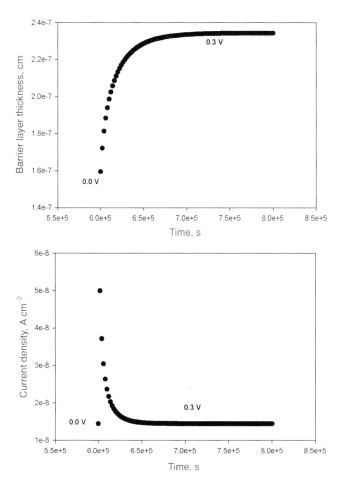

Fig. 6 *Transients in the barrier layer thickness and current density for a step decrease in potential from 0.0 V to 0.3 V(SHE) for Alloy C-22, T = 25°C, pH = 6.*

sharply with time, as film growth occurs at the metal/film interface due to the production of oxygen vacancies (Reaction 3, Fig. 1). The increase in film thickness on stepping the voltage in the positive direction is not instantaneous, as shown by the calculated transient thickness vs time in Fig. 6. Thus, for $\Delta V > 0$, the first term on the right side of eqn (1) becomes greater than the second term and the film grows. However, as the film thickens, the voltage drop across the metal/film interface decreases, such that the rate of film growth decreases. Ultimately, the values of the two terms on the right side of eqn (1) become the same and a new steady state is achieved. An important point to note is that the change in film thickness on cycling the applied voltage is not symmetrical, because the decrease in film thickness and the subsequent increase occur via fundamentally different processes (dissolution and oxygen vacancy generation, respectively).

We show in Fig. 7 calculated transients in thickness and current density for potential step polarisation of Alloy C-22 in saturated brine at 90°C, a temperature that is more

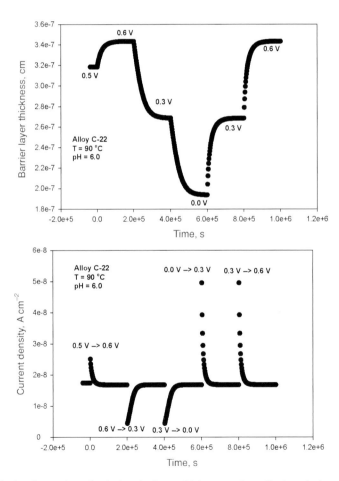

Fig. 7 Calculated transients in the barrier layer thickness and anodic (passive) current density for Alloy C-22 upon stepping the potential in the negative and positive directions. Alloy C-22, T = 90 °C, pH = 6. Voltages are with respect to the Standard Hydrogen Electrode (SHE).

representative of the conditions that are predicted to exist in the repository. Except for minor shifts in the values of the current and thickness, the predicted transients at 90°C differ little from those at 25°C. This is to be expected, because available experimental data on the passive current density suggests no strong dependence on temperature. Finally, while the transients are predicted to persist for considerable time, the times are short compared with the design life of the repository (10 000 years or about 3×10^{12} s). Accordingly we conclude that the accumulated damage due to general corrosion is readily predicted by using quasi steady-state models. This is an important finding, because a more rigorous treatment of the transient film growth problem requires solving the moving boundary (Stefan's) problem, which to date has only been achieved numerically [10].

4. Summary and Conclusions

The findings of this work may be summarised as follows:

1. A model, based on the Point Defect Model (PDM) for the growth and breakdown of passive films on metal surfaces, has been developed for predicting transients in barrier layer thickness and film growth current in response to step changes in potential. The current model is restricted to barrier layers in which the principal defect is the oxygen vacancy, but it is believed to be also applicable to systems where the principal defect is the cation interstitial.

2. The transients in barrier layer passive film thickness can be explained in terms of the relative impact of the voltage step on the rates of the processes that occur at the metal/film and film/solution interfaces.

3. The High Field Model, which is used extensively to describe the growth of passive films on metal surfaces is incapable of accounting for the experimental data obtained in previous studies and must be discarded as a model for predicting the growth of passive films on metal surfaces in HLNW repositories.

4. While the transients are predicted to persist for considerable time, the times are short compared with the design life of the repository and we conclude that the accumulated damage due to general corrosion is readily predicted by using quasi steady-state models.

5. Acknowledgements

The authors gratefully acknowledge the support of this work by the Department of Energy's Yucca Mountain Project through Subcontract A20257JG1S from Innovative Design Technologies, Inc.

References

1. L. Zhang, D. D. Macdonald, E. Sikora and J. Sikora, *J. Electrochem. Soc.*, 1998, **145**, 898.
2. E. J. W. Verwey, *Physica*, 1935, **2**, 1059.
3. N. Cabrera and N. F. Mott, *Rep. Prog. Phys.*, 1948, **12**, 163.
4. L. Young, *Anodic Oxide Films*, Academic Press, New York (1961).
5. D. D. Macdonald, *Pure Appl. Chem.*, 1999, **71**, 951.
6. D. D. Macdonald, M. Al-Rifaie and G. R. Engelhardt, *J. Electrochem. Soc.*, 2001, **148**(9), B343.
7. D. D. Macdonald, E. Sikora and J. Sikora, *Electrochim. Acta*, 1998, **43**, 2851.
8. G. T. Burstein and A. J. Davenport, *J. Electrochem. Soc.*, 1989, **136**, 936.
9. D. D. Macdonald, *Corrosion*, 1992, **48**(3), 194–205.
10. L. Zhang and D. D. Macdonald, *Electrochim. Acta*, 1998, **43**, 2673.

13

A Contribution to the Modelling of Atmospheric Corrosion of Iron

S. HŒRLÉ and F. MAZAUDIER

Laboratoire d'Etude de la Corrosion Aqueuse DEN/SCCME, CEA Saclay F-91191 Gif Sur Yvette Cedex, France

ABSTRACT

With the aim of predicting the long term atmospheric corrosion behaviour of iron, the characteristics of the rust layer formed during this process and the mechanisms occurring inside the rust layer during a wet–dry cycle are considered. A first step in modelling the behaviour is proposed, based on the description of the cathodic reactions associated with iron oxidation : reduction of a part of the rust layer (lepidocrocite) and reduction of dissolved oxygen on the rust layer. The modelling, by including some composition and morphological data of the rust layer as parameters, is able to account for the metal damage after one Wet–Dry cycle.

1. Introduction

Nuclear waste could be packaged in metallic containers and disposed in very long term interim dry storage (about 10^2 years). Nevertheless, in such conditions, condensation on the metallic containers cannot be completely prevented (e.g. thermal powerless containers). The containers' walls may then be exposed to cyclic wet and dry periods and will suffer from indoor atmospheric corrosion at room temperature.

Based on the well-known bilogarithmic laws [1,2], the atmospheric corrosion behaviour of iron based materials (e.g. low alloy steels) has generally been well predicted for periods of a few tens of years. For longer time periods, of one to a few centuries, which could be the case for some nuclear waste containers in very long term storage, predictions are rare and probably more uncertain. The main reason of this lack of reliability is the complex mechanisms involving the rust layer during atmospheric corrosion cycles and the subsequent modifications of the protective properties of the rust scale. Mechanistic modelling of these phenomena appears then to be necessary to establish more relevant and robust prediction of corrosion allowance in the behaviour of materials such as iron or low alloy steels.

Basic considerations on atmospheric corrosion are presented in Section 2, before details of the different stages of atmospheric corrosion process and the corresponding electrochemical phenomena in Section 3. Finally, the modelling of these phenomena is presented in Section 4.

2. Atmospheric Corrosion

Contrary to dry oxidation, atmospheric corrosion in nature is an electrochemical process needing aqueous conditions for its occurrence. As far as iron or low alloy steels are concerned, atmospheric corrosion can be summarised by the stoichiometric equilibrium:

$$4Fe + 3O_2 + 2H_2O = 4FeOOH$$

An aqueous layer, which acts as an electrolyte, is formed in indoor conditions by water condensation. The time of wetness, which defines the duration of the electrochemical process, is strongly dependent on many parameters which include the water vapour content of the atmosphere, i.e. the relative humidity (RH) at a given temperature. It is generally considered that atmospheric corrosion of carbon or weathering steels begins at about 60% RH with a very slow rate and increases sharply at 80% ≤ RH < 100% [3]. When 100% RH is reached, an electrolyte thickness of ~100 μm can be observed.

The RH and temperature variations lead to cyclic wet and dry periods, the so-called Wet–Dry cycles (Fig. 1). The Wet–Dry cycles are the key feature of atmospheric corrosion as the alternating wet and dry periods drastically change the rusting mechanisms from those obtaining in bulk aqueous corrosion. Indeed, earlier studies have shown that during a Wet–Dry cycle, the atmospheric corrosion of iron or low alloy steel can be divided into three stages. As initially proposed by Evans [4] and then experimentally investigated by Stratmann *et al.* [5,6], during the first stage (wetting), the anodic dissolution of iron is mainly balanced by the reduction of ferric species within the rust layer and little oxygen is reduced. After the reducible species are used up, the second stage (wet surface) begins, characterised by oxygen reduction

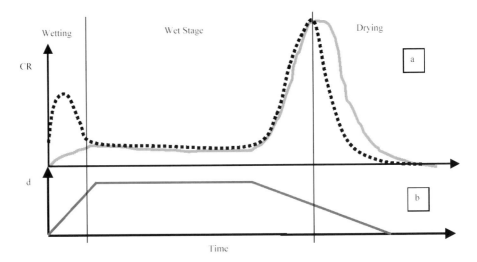

Fig. 1 *The Wet–Dry cycle. In (b) the schematic variations of electrolyte thickness (d) with time. In (a) the corresponding variations of Iron (dotted line) and Oxygen (solid line) consumption rates (CR) (from Stratmann [7]).*

as the major cathodic reaction. Eventually, at the end of the drying, the third stage of the cycle, the species reduced during the first stage and other ferrous compounds produced by the electrochemical corrosion processes are reoxidised by oxygen. Finally, the electrolyte film fully evaporates, so stopping completely the corrosion process. The corrosion rate and the rust layer modifications are thus correlated to the number and frequency of the Wet–Dry cycles. The phenomena occurring during the different stages of the cycle will be described in more detail in the Wet–Dry cycle section.

As shown by this short description of the Wet–Dry cycles, it is mainly the rust layer that is involved in the electrochemical phenomena so determining the transport processes that are indispensable for corrosion to occur. The composition and structure of the rust layer are then of some importance for the development of atmospheric corrosion. One can refer to the paper of Dillmann *et al.* [8] for a more detailed description of the rust layer and a presentation of the techniques used for its characterisation. A real rust layer is a very complex object (Fig. 2a). It is mainly composed of iron oxy-hydroxides and some other oxides such as lepidocrocite (γ–FeOOH), goethite (α-FeOOH) and magnetite (Fe_3O_4) — to mention only the most common. The relative fraction of these compounds is an important factor in determining the protective ability of the rust layer. Lepidocrocite, γ-FeOOH, is a semi-conducting and electrochemically active species, α-FeOOH is insulating and non-active [9,10] and Fe_3O_4, although a good conductor is considered as protective because of its compactness and thermodynamic stability [11,12]. For purposes of simplicity, a rust layer composed of only two species, a protective (say α-FeOOH) and an active (say γ-FeOOH) will be considered for the modelling. The composition of the rust layer and, to a certain extent, its protective ability, are characterised by the composition ratio α/γ, which we call the Protective Ability Index. For ancient rusts, the Protective Ability Index ranges from about 1 to 5 [8]. Obviously, some pollutants are also present in the rust layer, either from the external atmosphere (dust particles incorporated during the rust growth or sulfur compounds coming from industrial fumes, e.g. SO_2) or from the underlying metal (Cu, P, Cr,...). To begin with, they will

(a)   (b)

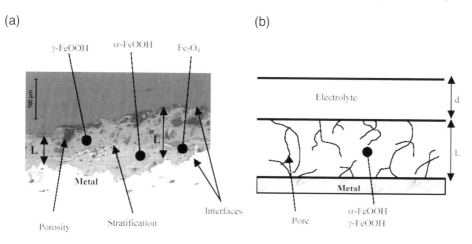

Fig. 2 The rust layer. (a) SEM micrograph of a 150 years old rust layer (CC01 sample, see ref. [8] for more detail). (b) Schematic representation of the rust layer used for modelling. L and d are the rust layer and electrolyte thickness, respectively.

not be considered, although they may have dramatic impacts on rusting [3, 4, 13–17]. The modelling will therefore be restricted to atmospheric corrosion of pure iron in an atmosphere free from any pollutant.

The protective ability of the rust layer is also determined by its structural properties. Indeed, the corrosion processes are governed by mass transport through the rust layer and by the surface area of the pores. Given the mean thickness derived from a few 10-year-old rust layers (~100 μm), it would seem that the transport phenomena probably depend mainly on diffusion and migration in the electrolyte through the pore system of the rust. The measured porosity of old rust layers is a few percent (~10%), made mainly of tortuous pores with nanometric widths. These old rust layers are quite protective as a result of their compactness, which is not necessarily the case with fresh rust layers which are far more likely to be porous. When active species, such as dissolved oxygen, have penetrated the rust layer, then the electrochemical reactions take place on the surface of the pores (see Wet–Dry cycle section). The specific area of the rust layer is then another important parameter. The measured specific area of rust layers is very large, i.e. about 20 m^2 g^{-1} (BET measurements on 420 year old rust powder), providing a potentially very large area for electrochemical reactions (the cathodic, see Wet–Dry cycle section). Thus, the commonly used porosity description (e.g. the cylindrical pores system) is unable to represent well the complex geometrical shape of the pores (Fig. 3) which is actually that of a large specific area associated with nanometric scale pores and low porosity volume (10%).

Just to complete the picture, since it will not be exploited further in the modelling, it can be mentioned that often 2 or 3 layers may be distinguished in a rust scale, differing in their relative amounts of oxides and oxy-hydroxides and porosity. Roughly, the first layer being more rich in α-FeOOH and compact (thus protective) and the second layer more γ-FeOOH rich and porous (thus non-protective). Sometimes, a third layer, mainly composed of incorporated dusts and very porous (with pores width many orders of magnitude greater than those of the rust strictly speaking) is also observed. Thus, the main parameters that appear to determine the properties of the rust layer with respect to atmospheric corrosion are: the composition

Fig. 3 *SEM surface micrographs illustrating the complex geometry of the rusts layers.*

(Protective Ability Index α/γ), the porosity (ε), the tortuosity factor of the pores (τ)* and the specific area of the rust (Sa). The parameters α/γ, ε and Sa can be measured [8,7,12,18], whereas τ is evaluated ($\tau \sim 0.1$) [32]. For the modelling, the rust layer will be considered as a layer of constant thickness L, with an homogeneous composition given by α/γ and morphological properties ε, τ and Sa that are independent of space variables (Fig. 2b).

3. The Wet–Dry Cycle

As presented in the preceding section, the main feature of atmospheric, compared to aqueous bulk corrosion, is the succession of Wet–Dry cycles. The cycles are divided into the 3 stages: wetting, wet stage and drying. These stages are characterised by their state of wetness and by their different electrochemical and chemical behaviour.

3.1. Wetting Stage

During the wetting, i.e. in the electrolyte build up, the anodic dissolution of iron begins. The electrons produced by this oxidation have to be consumed by a reduction reaction. Dissolved oxygen is obviously available as an oxidiser. Nevertheless, oxygen cannot be reduced on the rust layer, as it is not conducting (see Atmospheric Corrosion section). Furthermore, the metal surface left in contact with electrolyte by the rust layer is, in one hand, very small and, in other hand, difficult to access by oxygen, which has to diffuse through the thick rust layer (~ 100 μm) via tortuous and nanometric pores. So oxygen reduction cannot provide the large corrosion rates observed during the wetting stage [5]. Another available oxidiser is γ-FeOOH, present in the rust layer. One of the proposed reduction reaction [4,19–21] for γ-FeOOH is:

$$\gamma\text{-FeOOH} + e^- + H^+ \rightarrow \gamma\text{-Fe.OH.OH} \tag{I}$$

According to Stratmann [19,20], the reduced lepidocrocite (γ-Fe.OH.OH), keeps the same crystallographic structure as its mother phase γ-FeOOH, Fe[II] being in Fe sites and OH^- being in O sites, as long as the percentage of Fe^{2+} in the lattice is less than 2–4% [19] at which point Fe_3O_4 begins to form. Reaction (I) requires electrons and can start at the metal/rust interface with the reduction front propagating through the rust layer as γ-Fe.OH.OH is an electronic conductor. Reaction (I) also needs H^+ ions from the electrolyte and these reach only a few monolayers on the γ-FeOOH crystals surface inside the rust layer [19,22]. The species γ-Fe.OH.OH is also an ionic conductor that permits rather rapid movement of H^+ ions and, as the only mobile species involved are H^+ ions, the reduction reaction may be reversible [19,23]. The reduction reaction goes on over all the γ-FeOOH reducible surface and reduced γ-FeOOH finally coats the rust layer (Fig. 4). This depends on the porosity of the rust layer, the surface of the pores and the amount of γ-FeOOH on this surface. It should

---

*In this paper, following [32], the tortuosity factor is defined as $\tau = \left(\dfrac{l_0}{l}\right)^2$, where l_0 is the straight line length of a pore and l its actual length.

be pointed out here that the fraction of lepidocrocite in the rust layer and thus the γ-FeOOH amount on the surface of the pores (which is described by γ/α when the rust layer is considered to be formed of only γ-FeOOH and α-FeOOH) characterises the ability of the rust layer to be reduced, and conversely its protective ability. During the wetting stage, it is the rust layer itself that is responsible for corrosion, the anodic dissolution being balanced by the reduction of γ-FeOOH within the rust layer.

3.2. Wet Stage

When a large part of the available γ-FeOOH is reduced to the conducting γ-Fe.OH.OH, it coats the surface developed by the rust layer porosity. If a connection between the underlying metal exists then the reduced γ-Fe.OH.OH on the surface of the pores can act as a cathodic area. Oxygen can then be reduced on that surface. Cathodic and anodic areas are decoupled, the oxidation of iron taking place on the small metal area in contact with the electrolyte at the bottom of the pores and the reduction reaction on the large cathodic area formed by the reduced γ-FeOOH (Fig. 5). Moreover, there is experimental evidence that this corrosion process is under diffusional cathodic control [6,21,24–26]. As long as oxygen is provided (which is obviously the case for atmospheric corrosion), this process persists. However, it requires that the first stage is completed to a minimal extent. The actual mechanism of oxygen reduction is somewhat complicated and has not yet been clearly resolved in the presence of iron oxides [22,27,28]. But it can be said that oxygen reduction efficiency depends on the doping level of γ-FeOOH with Fe^{2+} ions. For modelling purposes, the detailed mechanisms will not be considered, but only the overall reduction reaction.

3.3. Drying Stage

When it dries, the electrolyte thickness decreases, which reduces the oxygen diffusion path and thus leads to an increasing cathodic current and an enhanced iron dissolution rate. But, in the same time, the decrease in electrolyte thickness increases the

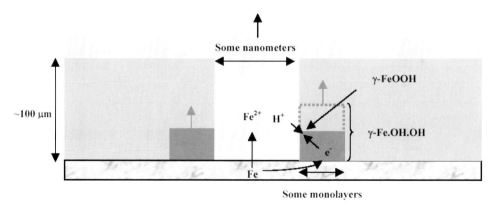

Fig. 4 *Schematic representation of the phenomena occurring during wetting.*

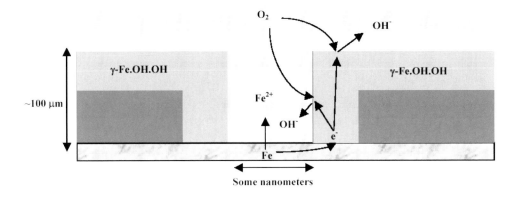

Fig. 5 *Schematic representation of the phenomena occurring during the wet stage.*

concentration of species dissolved during corrosion, namely Fe^{2+} ions, which can precipitate and block anodic and some cathodic sites and thereby lead to a kind of passivation-like state. These phenomena are observable as soon as the electrolyte layer thickness is below about 10 µm [6,25]. At the same time, the increase in cathodic current polarises the rust layer to more positive potentials, allowing oxidation of ferrous species. The γ-FeOOH species is regenerated from γ-Fe.OH.OH by the reverse (I) reaction, thus allowing another cycle to occur when the rust is wetted again. Some γ-FeOOH may also be formed from $Fe(OH)_2$ precipitates, which increases the protective ability of the rust, cycle after cycle. Moreover, by diminishing the amount of surface γ-Fe.OH.OH sites, the oxidation process also decreases the cathodic area, which lessens the current (Fig. 6).

To sum up, there is an activity peak during the drying. First, the corrosion rate is increased by the diminution of the dissolved oxygen diffusion path, and then it is

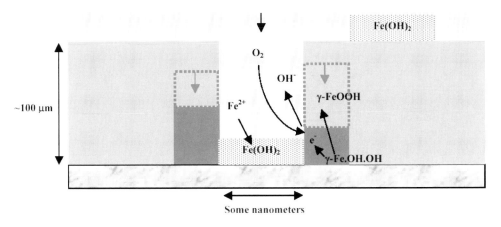

Fig. 6 *Schematic representation of the phenomena occurring during drying.*

decreased both by precipitation of dissolved species and re-oxidation of conducting γ-Fe.OH.OH into insulating γ-FeOOH. During the last instant of the drying stage, there is no more corrosion, but modifications of the rust layer composition and structure can occur that may dramatically change its protective properties and this can be critical for long term behaviour.

It appears that the rust layer takes an important part in the atmospheric corrosion process. Composition (particularly γ-FeOOH to α-FeOOH ratio) and morphological properties (porosity, that controls transport processes and the area where electrochemical reactions take place) mainly determine the protective ability of the rust layer and so have been thoroughly investigated [8] to allow robust modelling. Whereas all the parameters are highly correlated and the phenomena described above may, during intermediate stages, occur simultaneously, a typical Wet–Dry cycle will be considered for the modelling. This cycle is composed of a wetting stage with only γ-FeOOH reduction, a wet stage when only oxygen is reduced (including then a part of the actual drying) and a drying stage, more complex, that corresponds to the competition between increased oxygen reduction rate and precipitation/passivation leading to the extinction of corrosion processes.

4. Modelling

The modelling of atmospheric corrosion that is proposed is established on a modular concept. A first phenomenological module based on thermohydraulic considerations is described in earlier publications [29,30]. The outputs of this modelling are the Wet–Dry cycle main features (electrolyte layer thickness, time of wetness, wetting and drying average duration) of a stored well-defined metallic container according to climatic fluctuations. The other module is a mechanistic one, which is devoted to describing the phenomena occurring within the rust layer during a cycle. To begin with, only corrosion of pure iron is investigated (without taking into account specificities due, for instance, to alloying elements) and a typical Wet–Dry cycle is considered. The link between the phenomenological and mechanistic modules will be done later. The final aim of this modelling is to provide cycle characteristics given by the phenomenological module, for a number of cycles (corresponding to a certain rusting time). These characteristics are then used in a mechanistic model in order to predict the container walls' damage and the rust layer evolution.

First developments, that are presented here, deal with the processes connected with the wetting and the wet stages.

4.1. Wetting Stage

It is assumed that the rust layer is composed only of γ-FeOOH as the electrochemically active species and α-FeOOH as the inert and protective one. The corrosion behaviour during the wetting stage is controlled by the reduction of γ-FeOOH. The amount of reducible γ-FeOOH depends on the area of the pores, the composition of the rust layer (fraction of γ-FeOOH) and the thickness of γ-FeOOH that can be reduced. The area of the pores is given by

$$S = Sa(1-\varepsilon)\rho L \tag{1}$$

where Sa is the measured specific area, ε the porosity, ρ the mean density of oxides and L the thickness of the rust layer.

The density of γ-FeOOH surface sites can be evaluated as follows:

$$S_\gamma = \beta.\left(\frac{\rho}{M_\gamma}N_{Av}\right)^{\frac{2}{3}} \tag{2}$$

with M_γ the molar weight of γ-FeOOH and α-FeOOH and β the surface fraction of γ-FeOOH.

The available amount of γ-FeOOH for reduction is thus given by $\dfrac{S_\gamma.S.n}{N_{Av}}$, where n is the number of γ-FeOOH monolayers that are reduced. Assuming all the available γ-FeOOH is reduced when the electrons produced by the oxidation of iron: Fe $\rightarrow$ $Fe^{2+} + 2e^-$ are consumed, the damage (or corrosion depth) is given by:

$$p_\gamma = \frac{S_\gamma.S.n}{2N_{Av}}V_{Fe} \tag{3}$$

(with V_{Fe} the molar volume of iron) and the mean current density for a wetting stage that lasts Δt_w is then:

$$i_\gamma = \frac{F.S_\gamma.S.n}{\Delta t_w N_{Av}}V_{Fe} \tag{4}$$

Relations (3) and (4) are clearly overestimations as they consider the worst case. In real life, it is likely that γ-FeOOH is not homogeneously dispersed in the rust layer (thus the surface density is different from the inner one) and all the surface γ-FeOOH is not in electrical contact with the metal and thus will not be reduced (there are more likely to be γ-FeOOH islands than a continuous coating). When taking ρ as the mean density of α-FeOOH and γ-FeOOH (i.e. $\rho = 4.2 \; 10^6$ g m^{-3}) and assuming the surface γ-FeOOH fraction equals the bulk one given by characterisation methods [8] and that there are 2 γ-FeOOH monolayers reduced (according to some authors [22]), a maximum damage of 0.33 μm per cycle and a mean current $i_\gamma = 0.12$ mA cm^{-2} (for a wetting stage of 2 h) are obtained. These are quite reasonable values and accord well with published values [14,17]. No kinetic considerations were made for this description. However, the time necessary to reduce a sufficient amount of γ-FeOOH is critical as it determines the onset of oxygen consumption (see Atmospheric Corrosion section). The point is to find out the rate determining step of reaction (I). There are several possibilities, due to H$^+$ ions (penetration in the rust, diffusion in the γ-FeOOH lattice and reaction in an O site) or electrons (leap 'diffusion' in γ-Fe.OH.OH or electron transfer for the reductions of Fe[III]). This has not yet been resolved.

4.2. Wet Stage

When a sufficient amount of surface γ-FeOOH is reduced (wetting stage), the oxygen reduction becomes the prevailing cathodic reaction [31]. Indeed, oxygen can now be reduced on the conducting γ-FeOH•OH that coats the surface of the pores. Atmospheric oxygen is dissolved at the atmosphere/electrolyte interface. It is then supposed that the dissolved oxygen diffuses first through the electrolyte and then through pores of the rust (Fig. 7).

Oxygen is then reduced on the γ-Fe.OH.OH surface following the cathodic reaction:

$$O_2 + 4e^- + 2H_2O \rightarrow 4OH^- \tag{II}$$

Reaction (II) leads to an oxygen consumption which depletes oxygen concentration in the pores of the rust layer. The modelling is established as follows. Only diffusion of dissolved oxygen, first through the electrolyte and then the pores, where there is an oxygen sink (the consumption of O_2 by reaction (II)) is considered. Assuming that ε, τ and D (oxygen diffusion coefficient) are independent of space variables, the 1-D mass balance equation for diffusive oxygen mass transport is, in the electrolyte:

$$\frac{\partial C}{\partial t} = D\frac{\partial^2 C}{\partial x^2} \tag{5}$$

and in the rust layer [32]:

$$\varepsilon\frac{\partial C}{\partial t} = \tau\varepsilon D\frac{\partial^2 C}{\partial x^2} - \beta.s_a V_R \tag{6}$$

V_R is the rate of oxygen reduction on γ-Fe.OH.OH surface

$$(V_R(x) = k.C(x) \tag{7}$$

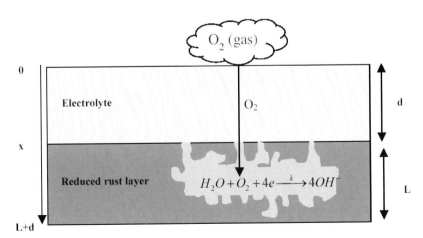

Fig. 7 Schematic representation of the oxygen reduction modelling. The electrolyte and the rust layer thicknesses are d and L respectively.

with the approximation of a first order reaction) and s_a the surface area developed by the pores per volume unit

$$s_a = Sa(1-\varepsilon)\rho \tag{8}$$

When it is assumed that all the γ-FeOOH was reduced during the wetting stage, then the area on which oxygen is reduced will be $\beta\, s_a$ per volume unit. When the stationary state is reached, (5) and (6) are solved for the oxygen concentration in the pores:

$$C(x) = \frac{C_0}{1+d/\lambda}\exp(-\frac{1}{\lambda}(x-d)) \quad x \in [d,\ L+d] \tag{9}$$

With the boundary condition: C_0 is the concentration of dissolved oxygen at the electrolyte/atmosphere interface (C_0 is given by Henry's law), and continuity conditions at the electrolyte/rust layer interface. As with the parameter λ, it characterises the depth of oxygen penetration inside the rust layer and is defined by:

$$\lambda = \sqrt{\frac{\tau\varepsilon\ D}{\beta\ s_a\ k}} \tag{10}$$

Then, integrating (7) with (9) through all the rust layer and using Faraday's law, the oxygen reduction current is given by the relation:

$$i_o = \frac{4F\tau\varepsilon\ DC_0}{d+\lambda}(1-\exp(-\frac{L}{\lambda})) \tag{11}$$

Equation (11) evaluates the oxygen reduction current as a function of parameters describing the morphology of the rust layer (ε,τ,L) and parameters that are more related to the electrolyte properties (D,C_0,d). A value of oxygen reduction current of 0.018 mA cm^{-2} is obtained using literature and experimental values of the parameters. It is of the same order as published values [14,17,25]. The corresponding damage p_o

$$p_o = \frac{i_o}{2F}V_{Fe} \tag{12}$$

for a wet stage of 5 h is 0.12 μm, which is also quite reasonable.

As expected, λ is governed by competition between transport ($\tau.\varepsilon.D$) and oxygen consumption processes ($\beta.s_a.k$). From experimental and literature values of these parameters, λ is evaluated as about 0.1 μm. The value of λ is then very small compared to the rust layer thickness ($L\sim100$ μm) [8]. The dissolved oxygen almost does not penetrate inside the rust layer and is reduced at the extreme surface. Furthermore, λ is also small compared to the electrolyte thickness. Indeed, for the validity of the modelling, the electrolyte thickness should be greater than about 10 μm; first to prevent precipitation processes and passivation (see Wet-Dry cycle Section 3) and secondly, because under a very thin electrolyte layer it appears that the rate determining step of oxygen reduction is no longer diffusion but solvation [25]. Thus, in the validity range of the modelling, λ remains very small compared to the electrolyte layer and rust scale thicknesses and (11) can be simplified to:

$$i_o = \frac{4F \ \tau\varepsilon \ DC_0}{d} \tag{13}$$

which is, except for the morphological factor $\tau\varepsilon$, the expression of the limiting current that would have been obtained when considering diffusion limited oxygen reduction

Table 1. *Values used for the parameters*

Parameter	τ	ε	D	β	s_a	k
Value	0.1	0.1	$1.9 \ 10^{-9} \ m^2 \ s^{-1}$	0.2	$7.38 \ 10^7 \ m^{-1}$	$10^{-5} \ m \ s^{-1}$

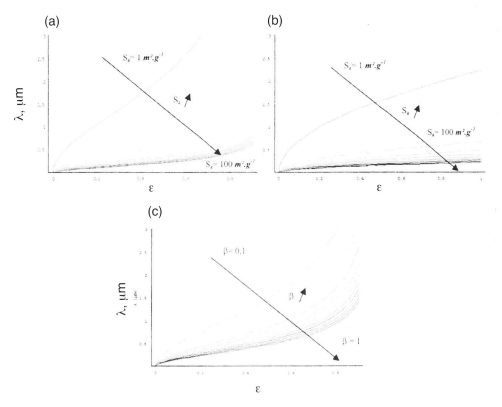

Fig. 8 *Effect of different parameters on the depth of oxygen penetration (λ). (a) Variation of λ with ε for different values of S_a. (b) Variation of λ with ε for different values of S_a. (c) Variation of λ with ε for different values of β (with $\varepsilon = 0.1$, $\tau = 0.1$, $S_a = 20 \ m^2 \ g^{-1}$, $\beta = 0.2$ when they are considered as parameters).*

on a plane. It agrees well with some experimental results showing a linear variation

of oxygen reduction current with $\dfrac{1}{d}$ [6, 21, 24–26]. It should be noted that for a range

of reasonable values of the parameters, the value of λ remains very small compared to both d and L (Fig. 8). The rate constant k, obviously depends on potential, as (2) is an electrochemical reaction. Nevertheless, it has been shown [27] that, for the range of potential at which atmospheric corrosion occurs, the value of k remains of the same order of magnitude ($k \sim 10^{-5}$ m s^{-1}). Thus the approximations leading to eqn (13) are applicable for typical atmospheric corrosion conditions.

To be more accurate, as long as the electrolyte thickness is greater than the diffusion layer width δ (Nernst diffusion layer), it is not d, the value of the electrolyte thickness that should be used in (11) and (13) [21,24–26,33], but δ. The current then remains constant during the wet stage and at the beginning of the drying. The width δ is replaced by d when the electrolyte thickness becomes smaller than δ and leads to an increase of the reduction current. Equations (11) and (13) are then applicable for the beginning of the drying, until d reaches values of about 10 μm, when precipitation/passivation begins and the cycle enters its last stage, which is not described by this modelling. The main conclusion of this first approach of oxygen reduction modelling is that there is nearly no penetration of the dissolved oxygen inside the rust layer. In fact, as long as all the rust layer is reduced, it is only the extreme surface that is involved in oxygen reduction. When the rust layer is partially reduced (only the inner part), then the diffusion of oxygen through the pores has to be examined in more detail as their morphological properties are very complex [31].

5. Conclusions

A first step in modelling the complex processes occurring during a Wet–Dry cycle is proposed. It is based on mechanistic considerations, so as to provide more reliable support than empirical extrapolations for long term predictions. The guiding principle in the modelling is to focus on the different cathodic reactions associated with the wetting and wet stages of the Wet–Dry cycle, namely γ-FeOOH and oxygen reduction. This modelling, although quite basic, gives some pointers for predicting the long term behaviour of the atmospheric corrosion of iron and, on the other hand, is also a very fruitful tool for investigating the mechanisms involved. Moreover, the description of the rust layer, necessary for establishing the modelling, has led to a wide characterisation effort [8] which has provided a better knowledge of the composition and structure of the rust layer.

The next step that will be considered is to conduct some electrochemical experiments on rust layers in order to acquire data about kinetics of electrochemical reactions that occur during Wet–Dry cycles and mass transport as well as to test this first modelling.

6. Table of Symbols

For calculated values, the numbers in brackets give the formula used.

LATIN LETTERS

C	O_2 concentration	calculated (9)	mol m^{-3}
C_0	O_2 concentration at atmosphere/electrolyte	table	0.25 mol m^{-3}
D	O_2 diffusion coefficient in the electrolyte	table	1.9 10^{-9} m^2 s^{-1}
d	electrolyte thickness	measured [33]	~100 µm and > 10 µm
F	Faraday constant	constant	96 500 C mol^{-1}
i_o	O_2 reduction current	calculated (11,13)	0.018 mA cm^{-2}
i_γ	γ-FeOOH reduction current	calculated (4)	0.12 mA cm^{-2}
k	rate constant of O_2 reduction on γ-Fe.OH.OH	measured [27]	10^{-5} m s^{-1}
L	rust layer thickness	measured	~200 µm
N_{Av}	Avogadro constant	constant	6.023 10^{23}
$M\gamma$	molar weight of γ-FeOOH	constant	89 g mol^{-1}
n	number of reduced monolayer of γ-FeOOH	evaluated	2
p_o	damage due to O_2 reduction	calculated (12)	0.12 µm
p_γ	damage due to γ-FeOOH reduction	calculated (3)	0.33 µm
S	surface of the pores per surface unit	calculated (1)	15 120
S_a	specific area of the rust	measured	20 m^2 g^{-1}
s_a	surface of the pores per volume unit	calculated	7.38 10^7 m^{-1}
$S\gamma$	density of γ-FeOOH surface site	calculated (2)	1.86 10^{18} site m^{-2}
V_{Fe}	molar volume of iron	constant	7.1 10^{-6} m^3 mol^{-1}
V_R	rate of oxygen reduction	calculated (7)	mol.m–2.s^{-1}

GREEK LETTERS

β	fraction of γ-FeOOH	measured	0.2
Δt_w	duration of the wetting stage	evaluated	2 h
ε	porosity of the layer	measured	0.1
λ	depth of O_2 penetration	calculated (10)	0.1 µm
ρ	mean density of the oxides	table/evaluated	4.2 10^6 g m^{-3}
τ	tortuosity of the rust layer	evaluated	0.1

References

1. M. Pourbaix, The linear bilogarithmic law for atmospheric corrosion, in *Atmospheric Corrosion* (W.H. Ailor, ed.), p.107–121. John Wiley & Sons, New York, USA, 1982.

2. M. Pourbaix and A. Pourbaix, *Corrosion*, 1989, **45**, 71–83.

3. T. E. Graedel and R. P. Frankenthal., *J. Electrochem. Soc.*, 1990, **137**, 2385–2394.

4. U. R. Evans and C. A. J. Taylor, *Corros. Sci.*, 1972, **12**, 227–246.

5. M. Stratmann, *Corros. Sci.*, 1987, **27**, 869–872.

6. M. Stratmann *et al.*, *Corros. Sci.*, 1990, **30**, 681–696, 697–714, 715–734.

7. M. Stratmann, *Ber. Bunsenges. Phys. Chem.*, 1990, **94**, 626–639.

8. Ph. Dillmann, V. Vigneau, F. Mazaudier, C. Blanc and S. Hœrlé, this volume, pp.316–333.

9. T. Misawa, K. Hashimoto and S. Shimodaira, *Corros. Sci.*, 1974, **14**, 131–149.

10. D. C. Cook, S. J. Oh, R. Balasubramian and M. Yamashita, *Hyperfine Interact.*, 1999, **122**, 59–70.

11. T. Misawa, K. Hashimoto and S. Shimodaira, *Corros. Sci.*, 1974, **14**, 131–149.

12. K. Kaneko and K. Inouye, *Bull. Chem. Soc. Japan*, 1974, **47**, 1139–1142.

13. T. Misawa, T. Kyuno, W. Suëtaka and S. Shimodaira, *Corros. Sci.*, 1971, **11**, 35–48.

14. M. Stratmann, K. Bohnenkamp and T. Ramchandran, *Corros. Sci.*, 1987, **27**, 905–926.

15. T. Nishimura, H. Katayama, K. Noda and T. Kodama, *Corros. Sci.*, 2000, **42**, 1611–1621.

16. M. Yamashita, H. Miyuki, Y. Matsuda, H. Nagano and T. Misawa, *Corros. Sci.*, 1994, **36**, 283–299.

17. T. Kamimura and M. Stratmann, *Corros. Sci.*, 2001, **43**, 429–447.

18. M. Yamashita, H. Miyuki, Y. Matsuda, H. Nagano and T. Misawa, *Corros. Sci.*, 1994, **36**, 283–299.

19. M. Stratmann and K. Bohnenkamp, H. J. Engell, *Corros. Sci.*, 1983, **23**, 969–985.

20. M. Stratmann and K. Hoffmann, *Corros. Sci.*, 1989, **29**, 1329–1352.

21. A. Cox and S. B. Lyon, *Corros. Sci.*, 1994, **36**, 1177–1192.

22. E. R. Vago, E. J. Calvo and M. Stratmann, *Electrochim. Acta*, 1994, **39**, 1655–1659.

23. A. Hicklin and D. J. G. Ives, *Electrochim. Acta*, 1975, **20**, 63–69.

24. S. H. Zhang and S. B. Lyon, *Corros. Sci.*, 1993, **35**, 713–718.

25. A. Nishikata *et al.*, *J. Electrochem. Soc.*, 1997, **144**, 1244–1252.

26. M. Yamashita, H. Nagano and R. A. Oriani, *Corros. Sci.*, 1998, **40**, 1447–1453.

27. E. R. Vago and E. J. Calvo, *J. Electroanal. Chem.*, 1995, **388**, 161–165.

28. E. R. Vago and E. J. Calvo, *J. Chem. Soc. Faraday Trans.*, 1995, **91**, 2323–2329.

29. M. Baklouti, N. Midoux, F. Mazaudier and D. Feron, *J. Hazard. Mater.*, 2001, **B85**, 273–290.

30. F. Mazaudier, D. Feron, M. Baklouti and N. Midoux, *J. Phys.*, 2001, **11**, 259–266.

31. M. Stratmann and J. Müller, *Corros. Sci.*, 1994, **36**, 327–359.

32. F. King, M. Kolar and D. W. Shoesmith, *Corrosion '96*, Paper No. 380, NACE, Houston, TX, USA, 1996.

33. H. Nagano, T. Doi and M. Yamashita, *Mater. Sci. Forum*, 1998, **289–292**, 127–134.

14
Model for Low Temperature Oxidation during Long Term Interim Storage

C. DESGRANGES, A. ABBAS and A. TERLAIN

Service de la Corrosion et du Comportement des Matériaux dans leur Environnement CEA-Saclay/Bâtiment 458, 91 191 Gif sur Yvette Cedex, France

ABSTRACT

Low-alloyed steels or carbon steels are considered as candidate materials for the fabrication of some nuclear waste package containers for long term interim storage. The containers are required to remain retrievable for centuries. One factor limiting their performance on this time scale is corrosion. The estimation of the metal thickness lost by dry oxidation over such long periods requires the construction of reliable models from short-time experimental data.

In a first step, models based on simplified oxidation theories have been derived from experimental data on iron and a low-alloy steel oxidation. Their extrapolation to long oxidation periods confirms that the expected damage due to dry oxidation could be small. In order to improve the reliability of these predictions advanced models taking into account the elementary processes involved in the whole oxidation mechanism, are under development.

1. Introduction

1.1. Context

For high level nuclear waste containers in long term interim storage, dry oxidation will be the main degradation mode. The reason is that, for this kind of waste, the temperature at the surface of the containers will be high enough to avoid any condensation phenomena for several years. Even if the scale growth kinetics is expected to be very slow since the temperature will be moderate at the beginning of the storage and will keep on decreasing, the metal thickness lost by dry oxidation over such a long period must be evaluated with a good reliability. To achieve this goal, modelling of the oxide scale growth is necessary and this is the aim of the dry oxidation studies performed in the frame of the COCON programme.

1.2. Oxidation Theories

All existing oxidation models are based on the two main oxidation theories developed by Wagner [1] between the 1930s and 1970s on the one hand and by Cabrera and Mott [2] in the 1960s and next by Fromhold [3] on the other hand. These used to be associated with high temperature behaviour for Wagner's theory and with low temperature for the second. Indeed it is certainly more relevant to consider their range of application in terms of the oxide scale thickness rather than in terms of temperature.

Wagner's theory relies on the assumption that the scale growth is governed by thermo-activated diffusion phenomena in the scale. It leads to a simple law for the scale growth kinetics (parabolic law) only for the stationary state. Wagner shows that the parabolic constant can then be evaluated from the self diffusion coefficient assuming that transport in the scale is dominated by one kind of defects. Models based on Wagner's theory have been successfully applied in the case of high temperature oxidation and for scales of sufficient thicknesses. For thinner scales, electric field created by oxygen adsorbed at the surface becomes very high and Cabrera and Mott's theory is more suitable. In fact, in this theory, the electron tunnel current through the oxide associated with the cation injection into the oxide is considered as the limiting reaction rate for the partial steps for oxidation. Fromhold [3] has developed further the Cabrera and Mott's theory by taking into account the effect of space charge in the growing film. He has demonstrated that it can enhance or retard the transport of species through the oxide film, and hence modify the growth rate.

1.3. Modelling for Long Term Interim Storage

Some basic models can be straightforwardly deduced from the application of oxidation theories with various simplifying assumptions such as considering a stationary state and/or assuming only one elementary process as the oxide growth limiting step. In such conditions, the oxidation rate can often be described by an analytical equation [4]. As mentioned above, in the case of high temperature oxidation models parabolic kinetics are assumed according to Wagner's resolution for the stationary state. These kinds of basic models well explain various observed types of behaviour. They are useful tools to correlate experimental kinetic measurements with 'theoretical values' of interest (like an activation energy for example) by constant rate determination. However, experimental results sometimes suggest that taking into account only one elementary mechanism to describe the kinetics of the whole oxidation process is not always sufficient to explain some types of behaviour. This is not surprising in view of the complexity of the scale morphologies sometimes encountered even in the simplest systems like pure Ni [5].

Advanced models based on numerical simulations have been developed in the 1990s in order to take into account several growth processes. Martin and Fromm [6,7] have developed a powerful model for a thin scale formed at low temperature. Different partial steps are described: reaction rate and electronic flux are calculated. For higher temperature oxidation models, various attempts at connecting microstructure and oxidation rate have been made considering processes like the effect of stress on diffusion [8], grain boundary diffusion [9], etc. One should notice here that these advanced models are very powerful tools but still in development and that their implementation cannot be compared to that of the basic models.

In the case of dry oxidation of waste containers, which theory should an appropriate model rely on? It can be expected that the oxide scale could have a thickness ranging from a few tens of nanometers up to a few tens of micrometers depending on temperature and duration. This wide range of oxide thicknesses suggests that an appropriate model will have to include the processes involved not only in the growth of a thin film but also those occurring in the thick oxide scale growth. For the thicker scales, Wagner's theory should be applied, as the rate limiting step will surely be the transport through the scale. At these temperatures, the diffusion along grain boundaries is often advanced as the main transport mechanism. However, growth mechanisms involved in the Cabrera and Mott theory , as well as space charge effects could be the dominant processes for thinner scales.

We present in this paper our attempt to apply two quite different approaches to model the dry oxidation of waste package containers in long interim storage. We shall first present some basic models developed from experimental data and then the basis of an advanced model and its first developments.

2. Basic Model

As mentioned above, for a scale of a sufficient thickness, a basic model consists in assuming that the oxidation rate is parabolic. Thanks to experimental data, parabolic constant as a function of the temperature can be evaluated and used to build a model through simple extrapolation.

2.1. Adjustment of Model Parameters on Experimental Data

Only a few studies dealing with iron and steel oxidation in the temperature range concerned by containers in long term interim storage (between 373 and 623K) have been reported in the literature. Moreover, most of these studies are focused on the first stages of the film growth and are therefore performed under a very low oxygen partial pressure and for only few hours. To handle this lack of data, a part of the COCON program presented elsewhere in this review [10] has consisted of some oxidation tests with iron and a low-alloy steel performed in wet air (2%vol. H_2O) (reference atmosphere) under atmospheric pressure for several hundred hours. We used these results to build a first basic model.

Hence, assuming that the oxide scale growth follows a parabolic law ($w^2 = K_p\, t$), the constant K_p has been evaluated from mass gain measurements (w) or oxide scale thickness evaluation (assuming a dense oxide scale with density of 5.2 mg cm^{-3}, typically Fe_2O_3 is formed) after 260 h oxidation at four temperatures in the 373–673K temperature range. An apparent activation energy E_a is deduced from exponential adjustment of the K_p variation with inverse temperature:

$$K_p = \alpha \exp[-E_a/kT].$$

For iron, this leads to an apparent activation energy of 97.5 kJ mol^{-1} if all the results are taken into account and 127 kJ mol^{-1} without the 373K value (see Fig. 1). This latter value is in good agreement with activation energies reported in the literature for shorter time measurements.

The discrepancy between the two adjusted values of the activation energy is not surprising. In fact, at the lowest experimental temperature (373K), a scale thickness of about 13 nm has been evaluated by XPS measurement. With a depth of this order

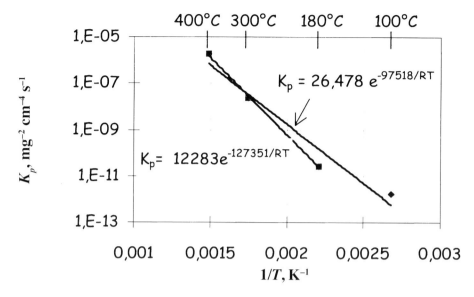

Fig. 1 *Variations of the parabolic constant of iron oxidation with inverse temperature.*

of magnitude, the electric field in the scale and interface phenomena would play an important role in the oxidation kinetics and could not be neglected in the interpretation of the oxidation tests performed at this temperature. This is the case in the basic model proposed here.

2.2. Extrapolation over Long Time Interim Storage Duration

For iron, the kinetic data obtained at 573K have been extended to a 100 year period. This calculation leads to the formation of a 60 μm thick oxide. However, as the oxide layer grows, mechanical stresses build up and lead to the formation of defects like cracking or spalling which decrease the protectiveness of the layer. If we assume that the layer completely spalls away as soon as its thickness reaches a critical value of 10 μm (which is a reasonable value because dense and adherent iron oxide layers with such a thickness have already been observed) and the oxidation kinetics remains the same, the total quantity of oxide formed for 100 years corresponds to a 130 μm oxide thickness. Moreover, from our data, we can expect the formation of an oxide layer thinner than 1μm due to dry oxidation at 373K under air with 2% H_2O. These calculations show that dry oxidation of iron could be very low in a 100-year period but they assume that the corrosion mechanism is the same for all the storage period and in particular that there are no changes in the temperature and atmosphere composition (no pollutant).

2.3. Assessment of the Basic Model Reliability

For the COCON programme, oxidation kinetics have also been measured by thermogravimetry during long-time experiments (about 700 h for iron and 300 h for the low-alloyed steel) using a microbalance. For iron, for the first 300 h the mass gain variations for the oxidation at 573K under the reference atmosphere obeys a parabolic law in good agreement with the previous adjusted model. However, the whole experiment cannot be correctly described by a parabolic fit. We can see that the extrapolation of the parabolic law deduced from the first 300 hours of 688 h leads to a overestimation of the mass gains (Fig. 2) but that a direct logarithmic plot well reproduces the entire measured oxidation kinetics.

In the literature, the reported mass gain variations of iron during oxidation in dry air or in oxygen at temperatures higher than 573K often follow parabolic laws [11–13]. However, in the 453–573K temperature range, these are also logarithmic [14–16] for the shortest time experiments. Sometimes, a change of law with time has also been observed. One interpretation for this change, is that an Fe_3O_4 film is first formed and then αFe_2O_3 nucleates and grows to form a film. As soon as αFe_2O_3 covers the Fe_3O_4 film, the oxidation rate decreases. However, this change has only been observed in the first hours of oxidation. Direct logarithmic ($x = k_0 \ ln(t/\tau+1)$) behaviour can also be explained by the Mott theory associated with electron transfer into the scale as the limiting step for oxidation. Hence, from results of Davies and Evans [17] obtained for iron oxidation performed at 423 and 523K for 15 min to 8 h, Uhlig [18] has calculated an activation energy of 29 kJ mol^{-1} for the Arrhenius dependence of the k_0 parameter. Nevertheless, whatever the supposed mechanism advanced in the literature to explain a direct logarithmic law, it always presumes a thin scale, and it therefore cannot explain the behaviour observed here for 700 h oxidation leading to a scale thickness of more than 1μm.

For the low-alloy steel, the measured oxidation kinetics by microbalance under the reference atmosphere at 523 and 573K cannot be reproduced by a parabolic law even for a 300 h long experiment. Indeed they follow a logarithmic law. Extrapolation of this law over a 100-year period for a 573K temperature leads to the formation of a layer about 1 μm thick, which is very low. In a study concerning the repository site

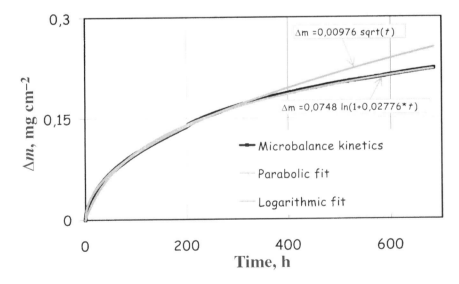

Fig. 2 Weight variations of an iron specimen oxidised at 573K under air with 2% vol. H_2O.

of Yucca Mountain, Larose and Rapp [19] have calculated following a procedure similar to our basic model, the thickness of the oxide scale on carbon steels in dry oxygen at temperatures between 473 and 573K for a 1000 year period from the data of Runk and Kim [20]. They obtained 50 and 3.6 μm thick oxides on a Fe–0.8wt%C steel for oxidation temperatures respectively equal to 573 and 373K. These values become 29 μm and 2.1 μm for a Fe–0.2wt%C steel. These estimations are higher than those predicted at 573K by our logarithmic law but they have been calculated assuming a parabolic law.

In any case, the expected damage calculated from extrapolations following those simple analytical laws is small. However, the reliability of this kind of model is poor, since it is based on the assumption that a single elementary process controls the oxidation rate which, furthermore, is not always in accordance with the measured kinetics. The reason is that in reality, several other phenomena play a major role in the oxidation rate at this temperature. First, as diffusion along grain boundaries can be the major transport mechanism, grain size distribution in the oxide should be taken into account. Second, it is well known that whether the metal is cold-worked or not can significantly affect the oxidation rate. Indeed, a larger number of dislocations leads to a higher vacancy annihilation capability. As a consequence, the decrease in the number of defects not only lowers the diffusion rate in the metal but also leads to a reduced tendency for vacancies to accumulate and cause local scale detachment. In their study for Yucca Mountain [19], Larose and Rapp outline the importance of these latter phenomena. Moreover, they suggest that a preliminary introduction of a proper amount of reactive elements at the surface of a steel prior to its oxidation at low temperature could induce an important reduction of its oxidation rate by pinning of the misfit dislocations at the metal/oxide interface so preventing the scale growth by cation diffusion. This effect is further investigated in [21]. In addition, experimental results on the low-alloy steel [22] show that even if alloying elements could not be found in the oxide scale, their distribution in the alloy near the metal/oxide interface certainly would affect the diffusion transport of oxidised species.

3. Advanced Model

We propose to build a numerical model able to reproduce all the major mechanisms described above in order to have a more powerful tool to describe the oxidation rate over long periods. This model is currently under development. So we present here mainly its general description and the way we manage to solve numerically the set of equations, and only a few preliminary results.

3.1. General Description

The system under consideration can be described by a 'sandwich-like' schematic image (Fig. 3) composed of a semi-infinite metal, the metal/scale interface, the oxide scale, the oxide/gas interface and a semi-infinite layer of gas.

In each layer we define the different species concentrations. The chemical species are the metallic species A and B, and an oxidant species O. In addition, in order to

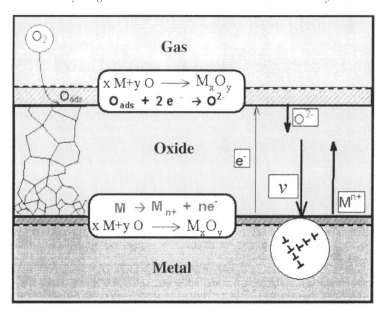

Fig. 3 *Schematic representation of mechanisms involved in the scaling rate.*

introduce effects like the influence on the scaling rate of defects annihilation (or creation), defects profiles will be calculated. In fact, the diffusion coefficient of the chemical species being proportional to the concentration defects, the scaling rate will be affected by the rate annihilation (or creation) of defects on sinks (and sources) in the metal which can be attributed to the dislocation density. Furthermore, explicitly calculating vacancy profile allows attention to be drawn to the risk of scale detachment in the case of local over-saturation of vacancy concentration near the interface. An overview of the different steps that we wish to treat in the final model is as follows:

(i)   the transport of chemical species by atomic diffusion via point defects but also by diffusion along short-circuits like grain boundaries or porosity in the oxide scale;

(ii)   the interface movement, considered either as an interface reaction taking place under conditions of a local equilibrium, or, by introducing a step reaction rate, that is, whether the oxidation reaction is considered to be instantaneous or not with respect to the diffusion;

(iii) the annihilation or creation of point defects;

(iv) the electronic fluxes, effect of high field on migration, and electron tunnelling.

At the present stage of development, only vacancies have been taken into account as defects. In the oxide we can distinguish two types of vacancies, anionic and cationic. Only the point defect diffusion is treated. Finally, we shall not at this stage introduce mechanisms described in (iv) that will be rate determining steps for thin oxide scales. This has been done in work by Martin and Fromm [6,7].

3.2. Set of Equations Governing the Evolution of the Concentration Profiles

To calculate the evolution of the system, we consider the profile evolution of the different species. To consider diffusion with non-conservative defects, we use Martin's description (see [23]). Hence, as shown in Fig. 4, concentration profiles are described by N labelling slabs of contiguous lattice plane with a thickness p^n.

We define in each slab n the concentration of the different species

$$C_k^n = \frac{N_k^n}{N_{sites}}$$

Following Fick's first law, using reduced units, the flux of chemical species i passing from slab n to $n+1$ is given by:

$$J_i^{n \to n+1} = -\sum_k d_{ik} \frac{C_k^{n+1} - C_k^n}{\frac{1}{2}\left(p^n + p^{n+1}\right)} \tag{1}$$

with

$$d_{ik} = -\sum_j \frac{M_{ij}}{kT} \frac{\partial(\mu_i - \mu_v)}{\partial C_k} \tag{2}$$

and

$$M_{ij} = \delta_{ij} z v_i C_v C_i \exp(-Q/kT) \tag{3}$$

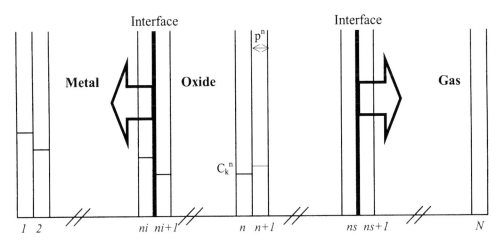

Fig. 4 *Schematic representation of the concentration profile.*

Note that the diffusion coefficient varies with respects to vacancies concentrations in the slab. In the oxide, we distinguish two sub-lattices for the exchange mechanism. Hence, the vacancy flux is nothing but:

$$J_{v_o}^{n\to n+1} = -J_o^{n\to n+1} \tag{4}$$

for anionic vacancies, and

$$J_{v_M}^{n\to n+1} = -\sum_{k=A,B...} J_k^{n\to n+1} \tag{5}$$

for cationic vacancies.

Assuming that vacancy source and sink are formed by a uniform dislocation density ρ the local rate of change in vacancy composition is given by:

$$\frac{\partial C_v}{\partial t}\bigg|_{sinks} = -\rho d_{vv}\left(C_v^{eq} - C_v\right) \tag{6}$$

For the metal, assuming that the vacancy creation or annihilation in slab n only increases or decreases the number of planes in the latter slab (without affecting cross section) the conservation equations are now written:

$$\dot{C}_i^n = \frac{\partial C_i^n}{\partial t} = -\frac{J_i^{n\to n+1} - J_i^{n\to n+1}}{p^n} - C_i^n \frac{\dot{p}^n}{p^n} \tag{7}$$

with

$$\frac{\dot{p}^n}{p^n} = \frac{\partial C_v}{\partial t}\bigg|_{sinks} \tag{8}$$

In eqn (7), the second term of the right hand side accounts for the change in the site concentration of species i resulting from the change in the number of sites in slab n keeping the atom fraction constant. This term disappeared for slabs in the oxide layer and in the gas. To compute time evolution we used a classical explicit finite difference algorithm with a small time increment Δt given by the stability criterion: $d_{max} \Delta t <^1/4 \Delta x$.

3.3. Algorithm for Moving Boundaries

To authorise growth of the scale by cationic and anionic transport we define two moving boundaries: between the slab ni and $ni+1$ for the interface metal/oxide and between ns and $ns+1$ for the oxide/gas interface. For each time step the thickness p^{ni} (p^{ns}) and $p^{ni}+1$ ($p^{ns}+1$) as well as concentrations of the corresponding slabs are recalculated in order to achieve local equilibrium over the two slabs around the interface. The complete algorithm is described in [24] and allows the movement of the interface in the entire system by incrementing variable ni (and ns) which marks the interface position when one of the slab becomes too large. Equilibrium can be for example simply defined by fixed equilibrium concentrations ($^{eq}C^{oxide}$; $^{eq}C^{metal}$) as in a binary phase diagram. In this case the interface position

$$\xi = \sum_{n=1}^{ni} p^n$$

obeys the classical rate equation:

$$\dot{\xi} = \frac{\partial \xi}{\partial t} = \frac{J^{ni-1 \rightarrow ni} - J^{ni+1 \rightarrow ni+2}}{^{eq}C^{oxide} - ^{eq}C^{metal}} \tag{9}$$

The algorithm for interface motion can also be used for more complex equilibrium phase diagram. For a multi-component systems, minimisation of the free energy over the two slabs around the interface gives the proportion for each phase from which the thickness p^{ni} (p^{ns}) and $p^{ni}+1$ ($p^{ns}+1$) are deduced, hence changing the

interface position $\xi = \sum_{n=1}^{ni} p^n$.

3.4. First Results

As the first step of this work, the model has been used considering vacancies as conservative species ($\rho = 0$; the second term of the RHS in the eqn (7) is null). For initial conditions, we define the concentration of each species in order to describe of the system: for slabs 1 to ni, the metal layer, the oxygen concentration is null (i.e. concentration of oxygen vacancies is 0.5, all the oxygen sub-lattice sites are occupied by vacancies), for slabs $ni+1$ to ns, the oxide layer, the concentrations of all the species is non-null, and finally, for slabs $ns+1$ to N, the gas layer, the concentrations of oxygen vacancies and metal null.

For boundary conditions, we chose fixed equilibrium vacancy concentrations at the interfaces. Notice that at the surface, the concentrations of the equilibrium vacancies in the oxide can be directly related to a fixed oxygen partial pressure [25]. Kinetic parameters (d_{ik}) have been chosen to reproduce an oxide scale with a larger cationic mobility than anionic mobility. With the set of parameters described above, the model has been numerically solved using $N = 100$ slabs. Figure 5 shows the calculated evolution of the oxide thickness obtained. As expected from the chosen initial conditions, the scaling rate is parabolic.

Figure 6(a) shows the evolution of profiles of metallic vacancies in the oxide during the first time steps. It can be seen that the concentration profile quickly develops into a linear gradient which is the expected stationary profile in respect to the chosen equilibrium conditions at the interfaces. For long time simulation leading to large scale growth, the concentration profiles of metallic vacancies in the oxide layer shown in Fig. 6(b) are still linear since stationary state has been reached. We can also see on this figure that the growth of the oxide layer is more important at the oxide/gas interface than at the metal/oxide interface, as expected from the chosen set of kinetics parameters.

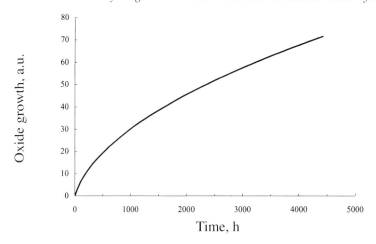

Fig. 5 Evolution of the oxide thickness with time.

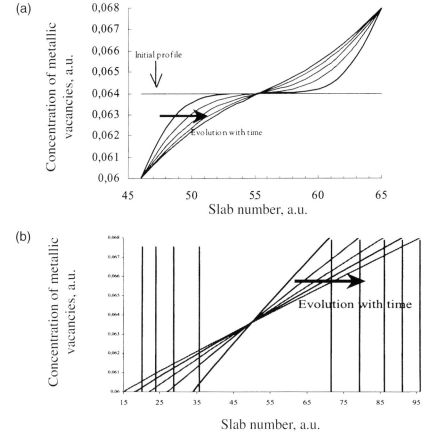

Fig. 6 Evolution of the metallic vacancies concentration in the oxide (a) for the first time steps (transition stage) and (b) for long time simulation (stationary state).

3.5. Future Development

The next step will be to use the model with a non null density of dislocations to evaluate the influence of the non-conservation of vacancies on the metal oxidation rate. In a second phase, diffusion via short-circuits will be introduced by defining in each oxide slab a density of short-cuts, following a description similar to that used to introduce a dislocation density in the metal.

4. Conclusions

In the framework of the COCON programme, two different approaches to the modelling of dry oxidation kinetics are proposed. A basic model consists of some extrapolations of available experimental data in the temperature range of interest following simple analytical laws deduced from classical oxidation theories. This leads to a very small thickness for the oxide scale and thus to the loss of very small amounts of metal even for extrapolations to over 100 years. However, the reliability of this kind of basic model is very poor since it is based on the assumption that a single elementary process controls the oxidation rate. Indeed, in the temperature range concerned by long term interim deposit of waste containers, several mechanisms can control the oxidation rate. A numerical model able to take in consideration several growth mechanisms is now in progress. At this stage of development the originality of the proposed advanced model consists in explicitly calculating the vacancy profiles and treating these as non-conservative species. Hence, the model should be able to reproduce various phenomena proposed to explain well known effects on oxidation behaviour but which are often not introduced in other models as, for instance, cold working of metal.

5. Acknowledgements

Stimulating discussions with G. Martin (CEA/SRMP) and D. Monceau (ENSCT) are gratefully acknowledged.

6. Notations

$J_i^{n \rightarrow n+1}$	flux of the species i from the slab n to the slab $n+1$
N_k^n	number of species k in slab n
C_k^n	concentration of species k in the slab n
$^{eq}C^{oxide}$, $^{eq}C^{metal}$	equilibrium concentration respectively in the oxide, in metal

$d_{ik} = \dfrac{D_{ik}}{(system\ length)^2}$ reduced coefficient of diffusion (D_{ik}: coefficient of diffusion)

p^n reduced length of the slab n

M_{ij} mobility

ν_i attempt frequency

ρ dislocation site density

$\delta_{ij} = 1$ (for $i = j$) $= 0$ (for $i \neq j$)   Kronecker's symbol

z coordinence

Q activation energy

μ_i chemical potential of i species

ξ interface positiontime derivative of x

$\dot{x} = \dfrac{\partial x}{\partial t}$ time derivative of x

$V_o V_M$ oxygen (anionic) and metallic (cationic) vacancies

References

1. C. Wagner, *Z. Physik. Chem.*, 1936, **B21**, 447.
2. N. Cabrera and N. F. Mott, *Rep. Prog. Phys.*, 1948-49, **12**, 163.
3. A. T. Fromhold, *Theory of Metal Oxidation*, Vol. II. North Holland Publishing, Amsterdam, 1990.
4. P. Kofstat, *High Temperature Corrosion*. Elsevier Applied Science, 1988.
5. R. Peraldi, Relations entre cinétique de croissance des couches d'oxyde et leurs microstructures lors de l'oxydation à haute température du nickel de haute pureté, Ph. D, Toulouse 2000.
6. M. Martin and E. Fromm, *J. Alloy. Comp.*, 1997, **258**, 7.
7. E. Fromm, Kinetics of metal–gas interactions at low temperature. Springer Series in Surface Sciences 1998.
8. H. Umimoto *et al.*, *IEE Trans. on Computer-aided Design*, 1989, **8**, 599.
9. H. Li *et al.*, *Corros. Sci.*, 1997, **39**, 1211.
10. C. Desgranges, F. Mazaudier, D. Gauvain, A. Terlain , D. Féron and G. Santarini, this volume, pp.49–61.
11. D. E. Davies, U. R. Evans and J. N. Agar, *Proc. Roy. Soc. Lond.*, 1954, **A225**, 443.
12. E. J. Caule, K. H. Buob and M. Cohen, *J. Electrochem. Soc.*, 1961, **108**, 829.
13. H. Sakai, T. Tsuji and K. Naito, *J. Nucl. Sci. Technol.*, 1989, **263**, 21.
14. A. B. Winterbottom, *J. Iron Steel Inst.*, 1950, **165**, 9.

15. M. J. Graham, S. I. Ali and M. J. Cohen, *J. Electrochem. Soc.*, 1970, **117**, 137.

16. W. Swanson and H. H. Uhlig, *J. Electrochem. Soc.*, 1974, **121**, 1551.

17. D. Davies, U. R. Evans and J. Agar, *Proc. Roy. Soc. Lond.*, 1954, **A225**, 443.

18. H. H. Uhlig, *Compounds*, **258**, 7–16, 1997.

19. S. Larose and B. Rapp, Review of low-temperature oxidation of carbon steels and low-alloy steels for use as high-level radioactive waste package materials. Report CNWRA97-003, Center for Nuclear Waste Regulatory Analyses, San Antonio, Tx, USA, February 1997.

20. R. B. Runk and H. J. Kim, *Oxid. Met.*, 1970, **2**, 285.

21. B. Pieraggi *et al.*, Oxid. Met., 1995, **44**, 63.

22. A. Terlain *et al.*, *Corrosion '01*, Paper No. 0119, NACE International, Houston, Tx, USA, 2001.

23. G. Martin and C. Desgranges, *Europhys. Lett.*, 1998, **44**, 150.

24. C. Desgranges, PhD Thesis, Université Paris-Sud, also published as report CEA-R-5805 or DE99619878, CEA -Saclay, Gif-sur-Yvette, France, 1998.

25. T. Narita *et al.*, *J. Electrochem. Soc.: Solid State Sci. Technol.*, 1982, **129**, 209.

15

Modelling Corrosion of Alloy 22 as a High-Level Radioactive Waste Container Material

D. S. DUNN, O. PENSADO, C. S. BROSSIA, G. A. CRAGNOLINO, N. SRIDHAR
and T. M. AHN*

Center for Nuclear Waste Regulatory Analyses, Southwest Research Institute, 6220 Culebra Road,
San Antonio, TX 78238-5166, USA
*U.S. Nuclear Regulatory Commission

ABSTRACT

The long-term performance of Alloy 22 waste package (WP) outer container is modelled using a total system performance assessment (TPA) code. The TPA code considers the susceptibility of the WP outer container to localised corrosion and the passive corrosion rate in the absence of localised corrosion. A description of the approach is provided along with an example output and a summary of ongoing activities to support the evaluation of Alloy 22 as a container material under geological disposal conditions.

1. Introduction

In the USA, the U.S. Department of Energy (DOE) is responsible for safe disposal of the nation's spent nuclear fuel (SNF) from commercial power reactors, high-level waste (HLW) from reprocessing and DOE-owned SNF including U.S. Navy SNF. Currently, the U.S. code of federal regulations (10 CFR Part 63) lists the regulatory requirements for the disposal of SNF and HLW in the proposed repository at Yucca Mountain, NV. The DOE strategy [1] for the disposal of HLW is focused on providing the safe containment of the HLW within an engineered barrier system (EBS) limiting the dose to the reasonably maximally exposed individual for the 10 000-year regulatory period. The proposed repository at Yucca Mountain and the current design of the EBS have several attributes that the DOE has suggested [1] would contribute to the isolation of the nuclear waste. The proposed repository horizon at Yucca Mountain will be located in the unsaturated zone that has a low water infiltration rate. In addition, the current WP design is intended to provide containment of the radionuclides for thousands of years. After the WPs are breached, the combination of the EBS and the limited availability of water at the proposed site may slow the mobilisation of the radionuclides and limit radionuclide transport away from the EBS.

The current DOE WP design [2,3] has an Alloy 22 (58Ni–22Cr–13Mo–3W–3Fe) outer container surrounding a type 316 NG (nuclear grade) stainless steel inner container to provide structural integrity. Complete descriptions of the fabrication sequence and non-destructive evaluation methods used to inspect the disposal container (DC) are provided in DOE documents [4,5] There are several variants of the WP design to accommodate fuel from boiling water reactors (BWR), pressurised water reactors (PWR), U.S. Navy SNF, and defence HLW glass canisters [2]. The Alloy 22 DCs will be constructed using roll formed plates welded into cylinders. The length of the DC requires that two cylinders be welded together. The bottom lid of the cylinders will also be welded to complete the DC. During fabrication, the longitudinal and circumferential welds of the Alloy 22 outer barrier will be solution annealed at 1125°C and quenched in water. Following final machining, the stainless steel inner container will be installed into the Alloy 22 outer container with a 1 to 4 mm gap between the inner and outer DCs.

The completed DC will be shipped to the proposed repository site where SNF and defence HLW will be loaded. An inner type 316 NG stainless steel lid will then be welded to the inner container. A dual lid design is proposed for the closure of the outer Alloy 22 container [3]. The inner lid will be 10 mm thick. Laser peening is specified for the inner lid closure weld to eliminate tensile stresses in the weld region. The outer lid will be secured with a deep U-groove weld joint. Induction annealing is specified for the outer lid closure weld to eliminate residual tensile stresses in the weld region. Following the completion of remote non-destructive inspection of the final closure weld using ultrasonic testing, the loaded and sealed DC will become a WP ready for emplacement in the repository drifts.

In addition to the WP, the EBS will include several other components. The WP pallet constructed with Alloy 22 V-shaped supports, connected together with stainless steel beams, will contact the WP [5]. The WP pallets will be placed with the WP on a bed of crushed rock. An inverted U-shaped Ti–Pd alloy drip shield (DS) will be installed prior to final closure of the repository to divert water away from the WP. Backfill has been retained as an option in the repository design. Other EBS components include carbon steel supports for the emplacement drifts and rock bolts.

A risk-informed, performance-based evaluation of the proposed repository is being conducted using the Total System Performance Assessment (TPA) code [6]. The U.S. Nuclear Regulatory Commission (NRC) plans to use the TPA code as a tool to evaluate any DOE license application for a proposed HLW repository. Corrosion related failure processes are the primary degradation processes of the WP and engineered barriers modelled in the TPA code. In a previous paper [7], the Engineered Barrier System Performance Assessment Code (EBSPAC) Version 1.1 was used to evaluate an earlier DOE WP design using an Alloy 825 inner container with an A516 steel outer container. The current EBSFAIL module of the TPA code is used to model uniform corrosion, corrosion potential (E_{corr}), localised corrosion susceptibility, and localised corrosion propagation rate. Future plans to improve the module include assessment of the effects of material performance variations, fabrication, thin water films, thermal oxides, anionic species such as nitrate, variations of the conditions within the emplacement drifts, and consideration of stress corrosion cracking (SCC).

2. Model Descriptions

Models abstracted from laboratory tests are implemented in the EBSFAIL module of the TPA code [6] to evaluate the performance of the WP. The abstracted models encompass a range of possible degradation modes and capture the important relationships that determine the mode and rate of the degradation processes.

For the proposed repository, alteration of the physical and chemical environment within the emplacement drifts may have a significant effect on EBS and WP performance. Prior to the formation of a water film on the WP, the conditions inside the drift are expected to be dry because of the combination of the heat generated through radioactive decay and either forced or natural ventilation of the repository drifts. This results in the formation and growth of a thermal oxide film. Although the thickness of the oxide film is likely to be a function of temperature, oxide penetration of the metal will be very small and is not anticipated to significantly affect WP performance. Heat generation and ventilation will also result in dryout of the surrounding host rock. The extent of dryout is likely to be determined by the thermal loading strategy and ventilation parameters. An extended dry period is desirable to prevent the onset of aqueous corrosion; however, elevated temperature in the adjacent rock may promote premature collapse of the drifts.

Water contact with the WPs is possible through a variety of mechanisms. Rockfall and drift collapse may damage the DS and allow dripping water to contact the WPs. Even without damage to the DS, condensed water is anticipated to contact the WPs when the relative humidity (RH) in the emplacement drifts exceeds the critical RH for the formation of a water film on the WP surface [7]. The critical RH for water film formation is dependent on a variety of factors including the presence of dust and hygroscopic salt deposits on the WP surface. After water contact with the WP, aqueous corrosion will become an important degradation mode. The rate of aqueous corrosion depends on the stability of the Alloy 22 passive oxide film. As long as a passive film is continuously maintained, the rate of corrosion is expected to be low and dictated by the slow dissolution of the alloy that is controlled by the oxide film. However, localised breakdown of the passive film could lead to the initiation of crevice corrosion and SCC causing rapid penetration of the WPs, if localised corrosion or crack propagation prevails. A combination of elevated temperature and solution compositions that contain critical concentrations of aggressive species, such as chloride, may promote localised corrosion. The presence of oxidising species, such as Fe(III) or H_2O_2, may facilitate the initiation of localised corrosion, whereas the presence of inhibiting species such as nitrate may both prevent localised corrosion initiation and promote repassivation of active localised corrosion. For the expected repository environment, which will have an oxygen partial pressure equivalent to that in air at the repository horizon, the corrosion mode will be determined by the composition of the water in contact with the WPs and temperature.

2.1. Temperature and Relative Humidity

The temperature and RH in the emplacement drifts are dependent on the thermal loading strategy, which can be manipulated by altering the drift spacing and WP spacing. The effects of ventilation, backfill, and DS can alter the temperature and RH within the emplacement drifts. At present, temperature and RH are calculated independently in the TPA code. The repository horizon average rock temperature is computed using an analytic conduction-only model for mountain-scale heat transfer based on line thermal sources, representing drifts, separated by a drift spacing and residing in a semi-infinite medium. The temperature increase in this semi-infinite medium is the sum of contributions from each line source. The WP temperature is computed as a function of the average rock temperature and the heat output per metric ton of heavy metal (MTHM) from the SNF and HLW. This function is derived via a multimode (i.e. conduction, convection, and radiation) heat transfer analytical model. Although the temperature of the WP depends on the specific characteristics of the WP, the thermal model only considers uncanistered WPs containing either 21 PWR or 44 BWR fuel assemblies. The model output provides a distribution of temperature and RH as a function of time and position in the emplacement drifts and the surrounding rock.

The thermal model assumes uniformly distributed WPs in parallel 5-m diameter emplacement drifts. The spacing between WPs and the drift spacing can be varied to accommodate changes in the thermal loading strategy. The thermal model calculates the temperature using inputs that include:

(i)   heat generated from the radioactive decay of the SNF,

(ii) heat removed by the heating of air, and

(iii) heat removed by latent heating of vapourised groundwater.

The groundwater vapourisation rate depends on the extent of dryout in the host rock surrounding the emplacement drifts. In addition, some of the heat may be removed by forced air ventilation. Ventilation will bring a steady flow of relatively cool dry air into the emplacement drifts. The air would be exhausted through ventilation shafts to remove heat and moisture from the repository horizon [8]. During the final closure phase, ventilation will be terminated and the ventilation shafts sealed to prevent the ingress of water to the repository horizon.

The WP temperature and RH in the emplacement drifts are required inputs as a function of time for both the corrosion models used in the EBSFAIL module and EBSREL, the radionuclide release module of the TPA code. The RH is calculated as the ratio of the actual vapour pressure to the maximal vapour pressure at the WP surface.

$$RH = \frac{P_v[\min(T_b, T_w)]}{P_v(T_{WP})} \qquad (1)$$

where P_v is the vapour pressure which is a function of temperature, T_b is the boiling point temperature at the repository horizon, T_w is the temperature of the drift wall, T_{WP} is the temperature of the WP, and $\min(T_b, T_w)$ is the minimum of the two values. Note that eqn (1) is a thermodynamic equilibrium expression, which may not be accurate under forced ventilation conditions. However, in this latter case, it is expected that eqn (1) would overestimate the RH of the EBS environment.

2.2. Emplacement Drift Environment

The environment model provides the chemical environment as a function of time in the immediate vicinity of the WP. Chemical parameters, including solution pH, oxygen partial pressure, chloride and bicarbonate concentrations, and dissolved silica, can have an important influence on the rate of corrosion of the WP, dissolution of SNF, and the formation of SNF alteration products.

The computer code MULTIFLO [9] is used to provide an estimate of the water composition contacting the WPs for the EBSFAIL module of the TPA code. MULTIFLO accounts for transport of reacting chemical constituents coupled to evaporation and condensation processes involving two-phase fluid flow. The code is capable of sequentially coupling two-phase fluid flow of liquid water, water vapour, and air with reactive transport of aqueous and gaseous species. For the sake of WP degradation computations, the repository chemistry was abstracted as a single component (chloride only) system, with an initial chloride concentration consistent with the composition of well water from the vicinity of the Yucca Mountain site. The MULTIFLO code was used to derive the chloride concentration at the top of the drift boundary during wet periods. During dry periods (i.e. temperatures above boiling), it is assumed that the chloride concentration of the solution in contact with the WP ranges from saturation in well water from the saturated zone (1.8M) to saturation of sodium chloride in pure water at ambient temperature (6.6M). This approach accounts for evaporation and potential accumulation of salts on top of the WP.

2.3. Humid Air and Aqueous Corrosion Models

In the EBSFAIL module of the TPA code, the corrosion models consider only passive dissolution and localised corrosion. Failure processes that are not currently considered include thermal oxidation of Alloy 22 and SCC. Thermal oxidation of Alloy 22 during the dry period may have a significant effect on the E_{corr} and subsequently, on the initiation of localised corrosion. However, significant loss of the WP wall thickness or penetration of the WP by thermal oxidation is not expected. SCC may be a valid failure mode for the WPs. Abstracted models of SCC will be included in future versions of the TPA code depending on the design and fabrication processes used for the construction and closure of the WP, as well as the components of the EBS included in the final repository design.

Another corrosion process that is not uniquely considered is humid air corrosion. In the EBSFAIL module of the TPA code, humid air corrosion is treated as uniform passive dissolution. This is a conservative assumption, since the critical RH above which humid air corrosion of most metals occurs is approximately 65%. Under these conditions, humid air corrosion is governed by the same electrochemical laws applicable to the corrosion of metals immersed in an aqueous electrolyte [10]. Empirical data obtained from long-term atmospheric exposures suggest that the rate of aqueous corrosion may be 10 times higher than that of the corrosion rates measured under atmospheric or humid air conditions [11].

2.3.1. Corrosion potential calculation.
The corrosion models calculate the rates of uniform corrosion and localised (crevice or pitting) corrosion. The dominant corrosion process at any given time is dictated by the E_{corr} and the appropriate critical potential for that process. The E_{corr} is the mixed potential established at the metal/solution interface when a metal is immersed in a given environment. For example, if the E_{corr} exceeds the critical potential for crevice corrosion initiation, active crevice sites are assumed to initiate and grow. If

the E_{corr} falls below the crevice corrosion repassivation potential (E_{rcrev}), active crevice sites are assumed to cease growing and the material passivates, corroding uniformly at a very low rate through a passive film. On the basis of experimental evidence [12,13] obtained for Alloy 825 in simulated groundwater containing 1000 ppm chloride at 95°C, it can be concluded that the potential required for the initiation of localised corrosion tends to decrease with increasing test time and the E_{rcrev} can be considered a lower limit for localised corrosion initiation. For this reason, the E_{rcrev} is conservatively adopted in this analysis as the critical potential for the initiation of localised corrosion, either as pitting or crevice corrosion. Both localised corrosion processes tend to coincide at temperatures higher than 20°C [14]. The concept of E_{rcrev} as a threshold condition for the initiation of localised corrosion has been shown to be consistent with a number of field observations [13,15].

The E_{corr} is defined as the potential at which the current due to all the cathodic processes is equal to the current due to all the anodic processes including the electrochemical dissolution of the metal [16], as indicated by eqn (2)

$$\sum_{j=1}^{n} I_{a,j} - \sum_{j=1}^{n} I_{c,j} = 0$$

(2)

where $I_{a,j}$ represents anodic (oxidation) currents including that of the metal dissolution, which is the corrosion current (I_{corr}), and $I_{c,j}$ represents cathodic (reduction) currents.

If the anodic and cathodic processes occur uniformly throughout the surface of interest, then eqn (2) can be written in terms of appropriate current densities. However, once localised corrosion initiates, this assumption is no longer valid because physical separation of anodic and cathodic areas occurs. The areas of passive corrosion, in which cathodic reactions take place, must be considered separately from the areas of active corrosion. In the EBSFAIL module of the TPA code, prior to the occurrence of localised corrosion, uniform distribution of cathodic and anodic areas is assumed for simplicity. Once localised corrosion occurs, it is reasonable to assume that the cathodic reaction takes place on the whole available surface because the pitted area is comparatively small. The two cathodic reactions assumed in the EBSFAIL module are the oxygen reduction equation (3) and the hydrogen evolution equation (4) reactions. Oxygen, diffusing through the electrolyte layer, is reduced at the oxide/solution interface according to the following reaction

$$O_2 + 2H_2O + 4e^- = 4OH^-$$

(3)

Under neutral and alkaline acidic conditions, the reduction of water takes place as follows

$$2H_2O + 2e^- = H_2 + 2OH^-$$

(4)

The oxygen reduction reaction is assumed to be a mixture of activation-controlled charge transfer and molecular diffusion-controlled transport processes. As discussed elsewhere [17,18] the resulting current density for oxygen reduction, i_{O_2}, can be written as eqn (5)

$$i_{O_2} = \frac{-k_{O_2} C_{O_2}^{bulk} \exp\left(-\dfrac{z_{O_2} \beta_{O_2} F E_{corr}}{RT}\right)}{1 + \dfrac{k_{O_2} \delta \exp\left(-\dfrac{z_{O_2} \beta_{O_2} F E_{corr}}{RT}\right)}{4FD_{O_2}}} \tag{5}$$

where k_{O_2} is the reaction rate constant for the oxygen reduction reaction, F is the Faraday constant, β_{O_2} is the charge transfer coefficient, R is the gas constant, T is the absolute temperature, z_{O_2} is the number of electrons involved in the reduction reaction, D_{O_2} is the diffusivity of oxygen in aqueous solution, δ is the thickness of the diffusion layer (e.g. water film), and $C_{O_2}^{bulk}$ is the bulk concentration of oxygen in solution. For the oxygen reduction reaction, the bulk concentration is related to the partial pressure of oxygen over the solution through Henry's law

$$C_{O_2}^{bulk} = K_H p_{O_2} \tag{6}$$

where K_H is the Henry's law constant for oxygen solubility.

The reaction rate constant is not usually measured but can be calculated from the exchange current density [16,19] for oxygen reduction, using eqn (7)

$$i_{O_2}^0 = k_{O_2} C_{O_2}^{bulk} \exp\left(-\frac{z_{O_2} \beta_{O_2} F}{RT} E_{eq}^{O_2}\right) \tag{7}$$

where $i_{O_2}^0$ is the exchange current density and $E_{eq}^{O_2}$ is the equilibrium potential for the oxygen evolution reaction. The equilibrium potential is given by the Nernst equation shown as eqn (8)

$$E_{eq}^{O_2} = E_0^{O_2} - 2.303 \frac{RT}{F} pH + 0.576 \frac{RT}{F} \log p_{O_2} \tag{8}$$

where $E_0^{O_2}$ is the standard electrode potential (V_{SHE}).

For the water reduction reaction, the cathodic current is assumed to be dictated only by the charge transfer process [17,18]. The cathodic current density then is given by eqn (9)

$$i_{H_2O} = k_{H_2O} \exp\left(-\frac{z_{H_2O} \beta_{H_2O} F}{RT} E_{corr}\right) \tag{9}$$

where k_{H_2O} is the reaction rate constant for the water reduction reaction.

The reaction rate constant can be calculated from the appropriate exchange current density for the hydrogen evolution reaction similar to that shown in eqn (7) but without the pre-exponential concentration factor. The equilibrium potential for the water reduction/hydrogen evolution reaction will depend on the bulk concentration of the H$^+$ ion, which is dependent on the pH. The temperature dependence of the reaction rate constants for the oxygen evolution and the water reduction reactions are assumed to be given by an Arrhenius-type relationship [7].

The E_{corr} is then calculated by solving simultaneously for E_{corr} using the passive current density, which is assumed to be independent of potential and temperature, and eqns (5) and (9). The effect of γ-radiolysis on the E_{corr} can be incorporated by including expressions for the cathodic reduction of oxidising species such as H$_2$O$_2$. In the EBSFAIL module of the TPA code the effect of γ-radiolysis is calculated according to eqn (10)

$$E_{corr}^{\gamma-radiolysis} = E_{corr} + \Delta E \tag{10}$$

The value of ΔE is calculated according to eqn (11)

$$\Delta E = \Delta E_0 e^{-\lambda t} \tag{11}$$

where t is time, ΔE_0 is the increase in the E_{corr} at $t = 0$, and λ is a decay constant.

2.3.2. Localised corrosion model.
Empirically derived equations are used in the EBSFAIL module of the TPA code for the dependence of the E_{rcrev} on environmental parameters. The E_{rcrev} is assumed to depend only on chloride concentration and temperature. Experimental results [15] have shown that the dependence of critical potentials on pH, bicarbonate, and silica is negligible for Alloy 825. Both nitrate and sulfate can act as inhibitors if the ratio of nitrate to chloride or sulfate to chloride is sufficiently high [14]. However, in the present calculations, the effects of these anionic species are not considered. The dependence of the critical potentials on chloride concentration and temperature for Alloy 22 was determined using the E_{rcrev} measured in solutions containing 0.5 to 4 molar chloride. Tests were conducted in autoclaves at temperatures in the range of 80 to 125°C [20,21]. The equation determined based on the experiments [21] for the E_{rcrev} of Alloy 22 is given by eqn (12)

$$E_{rcrev} = E_{rcrev}^0(T) + B(T)\log\left[Cl^-\right] \tag{12}$$

where the constants $E_{crev}^0(T)$ and $B(T)$, which are dependent on the material, are considered linear functions of temperature according to eqns (13) and (14)

$$E_{rcrev}^0(T) = 1540 - 13.1T \tag{13}$$

$$B(T) = -362.7 + 2.3T \tag{14}$$

and where E_{rcrev}^0 is in mV(SHE).

Aqueous corrosion could be uniform or localised. If it is localised because the E_{corr} is above the critical potential, the calculation of penetration is determined by eqn (15)

$$P = At^n \tag{15}$$

where P is the depth of localised corrosion penetration, t is time, and A and n are experimentally determined constants. When the depth of the localised attack is greater than the initial thickness of the Alloy 22 WP outer container, the WP is considered perforated because no credit is taken for the corrosion resistance of the Type 316 NG inner barrier. This is accomplished numerically by failing the inner type 316 container in a single time step. Otherwise, localised corrosion growth continues until either perforation of the Alloy 22 WP outer container occurs or the E_{corr} of the WP decreases below the E_{rcrev}, and the corrosion mode is switched from localised attack to passive dissolution.

2.3.3. Passive corrosion model.
When the E_{corr} is less than the E_{rcrev}, and the RH is greater than the critical RH for aqueous corrosion, penetration of the WP is assumed to occur by passive dissolution. In this case the penetration is calculated using eqn (16)

$$CR = \frac{i_{pass}EW}{F\rho} \tag{16}$$

where CR is the corrosion rate, i_{pass} is the passive dissolution current density, EW is the equivalent weight, F is Faraday's constant, and ρ is the density. For Alloy 22 ρ is 8.69 g cm^{-3}, and under passive conditions EW is 26.04 g/equivalent [22]. It should be noted however, that the long-term prediction of the performance of the Alloy 22 WP outer container under passive conditions using eqn (16) is dependent on the continuous presence of a passive film. Some evidence has been presented to suggest that the non-uniform dissolution of alloying elements can occur with Ni–Cr–Mo alloys under passive conditions [23,24]. Recent attempts to model the non-uniform dissolution of alloying elements from Alloy 22 suggest that this process may lead to enhanced dissloution of the passive film [25].

3. Total System Performance Assessment (TPA) Example Output

An example of the temperature calculation used as an input to the TPA code is shown in Fig. 1. The temperature profiles were calculated assuming no forced ventilation. Prior to the closure of the proposed repository, natural ventilation is assumed to remove some fraction of the heat generated from the WPs. This limits the maximum WP temperature to approximately 100°C. The large increase in temperature observed 50 years after emplacement is a result of repository closure and the cessation of natural ventilation. At 100 years after emplacement, the maximum WP temperature of 160°C is observed. The WP temperature remains above boiling until approximately 1000 years after emplacement. The results shown in Fig. 1 are for illustration purposes and can be modified with minor variations in design and related parameters.

The calculated RH at the WP surface is shown in Fig. 2. Initially the RH is approximately 55%. Within a few years after the emplacement of waste the RH exceeds the critical RH for aqueous corrosion, which is normally distributed in the range of 60–65% RH. It is important to note that the calculation of RH does not consider the removal of humid air as a result of natural ventilation. By assuming a closed system, the calculations are conservative with respect to the anticipated conditions within the emplacement drifts prior to repository closure. The large decrease in RH at 50 years after waste emplacement occurs as a result of repository closure and coincides with the sharp temperature increase shown in Fig. 1. After the maximum WP temperature is obtained and the WPs start to cool, the RH increases and exceeds the critical RH for aqueous corrosion at approximately 1000 years after waste emplacement. The RH remains greater than 95% for all times greater than 2000 years.

Figure 3 shows the results of the E_{corr} and E_{rcrev} calculations for Alloy 22. The inputs for the calculations include both the chloride concentration and temperature. The E_{corr} is calculated as described in Section 2.3.1. Two periods where aqueous corrosion is active are identified in Fig. 3. These periods are determined by the RH. During the initial repository operation after waste emplacement, the RH was calculated to be above the critical RH for aqueous corrosion. This period is relatively short, lasting from 5 to 50 years after waste emplacement prior to repository closure. The second period of aqueous corrosion occurs after 1000 years when the WPs cool to below boiling.

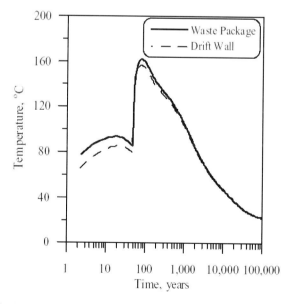

Fig. 1 *Calculated temperature of waste package surface and drift wall assuming an average thermal loading of 60 MTU (metric tons of uranium) per acre (0.015 MTU/m²).*

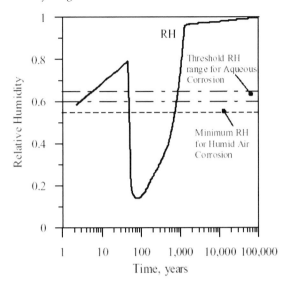

Fig. 2 *Relative humidity at waste package surface as a function of time after waste emplacement in horizontal drifts. The critical values of relative humidity assumed for humid air corrosion and for aqueous corrosion are indicated.*

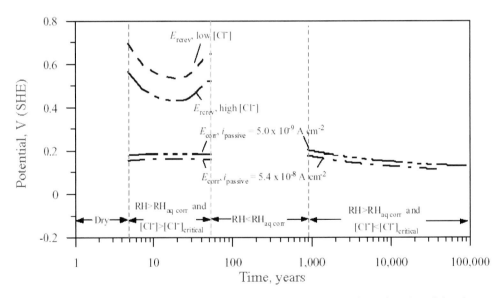

Fig. 3 *Corrosion potential and crevice corrosion repassivation potential as a function of time for the Alloy 22 waste package outer container.*

The E_{corr} of the Alloy 22 WP is calculated using both a low corrosion rate of 5.0×10^{-9} A cm^{-2} and a high corrosion rate of 5.4×10^{-8} A cm^{-2}. The variation of the E_{corr} with these passive dissolution rates is less than 50 mV, and the maximum E_{corr} is approximately 0.2 V(SHE). The E_{rcrev} is calculated when the RH is greater than the critical RH for aqueous corrosion. During the initial time after waste emplacement, the E_{rcrev} is determined for both a low and a high chloride concentration. The low chloride concentration is based on a J-13* water saturated with respect to chloride (1.8M Cl⁻). Saturated NaCl in water with no other dissolved ions is used for the high chloride concentration (6.6M Cl⁻). Localised corrosion is never initiated, and corrosion of Alloy 22 WPs occurs only by passive dissolution because for either the low or the high chloride concentration, the E_{corr} never exceeds the E_{rcrev}. After 1000 years when the WPs cool to below boiling, the value of E_{rcrev} is very high because the chloride concentration is assumed to decrease rapidly when water contacts the WPs. Figure 4 shows the remaining WP wall thickness as a function of time and passive dissolution rate. It is apparent that the high value of the E_{rcrev} for Alloy 22 results in a long WP lifetime by the successful prevention of localised corrosion. For the fast dissolution rate of 5.4×10^{-8} A cm⁻², penetration of the WP occurs after 37 000 years. For the lower passive dissolution rate of 5.0×10^{-9} A cm⁻², the lifetime of the WPs exceeds 100 000 years.

4. Uncertainty in Modelling Alloy 22 Corrosion

Uncertainties in the parameters used in the EBSFAIL module of the TPA code exist with respect to the performance of the Alloy 22 WP outer container. Factors that may contribute to parameter uncertainties include the evolution of passive film chemistry, fabrication processes, and the composition of the environment in the repository drifts. These factors are presently being evaluated using both laboratory testing and modelling.

The estimated Alloy 22 container lifetime shown in Fig. 4 assumes that the passive current density (i_{pass} in eqn (16)) is constant over thousands of years. The range of passive current densities used in the calculations plotted in Fig. 4 are based on measurements conducted at 20 and 95°C. Additional tests indicate that the passive current density is not strongly dependent on either the chloride concentration or pH [25]. Changes in the chemistry of the passive film that may occur by the preferential dissolution of Ni with respect to Cr are not accounted for in the present passive dissolution model. In the short term, preferential dissolution of Ni may result in a passive film enriched in Cr and consequently, in a lower the passive dissolution rate. In the long term, heuristic modelling of passive film dissolution based on the point defect model suggests that preferential dissolution may result in spalling of the passive film [25] and increase the susceptibility of the alloy to localised corrosion. The effects of fabrication processes on WP performance have been investigated by measuring the E_{rcrev} for both welded and thermally aged specimens. These results are compared to measurements for the as-received specimens from which the E_{rcrev} regression equations were generated (eqns (12–14)). As-received Alloy 22 was found to be resistant to crevice corrosion in solutions with chloride concentrations less than 0.5M [21,25]. Figure 5 shows the E_{rcrev} for welded Alloy 22 in chloride-containing solutions at 95°C. In 4M sodium chloride, the E_{rcrev} for the welded material is

---

*J-13 water is the water of a well in the proximity of Yucca Mountain, which is used as a reference water for the repository studies.

80 mV(SHE), well below the E_{corr} calculated by the EBSFAIL module of the TPA code (Fig. 3). In contrast to the results obtained for the as-received material, localised corrosion of welded Alloy 22 was observed in 0.005M chloride solutions (Fig. 5), indicating the welded material is more susceptible to localised corrosion. Similar

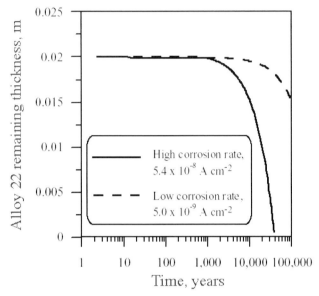

Fig. 4 *Remaining wall thickness for the Alloy 22 waste package outer container as a function of time after waste emplacement.*

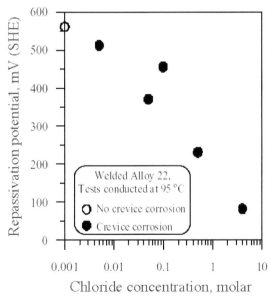

Fig. 5 *Crevice corrosion repassivation potential for welded Alloy 22 in chloride solutions at 95°C.*

results were obtained for as-received Alloy 22 after thermal aging for 5 min at 870°C (Fig. 6). The effects of exposure to temperatures where precipitation of secondary phases can occur may be significant because local induction annealing has been proposed to mitigate stress corrosion cracking of the WP closure weld [26].

The chemistry of the groundwater in contact with the WPs is presently modelled considering only the concentration of chloride. This simplifying assumption does not consider the effects of other dissolved species. Chloride concentration of groundwater contacting the WP will be dependent on the starting water composition. The J-13 water chemistry presently used has significant concentrations of sulfate, bicarbonate, and nitrate whereas the pore water obtained from the host rock is a chloride–sulfate type groundwater. The complex evolution of the water chemistry during heat up and cool down has not been determined, although modelling and experimental work is being conducted by the DOE [27].

The chemistry of the environment on the WP can have a significant effect on the localised corrosion resistance of the WP material. As a consequence of simplifying assumptions, the present TPA calculations do not consider the potential inhibitive effects of nitrate. The effect of nitrate to chloride concentration ratio on the E_{rcrev} of welded Alloy 22 is shown in Fig. 7. In pure 0.5M sodium chloride, the E_{rcrev} of welded Alloy 22 is 240 mV(SHE). The E_{rcrev} of welded Alloy 22 is not significantly altered in 0.5M chloride with the addition of 0.05M nitrate. With 0.5M chloride and nitrate concentrations of 0.1M or greater, no localised corrosion was initiated in short-term tests. An additional test was also conducted to determine the effect of nitrate on active localised corrosion by initiating crevice corrosion in a solution containing only 0.5M chloride and measuring the E_{rcrev} after the addition of nitrate to the solution. As shown in Fig. 7, the addition of nitrate increased the E_{rcrev} of welded Alloy 22 by

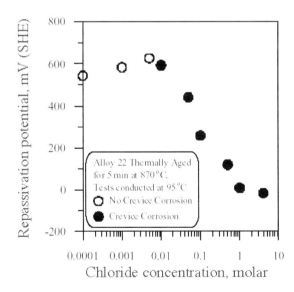

Fig. 6 *Crevice corrosion repassivation potential for thermally aged Alloy 22 as a function of chloride concentration at 95°C.*

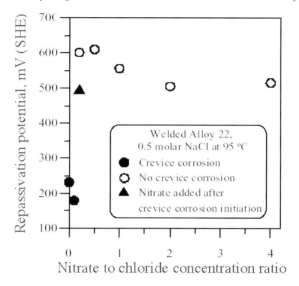

Fig. 7 *Crevice corrosion repassivation potential for welded Alloy 22 as a function of nitrate to chloride ratio in 0.5M chloride solutions at 95°C.*

more than 200 mV, indicating that nitrate has a significant effect on the inhibition of localised corrosion initiation and also promotes repassivation of localised corrosion.

5. Summary and Conclusions

The long-term performance of the EBS for the proposed HLW repository at Yucca Mountain, NV was evaluated using abstracted models based on the expected conditions within the repository drifts and data from laboratory investigations. For the Alloy 22 WP outer barrier, the performance assessment model calculates the E_{corr} based on anodic dissolution kinetics and the kinetics of water and oxygen reduction when the calculated RH exceeds the critical RH for the formation of a liquid water film. The E_{corr} at each time step is compared to the crevice corrosion E_{rcrev} selected as the critical potential for localised corrosion initiation. When the E_{corr} exceeds the E_{rcrev}, localised corrosion is assumed to initiate and propagate at an empirically determined rate which is a function of time. If the E_{rcrev} is greater than the E_{corr}, the Alloy 22 WP outer barrier corrodes at a rate determined by the passive dissolution of the material.

Performance assessment calculations suggest that the Alloy 22 WP outer barrier is not subject to localised corrosion in the environment expected to be encountered at the proposed repository site. The failure of the Alloy 22 WP is dependent on the passive dissolution rate. The shortest failure time occurs at 37 000 years, well beyond the 10 000-year performance period.

Uncertainties in the parameters used in the modelling of Alloy 22 performance are presently being investigated. Initial results indicate that variations in the chemistry of the Alloy 22 passive film may alter both the long-term passive corrosion rate and the susceptibility of the material to localised corrosion. Fabrication processes such as welding and exposure to elevated temperatures during induction annealing may reduce the localised corrosion resistance of Alloy 22. Environments that contain nitrate may mitigate the aggressive effects of chloride. The effect of these parameter uncertainties on the long-term performance of Alloy 22 will be evaluated using sensitivity analyses.

6. Acknowledgements

This paper was prepared to document work performed by the Center for Nuclear Waste regulatory Analyses (CNWRA) for the NRC under Contract No. NRC-02-97-009. The activities reported here were performed on behalf of the NRC Office of Nuclear Material Safety and Safeguards, Division of Waste Management. The paper is an independent product of the CNWRA and does not necessarily reflect the views or the regulatory position of the NRC.

References

1. CRWMS M&O, Repository Safety Strategy: Plan to Prepare the Safety Case to Support Yucca Mountain Site Recommendation and Licensing Considerations, TDR-WIS-RL-000001, Revision 04, ICN 01, Civilian Radioactive Waste Management System, Management and Operating Contractor, Las Vegas, NV, 2000.
2. CRWMS M&O, Update to the EIS Engineering File for the Waste Package in Support of the Final EIS, TDR-EBS-MD-000010, Revision 00, ICN 01, Civilian Radioactive Waste Management System, Management and Operating Contractor, Las Vegas, NV, 2000.
3. CRWMS M&O, Uncanistered Spent Nuclear Fuel Disposal Container System Description Document, SDD-UDC-SE-000001 Revision 01, Civilian Radioactive Waste Management System, Management and Operating Contractor, Las Vegas, NV, USA, 2000.
4. CRWMS M&O, Waste Package Operations FY-01 Closure Welds Technical Guidelines Document, TER-EBS-ND-000001, Revision 00, Civilian Radioactive Waste Management System, Management and Operating Contractor, Las Vegas, NV, USA, November 2000.
5. CRWMS M&O, Waste Package Operations Fabrication Process Report, TDR-EBS-ND-000003, Revision 01, Civilian Radioactive Waste Management System, Management and Operating Contractor, Las Vegas, NV, USA, September 2000.
6. S. Mohanty, T. J. McCartin and D. W. Esh, Total-System Performance Assessment (TPA) Version 4.0 Code: Module Descriptions and Users Guide, Center for Nuclear Waste Regulatory Analyses, San Antonio, TX, USA, 2000.
7. G. A. Cragnolino, S. Mohanty, D. S. Dunn, N. Sridhar and T. M. Ahn, *Nucl. Eng. Des.*, 2000, **201**, 289–306.
8. CRWMS M&O, Total System Performance Assessment for the site Recommendation, TDR-WIS-PA-000001 Revision 00, ICN 01, Civilian Radioactive Waste Management System, Management and Operating Contractor, Las Vegas, NV, USA, 2000.
9. P. C. Lichtner and M. S. Seth, User's Manual for MULTIFLO, Part II-MULTIFLO 1.0 and GEM 1.0, Multicomponent-Multiphase Reactor Transport Model, CNWRA 96-010, Center for Nuclear Waste Regulatory Analyses, San Antonio, TX, USA, 1996.
10. M. Stratmann, H. Streckel, K. T. Kim and S. Crockett, *Corros. Sci.*, 1990, **30**(6/7), 715–734.
11. G. D. Gdowski and D. B. Bullen, Survey of Degradation Modes of Candidate Materials for High-Level Radioactive-Waste Disposal Containers: Volume 2 — Oxidation and Corrosion, UCID-21362, Lawrence Livermore National Laboratory, Livermore, CA, USA, August 1998.

12. D. S. Dunn, G. A. Cragnolino and N. Sridhar, *Corrosion '96,* Paper No. 97, NACE International, Houston, TX, 1996.

13. D. S. Dunn, G. A. Cragnolino and N. Sridhar, *Corrosion,* 2000, **56**(1), 90–104.

14. N. Sridhar, G. A. Cragnolino and D. S. Dunn, Experimental Investigations Localized Corrosion of High-Level Waste Container Materials, CNWRA 93-004, Center for Nuclear Waste Regulatory Analyses, San Antonio, TX, USA, 1993.

15. N. Sridhar, G. A. Cragnolino and D. S. Dunn, Experimental Investigations of Failure Processes of High-Level Radioactive Waste Container Materials, CNWRA 95-010, Center for Nuclear Waste Regulatory Analyses, San Antonio, TX, USA, 1995.

16. Vetter, K.J., Electrochemical Kinetics, Academic Press, New York, 1967.

17. N. Sridhar, J. C. Walton, G. A. Cragnolino and P. K. Nair, Engineered Barrier Performance Assessment Codes (EBSPAC) Progress Report-October 1, 1992, through September 25, 1993, CNWRA 93-021, Center for Nuclear Waste Regulatory Analyses, San Antonio, TX, USA, 1993.

18. G. A. Cragnolino, *et al.,* Substantially Complete Containment-Example Analysis of a Reference Container, CNWRA 94-003, Center for Nuclear Waste Regulatory Analyses, San Antonio, TX, USA, 1994.

19. J. O'M. Bockris, and A. K. N. Reddy, *Modern Electrochemistry.* Plenum Press, New York, 1970.

20. G. A. Cragnolino, D. S. Dunn, C. S. Brossia, V. Jain and K. Chan, Assessment of Performance Issues Related to Alternate EBS Materials and Design Options, CNWRA 99-003, Center for Nuclear Waste Regulatory Analyses, San Antonio, TX, USA, 1999.

21. G. A. Cragnolino, D. S. Dunn, Y.-M. Pan and O. Pensado, Corrosion processes affecting the performance of Alloy 22 as a high-level radioactive waste container material. *Scientific Basis for Nuclear Waste Management*, Materials Research Society Symposium Proceedings Volume 663 (K.P. Hart and G.R. Lumpkin, eds). Materials Research Society, Pittsburgh, PA, USA, 2001, pp.507–514.

22. American Society for Testing and Materials. G102-89. Standard Practice for Calculation of Corrosion rates and Related Information from Electrochemical Measurements. *Annual Book of Standards*. American Society for Testing and Materials, pp.416–422, West Conshohocken, PA, USA, 1999.

23. S. Boudin, *et al., Surf. Interface Anal.,* 1994, **22**, 462–466.

24. G. Lorang, N. Jallerat, K. Vu. Quang and J.-P. Langeron, *Surf. Interface Anal.,* 1990, **16**, 325–330.

25. C. S. Brossia, *et al.,* Effect of Environment on the Corrosion of Waste Package and Drip Shield Materials. CNWRA 2001-003, Center for Nuclear Waste Regulatory Analyses, San Antonio, TX, USA, 2001.

26. CRWMS M&O, Waste Package Degradation Process Model Report, TDR-WDS-MD-000002, Revision 00, ICN 01, Civilian Radioactive Waste Management System, Management and Operating Contractor, Las Vegas, NV, USA, 2000.

27. N. D. Rosenberg, G. E. Gdowski and K. G. Knauss, *Appl. Geochem.,* 2001, **16**, 1231–1240.

16

Long Term Prediction of Corrosion of Passive Materials for High-Level Waste Containers based on Protection Potential

A. POURBAIX

CEBELCOR, Avenue Paul Héger, grille 2, B 1000 Brussels, Belgium

ABSTRACT

The corrosion rate of corrodible (active) materials is determined from tests of limited duration and by modelling. One difficulty is to assess the changes of the environment in the long term due to diffusion, precipitation of corrosion products etc.

For passive materials, one major risk is localised corrosion. The analysis of the localised corrosion processes provides some solid grounds for the long term prediction. This paper presents recognised basic aspects of localised corrosion processes and discusses the meaning of the 'protection potential against the propagation of localised corrosion'.

The value of the 'protection' potential is related to steady state conditions inside pits or crevices and it can be predicted from equilibrium conditions. In that sense, the 'protection potential' has some thermodynamic significance and is therefore useful for very long term prediction. The limits of application of the concept of 'protection potential' are discussed: long term future of pits under cathodic polarisation, effect of temperature, of thiosulfate, sulfide and anaerobic environments.

1. Introduction

Long term prediction for corrodible (active) materials calls for a detailed understanding and modelling of kinetics of reactions, including the effects of changes due, for example, to modifications of temperature, diffusion, precipitation, the formation of gas channels, etc. These are some of the difficulties associated with the long term prediction of reaction rate and in the interpretation of short term experiments.

The major risk for passive materials is localised corrosion. It is realistic to consider that localised corrosion initiation will always occur, as a result of mechanical damages, inclusions etc. Therefore, the question to address is: are there any safe conditions where there is no risk of localised corrosion propagation?

If such conditions are too stringent, it could then be necessary to examine whether more accessible conditions could prevent the initiation of pitting or enable metastable pits to become stable.

2. Concept of Protection Potential (against Propagation of Localised Corrosion)

The concept of a 'protection potential' against the propagation of localised corrosion was first introduced in 1962 as a result of a study conducted by CEBELCOR for EURATOM, on stainless steels [1].

At that time, the protection potential was determined as the zero-current potential on the back scan of a E–i polarisation curve (see Fig. 1), after pits were initiated at potentials above the 'pitting potential'.

The meaning of that 'protection potential' was described at that time as a potential below which exisiting pits would not grow. The potential range for passive metals was thus divided into three zones:

(1) above the pitting potential pits would initiate and grow;

(2) between the pitting potential and the protection potential, existing pits would continue to grow but no new pit would form, and

(3) below the protection potential, even the pits already formed would stop growing.

The practical interest of this 'protection potential against the propagation of localised corrosion' was immediately apparent: pitting will always occur, because real structures have structural heterogeneities, or are inevitably exposed to mechanical damages or to transient conditions that may damage the oxide film. Real-life steels always contain sulfide inclusions that are sites for pit initiation. It is thus more realistic to concentrate on preventing the growth of existing pits than to try to prevent their initiation.

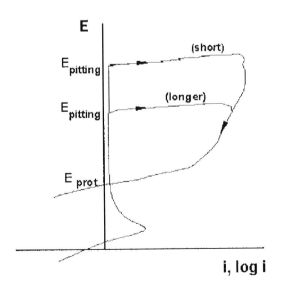

Fig. 1 Polarisation curve with pitting and protection potentials.

Furthermore, the value of the pitting potential determined as on Fig. 1 is lower with a slower scan rate or on longer exposure, which is of course important for our problem. The practical usefulness of the pitting potential is thus more limited than that of the protection potential.

Since the early work of 1962, knowledge on the local chemistry and electrochemistry of localised corrosion (pits, crevices and stress corrosion cracks) has progressed in a spectacular manner. As a result, a pseudo-thermodynamic approach of the control of localised corrosion was developed. The discussion below will cover first the meaning and the interpretation of the protection potential. Then, the methods to determine its value from basic electrochemistry and from specific laboratory studies and industrial experience are given.

Pits (and also crevices and stress corrosion cracks) are characterised by a restricted transport of reacting species, due to a barrier of precipitated corrosion products (oxide, salt film…on pits) or to a narrow and deep geometry (crevices, cracks). The term 'occluded corrosion cells' is used to underline the similarities between these different forms of localised corrosion.

The reactions that take place and their consequences are summarised below [2]:

(a) a small amount of metal is corroded in the recessed geometry, for example:

$$Fe \rightarrow Fe^{2+} + 2e^-$$

or

$$Cr \rightarrow Cr^{3+} + 3e^-$$

(b) The balancing reduction reaction is essentially oxygen reduction and takes place outside the pit, where access of dissolved oxygen is easy:

$$O_2 + 4H^+ + 4e^- \rightarrow 2H_2O$$

(c) When the solubility of the dissolved metal ions is exceeded in the pit, oxide or hydroxide precipitates (hydrolysis and hydrolysis and oxidation reactions):

$$Fe^{2+} + 2H_2O \rightarrow Fe(OH)_2 + 2H^+$$

or

$$2Fe^{2+} + 3H_2O \rightarrow Fe_2O_3 + 6H^+ + 2e^-$$

or

$$2Cr^{3+} + 3H_2O \rightarrow Cr_2O_3 + 6H^+ \; etc.$$

All these hydrolysis reactions produce acid. The pH in the pit decreases, and more metal ions go into solution. When the solubility for the given pH and potential conditions is attained again, more precipitation and acidification occur.

(d) As the concentration of H^+ ions and of metal cations increases in the pit, the ionic charge is balanced by diffusion of anions from the bulk solution towards the interior of the pit. Chloride is more mobile and diffuses preferentially, but other anions (SO_4^{2-} etc.) may also diffuse to the pit.

(e) Any additional corrosion in the pit decreases the pH further, increases the chloride and metal ion concentration and increases the local corrosion rate. This process goes on and on until, eventually, a metallic salt precipitates, instead of an oxide: then, the acidification process is stopped.

(f) The ultimate theoretical conditions in pits, crevices and cracks thus correspond to a saturated solution in the least soluble salt present (most often metal chloride, or metal sulfate). These conditions are seldom attained. However, metallic salts were occasionally observed in real pits.

(g) Localised corrosion sites (pits, crevice, cracks...) act as anion attractors: even in little concentrated chloride bulk solution, the final concentration inside the occluded cell will eventually become very concentrated in chloride ($FeCl_2$ or $CrCl_3$ etc.), possibly up to saturation.

(h) However, if there are no anions other than OH^-, the local acidification process cannot develop and there will be no localised corrosion.

(i) If differentiated sites for reduction and for oxidation do not exist, then localised corrosion will not occur.

(j) If oxygen (or another oxidant) is absent, then localised corrosion cannot occur. Some particular consideration must be given to possible local differences in kinetics for the hydrogen evolution, for example, as for differential sulfide concentration cells in H_2S media. This is discussed in Section 4.

(k) As the protection potential is related to the conditions in the pits, and as the conditions in the pit are independent of the chloride concentration in the bulk, the protection potential does not depend much on the bulk chloride concentration (see (o) below),

(l) for the same reasons, it is independent of the bulk pH,

(m) the conditions in the pits are largely influenced by the type of metal and of the major ions: the protection potential is expected to be more or less the same for AISI 316, AISI 304 or even AISI 430 stainless steels,

(n) the protection potential could be different (and higher) if nitrates or sulfates are present, together or without chloride,

(o) the protection potential differs slightly from the potential in the pits because of a diffusion potential. The protection potential thus depends within some small range on the geometry of the pit and on the bulk choride concentration,

(p) temperature should influence the protection potential, mostly through the solubility of the species present and less markedly through kinetics of reactions.

The active dissolution in pits and the relationship between the protection potential and the conditions in the pits are illustrated by several experiments.

Figure 2 shows an experimental arrangement for observing a progressing pit: the potential and pH are measured with electrodes inside the pit and on the surface exposed to the bulk solution. The current density is measured separately in the pit and on the passive surface. Here, the metal is pure iron and the bulk solution is 10^{-3}M NaCl + 10^{-3}M NaOH. Corrosion was initiated simply by deaeration of the pit.

Figure 3 shows the trajectory of E and pH in a pit on iron or carbon steel, as it progresses. The final conditions correspond to an active dissolution at low pH and low potential. The chemistry and electrochemistry of iron pits saturated or nearly saturated in chloride have been described in detail [3,4]: the ultimate composition of the pit correspond to the saturation in $FeCl_2$: 4.5M in $FeCl_2$, or 9M in Cl^-, pH = 3.5. In this solution, when deaerated, the potential is −570 mV(SCE). If oxygen reaches the pit, the pH drops further, the potential increases and the corrosion rate is much increased.

The polarisation curves of iron in nearly saturated or saturated $FeCl_2$ solutions are shown in Fig. 4. The limited current under anodic polarisation in the saturated solution looks like a passivation and is attributed to the precipitation of a salt film.

The individual polarisation characteristics of the pit and of the passive surface, plotted simultaneously under an overall cathodic polarisation, are shown in Fig. 5 [5]. The cathodic curve on the passive surface corresponds to oxygen reduction. At

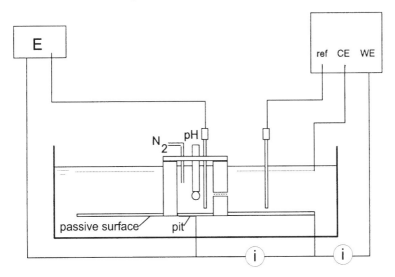

Fig. 2 *Experimental arrangement for the study of occluded corrosion cells.*

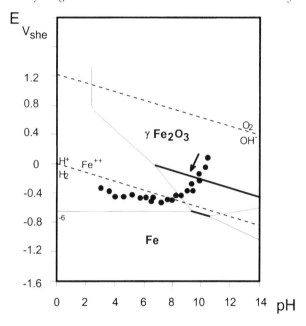

Fig. 3 *Potential and pH changes during the propagation of a crevice or a pit on iron. Experimental results (J. Van Muylder, 1969).*

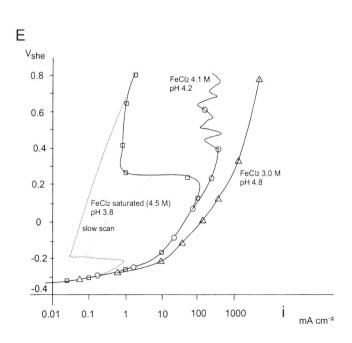

Fig. 4 *Polarisation curves for iron in saturated and nearly saturated FeCl$_2$ solutions.*

the free corrosion potential, the current in the pit is anodic and is clearly under activation control. As the cathodic polarisation is increased, the anodic current in the pit decreases and eventually becomes cathodic. The corrosion in the pit is stopped at a certain potential in the pit that corresponds to immunity, for the dissolved iron concentration in the pit. This potential in the pit corresponds to a higher potential measured outside the pit, which is the 'protection potential'. As the abscissa in Fig. 5 is a current density, the values for corresponding points inside and outside the pits differ by the ratio of the passive surface/pit surface area.

Figure 5 corresponds to a pit which is far from saturation (pH 5.5, compared to a $FeCl_2$ saturation pH of 3.5).

Figure 6 shows the slow evolution of E and pH inside and outside an iron pit, when the whole electrode is polarised in the cathodic direction. As the external potential decreases, the conditions in the pit become less acid, less concentrated, and the potential in the pits decreases towards the immunity region. Therefore, the term « repassivation potential » which is often used to describe the protection potential is not fully correct, at least for iron and carbon steel. The pit, in fact, deactivates rather than repassivates.

For stainless steels, there are many experimental data on the composition of the solution in pits: pH below 1.0 are reported by Suzuki [6]. Hakkarainen found 10M Cl^- and about 4.5M of 'metal ions' [7]. A chloride concentration up to 12M was observed in pits of 316 stainless steel at 25°C. The Cl^- concentration seems to decrease with an increasing size of the pit [8] and the corrosion rate is at a maximum for a concentration slightly below saturation. In all cases, the local conditions in active pits always correspond to conditions for active dissolution.

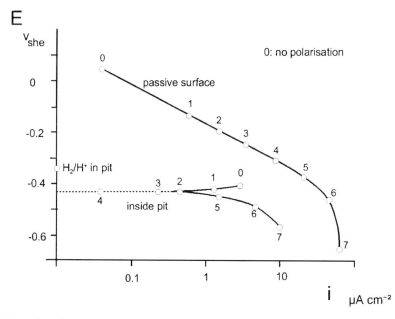

Fig. 5 *Cathodic polarisation curves for the passive surface and for a pit on iron, in NaCl 10^{-3}M + NaOH 10^{-3}M (pH$_e$ = 9.5, pH$_1$ = 5.5).*

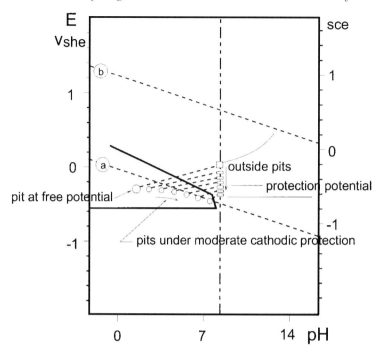

Fig. 6 *Changes of pH and potential in pits of iron as a function of the potential of the passive surfaces.*

Figure 7 represents the active pit and the passive surfaces of stainless steel in a schematic E–pH diagram: the black circle in the corrosion domain corresponds to the pit, and the vertical zone corresponds to the aerated passive surfaces.

The protection potential against the propagation of localised corrosion is slightly higher than the potential in the pit. The difference corresponds to the diffusion potential between the concentrated solution in the pit and the dilute solution outside.

It should be noted that the 'protection potential' (as defined in 1962 [1]) for existing pits of a reasonable size and the 'transition potential' for metastable very early pits [9,10] are not necessarily exactly the same.

The protection potential can thus be estimated from theory: it is the potential measured outside of the pit for which the corrosion in the pit comes to a stop. The ultimate potential in the pit is very near to the corrosion potential of the metal in solution saturated in the 'metallic salt' present. For stainless steel, when chloride is the major anion, the solution is saturated by a mixture of $FeCl_2$, $NiCl_2$ and probably of $CrCl_3$, with a very low pH [6]. For mixed sulfate/chloride solutions, the pH will be less acidic, since metal sulfates are generally less soluble than metal chloride. The detailed conditions in mixed salt environment requires experimental determinations or more development of mixed salt thermodynamics (Gouy, Milner, and Pitzer). The diffusion potential between the pit and the bulk, and the dissolved metal concentration in the pit have a slight effect on the protection potential, which is reflected by the width of the horizontal line for the protection potential in Fig. 7.

It is also interesting to determine the protection potential for a given metal/ environment system and for a given geometry of a crevice or a pit by conducting

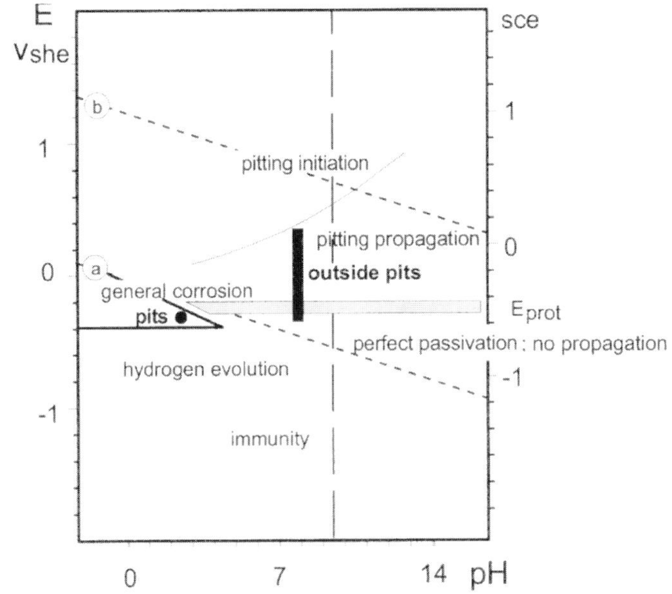

Fig. 7 Conditions inside and outside pits of stainless steel, with the protection potential range and the domain for perfect passivation.

some experiments, similar to those of Figs 2 and 5. Considering the importance of the protection potential for long term disposal of wastes, it is recommended to verify and confirm the values of the protection potential for passive materials both by calculation based on saturated solutions and by experimental verification on simulated pits.

The protection potential gives a solid indication of the conditions under which pits will not grow. The protection potential can be considered as a pseudo-thermodynamic concept, in the sense that it can be estimated from the equilibrium potentials of the hydrogen evolution reaction and of the corrosion reaction, for the pH and concentrations that correpond to nearly saturated metalic salt solutions. It is not influenced by time as is the pitting initiation potential. This protection potential concept has been considered in Japan (since 1984) and in USA (since 1993) for studies on the long-term behaviour of nuclear waste containers [11–17]. It is presently considered also in France.

A list of possible remedies against the propagation of localised corrosion cells has been presented elsewhere [18], for example, alkalisation, buffering, inhibitors that precipitate in acid solution, or potential control. The last is especially effective and valid in the long term: if the service potential is below the protection potential, then localised corrosion cannot propagate.

3. Values of Protection Potential

There is a large collection of data for the protection potential of carbon steel and for stainless steels in chloride solutions. Some were determined theoretically, some experimentally by the zero-current potential on the back scan polarisation curve, others by potentiodynamic or potentiostatic tests on samples with crevices in solutions simulating the chemistry of pits, or on arrangements simulating a local pit or crevice and the surroundng passive surfaces. Finally, a series of values obtained from industrial applications is available.

Most of the values given for austenitic stainless steels in chloride environment are in the range –200 to –550 mV(SCE) [19]. Some values of the potential in active pits for AISI304 and 316 stainless steels are higher (–100 to –150 mV(SCE)) [20]. We are inclined to play on the safe side and to consider the lowest range, i.e. between –450 and –550 mV(SCE). A number of industrial remedial applications have been quite successful in that range.

There are some opinions that this protection potential range is too conservative for austenitic stainless steels, especially for the molybdenum grade AISI316 or for stainless steels containing copper. To clarify these different viewpoints, we recommend that detailed experiments using simulated pit or crevice arrangements be conducted such as those reported in Figs 2 and 5. Several studies on stainless materials have used different simplified experimental approaches to determine the protection or repassivation potential. Considering the challenge of long term prediction, it is recommended that these values be confirmed or verified by a combination of basic calculation and experimental work on real geometries.

The same is even more true for the newer passive materials, such as Alloy 22.

4. Limitations and Further Studies

Some more work is needed to determine the value of the protection potential over the very long term: the propagation of a pit can be stopped by controlling the potential at a suitable low value. As a result, the potential in the pit decreases slightly, the pH increases and the concentrations of metal and of chloride decrease. The new protection potential, for these new conditions is then slightly decreased [18]. This is illustrated for iron in Fig. 8: the limits of the corrosion region are curved lines because the iron concentration in the pit decreases as the pH increases. The second abscissa scale represents the chloride concentration in the pit, which is directly related to the dissolved iron concentration and to pH.

This aspect is important for the very long term and deserves further study.

The situation in clay or backfilling material repositories would be satisfactory providing the absence of oxygen led to free potentials below the protection potential. If, however, low potentials are due not only to the absence of oxygen but also to the presence of sulfides, there could be a new concern. It is expected that sulfides would present no problem at slightly alkaline pH values, but we think that local acid formation in the presence of sulfide is a possibility. Previous studies on differential sulfide concentration cells provide some data on this, and a more detailed analysis on the role of sulfide is necessary.

Thiosulfate may be present in oxidised clay. It is thought that thiosulfate pitting may not occur at potentials in the range of the more conservative protection potentials, i.e. below –450 to –550 mV(SCE) [21]. This may not be true with the more optimistic

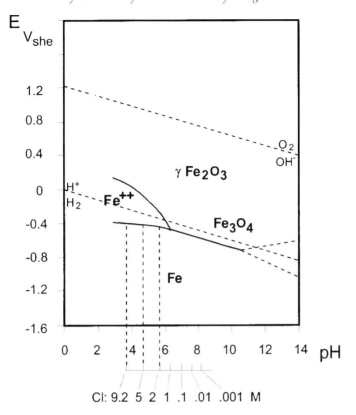

Fig. 8 *Schematic* E–pH *diagram for iron in chloride solution, with the corrosion domain represented by considering the relationship between pH, [Fe²⁺] and [Cl⁻] in pits.*

values of –100 to –150 mV(SCE). Again, a more detailed analysis on thiosulfate is recommended.

Another important subject for further studies is the possible dependence of the protection potential on temperature. It is known that the pitting initiation potential depends largely on temperature and the values of the 'critical pitting (initiation) temperature' has led to perhaps useless modifications of the design and the management of repositories. It is probably more relevant to analyse if and how the protection potential is influenced by temperature.

Pits and crevices may be considered as different in terms of risk for stainless materials because of their inherently different geometries. The experimental approach suggested above for the verification of the protection potential considers both types of geometries, and more especially a geometry that corresponds to contact with soils or backfills.

5. Provisional Conclusions

The concept of a 'protection potential' against the propagation of localised corrosion in a chloride environment is a useful tool for very long term prediction.

The interpretation is that if the zero-current potential of the passive materials considered for the waste containers is below their 'protection potential', localised corrosion due to chloride or sulfate environments will not propagate. This is true for pitting corrosion, crevice corrosion, and stress corrosion cracking.

Similarly, passive materials that exhibit a protection potential which is higher than their zero-current potential should be suitable for long term storage.

Considering the importance of the challenge, more detailed basic and experimental work is recommended to confirm the value of the protection potential for stainless steels and more alloyed passive materials. Examples of the basic approach and of experimental techniques have been given.

The value of the zero-current potential in the course of time, in the true storage environment, must be thoroughly documented and investigated. It is possible that just the absence of oxygen after some time may produce such low zero-current potentials.

A slight cathodic protection during the early years of aerated soils would be beneficial and might be considered.

Successive aerobic conditions with chloride and anaerobic conditions with sulfides must be analysed in detail, particularly for possible local acid formation.

The long term changes of the composition of pits at potentials below the protection potential have been little studied. Considering its importance for nuclear waste storage, it deserves more analysis.

The 'protection potential' is somehow a pseudo-thermodynamic value, in the sense that it is related to well defined electrochemical equilibria. This makes it especially interesting for solid prediction in the very long term.

References

1. L. Klimzack-Mathieu, J. Meunier, M. Pourbaix and C. Vanleugenhaghe, 'Sur le comportement électrochimique d'aciers alliés en solutions chlorurées — 1. Influence de chlorure et d'oxydants en présence de bicarbonate', Rapports Techniques CEBELCOR, Vol.86, RT.103 (1962); in English: *Corros. Sci.*, 1963, **3**, 239.

2. M. Pourbaix, 'Recherches en corrosion. Résultats de travaux récents, voyages aux Etats-Unis d'Amérique', Rapports Techniques CEBELCOR, 1969, Vol.109, RT.158.

3. L. Sathler, 'Contribution à l'étude du comportement électrochimique du fer en présence de solutions de chlorure ferreux, en liaison avec la corrosion localisée', Thèse Université Libre de Bruxelles, Fac. Sc. Appliquées, 1978.

4. L. Sathler, J. Van Muyler, R. Winand and M. Pourbaix, 'Electrochemical behaviour of iron in localized corrosion cells in the presence of chloride', in *Proc. 7th Int. Congr. Metallic Corrosion*, Rio de Janeiro, p.705–718, Abraco, 1978.

5. A. Pourbaix. 'Characteristics of localised corrosion of steel in chloride solutions', *Corrosion*, 1971, **27**, (11), 449–454.

6. T. Suzuki, M. Yamaka and Y. Kitamura, 'Composition of anolyte within pit anode of austenitic tainless steels in chloride solution', *Corrosion*, 1973, **29**, 18–22.

7. T. Hakkarainen, 'Anodic behaviour of stainless steels in simulated pit solution', in *Corrosion Chemistry within Pits, Crevices and Cracks*, Proc. Conf. NPL, Oct. 1984 (A. Turnbull, ed.). Publ. HMSO (1987) pp.17–26.

8. J. Mankowski and S. Szklarska-Smialowska, 'Studies on accumulation of chloride ions in pits growing during anodic polarization', *Corros. Sci.*, 1975, **15**, 493–501.

9. R. C. Newman, 'Dissolution and passivation kinetics of stainless steel alloys containing molybdenum. II: Dissolution kinetics in artificial pits', *Corros. Sci.*, 1985, **25**, 341–350.

10. N. J. Laycock and R. C. Newman, 'Localised dissolution kinetics, salt films and pitting potentials', *Corros. Sci.*, 1997, **39**, 1771–1790.

11. S. Tsujikama and Y. Hisamatsu, 'Repassivation potential as crevice corrosion characteristics for austenitic and ferritic stainless steels', in *Improvement of Corrosion Resistance of Structural Materials in Aggressive Media* (Ya.M. Kolotyrkin, ed.).Publ. Nauka, Moscow, 1984, p.119.

12. S. Tsujikama and Y. Kojima, 'Repassivation method to predict long term integrity of low alloy titanium for nuclear waste package', *Scientific Basis for Nuclear Waste Management XIV*, Materials Research Society Symposium Proceedings (T.A. Abrajano and L.H. Johnson, eds). Materials Research Society, Pittsburgh, 1990, **212**, 261.

13. G. Nakayama, H. Wakamatsu and A. Masatsune, 'Effects of chloride, bromide and thiosulphate ions on the critical conditions for crevice corrosion of several stainless alloys as a material for geological disposal package for nuclear wastes', *Scientific Basis for Nuclear Waste Management XVI*, Materials Research Society Symposium Proceedings (C.G. Interrante and R.T. Pabalan, eds). Materials Research Society, Pittsburgh, 1992, **294**, 294.

14. N. Sridhar, G. A. Cragnolino, J. C. Walton and D. Dunn, 'Prediction of Crevice Corrosion using Modeling and Experimental Techniques', in *Application of Accelerated Corrosion Tests to Service Life Prediction and Materials*, ASTM STP 1194 (G.Cragnolino and N.Sridhar, eds). ASTM, Philadelphia PA, 1994.

15. N. Sridhar and G. A. Cragnolino, 'Long-term life prediction of localised corrosion of Fe–Ni–Cr–Mo high-level nuclear waste container materials' in *Corrosion '93*, Paper 197, NACE International, Houston, TX, USA, 1993.

16. D. Dunn, N. Sridhar and G. Cragnolino, 'The Effect of Surface Conditions on the Localised Corrosion of a Candidate High-Level Waste Container', *12th Int. Corros. Congr.*, Houston, TX, 19–24 September, 1993, NACE International, Vol .5B, p.4021.

17. N. Sridhar and G. A. Cragnolino, 'Applicability of repassivation potential for long-term prediction of localised corrosion of alloy 825 and type 316L stainless steel', *Corrosion*, 1993, **49**, 11, 885–894.

18. A. Pourbaix, 'Corrosion localisée: fonctionnement et mécanismes de protection', Rapports Techniques CEBELCOR, Vol.148, RT.276 (1984). In english: in 'Corrosion chemistry within pits, crevices and cracks', Proc. Conf. NPL, Oct 1984 (A.Turnbull, ed.). Publ. HMSO, 1987.

19. A. Pourbaix, J. Kissel and I. Couzet, 'Electrochemical corrosion monitoring and corrosion prevention methods', in *Proc. Int. Symp. on Corrosion Science and Engineering*, Vol.2, ed. Cebelcor, Brussels, 12–15 March, 1989, Rapports Techniques CEBELCOR, vol.158, RT.298, pp.581–595 (1989).

20. T. Hakkarainen, personal communication, and 'Pitting of stainless steels. Basic requirements and effects of key variables', in *7th. Int. Symp. on Electrochemical Methods in Corrosion Research*, keynote 09, EMCR2000, Budapest 28 May 28–1 June, 2000.

21. A. Garner and R. C. Newman, 'Thiosulfate pitting of stainlss steels', in *Corrosion '91*, Paper 186, NACE International, Houston, TX, March, 1991.

Part 3

Analogue Session

17

Archaeological Analogue Studies for the Prediction of Long Term Corrosion on Buried Metals

D. DAVID, C. LEMAITRE and D. CRUSSET*

Université de Technologie, Laboratoire Roberval (UMR CNRS 6066), BP 20529, 60205 Compiègne, France
*ANDRA, 1–7 rue Jean Monnet, 92298 Châtenay-Malabry cedex, France

ABSTRACT

The prediction of the behaviour of metallic materials exposed to soil corrosion depends on many parameters. Any consideration of the effect of time, over long periods, can result only from a comparison with real cases of corrosion. It is the material–soil association which is the determining factor. This paper reports studies performed using modern techniques of materials science. The studies concern not only the characterisation of solid phase components, especially corrosion products since it is also necessary to conduct electrochemical evaluations of corroded layer properties.

1. Introduction

The development of nuclear energy, as of many other industrial activities, leads to the production of numerous types and quantities of wastes. These have the peculiarity of being radioactive for periods varying from some hours to millenia. The nuclear industry is not the only source of these wastes: other industries, as well as medicine, also generate waste. Most countries which have opted to dispose of their longer-lived products are exploring the possibilities of burial in suitable geological formations. In this way they are able to minimise the surveillance constraints for following generations.

The amount of radioactive waste annually produced in France represents about 28 000 m^3. The major part (90%) is of weak activity, with a half life of less than 30 years. The remaining 10% have a longer half life of up to several thousand of years [1]. In the case of underground disposal these radionuclides will be locked into metallic canisters, which will confine them for a few centuries. That is why one of the ANDRA programmes concerns metal corrosion.

Experimental corrosion studies in laboratories provide a good knowledge of corrosion mechanisms in a defined environment but only over a relatively short time scale. Their interpretation requires theoretical understanding of the processes and is achieved by modelling. Another approach is to use archaeological analogues and this is the subject of this chapter.

2. Natural and Archaeological Analogues: Definitions and Outline

"Natural analogues are defined more by the methodology used to study and assess them than by any intrinsic physico–chemical properties they may possess.(...) Studies about them have been performed, to date, on a wide range of phenomena, including ore deposits, natural fission reactors [2], marine sediments and man-made copper and iron artefacts. The latter, archaeological artefacts, are studied in the same way as natural analogues and are generally classed with them [3]."

The most relevant archaeological analogues for this purpose are those corresponding to the geologic and climatic conditions of France dating from 1700 to about 2000 years ago. They are Gallic or Gallo–Roman traces that can be found quite easily. In addition, other more recent metal artefacts also provide useful information, such as those of the First World War battlefields. We shall limit ourselves here to the case of conservation in soil as opposed to that in a maritime environment. These studies are conducted using modern techniques that have been developed by material science. This archaeological approach is recent, like paleometallurgy. However, the latter mainly concerns ancient techniques and does not take into account corrosion phenomena *per se*. The initial state of objects, as well as the conditions of their conservation over a very long period is often uncertain. It is therefore necessary to carry out many observations in order to give them a statistical value. In this way we shall be able to determine tendencies which underlie behavioural laws.

3. Mechanisms of Corrosion

3.1. Generalities

Corrosion in soils is mainly electrochemical, which means that it is a form of aqueous corrosion. Indeed, buried metals are in contact with an environment which globally acts like an electrolyte [4].

One of the characteristics of this environment is its electrical conductivity associated with the presence of salts in the water. The water contains, in solution, at least part of the soluble mineral salts present in the soil such as chlorides, sulfates, carbonates or nitrates. These dissolved and ionised salts give to the solution its electrical conductivity and thus its electrolytic character. All the examples that have been observed over very long periods were located between two extremes: either a liquid solution or a perfectly dry soil. The conductivity of the latter would be almost non-existent; such a soil would be insulating and would not produce electrochemical corrosion (there are, however, cases of corrosion in desert environments).

Heterogeneity is a key factor in this form of corrosion. Indeed, whatever the size of a corrosion cell, it is always the result of a lack of homogeneity of the metal or of the corrosive environment. In fact, soil is always heterogeneous, generally passing through the three following stages:

- **solid:** composed of mineral grains, which can have a heterogeneous composition and granularity;

- **liquid:** grains will be covered by an aqueous film in non-saturated soils, and entirely immersed in saturated soils;

- **gaseous:** mainly composed of air and carbon dioxide; in this stage, gases such as methane or hydrogen sulfide resulting from bacterial or chemical reactions can be present.

The liquid stage is that mainly responsible for corrosion since it allows electric currents to flow between anodic zones of the buried structure (oxidation — therefore corrosion phenomena) and cathodic zones (reduction of chemical species in the soil around the structure). The lateral extension of the initially localised anodic areas (pitting) will lead to a more uniform aspect of corrosion, at least on a macroscopical scale.

Two types of corrosion can therefore be observed: either general or localised, the latter being the more widespread among buried metals.

3.2. Thermodynamics

As seen above, the corrosion of metals, like iron, is an electrochemical process and is the result of two half-reactions which, in aqueous conditions, are [5]:

$$Fe \rightarrow Fe^{2+} + 2e^- \qquad (1)$$
$$\text{[oxidation, anodic reaction]}$$

$$O_2 + 4H^+ + 4e^- \rightarrow 2H_2O \qquad (2)$$

$$O_2 + 2H_2O + 4e^- \rightarrow 4\,OH^- \qquad (3)$$
$$\text{[reduction, cathodic reactions, in presence of air]}$$

In the absence of air the half-reactions are:

$$2H^+ + 2e^- \rightarrow H_2 \qquad (4)$$

$$2H_2O + 2e^- \rightarrow 2OH^- + H_2 \qquad (5)$$

The cathodic reaction (4), which is the reduction of hydrogen ions, does not occur significantly at pH values above 4, except perhaps in the presence of bacteria [6].

The electric circuit established between the anode and the cathode is completed by electronic transport in the solid phase and ionic transport in the aqueous phase. The solid corrosion products result from subsequent reactions involving on the one hand Fe^{2+} ions from the anode and, on the other hand, the components of the electrolyte in contact with the metal [7].

The classical way to represent the conditions of stability of these compounds is by

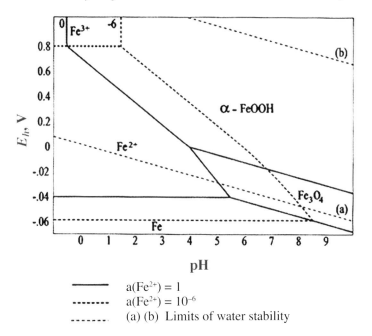

Fig. 1(a) Potential–pH diagram for the iron–water system (according to Turgoose).

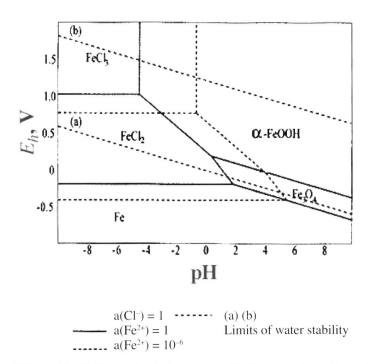

Fig. 1(b) Potential–pH diagram for the iron–water–chloride system (according to Turgoose).

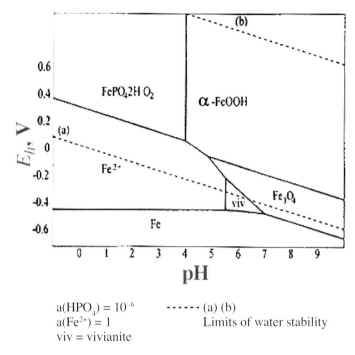

a(HPO$_4$) = 10^{-6} ----- (a) (b)
a(Fe^{2+}) = 1   Limits of water stability
viv = vivianite

Fig. 1 (c) *Potential–pH diagram for the iron–water–phosphate system (according to Turgoose).*

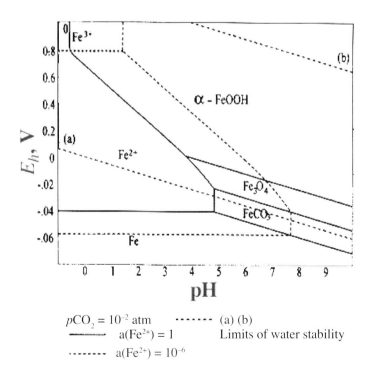

pCO$_2$ = 10^{-2} atm   ------ (a) (b)
——— a(Fe^{2+}) = 1   Limits of water stability
------- a(Fe^{2+}) = 10^{-6}

Fig. 1(d) *Potential–pH diagram for the iron–water–carbon dioxide system (according to Turgoose).*

the use of potential/pH diagrams. The diagram for the iron–water system is shown in Fig. 1. It shows the existence of two oxides, magnetite Fe_3O_4 and goethite α-FeO(OH), which in fact are those usually observed on archaeological artefacts. Hematite, α-Fe_2O_3, is more stable than goethite but the energy of the goethite/hematite transformation is weak and so, depending upon conditions, hydroxide or oxide can be found. Many authors have found that the corrosion products are those predicted by thermodynamics: magnetite, goethite and hematite. Akaganeite β-FeO(OH) was observed only once on roman nails, in zones with active corrosion and in the presence of chlorides [8].

The other main parameters are pH and Eh (redox potential). Their values depend on many factors. Their distribution as usually observed in natural waters is located, on the one hand, between the curves (a) and (b) of Fig. 1 and, on the other hand, between the pH values 4 and 9, although some values exceeding these limits have been observed.

The particular case of corrosion under corrosion products can drastically modify the characteristics of the environment in contact with the metal. In these conditions, the corrosion will continue by anodic or cathodic diffusion. In the first case, the anions released by the electrolyte will move towards the interface between metal and the corrosion product, where the oxidising process continues. In the second case, cations will move towards the electrolyte, and the corrosion process will continue on the external side of the corrosion product. These conditions can vary widely over the course of time. They can, at least initially, lead to the formation of pseudo-passivating layers without initiating a proper passivation process. Such formations are called blocking layers.

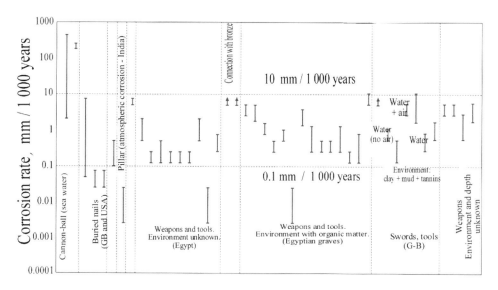

Fig. 2 *Corrosion rate data for iron artefacts (according to Johnson & Francis, reviewed by D. Crusset).*

3.3. Kinetics

A global estimation of the corrosion rate for archaeological artefacts was made by Johnson and Francis [9]. The results are shown in Fig. 2. These workers have listed 44 iron samples, but their origins were different and often not known precisely. A similar listing was provided for 33 copper artefacts.

We notice that, in spite of the variety of artefacts (origin, environment...), most corrosion rates are between 0.1 and 10 mm per millennium. The large number of samples confers a statistical value on this estimation. High values correspond to oxidising environments (presence of air or oxygen, in water or soil), sometimes aggravated by chlorides. Dry conditions are reducing (water or soil free from air or oxygen) and correspond to the lowest corrosion rates, such results being observed with artefacts preserved under corrosion layers. It should be noted that these corrosion rates are mean values and exclude pitting corrosion. Such data are not available for long-term localised processes which can sometimes lead to catastrophic failure. However, the numerous artefacts preserved show that this case is probably not the most frequent.

These assessments are confirmed by another study concerning more recent objects [10]. Workers have provided many corrosion cases on samples buried voluntarily, and correlated these with soil parameters. Most of the usual metals were studied, as well as their alloys: iron, copper, lead, zinc, aluminium, tin. In France, similar studies were performed by Laboratoire Central des Ponts-et-Chaussees [11].

Many authors report that corrosion rates decrease very quickly, at least in the first years. The losses observed in actual soils are always lower than those indicated by NBS (US National Bureau of Standards). For marine soils containing chlorides over 200 mg kg^{-1}, or soils with a resistivity slightly less than 1000 Ωcm, results are similar to those for aggressive soils reported by NBS.

Simulations over relatively long periods, i.e. a few decades, can give only very limited information so that extrapolation laws must then be defined and validated [12]. The parabolic law, often used for extrapolation, is:

$$x = k\, t^n \tag{6}$$

where x is the average thickness of corroded metal after time t, K a soil coefficient, and n a reducing parameter ranging from 0 to 1.

4. Soil Characterisation: A Specific Problem

4.1. Some Very Different Points of View

Soils are complex environments which, depending on the specific problem [13], can be studied by using different approaches including physics, chemistry, biology, or natural water cycles. Pedology, which is the science dedicated to soil studies and properties, is linked to geology. Both contribute to geography which is linked to human sciences. Humans have always left behind them many buried artefacts which nowadays provide many examples for the study of corrosion processes. The soils can either be at prehistoric sites or at modern battlefields.

The most significant soils in a corrosion context are:

- natural soils, without human influence; and

- homogeneous soils (clay, sand...).

Archaeological soils are generally stratified, showing human and natural influences throughout long periods of time (sediments, climatic precipitations). This is why archaeological soils containing artefacts are always different from their original condition. This is also why corrosion phenomena taking place in such environments are more difficult to understand since they result from numerous and interdependent parameters.

4.2. Solid Phase

A soil can be defined as the result of multiple interactions between natural minerals and a climatic environment in constant evolution. Minerals are generally crystallised particles enclosed in a colloidal matrix mainly inorganic but with organic compounds. The bulk is heterogeneous and is generally more physically and chemically unstable the closer it is to the surface. So, no soil is identical to another. Insoluble minerals are inert substances which form a skeleton. Dissolved compounds, minerals or organics and living ones (bacteria) are, on the contrary, active agents.

4.3. Fluid Phases

The main extraneous elements in natural waters are:

- dissolved gases (oxygen, nitrogen, carbon dioxide, sulfurous gas);

- mineral elements, including mineral salts (chlorides, sulfates, nitrates, bicarbonates), heavy metal compounds, silicates;

- organic matters, from animal or vegetable origin as well as agricultural effluents, oils and detergents; and

- micro organisms, such as algae and bacteria developing in muds and alluvia.

The chemical parameters are the levels of soluble salts and acidity (pH). The solution is an electrolyte which can lead to electrochemical corrosion. Mineral salt compounds contribute to electrical conductivity of the soils. They are able to take part in the formation of corrosion products, isolating the material from its environment. Their action is therefore complex. Some minerals will generate aggressive ions, such as Cl^-. Other ions, such as HPO_4^{2-}, Ca^{2+} and Mg^{2+} (in basic or neutral environments), will take part in the formation of compounds that will protect the metal.

Each modification of dissolved species can influence water aggressivity. For example, if the rates of supply of oxygen and carbon are lowered, the water will become less aggressive to metals. On the other hand, metallic ions in a water solution will tend to make it more corrosive. Some waters will corrode copper which will lead to localised corrosion of metals lower on the potential scale such as iron, galvanised steel and aluminium. This phenomenon can be explained by microscopic copper particles depositing on these metals to form electrochemical cells.

4.4. Global Evaluation of Soil Aggressivity

As well as metal corrosion, natural waters can be responsible for other adverse effects when metallic ions from a corrosion process are present. For example, at a toxicological level, lead pipes are worse than copper ones.

The concentrations of chemical species, present in solutions, colloidal or normal suspensions are relatively low, but can show important variations. For instance, water hardness ranging from 300 to 400 mg L^{-1} (CaCO$_3$) can be acceptable in drinking water, whereas an iron concentration of 1 mg L^{-1} is prohibited. (The usual unit is mass per volume except for gases.) Acidity is generally a corrosion factor. Soil pH is usually between 5 and 8.5, but its value compared to water is only approximate. The variety of corrosion factors make it difficult to establish a reliable relation between pH and the real aggressivity of the soil.

Contrary to what happens in liquid electrolytes, soils are not good removers of corrosion products which thus stay close to the corroded metal so that the chemical composition of the immediate environment is modified. The gangue* can act as a protection if they are compact and not too humid and will preserve metal almost forever. This is often so with archeological analogues. On the other hand, a water permeable product will tend to worsen the destructive process by forming occluded corrosion cells, which operate in a confined environment.

Fine mineral grains constitute yet another acceleration factor (clay elements made of aluminium silicate). Such grains will tend to fix metallic ions resulting from the start of the corrosion process, thus increasing it.

Basing his argument on this notion of total acidity (in mg-equivalents of hydrogen ions for 100 g of soil), Romanoff published a diagram showing that:

- for soil resistivity ranging from 4×10^3 to 5×10^3 Ω.cm, the aggressivity increases is proportional to total acidity;

- for total acidity ranging from 15 to 18 mg-equivalents, aggressivity is inversely proportional to resistivity.

The aggressivity criterion is used in deciding the need to repair pipelines and is above all a qualitative test. The data require values of electrical resistivity which is a major parameter and will be discussed below.

The huge economic importance of underground pipework systems has triggered numerous studies on the specific forms of corrosion. Comparable studies have been conducted on underground metallic parts of civil engineering buildings. For instance, pH data show a corrosion risk in acid environments, whereas E_h data are related to bacterial corrosion. Low E_h values indicate a deaerated soil which favours anaerobic

---

*Gangue is the material adhering to the metal object and is composed of mixture of corrosion products and clay (or other types of soil).

bacteria such as the sulfate-reducing. Resistivity values, which can give information on humidity of a soil, are also important.

Furthermore, all forms of corrosion do not have have the same effects. Pitting corrosion which is not very important in terms of mass losses, can be more dangerous than uniform corrosion. It must therefore be assessed according to another criterion, which is the pitting depth.

Therefore, it is not possible to define soil corrosiveness by only one parameter. On the contrary a global index is needed to simplify applications. Results should then be re-examined more thoroughly and precisely. This method, developed by Steinrath, is called 'global index approach'. The Commission for Western Europe (CEOCOR) has opted for it, for studying buried pipes corrosion (DVGW GW9 classification) [14]. It seems to give generally satisfactory results (see Annex).

5. Experimental Studies

Experience shows that buried artefacts generally evolve towards an equilibrium. Metal undergoes a loss of mass and a corroded interphase develops near its surface. The compounds in this interphase play an important role in chemical processes. Their characterisation, therefore, is an important step for the life prediction of such materials since it allows a determination to be made of the reactions that can occur. It can be complemented by loss of mass, or loss of thickness measurements, versus time. Finally, this passive study must be complemented by active determinations, mainly electrochemical.

5.1. Metal and Soil Characterisation

The phases present are the metal itself, the soil and the corrosion products interphase. The main analytical techniques that are the most useful for metal and compounds characterisation of metal and compounds are given below. Techniques used for soils are specific and do not come under these topics.

5.1.1. Metallography.
Metal is observed by means of usual metallography techniques: optical microscopy, scanning electron microscopy (SEM), X-ray diffraction. For example, gallic iron structures are ferritic with numerous grains having equiaxed orientations. Grain structure is hypoeutectoid. This material is not very different from modern grades of iron produced by the direct process (reduction without fusion). The metal is then very pure because impurities are not in the mass during fusion.

In Fig. 3 the corroded zone is rather rectangular in shape and can be seen on the right. On the left, the grey area corresponds to metal with several inclusions resulting from ancient metallurgical techniques. It is quite pure iron. In Fig. 4, ternary cementite is evident and can be seen as white lines between crystalline grains.

These observations are complemented by analytical techniques carried out in the same equipment: EDS spectroscopy, profilometry and element cartography using X-ray emission. This enables element distribution, for example, metal inclusions to be studied.

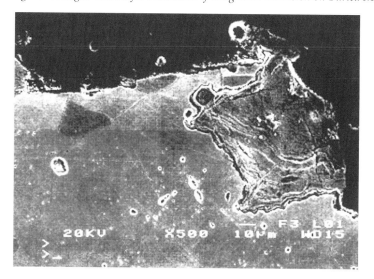

Fig. 3 *Iron nail, included in a Gallo–Roman wall. Technical purity, with numerous inclusions. Resin appears black. SEM microscopy, ×500.*

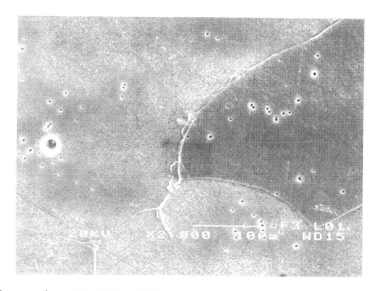

Fig. 4 *Same artefact as Fig. 3. The white lines between grains are ternary cementite.*

These are mainly silicates and iron oxides, and reveal the presence of elements usually contained in the ore gangue: silicon, oxygen, calcium, aluminium, potassium, magnesium, phosphorus, sulfur, as well as a considerable amount of chlorine. The quantitative variations in these can be explained by the origin of the ores which probably differs from one century to another.

Figure 5 shows an example of element cartography using X-ray emission. The artefact is a Gallo–Roman nail (1700 years old) and the choice of selected elements results from EDS analytical data. The top left represents an iron signal where metal with numerous cracks can be seen in the top right hand corner. The top right represents a calcium signal which is a marker for hydraulic mortar. Between these two, in the middle, can be seen a silicon map, whose distribution is less important. On the bottom row, from left to right, can be seen:

- the carbon signal from a component of the resin coating which penetrates the metal cracks; and

- the oxygen distribution which is mainly present in the corroded phase.

These analytical representations should be compared with the final map (bottom right) which represents the sample morphology.

5.1.2. Profilometry using nuclear microanalysis (RBS).

Another analytical method, quantitative and precise, is nuclear microanalysis. In its different configurations, it gives access to most of the chemical elements. Figure 6 presents an example of RBS (Rutherford Back Scattering) lateral profilometry of iron. The sample is a polished cutting of the same artefact as above. The signal plot corresponds to the surface image. Such measurements are also made on soil samples, to investigate the possible presence of iron arising independently of the artefact. These results contribute to the determination of the relative corroded mass.

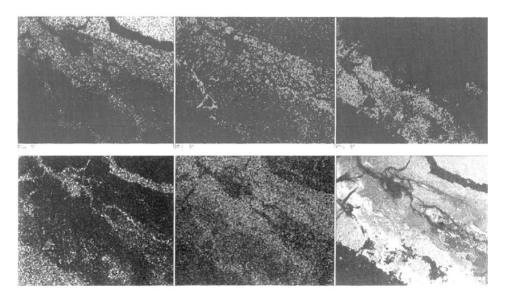

Fig. 5 *Element maps, using X-ray emission. From left to right, and upper to lower: Fe, Si, Ca, C, O, and SEM view of the sample surface.*

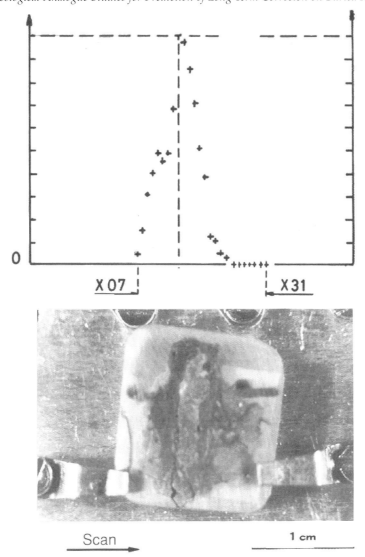

Fig. 6 Fe lateral profiling, using RBS. Upper: Fe signal, normalised scale. Lower: sample section, showing the metal nucleus, the corroded interphase and the soil. (Analytical apparatus: Groupe de Physique des Solides, UMR CNRS 75-88, Paris.)

5.1.3 Molecular analysis.

The methods described above do not provide information about chemical bonds and do not allow phases present to be identified. This information is however indispensable to the understanding of reaction processes. Among techniques that allow this identification, may be mentioned X-ray diffraction which is well known, and Raman laser spectroscopy [15].

As an example, Fig. 7 shows the Raman spectrum from the surface of the same corroded iron artefact shown in Fig. 6. Mortar overlapping metal is revealed by the

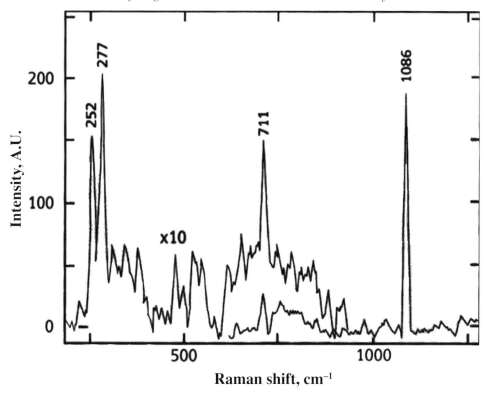

Fig. 7 Raman laser spectrosccopy. Gallo–Roman nail, embedded in concrete. Metal and mortaroverlap is shown by the presence of two compounds: the lepidocrocite γ-FeOOH and the calcite CaCO₃. (Analysis: A. Hugot Le Goff and S. Joiret, UPMC, Laboratoire de physique des liquides et d'électrochimie, UPR CNRS 15, Paris.)

presence of two compounds: lepidocrocite γ-FeOOH and calcite $CaCO_3$. This technique is very convenient for the identification of corroded phases. In the case of iron artefacts, the compounds are mainly magnetite Fe_3O_4, goethite α-FeOOH, akaganeite β-FeOOH, maghemite γ-Fe_2O_3 and hematite α-Fe_2O_3.

5.2. Measuring the Corrosion

The degradation of an archaeological analogue brings in a reference to time. The problem is how to observe the chronological variation of a parameter, which is a speed. There are three possibilities:

- measuring a loss of mass, or a loss of thickness;

- measuring the number of metal ions formed per unit of time; and

- measuring the number of electrons formed per unit of time, that is an electric current (this is the principle of electrochemical methods).

5.2.1. Loss of mass or thickness.

The first difficulty is to know the initial mass and this can only be estimated. This uncertainty does not deprive the subsequent measurements of any meaning. It is necessary that the artefact has initially a well-defined shape since the corrosion process usually increases the bulk appearance, while the non-corroded mass of metal decreases this. As a consequence, the original surface is located at an intermediate level, i.e. in the thickness of the corroded interphase. This level can sometimes be marked by the observation of soil silica particles. These grains are present outside the original surface but not inside this surface. There is also a possibility of measuring the loss of thickness and of calculating a corrosion rate. These data were listed by Johnson and Francis.

Another possibility is to estimate the loss of thickness by direct observation, using optical methods, on through sections of the artefact. Accuracy is improved if this observation is completed by analytical profiling to obtain the iron or corrosion compounds rate of concentration.

5.2.2. Metal ions in solution.

We measure the concentration of metal cations in solution with time. Experimental data can result from analyses of the corrosive environment. It is also possible to use specific electrodes that allow metal concentration to be obtained directly. These techniques give only mean values and do not reveal localised corrosion. Besides, if corrosion products are hardly soluble and precipitate, part of the metal ions will not be measured. The apparent rate is then lower than its real value. This defect limits the field of application of the method which is generally used for industrial controls.

5.2.3. Electrochemical methods.

Steady state electrochemical techniques allow us to measure the corrosion rate from the steady state current value. This is the resultant of the cathodic half-reaction (reduction, current I_c) and the anodic half-reaction (corrosion, current I_a). For the equilibrium potential, equation is :

$$I_a = -I_c = I_o \tag{7}$$

For that reason, it is necessary to move the potential away from the equilibrium, value in both directions, by polarising the system. We interpole the values of the observed current, which is:

$$I = I_a + I_c \tag{8}$$

The result is a polarisation curve. Current I and polarisation η are connected by the Buttler–Volmer's relation:

$$I = I_0 \left[\exp\left(\frac{\eta}{\beta_a}\right) - \exp\left(-\frac{\eta}{\beta_c}\right) \right] \tag{9}$$

This relation is in agreement with the cancellation of the current I if there is a lack

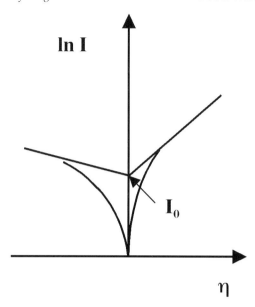

Fig. 8 *Schematic view of a polarisation curve: current vs potential. The tangents are Tafel lines.*

of polarisation. Figure 8 gives an example of its graphic representation. The two tangents are called Tafel lines. Their intersection allows us to determine the value I_0 of the exchange current (corrosion current), which is proportional to the corrosion rate.

Two experimental techniques are used for the construction of the polarisation curves:

(i)   varying the current and measuring the potential — the intensiostatic mode; and

(ii) varying the potential and measuring the current — the potentiostatic mode.

The latter is that most used, with the potential varied regularly with time.

Another electrochemical method is measurement of the impedance of the corroded layer. Current is varied using variable frequency with the potential following the same variation, but with a phase shift. The impedance value depends on the frequency. A diagrammatic representation allows equivalent electric circuit to be derived (Fig. 9). This technique provides a means for determining the corrosion mechanism. These techniques have been recently used for studying iron corrosion on artefacts that have been buried for very long times [16].

6. Conclusions

Corrosion studies on archaeological artefacts recovered from scientific studies are now developing appreciably. Their main purpose is to assist in understanding long

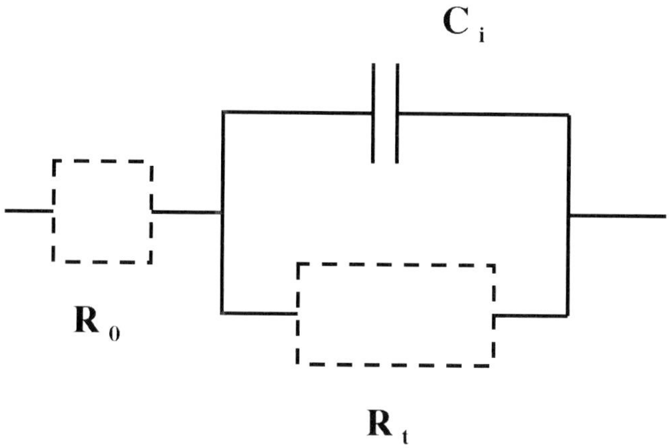

Fig. 9 *Randles' electrical equivalent network, with a resistance and a capacitance for interface. The material resistance is R_0.*

term corrosion behaviour in the geological disposal of radioactive wastes in metal containers. Data resulting from paleometallurgy, case by case, contains some imprecision. On the other hand, the variety of sites and times allows us to formulate predictive laws of behaviour and to estimate quantitatively the corrosion effects. Furthermore, natural and archaeological analogues, as well as samples drawn from the NBS study in natural soils, provide so many cases distributed at random that statistical values can be obtained.

Several general principles are now well established. Whatever the interest of researches that are concerned with materials and soils separately, it is the association of these two elements which is important for corrosion studies. Behaviour predictions, have to be concerned with material–environment combinations, including corrosion products.

Soil corrosion is an electrochemical process: a buried metals environment being generally similar to an electrolyte. Most workers recognise resistivity as a main parameter characterising soil aggressiveness. The pH then becomes important, which tends to give an advantage to slightly basic soils. As for the role of the other factors, the DVGW GW9 classification constitutes a useful basis for assessing the location of repositories.

Thermodynamics gives only a primary approach to the corrosion processes. It is advisable to complement it by kinetics to estimate adequately predictable behaviour. These laws are often not linear versus time and tend to be asymptotic when the equilibrium between metal and soil is reached. This stabilisation can last almost infinitely if the environment is stable. This condition would be credibly satisfied for geological repositories.

Concerning the choice of metal containers, archaeological studies are inevitably restricted to ancient materials.

Bronze would have been a possible choice. But, in spite of the survival of very numerous vestiges, some destructive forms of corrosion have been described [17]. So, these alloys, unlike pure copper, do not appear to be adequate.

On the other hand, ferrous metals appear to constitute the best choice, at least within the framework of this study. This opinion is based on the following observations:

- Numerous examples exist of conservation of objects buried in conditions that were not chosen with conservation in mind. The number of such cases must obviously be unknown but there are limited cases where dating is available by reference to a house or grave;

- The ability of iron or, rather, low carbon steel, to form a protective corrosion layer, under which a chemical equilibrium becomes established;

- A rather wide tolerance at the level of element compositions and structures is also proved by the conservation of gallic objects;

- Sacrificial over-sizing which is compatible with practical possibilities for manufacturing containers;

- A confirmation of paleometallurgical statements, by those made for the 20th century on numerous subterranean buildings;

- Use of soft steels for the reinforcements of reinforced-earth buildings, justifying the development of studies for operating modern techniques of characterisation and simulation;

- Possibility of improving temporarily the material resistance by galvanisation.

Furthermore, this property would tend to decrease the risk of pitting corrosion. These works, as experiments on modern materials and theoretical modelling, are a part of the numerous studies managed by ANDRA.

7. Acknowledgements

Work initiated and supported by ANDRA. The author wishes to thanks L. Uran (UTC), A. Hugot Le Goff and S. Joiret (UPMC) for their collaboration.

References

1. H. de Choudens, Le risque nucléaire, Tec & Doc, Paris, France, 2001.
2. R. Naudet, Oklo: des réacteurs nucléaires fossiles, Eyrolles, Paris, France, 1991.
3. W. Miller *et al.*, *Natural Analogue Studies in the Geological Disposal of Radioactive Wastes*, Elsevier, Amsterdam, 1994.
4. L. L. Shreir, R. A. Jarman and G. T. Burstein, (eds), *Corrosion* (3rd edition), Butterworth Heinemann, Oxford, UK, 1994.
5. D. David, Analogues archéologiques et corrosion, BIO Intelligence Service, Paris, France, 2001.
6. S. Turgoose, *Conservation of Iron* (R.W. Clarke and S.M. Blackshaw, eds), National Maritime Museum, Greenwich, London, UK, 1982.

7. D. H. Andrews and R. J. Kokes, *Fundamental Chemistry*, John Wiley & Sons, New York, USA, 1962.

8. Th. De Putter and J.-M. Charlet, Analogies naturelles en milieu argileux, ONDRAF, Bruxelles, Belgique, 1994.

9. A. B. Johnson Jr and B. Francis, Durability of metals from archaeological objects, metal meteorites and native metals, U.S. Department of Energy, Washington, USA, 1980.

10. M. Romanoff, *Underground Corrosion*, National Bureau of Standards, Circular 579, Washington, USA, 1957.

11. M. Darbin, J.-M. Jailloux and J. Montuelle, *Proc. Inst. Civ. Engrs, Part 1*, 1988, **84**, 1029–1057.

12. M. Kowaka, (ed.), *Introduction to Life Prediction of Industrial Plant Materials*, Allerton Press, New York, USA, 1994.

13. G. H. Bolt and M. G. M. Bruggenwert, *Soil chemistry — Basic Elements*, Elsevier, Amsterdam, 1976.

14. J. O. Harris and D. Eyre, in *Corrosion, op. cit.*, p. 2.86: German Gas and Waters Works Engineers' Association Standard, Merkblatt für die Beurteilung der Korrosiongefährdung von Eisen und Stahl im Erdboden, DVGW GW9, Frankfurt, DVGW (1971).

15. N. Boucherit, A. Hugot-Le Goff and S. Joiret, *Corros. Sci.*, 1991, **32**, (5/6), 497–507.

16. E. Pons, L. Uran, S. Joiret, A. Hugot Le Goff, C. Lemaitre and D. David, Long-term behaviour of iron in clay soils: a study of archaelogical analogues. This volume p.334–345.

17. L. Robbiola, N. Pereira, K. Thaury, C. Fiaud and J.-P. Labbé, in *Metal 98* (W. Mourey and L. Robbiola, eds). Publ. James & James, London, UK, 1998, pp. 136–144.

Annex

Corrosiveness of soils, from the DVGW GW9 classification. If the global index is ≥0, then the soil is not corrosive. From 0 to –4, slightly corrosive; from –5 to –10, corrosive; above –10, very corrosive. (according to Shreir, op. cit.)

Parameter	Nature of the soil	Index
Soil composition	Limestone, marl, marly sand, non-stratified sand	+2
	Vegetal earth, sandy earth (compost ratio ≤75%), marly earth, clayed sand (alluvium ratio ≤75%)	0
	Clay, marly clay, humus	–2
	Peat, coarse compost, swampy soil	–4
Water in contact with the metal	No water	0
	Always water	–1
	Sporadic water	–2
Resistivity	=10 000 Ω cm	0
	10 000 to 5000 Ω cm	–1
	5000 to 2300 Ω cm	–2
	2300 to 1000 Ω cm	–3
	=1000 Ω cm	–4

Continued overleaf (p.260)

Continued from p.259

Parameter	Nature of the soil	Index
Humidity ratio	≤20 %	0
	≥20 %	−1
pH	≥6	0
	≤6	−2
Sulphur and hydrogen sulphide	No	0
	Traces	−2
	Yes	−4
Carbonates	≥ 5 %	+2
	From 5 to 1%	+1
	≤1	0
Chlorides	≤100 mg kg^{-1}	0
	≥100 mg kg^{-1}	−1
Sulphates	≥200 mg kg^{-1}	0
	200 to 500	−1
	500 to 1000	−2
	≥1000	−3
Ashes and coke	No	0
	Yes	−4

18

Corrosion Resistance of the Delhi Iron Pillar — Scale Characterisation and Passive Film Growth Models

R. BALASUBRAMANIAM and P. DILLMANN*

Department of Materials and Metallurgical Engineering, Indian Institute of Technology, Kanpur 208 016, India
*Laboratoire Pierre Süe, CEA/CNRS CE Saclay 91191, Gif sur Yvette Cedex, France

ABSTRACT

The corrosion resistance of the Delhi pillar iron has been reviewed. The corrosion products on several ancient Indian irons have been characterised. The role of entrapped slag particles in aiding passive film formation in ancient Indian iron has been analysed by mixed potential theory. The protective rust formation process has been elucidated and possible models proposed. After an initial period of high corrosion rate, the initial corrosion resistance is conferred by the formation of protective amorphous compact layer of δ-FeOOH. The corrosion rate is further lowered by the formation of phosphates and their phase transformations.

1. Introduction

The Delhi pillar iron (Fig. 1) is testimony to the high level of skill achieved by the ancient Indian iron smiths in the extraction and processing of iron. The iron pillar at Delhi has attracted the attention of archaeologists and corrosion technologists as it has withstood corrosion for the last 1600 years. Several theories, which have been proposed to explain its superior corrosion resistance, can be broadly be classified into two categories: the environmental [1–3] and material [4–7] theories. These theories have been critically reviewed in detail in [6,7]. The proponents of the environment theory state that the mild climate of Delhi is responsible for the corrosion resistance of the Delhi pillar iron as the relative humidity at Delhi does not exceed 70% for significant periods of time in the year, which therefore results in very mild corrosion of the pillar. It is known from the classic researches of Vernon [8–10] that atmospheric rusting of iron is not significant for humidity levels less than 70%. Moreover, the climatic conditions of Udayagiri, where the pillar was originally located before it was brought to Delhi in the 11th century [11], should also be analysed in detail. On the other hand, several investigators have stressed the importance of the material of construction as the primary cause for its corrosion resistance. The ideas proposed in this regard are the relatively pure composition of the iron used [4], presence of phosphorus and absence of S/Mn in the iron [5], its slag enveloped metal grain structure [3], passivity enhancement in the presence of slag particles [6,7] and formation of phosphate film [5–7]. Other theories to explain the corrosion resistance are also to be found in the literature like the mass metal effect [3,12], initial exposure to an alkaline and ammonical environment [2], residual stresses resulting from the surface finishing operation [13], freedom from sulfur contamination both in the metal and in the air [13], presence of layers of cinder in the metal not allowing corrosion to proceed beyond the layer (cinder theory) [14] and surface coatings provided to the pillar after manufacture (treating the surface with steam [15] and slag coating [16]) and during use (coating with clarified butter) [3]. That the material of construction may be the important factor in determining the corrosion resistance of ancient Indian iron is attested by the presence of ancient massive iron objects located in areas where the relative humidity is high for significant periods in the year (for example, the iron beams in the Surya temple at Konarak in coastal Orissa [17] and the iron pillar at Mookambika temple at Kollur situated in the Kodachadri Hills on the western coast [18]). It is, therefore, obvious that the ancient Indians, especially from the time of the Guptas (300–500AD), produced iron that was capable of withstanding corrosion. This is primarily due to the high P content of the iron produced during ancient times, the reasons for which will be explored in detail later.

In order to understand the corrosion resistance of ancient Indian irons and the Delhi pillar iron, detailed characterisation of several ancient irons were performed. For example, a non-corroding iron clamp, exhibiting a thin adherent surface layer, produced during the time of the Guptas (the time period in which the Delhi pillar iron as constructed) was removed from one of the stone blocks in the ruined Gupta temple at Eran, Madhya Pradesh, dated to the 5th Century AD. Its microstructure [19], aqueous corrosion behaviour [19] and nature of its protective rust [20] have been earlier studied in detail. Similarly, the atmospheric rust from a non-corroding iron clamp from the Gupta temple at Deogarh (dated to the 6th Century AD) was characterised. As specimens from the Delhi pillar iron were not available for study, rust from several locations in the Delhi pillar iron, just below the decorative bell capital, was collected. This is the location where maximum rusting is observed on the exposed surface of the pillar [21] and, moreover, the rust from this location should be the oldest rust as this part of the pillar is inaccessible to the public [22]. The Delhi pillar iron rust was characterised [23] by XRD, FTIR and Mössbauer spectroscopy in order to clearly establish the nature of the protective passive layer that is responsible for the Delhi pillar iron's excellent corrosion resistance. The aim of the present paper is to understand the characteristics of protective rusts on the Delhi pillar iron, in particular, and on ancient Indian irons, in general. Possible passive film growth models that may be applicable to the understanding of the corrosion resistance of ancient Indian irons would be presented and discussed, in the light of the characterisation.

2. Typical Characteristics of Ancient Indian Irons

2.1. Composition

One of the notable features of corrosion-resistant ancient Indian iron is the relatively higher phosphorus content [24]. For example, Wranglen [2] utilised the available Delhi pillar iron compositions [5,25–27] (excluding that of Lal [26]) and estimated

Fig. 1 *Corrosion resistant Iron Pillar at Delhi.*

the average composition of the Delhi pillar iron iron as 0.15%C, 0.25%P, 0.005% S, 0.05%Si, 0.02% N, 0.05% Mn, 0.03% Cu, 0.05% Ni and balance Fe. A sample of Delhi pillar iron iron was also subjected to microprobe analysis in order to determine the composition of the elements Mn, Cr, Cu and Ni in the near surface regions and it was found that the composition of Cu (0.05%), Ni (0.05%), Mn (0.07%) and Cr (Nil) was uniform through several millimetres into the sample from the surface [3]. The Delhi pillar iron was originally located at Udayagiri in the Malwa region [11]. Interestingly, the composition of ancient iron from the Malwa region is similar to that of the Delhi pillar iron. For example, the composition of iron wedges found under the Gupta pillar at the northern gate of the Sanchi stupa was 0.05C, 0.09Si, 0.009 S, 0.303 P, 0.09 Mn. Hadfield [4] published detailed analyses of several iron specimens from Sri Lanka dated to the fifth century AD. For example, an iron nail (specific gravity of 7.69) contained traces of carbon, 0.11% Si and 0.32% P; an iron chisel (specific gravity of 7.69) contained traces of carbon, 0.12% Si and 0.28% P; and an iron bill hook (specific gravity of 7.50) contained traces of carbon, 0.26% Si and 0.34% P. Manganese and sulphur were practically absent in these irons, and the phosphorus content was high. The P contents of the iron clamps from Eran and Deogarh were relatively high. Wet chemical analysis of the Eran iron provided a P content of 0.25%.

2.2. Microstructure

The intimate relationship between structure and properties is well established in materials engineering. Nearly all ancient iron samples possess a non-uniform grain structure, arising essentially to the manufacturing process. In the un-etched condition, slag inclusions can be seen irregularly distributed in the microstructure. The etched specimens reveal a wide variety of structures from almost ferritic to high carbon structures. This has been already reported in the case of the Delhi pillar iron [3,5] and Eran iron [19,20]. The surface regions were generally free from pearlite, which seemed to increase towards the interior. Slip bands were sometimes seen in the surface ferrite grains due to the forging operation of manufacturing the iron products. Elongated slag particles and oxide particles were also seen in the deformed structure in the near surface regions and in the interior. The absence of an uniform distribution of pearlite in ancient Indian irons is indicative of segregation of P because, in such areas of P segregation, C diffuses out and the material becomes poorer in C content [28]. The interior portions of Delhi pillar iron were comparatively rich in C [5,27] indicative of surface decarburisation during the manufacturing stage [29]. Therefore, a wide variation in structure can be exhibited by ancient Indian iron. The non-uniform structure results in minor differences in the chemical analyses of the iron taken from different locations in the iron sample.

As a result of slag particles in the structure, the specific gravity of ancient iron is lower than that for the purest form of Fe. The slag particles in ancient iron are in fine microscopic form distributed unevenly in the microstructure. The entrapped slag results in the microstructure due to the processing method employed to obtain iron. Iron was produced in ancient India by solid state reduction of high quality iron ore using charcoal [30]. Once the reduction was complete, the iron lumps were removed from the furnace and hammered, in order to remove the liquid slag formed during the extraction process. Some of the slag invariably remained in the bulk of the material and this is the origin of the entrapped slag inclusions. The slags in ancient Indian irons are generally fayalite Fe_2SiO_4. Characterisation of entrapped slag inclusions using electron, photonic and nuclear microprobes [31] revealed the constituents of the slag particles. For example, detailed analysis of one slag inclusion in Eran iron revealed that the inclusion was made of wüstite dendrites entrapped in a polyphased matrix. One part of this slag appeared to be made of glassy phases but some other areas were crystallised. The micro-X-ray diffraction (µXRD) analyses revealed that these crystallised areas were made of fayalite Fe_2SiO_4. Energy dispersive spectroscopy analyses in the scanning electron microscope revealed that, in addition to Fe, Si and O, the inclusions also contained P. The absence of phosphates diffraction peaks on the µXRD spectrum could be explained by the fact that that the phosphates contained in the inclusions were glassy in nature [31]. Microstructural investigations on iron produced during the Gupta period show that in addition to slag, unreduced iron oxides were also present in the main body of Eran iron [19]. These unreduced iron oxides also contained carbon, presumably due the charcoal used for the extraction [19]. The presence of carbon in these unreduced oxides would render these oxides cathodic in nature with respect to the surrounding matrix. An interesting feature concerning the presence of slag particles in ancient Indian iron is that they are sometimes surrounded by a relatively higher fraction of pearlite compared to regions far removed from the slags. This is observed in the microstructures of DIP iron [3,4] and Eran iron [18,19]. Another typical example is seen in Fig. 2, which is the microstructure of the Dhar pillar iron. This pillar, which is about 900 years old, is nearly twice the length of the Delhi pillar iron, but lying presently in three broken pieces with a missing fourth piece and decorative capital [32]. Note that the pearlite volume fraction is relatively greater near the slag particles. The presence of a larger amount of pearlite near the slag inclusions is indicative of P depletion in these regions. Chemical analysis of the P content in DIP iron has revealed that the major portion of P content in the DIP is in solid solution while a smaller amount occurs in the slags. Out of 0.28% P in the material, 0.10% P was in the fixed state (i.e. in slags) with the rest being in solid solution in the metal [5]. The presence of P was also confirmed by micro-Particle Induced X-ray Emission (µPIXE) analysis near the entrapped slags of the Eran iron [31]. In the near vicinity of slag inclusions, the regions were depleted in P. These are the regions where carbon will concentrate [28]. Therefore, this is the reason for the presence of a larger fraction of pearlite near the slag-metal interfaces. The presence of more pearlite near the slag inclusions has an important implication as regards understanding potential cathodic sites in ancient Indian iron (i.e. slag regions would act as cathodic reaction sites due to the large volume fraction of cementite present at these locations).

2.3. Origin of Phosphorus in Metal

It will be shown later that the presence of P is crucial to the corrosion resistance of ancient India irons. As the ancient irons contain a relatively larger amount of P than modern-day iron (produced in the blast furnaces), the reason for the high P contents is briefly addressed. Modern steels cannot tolerate such high P contents, as they would be susceptible to cracking during the process of mechanical working. While it was earlier believed that P in ancient steels comes from slag inclusions [33], recent developments in slag chemistry help in understanding the probable reason. The relatively higher P content in ancient iron is related to the kind of slag that was created in the extraction process by solid state reduction. Lime was not added in the ancient Indian furnaces, unlike in modern blast furnaces, and therefore the slags generated in these ancient Indian furnaces were essentially fayalitic slags (i.e. consisting of iron orthosilicates Fe_2SiO_4). This is also corroborated by available

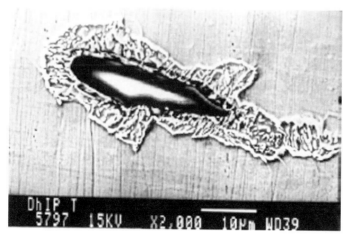

Fig. 2 *Microstructure of the Dhar pillar iron showing the presence of a larger fraction of pearlite near slag particles. The slag has been analysed in the electron microprobe analyser and it is essentially fayalite (Fe_2SiO_4).*

compositions of ancient Indian iron-making slags from archaeological excavation sites [26]. The slags do not contain lime [26]. The removal of P from the metal into the slag is facilitated by the basic components (for example, FeO and CaO) in the slag. The efficiency of removal of P from metal is much higher for CaO compared to FeO in the slag. These facts are well established in slag chemistry [34]. Therefore, the absence of CaO in the slags leads to a lower efficiency for removal of P from the metal, which invariably must have resulted in higher P contents in ancient Indian irons. Thermodynamic analysis of P removal from iron in the absence of CaO in the slag also provides the same answer [24]. As the entrapped slag seen in the ancient Indian iron is generally fayalitic without any CaO, thermodynamics dictates that a higher amount of P should remain in solid solution in iron. This must be one of the reasons for the presence of higher P in ancient Indian iron. It must also be noted that there are indications that P addition was intentional. For example, Buchanan[35], in his detailed description of steel making in Karnataka in the 18th century, describes one primitive furnace operated at Devaraya Durga. In this furnace, conical clay crucibles were filled with a specific amount of wood from the barks of a plant cassia auriculata, and pieces of wrought iron, then sealed and fired. Interestingly, the bark of this plant contains a high content of P, extracted by osmosis from the ground [6,7].

3. Role of Slag Particles in the Passivation Process

The method of extraction of ancient Indian iron resulted in the presence of fine slag particles and unreduced ore in the microstructure of the iron. The presence of these second phase particles in the microstructure would result in the creation of mini-galvanic corrosion cells when the iron is exposed to the environment. The metal in the matrix (which is almost pure iron [2,4,5]) would act as the anode and the second phase particles (slag and unreduced iron oxides) as sites for cathodic reactions. This is reasonable as the unreduced iron oxides contain carbon and the slag particles are surrounded by a large volume fraction of cementite (see Fig. 2). Initially, the cathodic reactions (that occur on these cathodic reduction sites) provide the sink for the electrons liberated by the corrosion of the metal to its ions. The slag particles would therefore accelerate corrosion on exposure to the environment. Relatively high corrosion rates were observed experimentally in the initial exposure period of Eran iron compared to mild steel of similar C content [19]. Therefore, it is anticipated that the two-phase structure should corrode at a faster rate compared to iron of similar composition as the ancient irons possess essentially a composite structure. The cathodic nature of the second phase particles in ancient iron should also be applicable in later stages of corrosion because the cathodic reduction of oxygen can be considered in atmospheric corrosion (as ferric oxyhydroxides are also good electronic conductors) in addition to the reduction of the corrosion product in the rust.

In order to explain the anomaly of the composite structure of Delhi pillar iron exhibiting excellent corrosion resistance, the oxidation and reduction processes occurring on the Delhi pillar iron were analysed using the mixed potential theory [6,7]. The analysis is briefly summarised by considering the Evans diagram presented in Fig. 3. The anodic polarisation behaviour of iron has been shown to exhibit active–passive behaviour as it is well known that rust layer(s) form on atmospheric corrosion

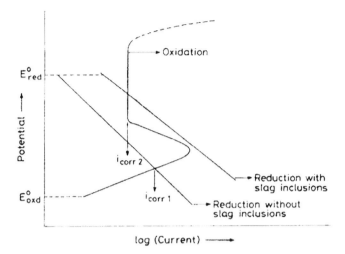

Fig. 3 *Mixed potential theory analysis for the enhancement in the passivity of the Delhi pillar iron in the presence of slag particles.*

of iron. Moreover, the Delhi pillar iron contains a relatively larger weight fraction of P (average composition 0.25% [2]) which will aid in inducing passivity in iron [36,37]. The formation of insoluble phosphates is thermodynamically favoured [38], even for P contents as low as 0.24 %P [39]. Potentiodynamic polarisation studies conducted on phosphated steels reveal that the critical current density for the formation of the phosphate layer can be higher by more than three orders of magnitude compared to that for steel without P [40]. Therefore, it is valid to indicate that the Delhi pillar iron will exhibit active–passive behaviour on anodic polarisation. It is important to stress that the exact nature of the passive film need not be known for this theoretical analysis and only the formation of a passive film is required while considering the mixed potential analysis.

If the iron did not contain slag particles, the cathodic activation polarisation line would have intersected the anodic polarisation curve in the active region. This is reasonable as relatively large currents are needed to induce passivity in iron containing P as the critical current density for passivation is relatively large in iron containing P [40]. However, in the presence of slag particles, the exchange current of the cathodic reaction(s) would be higher than in the case when there were no slag particles and the cathodic activation polarisation line would shift to the right as shown in Fig. 3. Notice that current is depicted in the X-axis and not current density because this representation is useful in elucidating the effect of second phase particles in inducing passivity on the iron. A similar representation is also employed in Evans diagrams to understand area effects in galvanic corrosion using the mixed potential analysis [41]. Therefore, in the presence of second phase particles, the higher current demand of the cathodic reaction(s) will demand a higher corrosion current and in the process, the critical current density for formation of the passive film should be attained. Once this is achieved, the passive film covers the surface and the corrosion rate is reduced to low rates. Therefore, the presence of slag particles may be indirectly beneficial in the case of the Delhi pillar iron as it helps in the induction of passivity

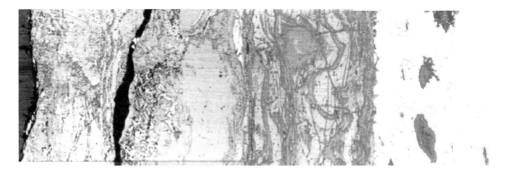

Fig. 4 *Optical microscopy micrograph showing the transverse section of the oxide scale on Eran iron. The P content was high in the metallic matrix adjacent to the scale–metal interface.*

on the surface. The above mixed potential analysis was experimentally validated by potentiodynamic polarisation studies using ancient Indian iron of different slag contents [6,7]. The above mixed potential theory analysis establishes that passivity can also be induced on the Delhi pillar iron due to the presence of second phase particles in the microstructure. The corrosion resistance of the Delhi pillar iron, therefore, has theoretical support. Although the above discussion clearly establishes that the second phase particles in the Delhi pillar iron may be beneficial in inducing passivity, it should be borne in mind that alternate wetting and drying conditions are implied while considering atmospheric corrosion. As shall be presented later, this is a very important factor in formation of the protective film at the metal–metal/oxide interface. In case of iron with such second phase particles (slags and oxides) is exposed to a more severe environment, for example complete immersion in acidic or alkaline solution, rapid localised attack occurs at the second phase-matrix interfaces, and the material corrodes at a much faster rate than normal mild steel. This has also been experimentally validated for ancient Indian iron by constant immersion corrosion testing and microstructural examination using a scanning electron microscope [19]. It is interesting at this juncture to note that the corrosion rates determined by constant immersion testing of DIP iron samples in 0.001% NaCl and 0.003% SO_2 solutions were 6 mg/sq dm/day (mdd) and 54 mdd, respectively [5]. Utilising these values, it can be easily seen that had the pillar been completely immersed in an aqueous solution, it would have been corroded much more severely.

4. Corrosion Resistance of Delhi Pillar Iron

Rust samples from the DIP were characterised by XRD, FTIR and Mössbauer spectroscopy [23]. The significant result of the XRD analysis was the identification of iron hydrogen phosphate hydrate in the crystalline form. The FTIR study clearly established that, in addition to iron hydrogen phosphate hydrate, the scale also consisted of γ-FeOOH (lepidocrocite), α-FeOOH (geothite), δ-FeOOH, magnetite and phosphates. The hydrated nature of these products was also indicated. It was also seen in the XRD pattern that there was a very small amount of iron oxide/oxyhydroxides present in the crystalline form. The identification of the oxide/

oxyhydroxides of iron by FTIR clearly established that they are present in the amorphous form. The FTIR study also indicated that phosphate ions were present in the rust, confirming the results of XRD. The Mössbauer spectroscopic study of the Delhi iron pillar rust samples proved conclusively that the oxyhydroxides and magnetite were present in the amorphous form and also that iron in the phosphate was in the +3 oxidation state. In summary, while XRD analysis proved the existence of crystalline iron hydrogen phosphate hydrate, FTIR and Mössbauer spectroscopy proved the presence of magnetite and several oxyhydroxides in the amorphous form.

The process of protective film formation on the exposed surface of the Delhi pillar iron can be outlined based on the detailed characterisation of Delhi iron pillar rust. It must be remembered that alternate wetting and drying conditions are implicit in the following discussion on atmospheric corrosion.

Initially, the corrosion of the matrix is relatively fast due to the presence of second phase particles in the microstructure. The usual corrosion products that are observed in the case of mild steels (exposed to atmosphere containing no chloride ions) are generated. It is well known that the corrosion products that form on iron on atmospheric exposure are α-FeOOH (geothite), γ-FeOOH (lepidocrocite), $Fe_{3-x}O_4$ (magnetite) and X-ray amorphous matter [36,37,42]. In the case of Delhi pillar iron, the formation of lepidocrocite and geothite was confirmed by XRD analysis of 1.5 year old rust from the Delhi pillar iron surface by Lahiri *et al.* [27]. It is also likely that the sample analysed by them could have contained amorphous δ-FeOOH [36,37], as the formation of this phase during the initial corrosion of iron has been conclusively proven by Misawa *et al.* [36,37, 43,44].

The initial enhanced corrosion of the matrix leads to the enrichment of P concentration at the metal-scale interface. In the presence of P at the interface between the metal and rust, the formation of a compact layer of amorphous δ-FeOOH layer next to the metal-metaloxide interface should be favoured like that observed in the case of P-containing weathering steels [36,37]. Moreover, experiments have shown that $H_2PO_4^-$ ions prevent crystal growth of the corrosion products [36,37]. The formation of amorphous δ-FeOOH confers the initial corrosion resistance to the iron. The δ-FeOOH phase forms in a discontinuous manner in normal mild steels while it forms as a compact layer next to the metal–metal/oxide interface in the case of P- or Cu-containing weathering steels due catalytic action [36,37]. It is important to note that the oxyhydroxides and magnetite present in the old Delhi pillar rust are nanocrystalline/amorphous in nature and not crystalline. As it is known that the initial oxide and oxyhydroxides that form on the DIP iron are crystalline in nature [27], the long-term conversion of the crystalline forms of these oxyhydroxides to the amorphous form is indicated. The crystalline oxide/oxyhydroxides of iron are converted to the amorphous state due to process of alternate wetting and drying, as has been shown in P-containing weathering steel [36]. The enrichment of P in the δ-FeOOH layer should continue with prolonged exposure and this has been observed in P-containing weathering steels [36,37]. This enrichment should be responsible for the precipitation of the insoluble phosphate, identified by XRD.

The process of formation of crystalline iron hydrogen phosphate hydrate can be understood based on the vast literature available on phosphating of steels [45]. The formation of a protective layer on Fe surface on phosphating involves several steps. In the first step, electrochemical attack of iron by orthophosphoric acid occurs over a wide range of concentration and temperature. In contact with phosphoric acid, the dissolution potential of iron becomes less noble and it causes the dissolution of Fe to sparingly soluble dihydrogen phosphate, $Fe(H_2PO_4)_2$. The transient oxides of iron (in which Fe is in the +2 oxidation state) will also be corroded by phosphoric acid to this phosphate. In the case of Delhi pillar iron, the formation of orthophosphoric acid next to the metal surface has to be first understood. The enrichment of P at this location was earlier established due to the initial corrosion of matrix. The concentration of P must be relatively higher near the surface regions of the Dehli pillar iron than in the bulk. It was earlier noted that the surface regions contain a higher amount of ferrite (i.e. depleted of carbon) which is due to the final surface finishing operation given to the pillar during its manufacture [29]. It is known that P diffuses and concentrates in regions where C is depleted [28] and therefore, concentration of P is relatively higher near the surface regions of the pillar. Therefore, enrichment of P in the δ-FeOOH layer is expected. This P enrichment has been experimentally determined by Ghosh [5] who mentioned that the P content in the DIP rust was 0.35% whereas the P content in the DIP iron was 0.18%. The enrichment of P in the rust would initially result in the formation of phosphoric acid at the metal–metal/oxide interface due to the presence of P at this location and also as free energy consideration favour its formation. The moisture for phosphoric acid formation obtains in the alternate wetting and drying cycles.

In the second stage, the contact of the metal shifts the equilibrium in such a way that massive precipitation of monohydrogen phosphate $FeHPO_4$ and tribasic iron phosphate $Fe_3(PO_4)_2$ occurs. Iron is in the +2 oxidation state in these phosphates. Both these phosphates are insoluble in nature. Moreover, these phosphates are also amorphous in nature and this is well corroborated by experimental evidence [45]. In normal phosphating processes, the phosphating reactions generally discussed are up to this point and other cations such as Zn or Mn are also added to phosphoric acid. In some cases, oxidising agents are added so that iron may appear in the coating as ferric phosphate. This has been proved to be more beneficial to corrosion resistance as the crystal reorganisation of the amorphous phosphates to ferric phosphates results in a large reduction in the porosity of the phosphate and subsequently, much improved protective properties [45]. The oxidation of iron to ferric phosphate can also be achieved by alternate wetting and drying cycles. Over time, $Fe_3(PO_4)_2$ is oxidised by atmospheric oxygen and H_3PO_4 to iron hydrogen phosphate hydrate $FePO_4.H_3PO_4.4H_2O$. The dissolution and reprecipitation reaction also leads to a change in the pH of the metal–solution interface, which leads to an crystalline reorganisation and followed, most importantly, by a large decrease in porosity. The crystalline reorganisation is a very deep-seated reaction (at the metal-phosphate interface) because it modifies the porosity of the passive layer and decreases markedly the exposed metallic surface [45]. Therefore, the formation of the crystalline modification of iron hydrogen phosphate hydrate from the amorphous phosphate is critical in providing excellent protection against further ingress of moisture and oxygen to the metal surface. It is important to also note that the crystalline modification of iron hydrogen phosphate hydrate was obtained in the XRD analysis [23], thus indicating that the phosphate in the rust is relatively old.

The continuous layer of crystalline iron hydrogen phosphate hydrate (formed at the metal–metal/oxide interface) is, therefore, responsible for the superior corrosion resistance of the Delhi pillar iron. Ghosh anticipated this mechanism in his excellent study of the DIP iron [5]. He stated that "...phosphorus accumulates in a new phase at the base of the main oxide layer in the oxide form in combination with iron" and "so long as the new phase at the base of the main oxide film was not formed, the metal behaviour was similar to that of exposure surface of ordinary irons" [5]. Ghosh performed some simple experiments with the DIP iron piece to assess the influence of P on rusting. The sample was polished and allowed to rust. When a thin and unequally distributed film of rust was formed, it was photographed and slightly polished to remove the rust from the surface. This surface was treated with Stead's solution to observe the distribution of P on the surface. In was found that P was generally low in the areas where rust appeared more intensely.

Alternate wetting and drying conditions play an important role in the case of atmospheric corrosion of the Delhi pillar because they accelerate the precipitation of protective crystalline phosphate and the amorphisation of the dehli pillar rust. The amorphisation of rust is also aided by the presence of $H_2PO_4^-$ ions [36,37]. The iron pillar's weight is estimated to be approximately 7 tonnes and therefore, the large mass of the metal plays a contributory role in aiding the alternate wetting and drying process. Sanyal and Preston [12] and, later, Bardgett and Stanners [3] proposed that the large mass of the pillar implies a large heat capacity for the iron and therefore, the pillar will heat up faster or cool down faster than the surroundings. This provides the right conditions of alternate wetting and drying of the iron pillar surface. The intensity of wetting and subsequent drying would be much more in the case of Delhi pillar iron because of the large mass of the pillar.

6. Rust Cross-Section

6.1. Eran Iron

The nature of the passivation process can be understood by analysing the rust cross sections on ancient Indian irons. They also provide ideas for modelling of the long term corrosion behaviour. Essentially, several different kinds of rust cross sections are encountered in atmospheric rusts in ancient iron. In the first case, there is no well marked differentiation in the rust as layers. The rust on Eran iron showed a two layered structure at some locations [20] and at other locations, such clear distinctions on the layered nature could not be made out (Fig. 4). The various constituents of the scale depicted in Fig. 4 were determined by microdiffraction methods. The P content in the metal adjacent to the surface oxide was also determined. P was not present uniformly in the scale but at certain places, the P concentration attempted 1 wt%. One such location is marked P in Fig. 2 of Ref. [54]. The μXRD pattern obtained from this location confirmed the presence of magnetite and goethite as major phases and crystalline phosphate $Fe_2(PO_4)(OH)$ as minor phase. In region A, in general, no diffraction was obtained indicating that these regions were amorphous in nature. Microdiffraction from region B indicated magnetite while microdiffraction from region C indicated goethite. The matrix near A zone possessed a higher P content. Therefore, these microdiffraction experiments provided significant insights into the nature of the protective rust on the ancient Indian Eran iron. The oxide scales formed near the high P containing metallic matrix consisted of a lower amount of crystalline compounds. On the other hand, when the P-content in the matrix was lower, the usual crystalline atmospheric corrosion products (magnetite and goethite) were detected. Finally, in the locations in the oxide where the P content was relatively high, crystalline phosphates were identified by µXRD. The formation of crystalline phosphate in the surface scales on iron generally occurs either in the presence of oxidisers or due to a large number of wetting-and-drying cycles (i.e. by the process of dissolution and re-precipitation). The alternate wetting and drying cycles experienced in the atmospheric corrosion process could probably explain the presence of these crystalline phosphates. Composition analysis in an EPMA indicated that there was a higher amount of P near the metal–scale interface [20]. The compositional variations of P across the atmospheric oxide scale at different locations on the sample were also analysed by µPIXE. The variation in P content as a function of distance into the oxide is presented in Fig. 5. In this Figure, three different profiles obtained at three different locations are provided. The profiles based on data points represented by the open and closed squares were obtained at locations where the P content was relatively low in the metallic matrix just below the scale. The profile based on the data points represented by triangles was obtained at a location where the P content was relatively high. These profiles indicated that significant enrichment of P could be obtained in the oxide scale. Notice also, that in certain regions, the oxide thickness could be as high as 800 µm. It is interesting that in the locations where the P content was relatively low in the metallic matrix (square data points in Fig. 5), there was a higher amount of P especially at the scale–metal interface and this continued into the scale. However, with increasing distance into the scale, the P content decreased.

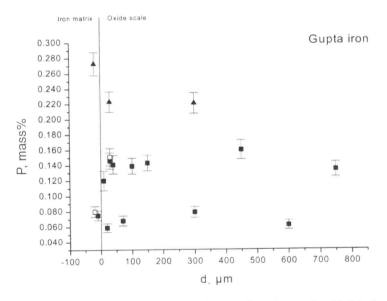

Fig. 5 *Variation of P content as a function of distance from the metal–oxide interface. Three different profiles obtained at three different places are provided. The profiles based on data points represented by the open and closed squares were obtained at locations where the P content was relatively low in the metallic matrix just below the scale. The profile based on the data points represented by triangles was obtained at a location where the P content was relatively high.*

The enrichment of P content at the metal-scale interface and in the near vicinity of the oxide scale, in regions both rich and poor in P, is proven.

6.2. Dhar Pillar Iron

Cross-sectional microscopy of the iron sample from Dhar iron pillar indicated that the atmospheric rust consisted of two layers, an optically dull inner thin layer and an optically bright thick outer layer (Fig. 6). The layered structure of the protective rust in the case of weathering steels is well established. Regarding the optical nature of the layers on weathering steels, it has been pointed out that the inner optically isotropic layer of the surface rust is composed of X-ray amorphous spinel type iron oxide which can protect the steel matrix [46,47]. Yamashita *et al.* [48] also observed that the rust layer present on a weathering steel exposed for 26 years could be divided into two parts: an outer layer which was optically active (i.e. illuminated) and an inner layer which was optically isotropic (darkened). On the other hand, the surface rust formed on mild steels consisted of the mottled structure consisting of the optically active and isotropic corrosion products [48]. It is also established that weathering steels obtain their protection due to the presence of the amorphous (dark) inner layer [48]. Therefore, based on the above studies, it is reasonable to state that the inner optically dull layer seen in the Dhar iron pillar rust microstructure is amorphous in nature while the outer layer is crystalline in nature. The innermost thin layer seen next to the metal surface in Dhar pillar iron must be the amorphous layer and this may be enriched with phosphorus or may contain crystalline phosphates. The outer layer must consist of the usual corrosion products expected on atmospheric corrosion of iron (i.e. magnetite, hematite, goethite and lepidocrocite) and can also be enriched in phosphorus or contain phosphates. As the Dhar pillar iron contains a significant amount of phosphorus (0.28% P), the formation of phosphates is expected. The golden reddish brown colour of the surface of the Dhar pillar pieces also indicates that phosphates must be present in the atmospheric rust.

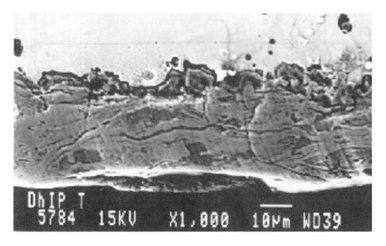

Fig. 6 Cross-section of the atmospheric rust of the Dhar pillar iron showing an optically dull thin inner layer and an optically bright thick outer layer.

6.3. Delhi Pillar Iron

In the case of the DIP, the long term rust cross section would provide not just two layers, but an additional third layer arising from the formation of insoluble phosphates at the metal-scale boundary. The process of protective film formation outlined earlier can be understood by analysing the only published cross section microstructure of the DIP rust [5]. The rust was obtained from the buried underground region when the pillar was excavated in 1961. The stone platform currently seen around the base of the pillar was not present till the last century [49]. It was erected by Beglar after he explored the underground regions of the courtyard area where the iron pillar is located in the 1860s. When the pillar was re-erected by Beglar, the stone platform was constructed and a coating of lead was provided on the buried underground surface of the pillar. This uneven coating of lead (of about 3 mm in thickness) was found to be in an excellent state of preservation when the buried regions of the pillar was again excavated in 1961 on the eve of the centenary of the Archaeological Survey of India [26]. However, the buried portion was found covered with rust layers ranging from a few mm to 15 mm [5,26]. Interestingly, numerous cavities and corrosion pits were also observed on the surface. The composition of the rust in the buried underground regions were chemically analysed and they were found to be similar to the rusts obtained above the ground [5]. The soil samples in the nearby vicinity was found loaded with soluble sulfates and chlorides [26]. The iron in the buried regions is corroded more than the exposed surface due to the galvanic action with the lead layer, as lead is cathodic with respect to iron [21]. Simulation experiments conducted with lead coated mild steel in soil environments [50] has clearly established the deleterious nature of the lead coating as the corrosion is intense in areas where coating defects exists. Under such circumstances, it is anticipated that cross sectional analysis of the rust layer in the buried underground would provide a picture of the rust nature, but on an amplified scale.

It is to be also noted that the surface film that forms on the iron pillar in the buried underground region would be slightly different compared to that seen in the atmospheric exposure region. This is due to the restricted nature of the buried soil environment. Under conditions of restricted oxygen supply, it is well established that the major corrosion product on iron is magnetite. Such a situation should obtain in the case of the Delhi pillar rust in the buried underground regions. Ghosh analysed a sample of rust from the buried underground region and it was seen that it was essentially magnetic in nature, based on the attraction of the rust to a magnet [5]. It is not possible to conclusively prove that the magnetic portion is entirely composed of magnetite as the other oxyhydroxides of iron are also paramagnetic and therefore will also be attracted to a magnet. The magnetic portion of the rust was analysed by XRD and it was shown to consist of magnetite in addition to lepidocrocite and geothite [5]. Ghosh did not mention the relative amounts of these phases based on the XRD study as he was quoting the results of the XRD work done at Bern. However, it can be reasonably predicted that a higher amount of magnetite must have been present in the rust obtained from the buried underground when compared to that obtained in the exposed regions. Apart from this major difference, the rust nature should be the same in the exposed and buried surfaces of the pillar. Therefore, in addition to the phases identified by Ghosh in the rust layers in the part of the pillar buried underground, the presence of amorphous phases like δ-FeOOH and amorphous phosphates is likely. XRD analysis would not have revealed these phases. Having established the general similarities in the rust that forms above and below the ground on the pillar, it is illuminating to view the published microstructure of the rust formed in the underground buried regions [5]. It is reproduced in Fig. 7. It is clearly noticed that the rust is composed of three distinct layers, with the otter layer (marked A) being optically active while the inner two layers (marked B and C) being optically dull. The inner most layer (marked C) is present as a thin layer next at the metal-metaloxide interface. It is reasonable to state that the inner two layers seen in the underground DIP rust microstructure of Fig. 7 are amorphous in nature while the outer layer is crystalline in nature. The innermost thin layer seen next to the metal surface must be due to precipitation of phosphates (most likely in the amorphous form as the rust did not undergo drastic alternate wetting and drying) because of the enrichment of P at this location. Of course, the major difference between the rust seen the buried underground and above the ground is that there should be a lower amount of crystalline oxide/oxyhydroxides in above ground rust due to alternate wetting and drying cycles. This is also corroborated by the characterisation of Delhi pillar rust [23]. Therefore, Fig. 7 literally provides the visual picture for the nature of protective passive layer that forms on the Delhi pillar.

7. Scale Growth Models

To understand and predict the long-term corrosion behaviour of ancient Indian iron, it is important to model passive film growth. Mathematical models dealing with atmospheric corrosion of several metals have been discussed in [51]. In the present section, the possible kinetics of scale growth would be addressed. Protective film growth is attributed to parabolic growth kinetics. This model has been previously

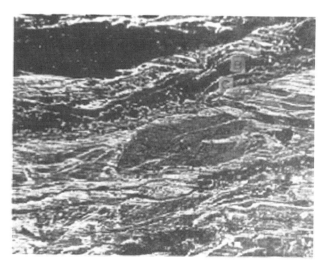

Fig. 7 Cross-section of DIP rust in the buried underground [5]. Note the presence of three distinct layers in the rust with the inner two layers being optically dull indicating their amorphous nature.

applied to the atmospheric rust on the Delhi pillar. Wranglen [2] utilised the weight loss data of Hudson [1] (who obtained the same for mild steel exposed to Delhi's environment) to estimate the thickness of protective film that forms on the surface of the pillar. Assuming parabolic growth kinetics, he predicted the thickness of the scale to be 200 µm after about 1600 years of growth [2]. He concluded that the good match of the estimated thickness with that experimentally estimated by Bardgett and Stanners [3], using a permanent-magnet type thickness gauge, was proof that the film that forms on the surface is protective in nature as it grows according to parabolic kinetics. However, the thickness measurements of Bardgett and Stanners [3] did show variations. For example, in the polished region, the film thickness was estimated to be 50 µm, while that in the rough portion above this polished region was estimated to be between 500 to 650 µm [3]. Moreover, the measurements were performed in the area assessable to the public and therefore, this may not be the state of the rust at the regions far away in the top sections of the pillar. Nevertheless, the applicability of the parabolic rate law indicates that thickness can be predicted assuming simple parabolic growth kinetics (curve 1 in Fig. 8).

The kinetics of the protective film process can also be understood based on the nature of the protective film, discussed in the earlier section. It was seen that the slag particles in ancient Indian iron indirectly aid in the passivation process. It was also pointed out while explaining the growth of the protective film form on the surface, that the corrosion rate is initially high on the surface leads in enrichment of P content on the surface. Therefore, enhanced corrosion is expected in the initial periods. This period has been shown in the kinetic diagram (curve 2 in Fig. 8) and represented by a high corrosion rate. The second stage would set in once the protective passive film has covered the surface. Due to the formation of the passive film, it is anticipated that the corrosion rate would be much lower than the initial stage. Therefore, the second stage has been represented by a lower kinetic rate (Fig. 8). Interestingly, an analogy may be

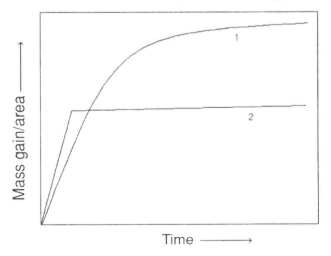

Fig. 8 *Some possible scale growth models to predict long-term atmospheric corrosion behaviour of ancient Indian iron.*

drawn with the kinetic behaviour of corrosion of metals that are passivated in the presence of cathodic alloying elements [52]. The kinetic regions of passive film growth on Delhi iron pillar have been analysed in Ref. [53].

It is also possible to envisage additional changes in the scale composition and structure which will further reduce the corrosion rate. This could be due to several reasons as has been outlined earlier for ancient Indian rusts. The first simple example is the precipitation of phosphates when P content is enriched in the region below the scale. Another example, described in the case of Delhi pillar iron rust, is the crystallisation of the amorphous phosphate due to alternate wetting and drying cycles. Of course, it is important to know the kinetics of growth of these various layers in order to obtain meaningful prediction about the growth of the scale. In order to achieve this, it is important to understand and study the scales on several ancient iron samples and based on the qualitative and quantitative analysis of the scale characteristics, it would be possible to obtain insights on the nature of long term scale growth.

8. Conclusions

The theories of corrosion resistance of the 1600 year old Delhi pillar iron have been briefly reviewed. The results of a detailed characterisation of the DIP's rust have been summarised. Rust characterisation clearly established that the major constituents of the scale were crystalline iron hydrogen phosphate hydrate ($FePO_4.H_3PO_4.4H_2O$), α-, γ-, δ-FeOOH and magnetite. The iron oxide/oxyhydroxides were present in the amorphous form. The role of slag particles in the matrix of the ancient Indian iron in enhancing the passive film formation has also been addressed. The process of protective rust formation and growth on DIP iron has been outlined based on the rust analysis. Initially, the corrosion rate of iron is high due to the presence of the slag particles. This results in enhancement of surface P content. In the presence of P, the formation of a protective amorphous compact layer of δ-FeOOH, next to the metal surface, is catalysed and this confers the initial corrosion resistance. The critical factor aiding the superior corrosion resistance of the Delhi pillar iron, however, is the formation of iron hydrogen phosphate hydrate, as a thin layer next to the metal–metal/oxide interface. The formation of the crystalline modification of this phosphate from the amorphous form is aided by alternate wetting and drying cycles (i.e. the environmental factor). The characterisation of protective passive films on other ancient Indian iron objects has also been presented. The passive film formation on the ancient iron will be contrasted with rusting of normal and weathering steels. Finally, possible kinetics models for the growth of the passive film on the surface of ancient Indian irons has been discussed.

9. Acknowledgements

The authors would like to thank the Archaeological Survey of India for their co-operation and help in studying ancient Indian iron.

References

1. J. C. Hudson, *Nature*, 1953, **172**, 499–500.

2. G. Wranglen, *Corros. Sci.*, 1970, **10**, 761–770.

3. W. E. Bardgett and J. F. Stanners, *J. Iron and Steel Inst.*, 1963, **210**, 3–10.

4. R. Hadfield, *J. Iron and Steel Inst.*, 1925, **112**, 233–235. 5. M. K. Ghosh, *NML Tech. J.*, 1963, **5**, 31–45.

6. R. Balasubramaniam, *NML Tech. J.*, 1995, **37**, 123–145. 7. R. Balasubramaniam, *Trans. Indian. Inst. Metals*, 1997, **50**, 23–35.

8. W. H. J. Vernon, *Trans. Farad. Soc.*, 1923–24, **19**, 839–934.

9. W. H. J. Vernon, *Trans. Farad. Soc.*, 1927, **23**, 113–204. 10. W. H. J. Vernon, *Trans. Farad. Soc.*, 1935, **31**, 1668–1700.

11. R. Balasubramaniam, *Bull. Metals Mus.*, 2000, **32**, 42–64.

12. B. Sanyal and R. Preston, Chemical Research Laboratory, London, 1952.

13. U. R. Evans, *The Corrosion and Oxidation of Metals*, First Supplementary Volume, Edward Arnold, London, England, 1968, pp. 191–192.

14. W. Rosenhain, *Trans. Faraday Soc.*, 1916, **11**, 236–243.

15. W. Rosenhain, *Physical Metallurgy*, Constable, London, England, 1915, pp. 329–330.

16. A. Herrero and M. de Zubiria, *J. Iron and Steel Inst.*, 1928, **118**, 109–125.

17. H. G. Graves, J. Iron. Steel Inst., 1912, 85, 187–202. 18. T. R. Anantharaman, *Current Sci.*, 1999, **76**, 1428–1430.

19. V. Puri, R. Balasubramaniam and A. V. Ramesh Kumar, *Bull. Metals Museum*, 1997, **28–II**, 1–10.

20. A.V. Ramesh Kumar and R. Balasubramaniam, *Corros. Sci.*, 1998, **40**, 1169–1178.

21. R. Balasubramaniam, *Bull. Metals Museum*, 1998, **29-I**, 19–39.

22. R. Balasubramaniam, *J. Metals*, 1998, **50**, (3), 40–47. 23. R. Balasubramaniam and A. V. Ramesh Kumar, *Corros. Sci.*, 2000, **42**, 2085–2101.

24. V. Kumar and R. Balasubramaniam, *Metals Materials and Processes*, 2002, **14**, 1–14.

25. R. Hadfield, *J. Iron Steel Inst.*, 1912, **85**, 134–174.

26. B. B. Lal, in *The Delhi Pillar Iron: Its Art, Metallurgy and Inscriptions*, (M.C. Joshi, S.K. Gupta and Shankar Goyal, eds). Kusumanjali Book World, Jodhpur, India, 1996, pp 22–58.

27. A. K. Lahiri, T. Banerjee and N. R. Nijhawan, *NML Tech. J.*, 1963, **5**, 46–54.

28. A. Bramley, F. W. Haywood, A. T. Coopers and J. T. Watts, *Trans. Faraday Soc.*, 1935, **31**, 707–734.

29. R. Balasubramaniam, *Bull. Metals. Museum*, 1999, **31**, 40–63.

30. B. Prakash, *Ind. J. Hist. Sci.*, 1991, **26**, 351–371.

31. P. Dillmann and R. Balasubramaniam, *Bull. Mater. Sci.*, 2001, **24**, 317–322.

32. R. Balasubramaniam, *Indian Journal of History of Science*, 2002, **37**, 115–151.

33. U. R. Evans, *Corrosion of Metals*, Edward Arnold, London, England, 1926, p.266.

34. J. J. Moore, *Chemical Metallurgy*, Second Edition, Chapter 5, Butterworth-Heinemann, Oxford, England, 1990, pp.152–192.

35. F. A. Buchanan, *Journey from Madras Through the Countries of Mysore,Canara and Malabar*, East India Company, London, England, 1807.

36. T. Misawa, T. Kyuno, W. Suetaka and S. Shimodaira, *Corros. Sci.*, 1971, **11**, 35–48.

37. T. Misawa, K. Asami, K. Hashimoto and S. Shimodaira, *Corros. Sci.*, 1971, **14**, 279–289.

38. T. L. Wood and R. M. Garrels, *Thermodynamic Values at Low Temperature for Natural Inorganic Materials: An Critical Summary*, Oxford University Press, Oxford, England, 1987, pp.100–106.

39. V. A. Urasova, N. P. Levenets and A. M. Samarin, *Izvest. Akad. Nauk SSSR, Metally*, 1966, **6**, 24–30.

40. R. L. Chance and W. D. France, *Corrosion*, 1969, **25**, 329–335.

41. M. G. Fontana and N. D. Greene, *Corrosion Engineering*, Second Edition, McGraw-Hill International Student Edition, New Delhi, India, 1978, pp.330–335.

42. I. Suzuki, Y. Kisamatsu and N. Masuko, *J. Electrochem. Soc.*, 1980, **127**, 2210–2215.

43. H. Kihira, S. Ito and T. Murata, *Corros. Sci.*, 1990, **31**, 383–388.

44. J. T. Keiser and C. W. Brown, *Corros. Sci.*, 1983, **23**, 251–259.

45. E. L. Ghali and R. J. A. Potoin, *Corros. Sci.*, 1972, **12**, 583–594.

46. H. Okada, *J. Soc. Mater. Sci. Japan*, 1968, **17**, 705–7709.

47. H. Okada, Y. Hosoi, K. Yukawa and H. Naito, *J. Iron Steel Inst. Japan*, 1969, **55**, 355–365.

48. M. Yamashita, H. Miyuki, Y. Matsuda, H. Nagano and T. Misawa, *Corros. Sci.*, 1994, **36**, 283–299.

49. R. Balasubramaniam, *Curr. Sci.*, 1987, **73**, 1057–1067. 50. R. Balasubramaniam, *Curr. Sci.*, 1999, **77**, 681–686.

51. C. Leygraf and T. E. Graedel, *Atmospheric Corrosion*. Electrochemical Society, Princeton, USA, 2000.

52. N. D. Tomoshov and G. P. Chernova, *Passivity and Protection of Metals Against Corrosion*. Plenum Press, New York, 1967, pp.151–180.

53. R. Balasubramaniam, *Curr. Sci.*, 2002, **82**, 1357–1365.

54. P. Dillmann, R. Balasubramaniam and G. Beranger, *Corros. Sci.*, 2002, **44**, 2231–2242.

19

Expansion Due to Anaerobic Corrosion of Steel and Cast Iron: Experimental and Natural Analogue Studies

N. R. SMART, A. P. RANCE, P. FENNELL and L. WERME*

Serco Assurance, Culham Science Centre, Abingdon, Oxfordshire, OX14 3ED, UK
*SKB, Box 5864, SE-10240, Stockholm, Sweden

ABSTRACT

An apparatus was constructed to measure the expansion caused by the anaerobic corrosion of steel and cast iron whilst under representative compressive loads. The detection of hydrogen and the identification of magnetite on the surface of the specimens demonstrated the occurrence of anaerobic corrosion, but no expansion was observed after over two years' exposure, suggesting that the corrosion product is too soft and deformable to cause jacking of the walls of canisters used for encapsulating spent nuclear fuel. The use of natural analogues to examine the potential for expansion caused by anaerobic corrosion in confined spaces over long time periods is discussed.

1. Introduction

Sweden has eleven nuclear reactors in operation at four different sites. These reactors produce about 50% of all the electricity used in Sweden. By 2010 the current nuclear programme will have produced approximately 8000 metric tons of spent nuclear fuel. After 30 to 40 years of storage in the intermediate storage facility, CLAB, the fuel will be encapsulated in corrosion-resistant disposal containers. After encapsulation, the fuel will be transported to a geological repository, where the containers will be deposited at a depth of 500 to 700 m in granitic rock and surrounded by a bentonite clay backfill material. The proposed design for a final repository for spent fuel and other long-lived residues in Sweden is based on the multi-barrier principle. The waste will be encapsulated in sealed cylindrical canisters, which will then be placed in vertical storage holes drilled in a series of caverns excavated from the granite bedrock at a depth of about 500m and surrounded by compacted bentonite clay. Groundwater in granitic rock in Sweden is oxygen-free and reducing below a depth of 100m to 200 m. The redox potential below this depth ranges between −200 and −300 mV on the hydrogen scale and the water has a pH ranging from neutral to mildly alkaline [1,2].

Resistance to corrosion can be achieved in several ways. SKB (The Swedish Nuclear Fuel and Waste Management Co.) has decided to approach the long-term corrosion problem by choosing a container material that is as close as possible to being immune to corrosion under the expected repository conditions. Copper has a wide stability range in oxygen-free water [3] and oxygen-free conditions are expected during most of the repository performance lifetime. Pure copper does not have sufficient mechanical strength to withstand the external overpressure of about 14 MPa at the disposal level in the repository. This pressure is composed of a water pressure of 7 MPa, corresponding to a depth of 700 m, and 7 MPa swelling pressure from the bentonite. A cast iron or carbon steel insert in the container is, therefore, used to give the waste package sufficient mechanical strength, as shown in Fig. 1. The insert will also keep individual fuel elements at a safe distance from one another, thereby minimising the risk of criticality.

As part of the safety case for the repository, one of the scenarios being addressed by SKB involves the early mechanical failure of the outer copper overpack, for example due to the formation of a small hole in the copper, allowing water to enter the outer container and corrode the inner one. If penetration by water occurred the inner cast iron insert would start to corrode. At this stage the groundwater surrounding the container would be devoid of oxygen (i.e. anoxic) and the ferrous insert would corrode by slow anaerobic corrosion [4] (i.e. corrosion in the absence of oxygen), a process which leads to the formation of hydrogen and a solid corrosion product. One possible consequence of this failure therefore would be the long-term build up of corrosion product in the annulus between the insert and the outer canister, which could in principle induce stresses in the spent fuel canister and ultimately deform the outer copper canister further. Further degradation of the outer canister is undesirable because it could lead to more rapid release of material from inside the canister. This issue is therefore one that should be taken into account in making safety assessments of the waste disposal system.

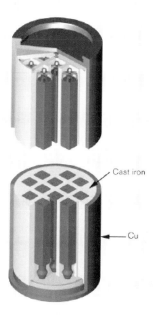

Fig. 1 *Schematic of SKB container design.*

In previous work [5,6] the effect of the accumulation of corrosion product on the inner container on the expansion of the outer copper canister was modelled, assuming certain mechanical properties for the oxide. The modelling showed that deformation of the outer copper canister would occur as corrosion product accumulated. The results of the modelling depended on the mechanical properties assumed for the oxide films and, since no data were available for the oxides formed at low temperatures, data for oxides formed at higher temperatures were used. In order to validate the predictions from the modelling, an experiment was set up to measure expansion that may be caused by the anaerobic corrosion of steel in a simulated repository environment, whilst under representative compressive loads. The apparatus was known as the 'stress cell'. The design of the cell and some preliminary results have been described previously [7]. In parallel, the mechanical properties of corrosion products formed on steel and cast iron by anaerobic corrosion were investigated using atomic force microscopy (AFM)[8].

In view of the very long timescales involved in the geological disposal of radioactive waste containers a number of investigators have studied the corrosion of materials which have been buried for long periods. This approach has been dubbed the 'natural analogue' concept. The materials examined hitherto have included archaeological artefacts. In the case of SKB's specific application most attention so far has been paid to the corrosion of copper-based natural analogues [8,9]. Corrosion of ferrous natural analogues is of interest in relation to the effect of anaerobic corrosion product formation on the generation of stresses in neighbouring materials. The advantage of examining natural analogue materials is that they are likely to have been exposed to the corrosive environment for many years and will therefore have developed a thicker oxide film than can be achieved on a reasonable timescale in laboratory tests. Therefore, in parallel with the stress cell experiment, the feasibility of using natural analogues to examine the potential for expansion caused by anaerobic corrosion in confined spaces over long time periods was investigated. This involved carrying out a literature survey and attempting to identify situations and types of artefacts that could be used as analogues for corrosion-induced expansion under anoxic conditions.

This paper summarises the results obtained using the stress cell and considers the use of natural analogues for the expansion of ferrous materials undergoing anaerobic corrosion.

2. Experimental

2.1. Stress Cell

The overall design of the apparatus was a stack of carbon steel or cast iron discs interleaved with copper discs, which were immersed in the test solution. A system of levers amplified the change in height of the stack and the displacement was measured using sensitive transducers. The lever system was also used to apply a compressive load to the test stack. As the anaerobic corrosion rate of steel was expected to be very low [4], typically <1–5 μm/year, the increase in volume due to the corrosion of a single specimen over a reasonable period of exposure would be too small to be measured accurately. It was therefore necessary to devise a method of amplifying any expansion due to corrosion so that it could be measured more accurately. The method adopted was to set up a stack of ferrous discs separated by copper discs and to measure the total expansion of the stack of discs. This design provided up to two hundred copper–iron interfaces at which corrosion could take place. The alternating copper–iron interfaces simulated the combination of materials that would occur if the surfaces of the inner and outer canisters were to come into contact. All the test specimens were 38 mm dia.; carbon steel discs were 1 mm thick, cast iron disks were 3 mm thick and they were separated by 1.5 mm thick copper discs. The surfaces of the carbon steel specimens were prepared by pickling in hydrochloric acid followed by thorough rinsing in demineralised water. This procedure was to activate the surface; it has been used previously to prepare samples for gas generation rate experiments.

A central hole was machined in each specimen and a locating copper rod was inserted to ensure stack stability. The stack of discs, which was nominally 251.5 mm high, was mounted in a rigid, aluminium alloy frame and the test environment was contained in a cylindrical stainless steel vessel which was sealed with a sliding O-ring system. A polysulphone viewing window was attached to the cylinder, using a butyl rubber gasket, to allow visual inspection of the corroding specimens. A photograph of the assembled apparatus is shown in Fig. 2. Three complete cells were set up.

To ensure entry of water to the interface between the test specimens and allow a free exit path for any gas produced during the corrosion process, the faces of the steel or copper discs were impressed with a matrix of concentric and radial lines approximately 100 µm deep. In addition, the surfaces of the copper discs were roughened by glass bead peening to provide a peak-to-peak height comparable in size to the thickness of the oxide expected on the surface of the corroding steel. The surface roughness for the glass bead peened surface, as described by the R_a value, was 0.5 µm. The test stack was inclined at an angle of ~10° to the horizontal to encourage any gas produced by corrosion to escape.

Fig. 2 *Photograph of assembled apparatus.*

The system of levers gave an amplification factor of 15.9 for the total expansion of the stack. The movement was measured by means of a high temperature linear variable displacement transducer (LVDT); cell 3 was fitted with two transducers, as this was seen as the most important cell because it was closest to the compressive loads expected in the repository. The amplification of the movement in the stack height, and the fact that two hundred corroding interfaces were present in the cell, represents a total magnification factor of 3200 times relative to the expansion at a single corroding steel face. The theoretical detection limit for a change in specimen thickness was ~0.001 μm. The equipment was calibrated using a specially constructed expanding calibration tool based on a micrometer screw thread. This enabled a known expansion to be produced in place of the specimen stack and the deflection at the end of the lever arms to be measured.

The lever arms were also used to transmit a compressive load to the stack of specimens, by placing lead shot in a holder at the end of the lever arms. High precision bearings were used and moving parts were lubricated. The load was monitored by means of load cells mounted directly in line with the specimen stack. All the wetted metallic parts that were in electrical contact with the test specimens were made from mild steel or copper only, in order to prevent any extraneous galvanic effects.

Inlet and outlet pipes were provided to each test cell (i) to enable the test solution to be drawn in to the test chambers, and (ii) to allow gas escape. A reservoir system containing the test solution was attached to the cell inlets and a manometer system was attached to the outlets to enable changes in gas pressure to be monitored. Each test cell contained electrical connections to a carbon steel wire and a silver–silver chloride reference electrode in the environmental chamber. The corrosion potential of the steel was measured to determine whether anaerobic corrosion was occurring. The oxygen concentration was not measured directly but it would be expected that any residual oxygen at the start of the experiment would be rapidly 'gettered' from the test chamber by the walls of the test cells and the corrosion specimens.

Display panels were provided for the load cell and LVDT outputs, and the load, position and temperature of the solutions were monitored automatically using a PC-based datalogging system. The expansion monitoring system was tested and calibrated before the experiments were started by applying a known expansion with a calibrated micrometer screw thread device.

All three rigs were mounted inside a separate environmental chamber, fabricated from stainless steel. It had a glass lid, which was sealed to the box using a butyl rubber gasket. The box was purged slowly with nitrogen, as a precaution to ensure that anoxic conditions were maintained within the test cells. The gas in the test chamber was heated and circulated using small electric fans attached to electric heaters. Platinum resistance thermometers were used to monitor and control the temperature of the gas in the environmental cabinet and three chromel–alumel thermocouples were inserted into pockets in each test cell to enable the temperature of the test solutions to be monitored.

The procedure for setting up and running the experiments was as follows. The test specimens were prepared (ferrous discs were pickled and copper discs were cleaned). The components of the test cells were assembled in the open atmosphere, mounted in their respective frames in the environmental chamber and connected to the transducer systems. The stacks were compressed by filling the lead shot containers to give the desired loads. The reservoir of test solution was deaerated by purging with high purity nitrogen for 24 h. The test cells were then evacuated and the deaerated test solution was allowed to enter the test chambers and brought to the test temperature of 69°C. This temperature was chosen to ensure that the pressure generated by heating the gas in the test chamber was not greater than the hydraulic head in the gas manometer. The following parameters were then monitored: electrochemical potential of the carbon steel electrode, load, displacement of the lever arms, temperature, and height of the liquid in the manometer columns.

2.2. Materials

Experiments were carried out with carbon steel (BS EN 10130:1991 Fe PO1 CR4), cast iron (BS2789 Grade 420/12, nodular spheroidal graphite) and copper (ASTM C101 OFE) discs. The compositions of the materials (mass%) are shown in Table 1.

The test solution was a simple artificial groundwater containing 31.56 gL^{-1} sodium chloride and 1.06 gL^{-1} sodium carbonate, with pH~10.4. This water simulated the pH and ionic strength of a granitic groundwater equilibrated with bentonite backfill.

2.3. Experimental Matrix

In the first set of experiments the three test cells in the rig were set up as follows. In two of the cells, stacks of copper discs and carbon steel discs were used. In one experiment the applied compressive load at the top of the specimen stack was 10 MPa, which was estimated to be the restraining force resulting from the outer copper canister plus the external hydrostatic head in the repository. The second cell was operated at a lower load (~1 MPa), to determine whether the applied load had any effect on the amount of expansion caused by corrosion product formation. The third cell was a control test at a compressive load of 10 MPa that used stainless steel discs rather than carbon steel, to ensure that any expansion measured in the other two experiments was due to corrosion and not some other unforeseen effects.

In a second series of experiments, stacks were set up with cast iron specimens in artificial groundwater, and with carbon steel in aerated demineralised water, under a load of approximately 1 MPa. The experiments with cast iron were chosen because previous work [7] had suggested that cast iron produces a thicker more voluminous corrosion product than carbon steel and so the expansion caused by cast iron needed to be measured and compared to that of carbon steel. The experiments under aerated

Table 1. *Composition of materials used in stress cell experiments*

	Fe	Cu	C	Mn	Si	P	S	Pb	Al	O_2
Carbon steel	bal.		0.035	0.18	0.011	0.011	0.012		0.046	
Copper		99.996				0.0005	<0.005	0.001		0.003
Cast iron	bal.		3.65		2.29	< 0.005	<0.005			

conditions were designed to test the behaviour of the apparatus when rapid corrosion rates were occurring.

2.4. Analysis of Corrosion Products

Measurements were carried out to identify the composition of the both the gaseous and solid products of the corrosion reactions proceeding in the stress cell. The gaseous products of the reactions were analysed by attaching an evacuated gas sample bomb to the outlet from cell number 3. The outlet valve was opened and a sample of the gas above the test fluid was collected and analysed by mass spectroscopy.

Two of the cells from the first set of stress cell experiments, namely the cell containing the stainless steel control stack and the cell containing carbon steel at 10 MPa were dismantled after they had been operating for 380 days and the samples were removed and inspected visually.

The composition of the oxide formed on the carbon steel specimens was determined using Raman spectroscopy, by the following procedure. The cell containing the specimen stack was rapidly removed from the stress cell apparatus and transferred to a nitrogen-purged glove box, where the specimen for Raman analysis was transferred to a gas tight glass cell containing an optical flat to permit access of the laser light for the Raman analysis. This procedure ensured that oxidation of the corrosion product was minimised. Raman spectroscopy is an optical technique whereby chemical and phase specific information can be obtained non-destructively and *in situ* with a spatial resolution of 2–4 μm. A Renishaw Raman Microscope was used. Spectra were recorded in the back-scattered geometry. The exciting laser wavelength was 632.8 nm (from an air-cooled HeNe ion laser). The power was approximately 20 mW. The intensity and position of the bands in a laser Raman spectrum depend both on the chemical composition and the crystal phase composition of the sample.

The composition of the surface layer on one of the carbon steel specimens was determined using X-ray photoelectron spectroscopy (XPS). The sample was transported from the glovebox in an air-tight transport vessel and transferred in air quickly from the transport vessel to the fast entry port of the analysis instrument. It is inevitable that some aerobic oxidation occurred during the transfer. XPS was performed in a VG ESCALAB Mk II instrument, using 250 Watt Al Kα radiation.

The morphology of the oxide on the corroded steel discs was examined using a Hitachi S-800 low voltage scanning electron microscope (LVSEM). This instrument has a high brightness field emission gun which gives imaging for accelerating voltages from 500 to 25 kV. The low voltage capability of the instrument is useful in examining delicate materials where beam damage effects are minimised. A JEOL 840 scanning electron microscope was used to determine the elemental composition of the corrosion product using energy dispersive analysis (EDX).

3. Results of Experimental Work

3.1. First Set of Stress Cell Experiments

Initially three cells were set up; two contained alternate mild steel and copper discs, at compressive loads of 1 MPa and 10 MPa. The third, control cell, contained interleaved stainless steel and copper discs. The change in the actual load on the specimen stacks and the change in the displacement of the transducers at the end of the lever arms of the three test cells are shown in Figs 3 and 4.

There was a relaxation in the load of all three cells. There was a contraction on the displacement transducer on the stainless steel stack of 1.38 mm, corresponding to a change in a single specimen surface of 0.43 ☐m. The height of the stacks containing the mild steel remained virtually constant.

The potential of the carbon steel electrodes in the test cells rapidly assumed a value more negative than –880 mV vs AgCl, indicating that anaerobic corrosion was occurring. The edges of the carbon steel specimens were seen to go black during the experiments, which is also consistent with anaerobic corrosion.

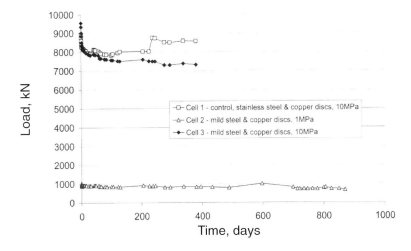

Fig. 3 *Change in load on stress cells — artificial groundwater, 69 °C.*

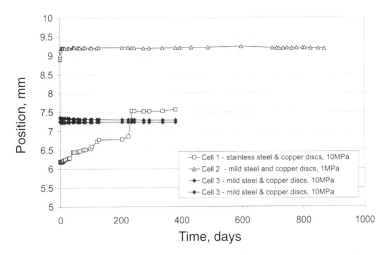

Fig. 4 *Change in displacement on stress cells — artificial groundwater, 69 °C (note: cell 3 had two position transducers).*

3.2. Analysis of Corrosion Products

3.2.1. Visual inspection.

After dismantling cells 1 and 3, the condition of the specimens was examined. A black corrosion product was visible on the surfaces of the carbon steel discs, both around the edge of the specimens and on the surfaces in contact with the copper. The stainless steel specimens in cell 1 remained shiny although some black corrosion product had formed on the carbon steel end plugs and this had spread on to the outer edges of the disks. The appearance of the specimens is shown in Fig. 5.

3.2.2. Gas analysis.

Mass spectral analysis of the gas sample taken from cell 3 showed that the gas contained approximately 3% hydrogen in nitrogen. The presence of hydrogen confirms that the carbon steel was corroding anaerobically.

3.2.3. Raman spectroscopy.

The Raman spectra obtained from the black areas of a corroded carbon steel disc from cell 3 were characteristic of magnetite Fe_3O_4.

3.2.4. X-ray photoelectron spectroscopy (XPS).

The XPS analysis detected oxygen, carbon, iron, chlorine, silicon and sodium. From the detailed scans the predominant species detected were FeO.OH and Fe_2O_3 (goethite and haematite). This result indicates that some oxidation of the corrosion product

Fig. 5 *Appearance of specimens from cells 1 (stainless steel control) and 3 (carbon steel) after exposure to artificial groundwater for 380 days at 69°C. The stack of specimens can be seen at the top and the cell end plugs at the bottom.*

surface had occurred during transfer of the specimen to the spectrometer. The presence of silicon is believed to be due to residual glass beads from peening the surface of the copper discs.

3.2.5. Low voltage scanning electron microscope (LVSEM).
The scanning electron micrographs (Figs 6 and 7) show the morphology of the deposit. The low magnification image (Fig. 6) shows an extensive deposit, although it was not a contiguous film.

Higher magnification (Fig. 7) imaging of the deposit shows that the deposit is composed of smooth 'particles' *ca.* 5–10 µm in size on a background of very small particles ($<< ca.$ 1 µm). The EDX spectrum obtained from the deposit showed that, in addition to iron, silicon, sulphur and chlorine were observed. Surface profilometry measurements on the surface of one of the copper specimens showed no flattening of asperities.

3.3. Second Set of Experiments

After 380 days' operation, test cells 1 and 3 (cell 1: stainless steel/copper, 10 MPa and cell 3: carbon steel/copper, 10 MPa) were dismantled and the specimens examined. The remaining cell (cell 2: carbon steel/copper, 1 MPa) was left running. Cell 3 was set up again using carbon steel and copper discs at ~1 Mpa load using demineralised water as the test solution, open to the atmosphere (i.e. the test solution was aerated), with the aim of demonstrating that rapid corrosion would lead to expansion in the test rig. The results to date of these experiments are shown in Fig. 8.

Fig. 6 LVSEM image of carbon steel sample from cell 3 (×200).

Fig. 7 LVSEM *image of carbon steel sample from cell 3 deposit showing smooth regions on a background of very small particles (×1700).*

The potential of the steel electrode with respect to the silver chloride reference electrode was compatible with the solution being oxygenated (*ca.* –30 mV vs Ag | AgCl). Although the test specimens in the naturally aerated demineralised water were clearly corroding, no expansion was observed after 49 days. The level of water in the test cell was reduced so that the specimens were exposed to a humid atmospheric phase, rather than an aqueous one, with the aim of reducing the

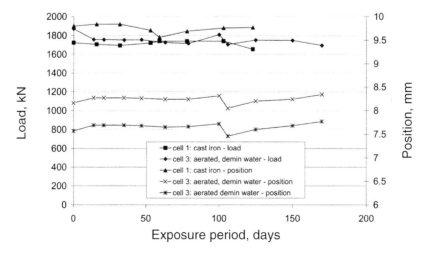

Fig. 8 Results of second series of stress cell experiments at 1 MPa load (note: cell 3 had two position transducers).

possibility of removing corrosion product in the aqueous phase. No expansion has been observed under these conditions. In the next phase of the experiment the load will be reduced to test whether expansion is possible under lower load conditions. No expansion has been observed with the cast iron specimens.

4. Natural Analogues

In view of the long timescales involved, the use of natural analogues has been postulated as a means of validating corrosion models for the failure of metallic waste containers [10]. The main natural analogues for the anaerobic corrosion of ferrous materials are archaeological artefacts (>1000 years) and material from modern industrial or domestic operations (<1000 years). Iron meteorites have also been viewed as natural analogues and there are a few examples of native iron in geological sites with a low redox potential and high pH [11]. The use of archaeological metal artefacts for predicting the long-term corrosion rates of metals for radioactive waste containers has been reviewed for Nagra [12] (the Swiss National Cooperative for the Disposal of Radioactive Waste).

Most studies of the corrosion of archaeological iron objects have been in relation to conservation [13]. The metal core of iron artefacts is typically covered with a layer of magnetite, several millimetres thick, overlain with a looser FeOOH layer [14]. The pore solution may contain dissolved ferrous ions and chloride and on exposure to air the ferrous ions are oxidised to FeOOH, which results in a volume expansion and spallation of the oxide film.

4.1. Archaeological Artefacts

Johnson and Francis [15] have published the widest ranging assessment of the corrosion rates of archaeological iron artefacts in relation to radioactive waste disposal. The samples they considered had been subjected to a wide range of environmental conditions and oxidising power; the most heavily corroded ferrous items were cannon balls removed from oxidising sea water and the least corroded were exposed to dry atmospheres. The corroded items typically exhibited corrosion rates of *ca.* 0.1 µm/year, which are consistent with long-term laboratory experiments on steel wire immersed in anoxic artificial groundwater.

The deposit of two thousand year-old Roman nails at the Inchtuthil legionary site in Scotland is one of the best known archaeological finds of iron [16,17], where over a million nails were buried in a pit. In this case the nails at the centre of the find were hardly corroded, because they experienced low oxygen concentrations and less exposure to water due to the impermeable crust formed by corrosion of the nails at the edge of the mound. Some of the nails at the centre were covered with a smooth black closely adherent scale [17]. Iron artefacts from Australian shipwrecks have been found embedded in coral concretions, which lead to the development of anoxic conditions [18] around the objects. It is probable that anaerobic corrosion occurs in these situations. Iron objects embedded in seabed silts would also be expected to corrode anaerobically.

In examining iron artefacts in relation to predicting the lifetime of radioactive waste containers, most emphasis has been placed on estimating corrosion rates rather than on examining corrosion product properties. There are no known reports in the literature regarding the mechanical properties of corrosion products formed on archaeological objects after long-term exposure to anoxic conditions. There is therefore a requirement to discover fresh objects and to obtain detailed information about the environmental conditions around them before carrying out thorough chemical and physical analyses of the corrosion products. To address the issue of corrosion product expansion, items are required where constrained geometries have been established, for example in crevices between two components of an object.

It should be recognised that although archaeological ferrous items may have suffered from anaerobic corrosion, they will also probably have been exposed to much higher levels of microbial activity than will be the case in a repository backfilled with bentonite. Further important differences between natural analogues and the repository situation are the lack of a radiation field and the low ambient temperatures, both of which could be expected to affect corrosion behaviour.

4.2. Industrial Analogues

In addition to using archaeological artefacts as natural analogues there is a range of analogous situations in modern industrial and domestic operations where anaerobic corrosion of ferrous items may proceed. These include:

- old hot water systems where copper pipes and steel radiators are exposed to near neutral waters, and black magnetite is commonly found as a deposit in the heating systems;

- steel corroding in old concrete, where the rate of diffusion of oxygen through the concrete is so slow that anaerobic conditions pertain at the surface of the reinforcement bars. The fact that spallation of concrete cover occurs due to corrosion of the rebar is an illustration of the expansive force which can be generated though corrosion processes; and

- oxide jacking, for example in civil engineering structures and heat exchangers.

5. Discussion

In the stress cell experiments with carbon steel, no expansion occurred, even though corrosion product analysis identified the presence of both hydrogen and magnetite, confirming that anaerobic corrosion was taking place. In the stainless steel–copper control cell some contraction of the stack took place. The contraction may have been due to creep of the copper at asperity tips on the roughened copper surface. No expansion was observed with cast iron specimens, where a thicker more voluminous oxide would be expected [7], or even under aerated conditions, where much more rapid corrosion (*ca.* 100 μm/year compared with 1 μm/year) would have occurred. The reason for the lack of any detectable expansion in the stress cell experiments is probably because the corrosion product was easily compressed or extruded from the copper-steel interface and it was not capable of withstanding the applied loads. Scanning electron microscopy showed that the anaerobic corrosion product was finely divided and it had deformed to fit into the available spaces between the discs in the stack of specimens, rather than forcing them apart. It is probable that the surface roughness and the grooves in the copper discs were filled with corrosion product from the anaerobic corrosion of the steel. This situation is in contrast to that experienced in concrete or in heat exchangers where jacking effects can cause severe mechanical disruption of the surrounding structures.

One possible reason for the low mechanical strength of the corrosion product formed at low temperature under anaerobic conditions is the higher proportion of water, in contrast to the situation in concrete where the supply of water is limited, or to the oxide films formed at high temperature which have a much higher hardness [19] and a lower water content. If water were to penetrate the outer canister wall (see Fig. 1) and enter the annulus it is likely that any anaerobic corrosion product formed would deform and spread around the annulus, rather than expand to force the copper outer canister away from the iron insert. Eventually the annulus may fill up with corrosion product, which would gradually be compressed by the confining walls of the canister. Subsequent corrosion would depend on whether water could penetrate the corrosion product in the annulus and reach the surface of the iron insert. It is possible that natural analogues may be found that will support this hypothesis, but the evidence is not available in the current body of literature and further careful work on industrial or archaeological objects is required involving fresh samples, with full environmental characterisation.

6. Conclusions

1. A 'stress cell' apparatus did not detect any expansion from the anaerobic corrosion of carbon steel or cast iron after more than two years' exposure.

2. The anaerobic corrosion product formed on iron-based materials is easily deformed and appears to be incapable of causing expansion of confining metal surfaces under compressive loads which simulate those expected in a repository.

3. Natural analogues to support the results of these measurements are not described in the open literature.

7. Acknowledgements

The authors are grateful for the financial support provided by SKB and wish to thank Drs C. Johnston and A. J. Crossley (AEA Technology) for providing the analytical data.

References

1. J. A. T. Smellie and P. Wikberg, *J. Hydrol.*, 1991, **126**, 129.

2. J. A. T. Smellie, M. Laaksoharju and P. Wikberg, *J. Hydrol.*, 1995, **172**, 147.

3. D. G. Brookins, *Eh–pH Diagrams for Geochemistry*. Springer-Verlag, Berlin, New York, 1988.

4. R. Grauer, B. Knecht, P. Kreis and J. P. Simpson, *Werkst. Korros.*, 1991, **42**, 637.

5. A. E. Bond, A. R. Hoch, G. D. Jones, A. J. Tomczyk, R. M. Wiggins and W. J. Worraker, Assessment of a Spent Fuel Disposal Canister. Assessment Studies for a Copper Canister with Cast Steel Insert, SKB Technical Report TR-97-19, (1997).

6. SKB Deep repository for Spent Fuel Nuclear Fuel, SR 97 - Post-closure safety, SKB Technical Report TR-99-06, (1999).

7. N. R. Smart, A. E. Bond, J. A. A. Crossley, P. C. Lovegrove and L. Werme, *Mechanical Properties of Oxides Formed by Anaerobic Corrosion of Steel*, MRS Symp. Proc. Vol. 663, 2000, Materials Research Society, Scientific Basis for Nuclear Waste Management XXIV, Sydney, 2001.

8. R. F. Tylecote, The Effect of Soil Conditions on the Long-Term Corrosion of Buried Tin-Bronzes and Copper, BNFL report, October 1979.

9. A. Bresle, J. Saers and B. Arrhenius, Studies in Pitting Corrosion on Archaeological Bronzes, SKB report 83-05, 1983.

10. M. B. McNeil and J. B. Moody, Corrosion Model Validation in High Level Nuclear Waste Package Research, MRS Symp. Proc. Vol. 294, 1992, Materials Research Society, Scientific Basis for Nuclear Waste Management XVI, Boston, 1993, p.549–556.

11. K-H Hellmuth, The Existence of Native Iron — Implications for Nuclear Waste Management, Finnish Centre for Radiation and Nuclear Safety report, STUK-B-VALO 68, 1991.

12. W. Miller, R. Alexander, N. Chapman, I. McKinley and J. Smellie, Analogues of Repository Materials, Chapter 4 in Natural Analogue Studies in the Geological Disposal of Radioactive Wastes, NAGRA Technical Report 93-03, Elsevier, 1994.

13. Corrosion and Metal Artifacts — Dialogue Between Conservators and Archaeologists and Corrosion Scientists, (B.F. Brown H.C. Burnett, W. Thomas Chase, M. Goodway, J. Kruger and M. Pourbaix, eds). NACE, Houston, USA, 1991.

14. N. Hjelm-Hansen, J. Van Lanschot, C. D. Szalkay and S. N. Turgoose, *Corros. Sci.*, 1993, **35**(1–4), 767–774.

15. A. B. Johnson and B. Francis, Durability of Metals from Archaeological Objects, Metal Meteorites and Native Metals, Battelle Pacific Northwest Laboratory, PNL-3198, 1980.

16. W. H. Manning, The Iron Objects, Chapter 28 in Inchtuthil, The Roman Legionary Fortress Excavations 1952–65 (L.F. Pitts and J.K. St Joseph, eds). Society for the Promotion of Roman Studies, British Monograph series No. 6, 1985.

17. N. S. Angus, G. T. Brown and H. F. Cleere, *J. Iron and Steel Inst.*, Nov. 1962, 956–968.

18. L. E. Samuels, *Mater. Charact.*, 1992, **29**(2), 69–109.

19. J. R. Nicholls, D. J. Hall and P. F. Tortorelli, *Mater. High Temp.*, 1994, **12**(2/3), 141–151.

20

An Analytical Study of Corrosion Products Formed on Buried Ferrous Archaeological Artefacts

D. NEFF, P. DILLMANN and G. BÉRANGER*

Laboratoire Pierre Süe (CEA/CNRS), CEA Saclay, 91191 Gif-sur-Yvette cedex, France
*Université de Technologie de Compiègne, Laboratoire Roberval, Centre de recherches Royallieu, BP 20529, 60205 Compiègne, France

ABSTRACT

A statistical approach to long term corrosion behaviour of iron in soil has been initiated with a study of nine archaeological iron artefacts. By combining classical materials science analytical techniques (optical and scanning electron microscopy and electron microprobe) with microdiffraction under synchrotron radiation, the three components of the corroded iron in soil have been characterised: the metallic substrate (structure and composition), the corrosion products (identification of three main phases: goethite, magnetite and siderite) and the soil (specific compositions close to the original surface of the artefact). An estimate of the corrosion rate has been made from thickness measurements of the corrosion products.

1. Introduction

Disposal of nuclear wastes within geological barriers is envisaged for the coming years in France. High-level radioactive wastes will be confined in a glass matrix and stored in a metallic canister. The French programme plans to use a low alloy steel overpack. Simulations of carbon steel corrosion in soil have therefore been carried out in order to understand the long term behaviour of this material. However, predictions made from these are not sufficiently reliable for very long periods (several thousands of years). One way to validate these experiments is by studying recently excavated archaeological iron artefacts.

In the past the corrosion of these kinds of artefacts in soil has been studied for different purposes. On the one hand, it provides information for preventing degradation after excavation and ensuring reliable restoration. On the other hand, some authors consider archaeological artefacts as analogues for the estimation of iron corrosion behaviour in soil, particularly for obtaining average corrosion rates.

Accary *et al.* and Miller *et al.* [1–3] have studied archaeological artefacts in the context of long term corrosion. Their results are not very detailed and no indications of the structure and composition of the metal and the corrosion products or of the nature of the soil are given. The corrosion rates they collected are between 0.01 and 1 μm/year with a value of 1 mm/year being obtained on one artefact. This extreme value of 1000 μm/year is very high when compared with the other studies (see Table 1).

No precise description of the corrosion rate measurement methods are given in these papers and no critical point of view about the obtained values is presented.

More recent studies aimed at conservation of archaeological artefacts have been undertaken. Soerensen *et al.* [4] underline the difficulty of this kind of work. Soil is in fact a complex environment and the following parameters play a role in the corrosion process:

- geology (soil types, electrical resistivity),

- hydrological factors (water flow through ground layers),

- chemical composition of the soil waters (pH, dissolved O_2, contents of salts such as SO_4^{2-}, S^{2-}, PO_4^{3-}, Cl^-, CO_3^{2-}), and

- bacteriological activity.

These authors measured uniform corrosion rates between 0.025 and 1.2 μm/year.

Gerwin *et al.* [5] worked on iron nails and measured the degree of corrosion by comparing the size of the metal core remaining with that of the original size of the object estimated from X-ray photographs. This method is not very precise because of the difficulty in determining the original size. But the results show general tendencies for the influence of soil composition on iron corrosion rate. For example, phosphates and especially carbonates led to the formation of a protective layer on the metal that slows down the corrosion process, but the type of the phase that was formed was not specified. Conversely a high content of sulfates or chlorides in soil water ensured the destruction of layers covering the metallic substrate that would have been protective.

To estimate corrosion rates, two simulation studies have been carried out of steel buried in soil. Galliano *et al.* [6] buried carbon steel samples in a sandy soil and estimated an average corrosion rate of 15 μm/year, after 500 days by measuring the metal mass loss.

Angelini *et al.* [7] studied a soil with a high content of organic matter and carbonates. These authors showed that the corrosion rate of samples decreases with burial time (from 0.035 g dm^{-2}/day after 30 days to 0.028 g dm^{-2}/day after 120 days). To complete these observations X-ray Diffraction (*XRD*) and Fourier Transform Infrared Spectroscopy (*FT–IR*) data were collected in order to identify the corrosion layers. They report the presence of goethite (α-FeOOH), akaganeite (β-FeOOH containing chloride), magnetite (Fe_3O_4) and siderite ($FeCO_3$) in the corrosion products.

These studies (summarised in Table 1) give some idea about the corrosion of iron in soil. Nevertheless, all the results are very heterogeneous because of the diversity of the materials of the artefacts (iron, carbon steel), the soils (sand, clay, natural or with associated elements) and the measurement methods.

Fell *et al.* [8] have analysed archaeological artefacts excavated from a waterlogged deposit. This particular environment leads to the formation of aggregates of sulfides identified by XRD. These phases are formed principally through biological activity. On some samples the major corrosion component was siderite: it is assumed that this phase is often associated with sulfides.

Table 1. *Summary of literature data on corrosion rates of ferrous alloys in soils*

Type of sample	Environment	Measurement technique	'Corrosion rate'	Ref.
A	Waterlogged soil	X Radiography	0.025–1.2 µm/year	[4]
A	3 types of soils: - urban (phosphates, carbonates, chlorides, sulfates) - sandy (acidic, low salts content) - loess (neutral or slowly alkaline)	X Radiography	Degree of corrosion: 80–100% 60–80% 40%	[5]
A and meteorites, ferrous alloys	?	Thickness measurement	0.01–1000 µm/year	[1]
A	?	Thickness measurement	0.1–10 µm/year	[2]
S (500 days)	Sandy	Weight loss	15 µm/year	[6]
S (30 and days at 25°C)	Soil: organic matters and chlorides water	Weight loss	After 30 days: 0.035 g dm^{-2}/day After 120 days: 0.028 g dm^{-2}/day	[7]

A: archaeological ; S: simulation

To find the original shape of archaeological artefacts, Bertholon [9,10] introduced the notion of original surface (OS). The author defined it as the limit between the 'matters' that belonged to the object (metal, mineral and organic parts) and the surrounding soil, before the beginning of the corrosion process. Then the difference is made between what is called 'inner corrosion products', under the original surface and 'outer corrosion products', above this surface. The inner are recognisable because they contain 'inner markers' coming out from the metallic substrate (slag inclusions for example), and the outer by the presence of 'outer markers', such as soil minerals (quartz grains...). The OS does not always coincide with the original level of the object because it can move during the formation of oxide scales.

All the studies on archaeological artefacts reviewed here are very dissimilar and never give a complete characterisation of the whole system represented by metal in soil. It appears that literature data alone are not sufficient to provide a complete and statistical overview. For this purpose a study on iron archaeological artefacts has been launched by ANDRA (Agence Nationale pour la gestion des

Déchets Radioactifs) and CEA (Commissariat à l'Energie Atomique). In this work a corpus of artefacts, taken from different excavation sites, has been studied. The first analytical results are presented below.

First it is important to describe briefly the ancient iron reduction processes in order to understand the metal structure of the sample artefacts. In Europe before the 12th century, iron ore was smelted in a shaft furnace. An iron bloom was obtained which contained the non reduced impurities of the ore and it was therefore necessary to separate the metal from this slag by hammering the bloom. An iron ingot was then obtained, so that this iron smelting process is called the 'direct process' [11].

During the Middle Ages the use of hydraulic force appeared for the ventilation that permitted the furnace temperature to be increased. The iron ore reduction took then place in a blast furnace. In this process carbon and other elements of the ore like silicon, phosphorus or manganese diffuse more easily in the metal. The obtained product was pig iron that was cast under the furnace. In order to decarburise it and obtain wrought iron, the metal was reheated in an open hearth. This iron ore reduction process is called the indirect process because it requires two steps (the first in the liquid state and the second in the solid state) to obtain carbon steel or iron.

Nevertheless, whether by the direct or the indirect process, the carbon steel is obtained in the solid state, so that impurities remain in the metal, and come out in the slag. Even after the forging operation these impurities are always present as slag inclusions. The better the forging quality is, the less the chance that these second phase particles will be present in the metallic matrix. The general alignment of these inclusions gives information on the direction of any cold working. As these slag inclusions can measure up to 800 µm in length they are observable with an optical microscope. Moreover, compared to contemporary low alloy steels, the metallic matrix of the ancient artefacts can contain very high amounts of additional elements. The most important is carbon but other elements like silicon, phosphorus or manganese, for example, can be detected. It is important to appreciate these points for the understanding of the further analysis presented in this study.

In the first part of this paper the experimental corpus and the sample preparation will be presented. The different analytical methods will then be described, from classical material sciences analysis (optical and electron microscope and electron microprobe) to microdiffraction under synchrotron radiation for local and structural characterisation. In the last part the results will be presented and discussed: firstly the metallic substrate determination, then the characterisation of corrosion products formed during the burial period, with separate considerations on interfaces, and finally the estimations of corrosion rates from thickness measurements.

2. Analysed Samples and Analytical Methods

A corpus of nine archaeological artefacts has been collected. These artefacts were taken from three different archaeological excavation sites in France: Glinet, Montbaron and Avrilly (see Table 2).

Sampling included the adhering soil in order to conserve the close environment. The first step of the sample preparation was drying: artefacts were put in a drying oven under a primary vacuum in order to withdraw the non-combined water from

Table 2. *Corpus of archaeological artefacts*

Sample	Excavation site	Dating	Artefact	Type of soil
GL 00 03	Glinet (Somme): forge	16th century AD	Nail	Mainly clay
MONT 1	Montbaron (Indre et Loire): seigniory	12–13th century AD	Horseshoe	Mainly clay
MONT 2			Ploughshare	
MONT 4			Plate	
MONT 5			Point	
MONT 6			Iron tool	
MONT 6-2			Nail	
AVRI 2-1	Avrilly (Eure et Loire): Artisanal zone	15th century AD	Nail	Mainly clay
AVRI 3			Piece of metal	

the soil. The artefacts were then mounted in epoxy resin and held for one day under primary vacuum to ensure good penetration into the soil and the corrosion products. Finally, the metal was cut to obtain cross-sections. Films thinner than 50 μm were also prepared from cross sections for micro X-ray diffraction (μXRD) experiments in transmission mode (see below). All the samples were polished with diamond paste (3 and 1 μm) for observation by optical microscopy (OM). Before etching, the slag inclusion distribution in the metallic matrix was observed for information on the metal heterogeneity. The OM was also used to observe the morphology of corrosion products around the metal. It is particularly interesting to observe the metal/oxides interface on the one hand and the oxide/soil interface on the other hand. The metallographic structure was then revealed (ferrite, pearlite...) after etching with 2% Nital. After repolishing, the Oberhoffer etching (cupric etching) was applied to study phosphorus distribution in the metallic matrix: this reagent deposits copper preferentially on the zones of lowest phosphorus content [12] so that they appear darker under OM.

For the composition analysis several techniques have been employed. The first is the Energy Dispersive Spectroscopy (EDS), directly coupled to the Scanning Electron Microscope. This technique was mainly used to analyse the element compositions of slag inclusions, before etching. It is also an efficient method to locate specific elements (Si, Ca, Al, P...) on the interfaces between corrosion products and soil by elementary X-ray mapping.

The Electron Probe Microanalysis (EPMA) was specially used to analyse minor elements in the metallic matrix. A $2 \times 2 \ \mu m^2$ beam with an accelerating voltage of 15 kV was focused on the sample. In the selected analysis conditions, the detection limit was about 200 ppm. Element concentration profiles were also measured from metal to soil. The quantification results given by EPMA enable phase stoichiometries to be deduced. Finally, X-ray microdiffraction (μXRD) experiments were conducted on the D15 beamline at the Laboratoire pour l'Utilisation du Rayonnement Electromagnétique (LURE) at Orsay, France. The white X-ray beam delivered by the DCI ring was focused and monochromatised by a carbon/tungsten (C/W) Bragg Fresnel Multilayer Lens (BFML) [13]. The BFML is a wide band pass monochromator that results in broadening of the diffraction peaks compared to the classical diffraction set-up. Photons centred around 14 keV (λ = 0.08857 nm) were selected and focused down to a 20 × 20 µm² beam. The diffraction patterns were collected with an image plate downstream of the sample. A detailed description of the experimental set-up is provided in [14,15]. 2-D diffraction patterns were obtained by circularly integrating diffraction rings using the FIT2D software developed at the European Synchrotron Radiation Facility (ESRF) [16]. The reference sample used was a silicon powder. The spectra were compared with the JCPDF database using the Diffract + program. Thus, crystalline phases can be identified and precisely located in the oxide scale: structural profiles are obtained by this method.

3. Results and Discussion

The corrosion of iron in the soil is a complex system composed of metal, corrosion products and soil. Because all the studied artefacts were sampled with the adhering soil it was possible, by preparing cross sections, to observe the successive parts of this system. This section will be divided into three paragraphs. Firstly the metallic substrate will be examined i.e. composition, structure and slag inclusions. Then, after the metal/oxide interface, the corrosion products under and on the original surface (OS) (following Bertholon's definition [9]) (see Fig. 1) and finally calculations of corrosion rates will be presented.

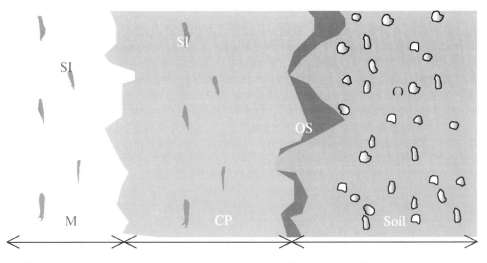

Fig. 1 *Schematic representation of a cross-section ; M: metallic substrate ; CP: corrosion products under OS ; Sl: slag inclusions ; OS: original surface ; Q: quartz grains.*

3.1. Metallic Substrate Examination

Except two artefacts which are entirely corroded and contain only metallic zones of few hundreds of micrometers (see Fig. 2), all the samples were analysed following the experimental method described above. All the analysed artefacts are hypoeutectoid carbon steels. The maximum carbon content that was estimated by quantitative metallography after Nital etching is about 0.5 mass% (see Table 3). On one sample, tertiary cementite can be observed at grain boundaries (see Fig. 3). The most carburised samples show pearlitic zones distributed heterogeneously. For example, Fig. 4 shows a ferrito–pearlitic zone observed on the MONT 1 sample. Only on the MONT 2 sample, could martensite be observed very locally. Generally, all the metallic microstructures are equiaxed. Nital etching on the AVRI 2-1 sample also revealed twinning zones, which were caused by rapid deformations at ambient temperature as a result of the use of the artefact [17].

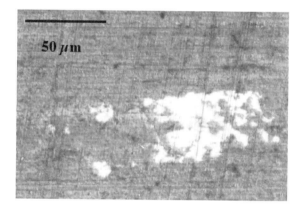

Fig. 2 *Isolated non corroded zone (light zone) in corrosion products, OM, MONT 4 sample.*

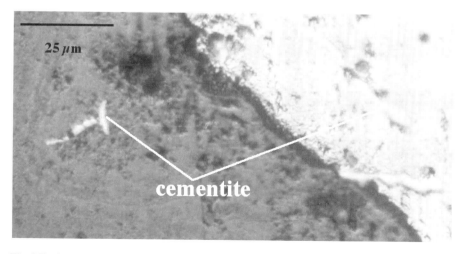

Fig. 3 *Tertiary cementite at grain boundaries in the metallic matrix and in the corrosion products, OM, MONT 6-2 sample.*

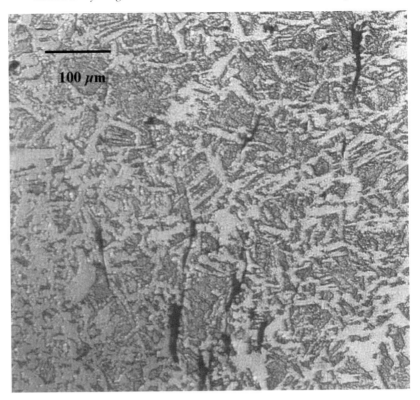

Fig. 4 *Ferrito–pearlitic matrix, OM, MONT 1 sample.*

Phosphorus is the only minor element detected by EPMA in the metallic matrix. Five artefacts (see Table 3) contain high phosphorus amounts (400–5500 ppm). The distribution of this element was studied using the Oberhoffer etching that reveals mesoscopic heterogeneity. In fact, the P distribution follows the cold work lines marked by the slag inclusions. It is particularly clear on the sample GL 00 03 (see Fig. 5(a)). Moreover, one can observe a P depletion, called a ghost structure [18], around the slag inclusions (see Fig. 5(b)), which is proof of the dephosphorating role of these second phase particles [12,19].

The main structural difference between contemporary carbon steels and ancient iron artefacts lies in the presence of slag inclusions which can be very numerous in the latter. Previous studies on slag inclusions in ancient iron artefacts [20,21] revealed that they are mainly composed of fayalite (Fe_2SiO_4), wüstite (FeO) and glassy phases. Their distribution in the metallic matrix varies from one sample to another and depends on the forging quality (see Table 4). Thus, even if the metallic microstructure presents only equiaxed grains, it is possible to deduce some information about the forging (see Table 4). For example, GL 00 03 shows numerous inclusions elongated along the cold work lines (see Fig. 5(a)).

These first observations on the metallic matrix of ancient iron artefacts reveal that these materials are often very heterogeneous. This heterogeneity is expressed in the composition (carbon and phosphorus contents) and the structure (deformation lines,

Table 3. *Structure and composition of the metallic matrix*

Sample	Carbon rate (mass%)*	Metallic structure	Phosphorus composition[†] and distribution
GL 00 03	< 0.02–0.1	Ferritic matrix Some pearlitic zones	1400–5200 ppm Mesoscopic distribution along cold work lines Impoverishment around slag inclusions Ghost structure
MONT 1	< 0.02–0.4	Ferritic matrix Some pearlitic zones with acicular ferrite	< 500 ppm
MONT 2	<0.02–0.5	Ferrito–pearlitic matrix Locally martensite	400–1800 ppm Ghost structure
MONT 4	Corroded	Some metal traces	Traces
MONT 5	<0.02–0.5	Ferritic and ferrito–pearlitic matrix	300–400 ppm
MONT 6	Corroded	Some metal traces	
MONT 6-2	0.02	Ferritic matrix Tertiary cementite at grain boundaries	1000–2000 ppm
AVRI 2-1	< 0.02	Twinning ferrite	1600–4500 ppm
AVRI 3	< 0.02–0.1	Ferrito–pearlitic matrix	1700–5500 ppm

*Quantitative metallography; [†]EPMA

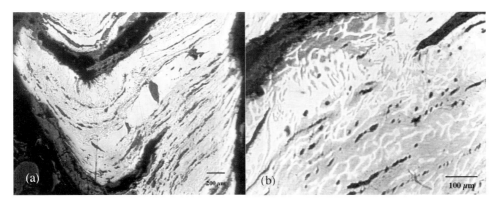

Fig. 5 *(a) Mesoscopic heterogeneity of the P distribution following the cold working lines; (b) P distribution around slag inclusions, OM, Oberhoffer etching, GL 00 03 sample.*

Table 4. *Slag inclusions in the metallic matrix*

Sample	Slag inclusion distribution	Type of slag inclusion	Cold work
GL 00 03	N	F + W	Elongated slag inclusions along shape lines Equiaxe ferrite
MONT 1	C	F + W	Elongated slag inclusions along shape lines Equiaxe ferrite
MONT 2	C Central alignment of some slag inclusions	F + W	Elongated slag inclusions along shape lines
MONT 4	Some slag inclusions present in the corroded matrix	F + W	
MONT 5	C	Iron oxides	Slag inclusions alignment between ferrite and carbon steel matrix
MONT 6	–	–	–
MONT 6-2	C	F + W	Elongated slag inclusions along shape lines Equiaxe ferrite
AVRI 2-1	N	Monophased Amorphous silicate	Twinning system
AVRI 3	N	Monophased Amorphous silicate	Elongated slag inclusions

N: numerous slag inclusions; C: relatively clean metallic matrix; F: fayalite (Fe_2SiO_4); W: wüstite (FeO)

second phase particles, carbide distribution...), and is a way to compare the influence of different metallic structures (ferritic, carburised zones...) on corrosion formation in the same burial conditions.

3.2. Corrosion Products Examination

The metal/oxide interface can be very different from one sample to another. It can be sometimes relatively regular, but in most of the samples the corrosion penetrates in the metallic matrix either along slag inclusions, (the direction depends on their alignment), (Fig. 6), or along grain boundaries (Fig. 7). This penetration of corrosion products can reach to 100 μm. In some cases, the corrosion perforates the artefact.

The MONT 6-2 sample is peculiar because it is the only one that presents an

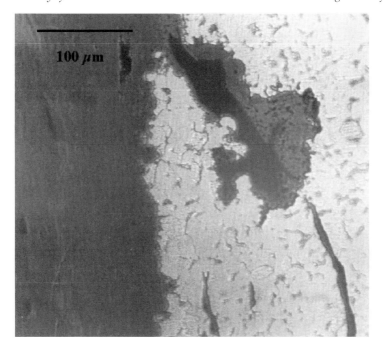

Fig. 6 *Corrosion penetration relatively parallel to the corrosion progression around a slag inclusion, OM, MONT 1 sample.*

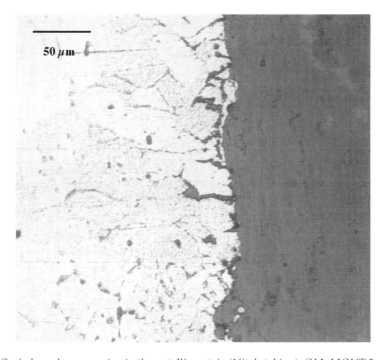

Fig. 7 *Grain boundary corrosion in the metallic matrix (Nital etching), OM, MONT 2 sample.*

anisotropic corrosion morphology. On the head of the nail, the corrosion looks fingered (Fig. 8), as opposed to the rest of the sample where the corrosion front is relatively linear. However, no anisotropic structure or composition of the metallic substrate was observed that could explain the corrosion product morphology.

The observations made on the metal/oxide interface show that the local penetration of corrosion in the metallic substrate seems to be highly influenced by the metal microstructure. The presence of slag inclusions could favour the migration of ions along defects generated by these second phase particles. Lastly, no clear explanation on the penetration at grain boundaries has been found. The difference in composition at these places (for example the phosphorus content) has to be investigated in further experiments.

To determine the nature of the corrosion products and their local distribution in the oxide scales, EPMA and μXRD analyses were combined. The quantitative analyses made by EPMA give the elementary distribution in the oxide scale. The μXRD spectra obtained at the same places lead to the clear identification of the constitutive phases of the corrosion products. Finally, the comparison of these compositional and structural data with OM observations precisely enable the different identified phases in the corrosion scales from metal to soil to be located.

No amorphous phases seem to be associated with the corrosion products under the OS (see Fig. 1). Indeed, every μXRD spectrum shows relatively well crystallised phases (very intense diffraction peaks). Moreover, the complete diffraction rings observed on the 2D-patterns reveal that the identified phases are not textured. Three different major phases were identified in the corrosion products (see Table 5). The main phases are magnetite Fe_3O_4 and goethite α-FeOOH (see Fig. 9). In some cases, siderite $FeCO_3$ (see Fig. 9) can be identified. EPMA analyses show that this phase contains various quantities of calcium (up to 4 mass%). Table 5 indicates for each sample the distribution of these phases in the corrosion scale.

Calcium is observed on five samples in the form of aggregates close to the original surface (see Fig. 10) or as fine strips in corrosion products under the original surface (see Fig. 11). The compositions of the Ca-rich phases were analysed by EPMA as

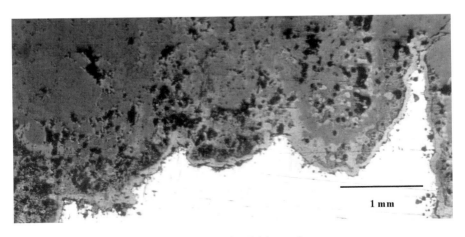

1 mm

Fig. 8 Fingered interface of corrosion, OM, MONT 6-2 sample.

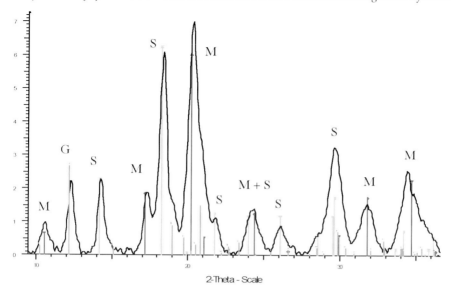

Fig. 9 µXRD spectrum obtained at a point of the corrosion scale of the AVRI 3 sample. Vertical lines: magnetite, M (JCPDF 79-419), siderite, S (JCPDF 83-1764) and goethite, G (JCPDF 81-464).

shown on Fig. 10, but the strips were too fine to be analysed correctly by this technique and this is the reason why Fe is overquantified.

Chlorine was detected by EDS and EPMA on one sample (see Table 5). Two FeOOH) and an iron chloride ($Fe_2(OH)_3Cl$). Figure 12 presents the µXRD spectra of

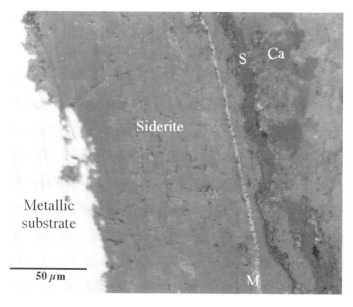

	mass%
O	42.14
Ca	18.93
Fe	28.34
Total	90.13

Fig. 10 Cross-section from metal to soil, (OM), M: magnetite, S: sulfur ; Ca: aggregates of Ca, and corresponding EPMA analyse of Ca-rich phase GL 00 03 sample.

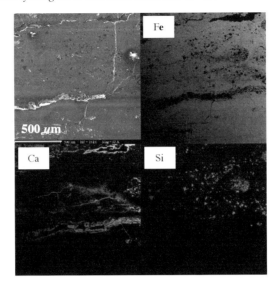

Fig. 11 *Calcium strips on X-ray mapping of Fe, Ca and Si around the OS, MONT 6 sample.*

these two phases. Akaganeite forms in chloride environment. Moreover, the iron chloride $Fe_2(OH)_3Cl$ has been identified by Refait *et al.* [22] as a precursor of either akaganeite or goethite, depending on the chloride environment. Thus, it seems that the akaganeite was formed from this iron chloride in the MONT 6-2 sample.

Sulfur has been detected near the original surface on the GL 00 03 sample in a fine brown strip on OM pictures (see Fig. 10). Its amount has been quantified by EPMA: up to 5 mass% of S is detected. But it is difficult to make a reliable quantification because S was found in a crack caused by the sample preparation.

Even for samples with a high phosphorus content in the iron matrix, the amount of phosphorus in the corrosion products is never higher than 1 mass%. The amounts of other elements like aluminium or silicon also do not exceed 1 mass%.

A last important point concerning the corrosion products under the original surface is the presence of 'inner markers' coming from the corroded substrate. These markers are either non-corroded slag inclusions that exist in the corrosion products (see Fig. 13) or tertiary cementite coming from the substrate (Fig. 3).

Except for two samples where insufficient adhering soil had been taken (MONT 2 and 5), it is always possible to locate the original surface (as described by Bertholon [9]). It is the frontier between the zone in which the corrosion products have entrapped inner markers and the zone where quartz grains or other markers coming from the soil can be found. Moreover, this OS can be characterised by the presence of particular phases or elements: in one case sulfur and a fine magnetite strip have been observed, and more generally an enrichment of Ca noticed near the OS.

Some analyses were performed on outer corrosion products (see Fig. 1). The first observations are EPMA concentration profiles. The profile of Fig. 14 shows a strong iron concentration heterogeneity up to the OS. Nevertheless, the average iron content is about 20 mass% which is higher than that of iron in soil. This means that some iron cations have migrated from the corrosion products to the environment. Moreover, μXRD spectra obtained in the soil reveal a progressive broadening of the peaks

Table 5. *Corrosion products under OS, structure and composition*

Reference	Interface M/O	Corrosion products from metal to OS (thickness, see Table 7)	Inner markers of metal in oxides
GL 00 03	PI ⊥ (metal perforated) PGB	S (100 μm) M (<10 μm) + sulfur (<10 μm) Ca-rich phases close to the OS	None
MONT 1	PI // (up to 500 μm in the metal)	G/M mix Strips of Ca rich phases (10 μm)	Slag inclusions
MONT 2	PI // (up to 350 μm in the metal) PGB up to 60 μm	G + M Fine strips of Ca rich phases (20 μm)	Slag inclusions
MONT 4		G + M Fine strips of rich Ca phases Locally, some needle structure of G + M	Slag inclusions
MONT 5		G G + M	None
MONT 6		G/M in inner oxides G in outer corrosion products	Slag inclusions
MONT 6-2	Anisotropic interface between metal and corrosion products: fingered overhang of corrosion	Fe$_2$(OH)$_3$Cl (50–200 μm) G + A S Spheroids partially recovering the G + A scale (3 mm) G + M Fine strips of rich Ca phase at the OS	Slag inclusions, cementite
AVRI 2-1	PGB up to 100 μm in the metal	M	None
AVRI 3	–	S close to the metal G/M mix	None

PI: penetration of corrosion products along inclusion lines; //: parallel to the metal/oxide interface; ⊥: perpendicular to the M/O interface; PGB: penetration of corrosion products along grain boundaries; L: linear interface; A: akaganeite (Cl β-FeO(OH)); G: goethite (FeO(OH)); M: magnetite (Fe$_3$O$_4$); S: siderite (FeCO$_3$); G/M: not discriminated by μXRD, G + M: discriminated by μXRD.

belonging to iron-containing phases, compared to corrosion products. This could be explained by the increase of the amount of amorphous phases (Fig. 15). Also, some crystallised goethite was identified on two samples by μXRD, more than 13 mm from the metal/oxides interface and in the outer part in contact with soil i.e. the

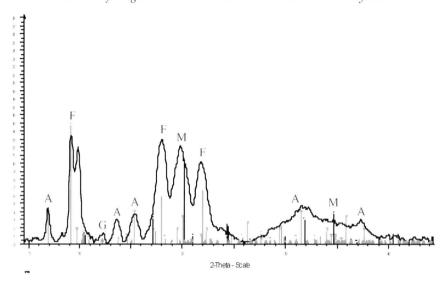

Fig. 12 μXRD spectrum obtained on MONT 6-2 sample; Vertical lines: akaganeite, A (JCPDF 13-157, magnetite, M (JCPDF 82-1533), $Fe_2(OH)_3Cl$, F (JCPDF 72-619) and goethite, G (JCPDF 81-464).

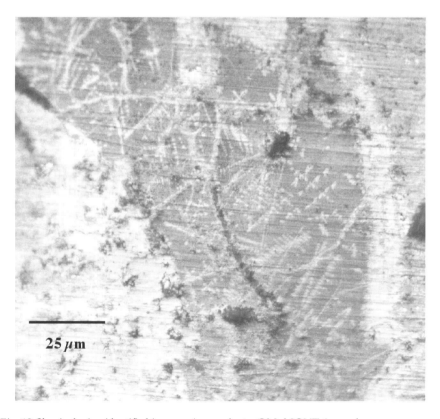

Fig. 13 Slag inclusion identified in corrosion products, OM, MONT 4 sample.

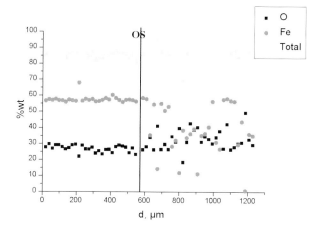

Fig. 14 *EPMA concentration profiles of Fe and O, MONT 6 sample.*

zone between OS and soil. Thus, at the moment it is impossible to conclude whether or not the phases formed in soil are crystalline.

These results lead us to make some general observations. The samples can be arranged in two groups depending on the corrosion product type: (i) the samples containing only magnetite and goethite, and (ii) those also containing siderite. This classification is chosen on the basis of literature results [5] that suppose carbonates to be protective corrosion products.

Magnetite and goethite phases are often present together or separated but in most cases they could not be discriminated by μXRD ($20 \times 20 \times 50 \ \mu m^3$ beam). When μXRD was able to discriminate these phases, goethite and magnetite are referenced

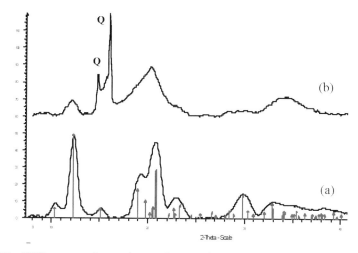

Fig. 15 *μXRD spectra of corrosion products under the OS (a) and in soil (b); Vertical lines correspond to the JCPDF file 81-464 of the goethite; Q: quartz (83-542 JCPDF file), MONT 4 sample.*

as G + M in Table 5. When these two phases are mixed in the 20 × 20 × 50 μm^3 volume, they are referenced as G/M mix in the Table 5. The mechanisms of siderite formation in the three artefacts seem to be different because of the morphology of this phase (spheroids in the MONT 6-2 sample and continuous layer in GL 00 03 and AVRI 3, see Table 5).

As mentioned above, the substrate microstructure and composition seem to have an influence on the metal/oxide interface morphology and on the local corrosion penetration, i.e. along grain boundaries or slag inclusions. Nevertheless, at this point of the study, no clear correlation can be made between the metallic substrate and the type of corrosion products. Indeed, the main influence on corrosion product formation seems to come from the soil environment.

3.3. Corrosion Rate Calculation

In order to evaluate approximately the corrosion rates and make some comparisons with literature data, the corrosion product thicknesses were measured on cross sections taken from the metal/oxide interface to the original surface, perpendicularly to the interface. This value represents a minimum of the total metal loss because iron cations that migrate over the original surface in the soil are not taken into account. Moreover, the porosity of the oxide scale is not included in the calculated corrosion rate. For this approximation, the corrosion has been assumed as uniform i. e. local overhang was not considered.

To take into account the expansion of the oxide layers compared to the metal volume during the corrosion process it is necessary to introduce the Pilling–Bedworth ratio (PBR) [23]. This is the ratio of the specific volume of the oxide and the metal. This parameter is important in determining whether an oxide film can be protective or porous. It can be theoretically calculated by the following formula:

$$PBR = \frac{Md}{nmD}$$

where M is the molecular weight and D is the density of the oxide, m and d are the atomic weight and density of the metal respectively, and n is the number of the metal atoms in the molecule of the oxide. The ratio has been calculated for the three main iron oxides or oxy-hydroxides identified in this study (see Table 6).

According to Gupta *et al.* [23] all these phases are supposed to have a protective

Table 6. Pilling–Bedworth ratio

Iron oxides	D	PBR
Magnetite: Fe_3O_4	5.20	2,09
Goethite: α-FeOOH	4.26	2,94
Siderite: $FeCO_3$	3.93	4,15

role for the metal during the corrosion process because the PBR is greater than unity: the most protective must be siderite. Only the thickest phase was taken for the corrosion rate calculation. When two oxides are mixed, the phase with the higher Pilling–Bedworth ratio was used (indicated in brackets). From the thickness values, corrosion rates were evaluated by the following calculation:

$$CR = \left(\frac{thickness}{dating} \right) / PBR$$

The measured thicknesses and the calculated corrosion rates are given in Table 7.

The corrosion rates calculated here are of the same order of magnitude as those found in the literature (0.01–10 μm/year). Nevertheless, our results are closer to the minimal values. This could be explained by the fact that, as stated before, the total metal loss is minimised in our calculation. Furthermore, there is no evident correlation between the nature of the corrosion products and estimated corrosion rates which means that these calculations need to be refined by taking more parameters into account.

4. Conclusions

To begin a statistical approach of the long term corrosion behaviour of iron, archaeological iron artefacts, originating from three different excavation sites, have been studied and the first results obtained on nine artefacts presented.

The metallic substrates of the samples were very heterogeneous because of the presence of slag inclusions and the variable composition of the metallic matrix. From the metal to the soil, the following profiles were identified: corrosion products containing 'inner markers' coming from the substrate (for example slag inclusions), an original surface, and then the soil, in which 'outer markers' like quartz grains are entrapped. The corrosion products formed during the burial period have been identified. Three main corrosion products have been characterised: goethite, magnetite and siderite, independently of the excavation site. Finally corrosion rates have been estimated. These first estimates lead to corrosion rates between 0.03 and 1 μm/year. These values are comparable to literature values. The use of various techniques revealed no influence of the metal structure on the nature of corrosion products which seems to depend mainly on the soil characteristics. In addition, the nature of the corrosion products seems to have no clear influence on the corrosion rate.

However, structure and composition variations seem to have an influence on the local corrosion penetration. This point is important to consider in the context of deep repository of radioactive wastes, where local perforations of the carbon steel container can be dramatic.

These results show that the metal/soil system can lead to very varied corrosion profiles. Thus, more samples have to be examined to confirm these general tendencies.

Table 7. Thicknesses of the corrosion products and calculated corrosion rates

Reference	Dating	Corrosion thickness (μm)	Corrosion rate (calculated from t_m) (μm/year)
GL 00 03	500 years	$100 < t < 350$ $t_m = 225$	$0.05 < CR < 0.17$ $CR_m = 0.11$ (S)
MONT 1	800 years	$60 < t < 900$ $t_m = 219$	$0.03 < CR < 0.38$ $CR = 0.09$ (G)
MONT 2	800 years	Original surface not visible on this sample	–
MONT 4	800 years	$450 < t < 550$ $t_m = 510$	$0.19 < CR < 0.23$ $CR_m = 0.22$ (G)
MONT 5	800 years	Original surface not visible on this sample	–
MONT 6	800 years	$650 < t < 1000$ $t_m = 825$	$0.27 < CR < 0.43$ $CR_m = 0.35$ (G)
MONT 6-2	800 years	Fingered zone: $2800 < t < 4700$ $t_m = 4100$ Isotropic zone: $1100 < t < 2400$ $t_m = 1890$	CR have been calculated with the goethite equivalent because of the complexity of the oxide scale $1.19 < CR < 2.00$ $CR_m = 1.74$ (G) $0.47 < CR < 1.02$ $CR_m = 0.80$ (G)
AVRI 2-1	600 years	Head: $350 < t < 450$ $t_m = 400$ Body: $250 < t < 350$ $t_m = 300$	 $0.28 < CR < 0.36$ $CR_m = 0.32$ (M) $0.20 < CR < 0.28$ $CR_m = 0.24$ (M)
AVRI 3	600 years	$1300 < t < 2300$ $t_m = 1560$	$0.52 < CR < 0.92$ $CR_m = 0.63$ (S)

t_m: mean thickness ; M: magnetite ; G: goethite ; S: siderite

5. Acknowledgements

This study was supported by ANDRA and CEA.

References

1. A. Accary and B. Haijtink, *Proc. Conf. Journées de Paleométallurgie*, 1983, 323.

2. B. Miller and N. Chapman, *Radwaste magazine*, 1995 (Jan.), 32–42.

3. A. B. Johnson and B. Francis, Durability of metals from archaeological objects, metal and meteorites and native metals, U.S. Department of Energy (1980), in A. Accary and B. Miller [1,2].

4. B. Soerensen and D. Gregory, *Proc. Metal '98 Conf.* (E.W. Mourey and L. Robbiola, eds). James and James (Science Publishers), 1998, 94–99.

5. W. Gerwin, W. Scharff and R. Baumhauer, *Proc. Metal '98 Conf.* (W. Mourey and L. Robbiola, eds). James and James (Science Publishers), 1998, 100–105.

6. F. Galliano, W. Gerwin and K. Menzel, *Proc. Metal '98 Conf.* (W. Mourey and L. Robbiola, eds), James and James (Science Publishers), 1998, 87–91.

7. E. Angelini, E. Barberis, P. Bianco, F. Rosalbino and L. Ruatta, *Proc. Metal '98 Conf.* (W. Mourey and L. Robbiola, eds), 1998, 106–110.

8. V. Fell and M. Ward, *Proc. Metal 98 Conf.* (W. Mourey and L. Robbiola, eds), 1998, 111–115.

9. R. Bertholon, in La conservation en archéologie (M. C. Berducou, ed.), Ch. 5, Masson, Paris, France, 1990.

10. R. Bertholon, *Proc. Surface Modification Technologies 14 Conf.*, 11–13th Sept., 2000, ASM International, 2001 (T. S. Sudarshan and M. Jeandin, eds).

11. R. F. Tylecote, *A History of Metallurgy*, The Metals Society, London, 1979.

12. J. W. Stewart, J. A. Charles and E. R. Wallach, *Mater. Sci. Technol.*, 2000, **16**, 283–290.

13. P. Chevallier *et al.*, J. *Trace and Microprobe Techn.*, 1996, **14**, 517.

14. P. Dillman *et al.*, J. *Trace and Microprobe Techn.*, 1997, **13**, 251.

15. P. Dillmann, R. Regad and G. Moulin, *J. Mater. Sci. Lett.*, 2000, **19**, 907.

16. A. P. Hammersley, *High Press. Res.*, 1996, **14**, 235.

17. I. Guillot, P. Fluzin, M. Clavel and G. Béranger, *Matér. Techniques*, 1987, **10–11**, 411–419.

18. V. F. Buchwald, *Mater. Charact.*, 1998, **40**, 73.

19. D. Neff and P. Dillmann, *Nucl. Instr. Meth. in Phys. Res. B.*, 2001, **181**, 675–680.

20. P. Dillmann and R. Balasubramaniam, *Bull. Mater. Sci.*, 2001, **24**, 3, 317–322.

21. P. Dillmann, PhD Thesis, Université de Technologie de Compiègne, 1998.

22. P. Refait and J. M. Génin, *Corros. Sci.*, 1997, **39**, (3), 539–553.

23. L. Chawla and K. Gupta, *Materials Selection for Corrosion Control*, ASM International, 1993.

Rust Characterisation of Ancient Iron Artefacts Exposed to Indoor Atmospheric Corrosion

P. DILLMANN*, V. VIGNEAU, F. MAZAUDIER, C. BLANC and S. HOERLÉ

LECA/SCCME/DEN, CE Saclay 91191, Gif sur Yvette, Cedex, France
*Laboratoire Pierre Süe CEA/CNRS, DRECAM/DSM, CE Saclay, 91191 Gif sur Yvette, Cedex 2, France

ABSTRACT

A corpus of eleven iron artefacts exposed several centuries to indoor atmospheric corrosion has been collected. The general morphology of the rust layers has been studied as well as the iron substrate. Second phase particles (slag inclusions) and minor elements (carbon, phosphorus) were detected. The compositions of the rust layers have been analysed by EPMA on cross sections. The distribution of endogenous and exogenous elements has been studied. An original analytical method (micro XRD under synchrotron radiation) was used to study the structure distribution on thin film transverse sections. Using this technique, the different constitutive crystallised phases in the rust layer have been localised. Goethite has been detected preferentially in inner layers, lepidocrocite in outer layers. The goethite to lepidocrocite content ratio in the rust layers, the so-called protective ability index α/γ, has been calculated and compared to values found in literature.

1. Introduction

As far as the behaviour of nuclear waste iron-based containers during long term interim storage (several centuries) is concerned, two types of corrosion have to be considered: dry oxidation for high level wastes and indoor atmospheric corrosion for low level wastes. Thus, in order to predict the long term corrosion behaviour of the radioactive waste canisters, a modelling approach as well as analytical studies have to be carried out.

Atmospheric corrosion of iron or low alloy steels is a complex and composite phenomenon that deals with a number of parameters among which are the environmental features and properties of the rust layers. Previous studies on the atmospheric corrosion behaviour of low alloy steels showed that the rust itself plays an important role in the oxidation mechanisms [1–3]. It is therefore critical to characterise precisely the chemical and structural properties of the rust layers in order to predict this long term behaviour. The characterisation of rust that is several years old and formed on iron artefacts exposed to indoor atmospheric corrosion can provide a useful set of data and an extended knowledge of atmospheric corrosion features. In fact, most of atmospheric rust characterisation studies deal with mild steels and weathering steels exposed for about 30 years to rural or marine atmospheres [4–12]. Very few studies available in the literature deal with the characterisation of ancient artefacts: some concern the excellent corrosion resistance of the Delhi Iron

Pillar (1600 years old) [13–16], whereas others are dedicated to isolated iron artefacts with an unknown historical and exposure context [10,17]. The main aim of the present study is to extend the characterisation data about the rust layers. This investigation has been conducted on a corpus of ancient iron artefacts (120–1600 years old) exposed to indoor atmospheric corrosion.

The main phases constituting the rust layers formed on ancient iron artefacts exposed to atmospheric corrosion are either amorphous or crystallised iron oxyhydroxides (lepidocrocite γ-FeOOH, goethite α-FeOOH, akaganeite β-FeOOH and feroxyhite δ-FeOOH) and iron oxides (magnetite Fe_3O_4). Using infrared spectrometry, Raman [15] identified γ-FeOOH, amorphous oxyhydroxides and quartz grains in rust from an old iron axe. Yamashita *et al.* [10] analysed the rust coming from a 400 year old Chinese cast iron and found only crystallised α- and γ-FeOOH. X-ray diffraction experiments made on the rust of the Delhi Iron Pillar [13,14] reveal, in addition to the presence of crystallised α- and γ-FeOOH and amorphous phases, the presence of a specific crystallised iron phosphate: $FeH_3P_2O_8 4H_2O$ as well as amorphous phosphates. The presence of this phosphate is attributed to the specific composition (high P-content) of the iron substrate.

Optical microscope observations made on weathering steels rusts [7,8,10] reveal sometimes a two-layer morphology of the rust scale: an inner layer made of amorphous iron oxyhydroxides with α-FeOOH and an outer layer consisting mainly of γ-FeOOH.

Some authors consider that lepidocrocite γ-FeOOH plays an active role in the corrosion mechanisms [1], on the other hand, goethite, α-FeOOH, is assumed to be a non-reactive phase with a more important protective ability. That is why some authors [4,7,10] have concentrated on the goethite-to-lepidocrite content ratio in the rust layers, the so-called protective ability index α/γ which, according to Yamashita [10], if greater than 2, means that the rust layer is quite stable — assuming it is dense enough.

To clarify all these points, this paper will present the first results of a study made on 11 iron samples exposed several centuries (120 to 1600 years) to indoor atmospheric corrosion. After presenting the analysed corpus and the experimental methods, the obtained results will be presented and discussed. First, the iron substrates will be carefully characterised (composition, second phase particles, structure) and then the general characteristics of 'old rust' layers (metal/oxide interfaces, morphology, thickness) will be presented. After this general overview, the composition in terms of specific elements will be investigated. Finally, to investigate the local structure variations of the different oxide scales, an original experimental method (microdiffraction under synchrotron radiation) was performed. Protective Ability Index, α/γ, values were determined and are discussed.

2. Samples and Experimental Methods

2.1. Analysed Corpus

Several objects coming from French old architectural monuments were chosen. Different criteria were taken into account: the indoor atmospheric exposure and the relatively good dating of the artefact as determined from the building history (see Table 1). Moreover, all the artefacts have been exposed to a rural atmosphere until

Table 1. *Location and description of the analysed samples*

Sample name	Literature reference	Age (years)	Location	Sample type and size
NO01	[18]	120	Noisiel mill — France	Sheet fragment — 1 cm^3
CC01	[18]	150	Conques church — France	Big nail
MAR01	[18]	220	Marly aqueduct — France	Circular rod length: 80 cm; dia.: 1 cm
MO01	[18]	250	Montbard castel — France	Rod length: 20 cm; section: 1 cm^2
CV01	[18]	420	Larrazet castle — France	'Iron' piece coming from a keystone
ME01		700	Meaux cathedral — France	Rod fragment — 1 cm^3
CBB01	[18]	750	Breuil Benoit church	Rod fragment — 1 cm^3
PP01		800	'Palais des Papes' in Avignon — France	Rod fragment 5 cm^3
CP02		800	Popes castle in Chateauneuf du Pape — France	Rod fragment 5 cm^3
GU01	[22]	1600	Temple of Eran — India	Rod fragment 1 cm^3
DE01	[14–16]	1600	Temple of Deogarth — India	Rod fragment 1 cm^3

the 19th century. Some may then have been exposed to semi-industrial atmosphere (NO01 for example). Figure 1 shows the localities of the different samples in France. Because of their very old age (1600 years), two samples coming from India were also analysed, with the support of Prof. R. Balasubramaniam. These two samples were also exposed to rural indoor atmospheres.

When possible the iron object itself was sampled but, in most cases, it was only possible to sample small volumes (some cm^3 of metal and rust). For this reason, not all the analyses were performed on all the collected samples.

2.2. Experimental Methods

First, for the largest samples available, a part of the rust was taken for XRD characterisation. Samples were mounted in epoxy resin. A cross section was made and carefully mechanically polished with SiC papers, grade 180-2400 and diamond paste (3 and 1 μm). A second part of these cross sections was prepared to a 50 μm

Fig. 1 *Sample locations in France.*

thin film for micro X-ray diffraction (μXRD) experiments in transmission mode (see below). After all rust observations on the different cross sections had been made, samples were etched with Nital 4% to determine the carbon content. Then, after repolishing, samples were etched with Oberhoffer reagent [24] in order to reveal the eventual phosphorus repartition.

All metallographic structure and rust cross section observations were made with a LEICA DMRXA optical microscope (OM) and a Stereoscan 200 Scanning Electron Microscope (SEM). The rust layer compositions were then investigated by Energy Dispersive Spectrometry (EDS) coupled to SEM. Minor elements in the metallic substrate and in the rust were analysed by Electron Probe Micro Analysis (EPMA) with a 2×2 μm^2 beam (energy 15 keV, current 40 nA).

The structure of the rust layers, was investigated by XRD powder analyses in grazing incidence with a Philips diffractometer. A Co anticathode was used (40 kV, 40 mA). Spectra were collected for a 2θ value between 5 and 130°.

To study the distribution of the different phases in the scales, μXRD experiments were performed to obtain micro-diffraction spectra at different points of the oxide thin film cross sections. The μXRD experiments were conducted on the D15 beamline at LURE (Laboratoire pour l'Utilisation du Rayonnement Electromagnétique — Orsay, France). The white X-ray beam delivered by the DCI ring was focused and monochromatised by a carbon/tungsten Bragg Fresnel Multilayer Lens (BFML) [18]. The BFML is a wide band pass monochromator resulting in broadening of the diffraction peaks. Photons centred around 14 keV ($\lambda = 0.08857$ nm) were selected and focused down to a 20×20 μm^2 beam. The diffraction patterns were collected with an image plate downstream from the sample. A detailed description of the experimental arrangement has been provided elsewhere [22,23]. 1-D diffraction patterns were obtained by circularly integrating diffraction rings using the FIT2D software developed at the European Synchrotron Radiation Facility (ESRF) [24]. The reference sample used was silicon powder. The spectra were compared with the JCPDF database using the Diffrac+ program.

Nishimura *et al.* [17] gave the measured intensities of some diffraction peaks corresponding to the 2θ angles (Cu anticathode) at 30.09°, 21.22°, 11.84° and 14.14° for respectively magnetite Fe_3O_4, goethite α-FeOOH, akaganeite β-FeOOH and lepidocrocite γ-FeOOH. These measurements were obtained in the same conditions for each separate phase mixed with the same quantity of the internal CaF_2 reference. We used these values to correct the measured intensity of each corresponding diffraction peak on the μXRD spectra. This calculation can be made because the rings observed on the μXRD patterns showed no texture effects. Moreover, the corrections were calculated for different phases but on the same spectrum, so that the absorption effect of the diffracted matrix is the same for each phase and can be ignored in a given spectrum. Relative concentrations of the different phases were obtained for each spectrum. Thus, it was possible, by comparing the corrected intensities, to obtain a relative percentage of the identified, i.e. crystallised, phases at a given location in the oxide scale.

3. Results and Discussion

3.1. Metallic Substrate

The metallographic observation reveals that all the examined samples are hypoeutectoid steels. The most carburised sample (PP01) contained 0.4 mass% carbon and most of the structures about 0.1 mass% carbon.

The grain sizes (a few tens of micrometres) are higher than in contemporary iron and steels and can sometimes reach several hundred micrometres for some samples containing phosphorus (see Table 2). All the observed microstructures are of equiaxed grains.

A peculiarity of ancient iron is that it can content high amounts of non-metallic slag inclusions coming from the iron ore reduction stage [18,20,21]. Actually, before the 19th century iron and steels were produced with solid state processes that generated more inclusions than the liquid states processes used later.

Figure 2 (a) shows a microphotograph taken on zone of the ME01 sample containing numerous slag inclusions. These inclusions are elongated perpendicularly to the hammering plane. The constitutive phases of these kinds of inclusion were studied by XRD and μXRD [18,22] and are mostly wüstite (FeO) dendrites entrapped in a glassy and fayalitic (Fe_2SiO_4) matrix. Some other elements could be present in solid solution in the inclusions, e.g. aluminium, calcium, phosphorus, etc.

Figure 2 (b) shows an optical microphotograph of the dendritic structure of slag inclusions. The number of inclusions entrapped in the iron artefact depends on the quality of the post reduction work, i.e. the forging. Therefore, on the one hand it is possible to find iron with very few slag inclusions, but on the other hand, because ancient irons are heterogeneous, clean zones can be located even when the iron

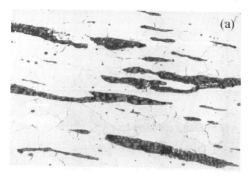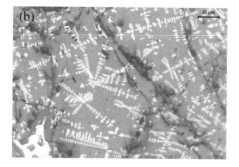

Fig. 2 (a) Slag inclusions in a ferritic matrix (ME01); (b) dentritic structure of a slag inclusion.

contains a lot of slag inclusions. Thus, in this kind of sample, the two zones and their corrosion products must be compared.

In addition to carbon, the only other minor element detected in high amounts in the samples is phosphorus. It is present in five samples with a content over 0.1 mass% (see Table 2). This element, often present in ancient artefacts [23,24], comes from the iron ore used to obtain the metal or from the charcoal used during the reduction process [22]. It is distributed in an heterogeneous manner. It is possible to observe mesoscopic combined with microscopic variations. These variations can be explained by the annealing and cooling operations during the forging stage [22] since at the forging temperature, ferrite and austenite can coexist. These two phases have different P concentration limits. After rapid cooling, this difference is retained in the ferritic matrix and can be revealed by Nital etching (see Fig. 3(a)) Because of their metallographical aspect these concentration heterogeneities are called ghost structures [24]. The Oberhoffer reagent also clearly reveals all the phosphorus heterogeneities. Figure 3(b) shows a mesoscopic (100–500 μm) variation of the phosphorus content observed on the GU01 sample. The dark zones correspond to the low P iron matrix and the clear zones to the high P contents.

Fig. 3 (a) Phosphorus ghost structures revealed by Nital etching on the MO01 sample. (b) Mesoscopic variation of the P content revealed by Oberhoffer reagent on the GU01 sample. Dark zone: low P content. Clear zones: high P content.

Even if sulfur was sometimes detected in the inclusions (see Table 2), in the experimental conditions, this element could not be detected by EPMA in the iron matrix. That means that its concentrations is below 200 ppm in the matrix of the analysed samples. The same remark can be made for manganese.

Table 2. *Metallographic observation and minor alloying elements*

Sample	mass%C[a]	mass%P[b]	Metallographic structure	Grain size (μm)	Inclusions
NO01	0.1	–	Ferrite and pearlite	10–20	Small sulfur
CC01	0.02	–	Equiaxed ferrite and some tertiary cementite	10	Multiphased slag inclusions
MAR01	< 0.02	0.05–0.25	Equiaxed ferrite Presence of phosphorus ghost structures	50	Numerous multiphased slag inclusions Presence of S
ME01	< 0.02	–	Equiaxed ferrite	30–40	Numerous multiphased slag inclusions
MO01	<0.02	<0.02–0.10	Equiaxed ferrite	100–200	Some rare multiphased slag inclusions
CV01	0.1	–	Ferritic–pearlitic structure	30–40	Multiphased slag inclusions
CBB01	0.1	<0.02–0.15	Ferritic–pearlitic structure	20	Rare multiphased slag inclusions
PP01	0.3–0.4	–	Acicular ferrite and pearlite	40	Numerous multiphased slag inclusions
CP02	0.02	–	Equiaxed ferrite	50	Multiphased slag inclusions
GU01	0.02–0.3	0.06–0.23	Ferritic pearlitic structure Presence of phosphorus ghost structures	30	A lot of multiphased slag inclusions
DE01	0.1	0.12–0.25	Ferritic pearlitic structure Presence of phosphorus ghost structures	70–100	A lot of slag inclusions

[a]Determined by quantitative metallography. [b]Determined by EPMA and Oberhoffer etching.

3.2. Rust layers

The general morphology of the rust layers will be considered in the following paragraph. Optical microscopy and SEM observations revealed three kinds of morphology. The first type is relatively homogeneous. The second type is made up of two different layers (see Fig. 4). Lastly, on some samples it is possible to observe on the outer zone of the rust, an 'overlayer' containing some impurities like quartz grains. In fact, this third layer may be contaminated by the dust coming from the stone and the cement of the building walls. Figure 5 shows an 'overlayer' observed by SEM on the CV01 sample. Quartz grains are clearly identified on the high magnification view.

There are various aspects of the metal/oxide interface. For example, on some samples, this interface is relatively irregular (see Fig. 6). On others, some local penetration can be observed (Fig. 7). These local penetrations are always larger than the grain size. On some samples containing zones with a high number of slag

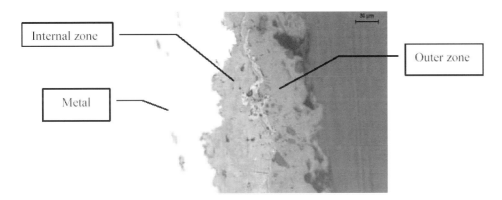

Fig. 4 *Internal and external zones observed on a transverse section of the CC01 sample.*

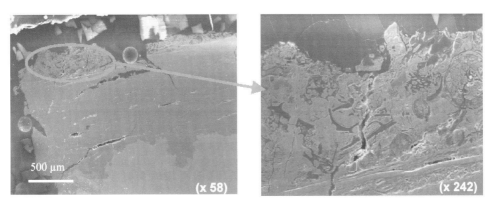

Fig. 5 *SEM microphotograph of a rust layer with an 'overlayer' containing quartz grains (sample: CV01).*

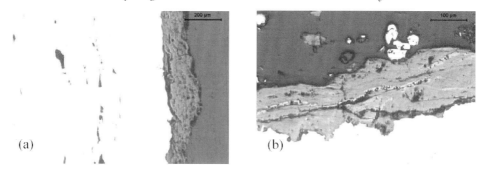

(a)   (b)

Fig. 6 (a) *Regular metal/oxide interface observed on sample CC01.* (b) *Irregular metal/oxide interface observed on sample PP01.*

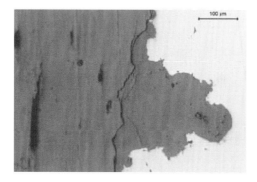

Fig. 7 *Local penetration at the metal/oxide interface observed on sample CV01.*

inclusions, it was possible to observe inclusions entrapped in the oxide scale near the interface (see Fig. 8), but no slag inclusions were seen by OM and SEM observation in the outer zone of the rust layers.

In all the samples, some cracks, sometimes of several hundred micrometers, could be observed. These cracks can be perpendicular or parallel to the layers. The general morphologies of rust layers on the different samples are summarised in Table 3.

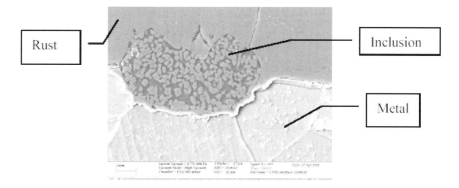

Rust   Inclusion

Metal

Fig. 8 *Inclusion entrapped in the rust inner layer near the interface (CV01 sample).*

Table 3. General description of the rust layers

Sample	Cracks and porosity*	Metal/oxide interface	Layers Internal (I), Outer (O) 'Overlayer' (OV)
CC01	10 µm Po in outer layer	Regular	I (50µm)/ O (50µm)/ OV
NO01	// and •cracks	Local penetration (LP) > 20 µm	I/O/OV
MO01	Numerous // and •cracks	Irregular	
MAR01		Very irregular LP •50 µm (equal to grain size)	I (50µm)/ O (60–70 µm)
CV01	// and •cracks	Irregular Inclusions LP > 100 µm	I (150 µm)/O (150 µm)/OV
CBB01	// and •cracks	Irregular LP •20 µm (equal to grain size)	No visible scales
PP01	// and •cracks	Very irregular LP	I (30 µm)/ O (70 µm)/OV
CP02	Numerous // and •cracks Po in several zones	Very irregular LP	
ME01	Po in several zones	Very irregular	No visible scales
GU01		LP	I/O
DE01	Presence of Po	Very irregular LP	No visible layers

*Po: porosities. // parallel to the metal/oxide interface. • perpendicular to the metal/oxide interface.
LP : local penetration

Figure 9 shows the rust layer thickness of the samples vs their ages. The maximal and minimal measured thicknesses are represented for each sample. Points presented in open squares are values found in literature [4–8,11]. The overall tendency is for an increase of the thickness with the age of the sample. Nevertheless, a high dispersion of the results was found for samples of about the same age. Thus, the CBB01, ME01, PP01 and CP02 have very different rust layer thickness. Several explanations can be

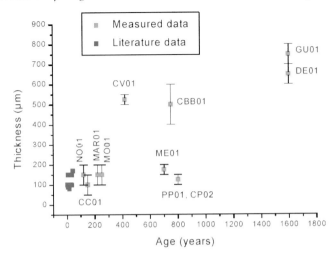

Fig. 9 *Rust layer thickness.*

proposed for this, such as different exposure atmospheres, different substrate compositions, material history and so on.

After this general overview of the morphology of rust layers observed on ancient iron artefacts, the composition and structure of these layers was considered. Quantitative concentrations of Fe, O, P, Cl, Si, S and Ca profiles were made on the rust layer cross sections by EPMA. No other element could be detected in the layer with this technique. These results are confirmed for major elements by EDS analyses. Hydrogen, of course, cannot be detected by these two methods. In the samples where no 'overlayer' was observed, the iron and oxygen contents were relatively constant across the whole thickness of the layer. Figure 10(a) shows an iron profile obtained on such a sample (MAR01). On the samples showing an 'overlayer', the concentration profiles show an increase of Ca, Si, S and sometimes of Cl in this zone and a decrease of the iron content. Figure 10(b) shows the iron content profile obtained on such a sample (CV01). This Figure shows clearly the difference of iron composition between the rust layer and the 'overlayer'. Table 4 shows the iron content analysed in the rust layers. 'Overlayers' are not considered.

For all the analysed samples, except DE01, the iron content is in good agreement with the iron content of the oxyhydroxides goethite or lepidocrocite (α- or γ-FeOOH) i.e. 62 mass% of iron. The DE01 sample agrees better with a magnetite (Fe_3O_4) iron content, i.e. 72 mass%.

Because phosphorus is an element that is well known to play a role in atmospheric corrosion [12], distribution profiles of this element were made in the rust layers of the samples (MAR01, MO01, CBB01, DE01 and GU01) that contain more than 1000 ppm of P in the metallic matrix. In all the examined samples, the distribution of P was very heterogeneous across the thickness of the rust layers (see Fig. 11(a)). In fact, the phosphorus content in the oxide follows the mesoscopic variation of phosphorus in the metallic matrix. This is corroborated by the analyses made on the GU01 sample (see Fig. 11(b)). Oxides situated near a high phosphorus iron matrix contain a higher quantity of this element than oxides formed near low phosphorus zones.

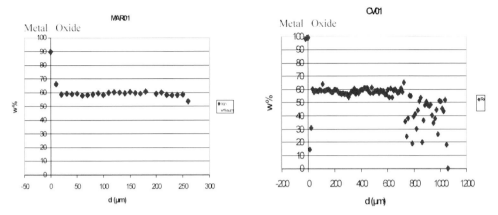

Fig. 10 *EPMA iron distribution in the rust layer cross section (a) MAR01 sample and (b) CV01 sample; d = distance from the interface.*

Table 4. *Iron content in the rust layers analysed by EPMA. Overlayers are not considered*

Sample	mass% Fe
MAR01	58–63
ME01	59–62
CC01	59–60
DE01	69–72
PP01	59–64
MO01	57–61
CV01	58–62
CBB01	58–62

Sulfur concentration profiles were obtained on cross sections of the rust layers. In all the analysed samples, this element was detected in relatively high quantities in the outer layer of the rust layer or in the overlayer. A typical example of this external sulfur enrichment is shown in Fig. 12. This enrichment could be explained by an exogenous pollution of the oxide scale by the SO_2 that has been present in the atmosphere since the middle of the 19th century. A more surprising result is the presence of sulfur (in lower quantities than in the outer layer) in the inner layers of some samples (see Table 5). The particular high level in the N001 sample (0.2 mass% in the inner layer) could be explained by the fact that this sample had always been exposed to industrial atmospheres since the middle of the 19th century. For some

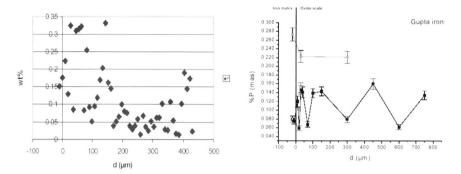

Fig. 11 *Phosphorus distribution in the oxide scales (EPMA); (a) CBB01, (b) GU01. d — distance from the metal/oxide interface.*

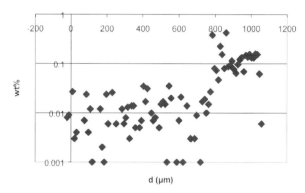

Fig. 12 *Sulfur concentration profile in the oxide layer of the CV01 sample (EPMA). d — distance from the metal/oxide interface.*

Table 5. *Sulfur concentration in the rust layers (EPMA)*

Sample	Average mass% S, inner layer	Max. mass% S, outer layer
CBB01	0.058	0.15
CV01	< 0.01	0.45
MO01	0.038	0.52
NO01	0.2	0.25
CC01	0.08	0.35
MAR01	0.35	0.85
PP01	0.25	0.35
DE01	0.02	0.15

samples (such as MAR01), this high concentration could be explained by the presence of sulfur in the substrate (solid solution in iron or S in the slag inclusions).

Chlorine was only found in the outer layer of the CV01 sample. Because no sample was exposed to marine atmospheres (see Fig. 1), the presence of this element could be attributed to the water dripping from the building walls and contaminated by the cement that could contain chlorine.

Silicon and calcium can have both endogenous and exogenous origins. On the one hand, slag inclusions of the substrate contain high amounts of silicon and sometimes several mass% of calcium. On the other hand, the dust coming from the building walls can be incorporated in the outer scales of the rust layers (see above). This can be an exogenous source of these two elements. Thus, the exogenous silicon can be identified on the EPMA profile by an important enrichment of the outer layers (see Fig. 13). In the inner layers, it is possible to observe some very localised increases of Si and / or Ca concentration. This could be explained by the presence of dissolved slag inclusions in the rust.

For some samples (CV01, PP01 and MA01), rust powder was mechanically withdrawn in order to conduct XRD analyses. Certain phases were identified in all spectra: magnetite Fe_3O_4, goethite α-FeOOH and lepidocrocite γ-FeOOH. On chlorine-containing samples, namely CV01, akaganeite β-FeOOH was also observed.

No XRD pattern seems to present the characteristic shape linked to the presence of amorphous phases.

Micro X-ray diffraction profiles were made on thin films of cross sections of four samples (CV01, DE01, NO01 and PP01). At different points of the rust layer μXRD spectra were collected. Figure 14 shows the 2D diffraction pattern obtained at a point of the rust layer of the CV01 sample and the corresponding $I = f(\theta)$ integrated spectrum. The complete diffraction rings observed on the 2D diffraction pattern showed that the constitutive phases of the rust layer are not textured and this was observed for all the analysed samples. After corrections of intensities (see chapter 2.2.), the relative proportion of the different identified phases is obtained and Fig. 15 shows these profiles.

Several general tendencies concerning the four samples can be observed. Goethite is present in more important proportions in the inner layers and lepidocrocite is present more in the outer layers. On three samples, (NO01 CV01 and PP01), the

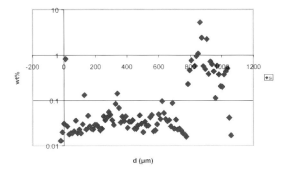

Fig. 13 *Silicon enrichment of the external zone (CV01) sample. EPMA. d — distance from the metal/oxide interface.*

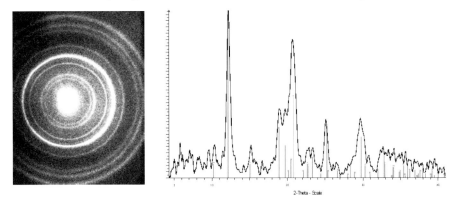

Fig. 14 *2D micro-diffraction pattern and corresponding integrated I = f(j) spectrum. PP01 sample. Vertical lines correspond to the JCPDF 81-462 reference file of goethite. Beam size: 20 × 20 μm. Energy: 14 keV.*

main phases are these two oxyhydroxides (α– and γ-FeOOH). For the Indian iron sample (DE01) it is the magnetite that is the main phase. These observations are in very good agreement with the EPMA quantitative analyses: DE01 is the only sample for which the Fe and O contents correspond to magnetite. The others had better agreement with the FeOOH stoichiometry. EPMA analyses and OM observations reveal the presence of an overlayer containing high amounts of chlorine on the CV01 sample. This is confirmed by the μXRD profile that shows the presence of akaganeite

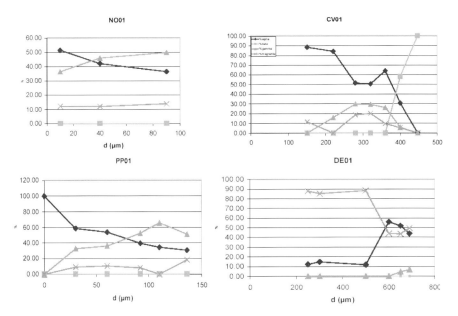

Fig. 15 *Relative concentrations of the crystallised phases obtained by μXRD experiments on thin films on cross sections. d — distance from the metal/oxide interface. Diamond: goethite; square: lepidocrite; disc: magnetite; triangle: akaganeite.*

(β-FeOOH) in the overlayer, which is an iron oxyhydroxide that contains high amounts of Cl.

This distribution of the corrosion products (more goethite present in the inner layer and more lepidocrocite in the outer layer) was also observed by two authors on weathering steels that had been exposed for 16 and 26 years [8,10]. However, these authors also found amorphous phases in the inner layer but it seems not to be the case here, even allowing for the fact that μXRD is not the most suitable method to detect amorphous phases.

From these μXRD profiles, the average composition of the entire rust layer was deduced by integration of the composition curve of each phase on all the layer thickness. It was then possible to obtain, for each sample, the protective ability index α/γ. Figure 16 shows the evolution of this ratio with the age of the analysed sample.

As a general tendency, the older the sample, the higher its protective ability index. This could be explained by the progressive transformation of lepidocrocite into goethite with the succession of wet–dry cycles during the atmospheric corrosion process. Yamashita *et al.* [8] consider that the rust layer is stabilised when this ratio is higher than 2. Considering this point of view, it seems that, although PP01 is older than CV01, its rust layer is not so stable. Because magnetite and akaganeite were also identified in the rust layers, we propose to introduce another ratio that is:

$$\alpha^*/\gamma^* = (\text{mass}\% \ \alpha\text{-FeOOH} + \text{mass}\% \ Fe_3O_4)/ \ (\text{mass}\% \ \gamma\text{-FeOOH} + \text{mass}\% \ \beta\text{-FeOOH})$$

The evolution of this new protective ability index α^*/γ^* which seems to characterise better the stability of the rust layer, is presented in Fig. 16. The general tendencies are the same that for the α/γ ratio. Nevertheless, the α^*/γ^* ratio is much higher for the DE01 sample because the rust layer is mainly composed of magnetite.

The protective ability index α/γ values are generally lower than those reported in the literature for much shorter times: Yamashita *et al.* [10] report an α/γ ratio of 1 for weathering steel exposed for one year and about 3 after 26 years. As far as a very long term (i.e. a few centuries) of exposure of samples is concerned, the only reference we have found in the literature deals with a pig iron [10] that has been exposed to the atmosphere for 400 years. The protective ability index of the rust layer of this old pig iron is about 4 and is in good agreement with our data.

To close this chapter, important points highlighted in this study will be summarised.

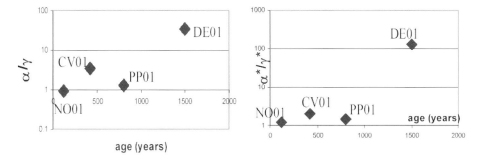

Fig. 16 *α/γ and α^*/γ^* ratio calculated from the μXRD data in function of the sample age.*

The iron archaeological analogues examined here are all hypoeutectoid steels containing ferritic or ferritic–perlitic structures. The grain sizes are higher than in contemporary low alloy steels. In some samples, high phosphorus contents were evident. The parculiarity of old iron is that it sometimes contains numerous slag inclusions coming from the reduction process used to obtain the metal.

As mentioned in the literature, the present study shows that the main crystalline phases composing the atmospheric corrosion products are: magnetite (Fe_3O_4), goethite (α-FeOOH) and lepidocrocite (γ-FeOOH). Micro X-ray diffraction experiments show that these phases are localised in a particular way in the rust layer: goethite in the inner layer and lepidocrocite more in the outer layer of the rust. This is also confirmed by other authors. This composition distribution appears to be not very sharp. Moreover, in agreement with the electrochemical processes connected with the wet–dry cycle, the older the rust is, the more magnetite and/or goethite is formed to the detriment of lepidocrocite, which is an active phase in the atmospheric corrosion process. This is verified by the evolution of the protective ability indexes α/γ and α^*/γ^*. No amorphous phases was detected with the analytical techniques that were used here. This point seems to be in contradiction with some authors who very often found amorphous phases in the inner layer of the rust. Nevertheless, the analysed samples found in the literature are far younger that those studied here. The older age of the rusts could be an explanation of the apparent absence of amorphous phases. This hypothesis is also proposed by Yamashita [10] who explains that during a long period of exposure, the amorphous inner layer can transform to a densely packed aggregate of nano-particules of goethite. It is important to note that some authors [14–16] mentioned that phosphates that could be present in the rust layers of high phosphorus irons, are mainly amorphous in very old layers — like in the Delhi iron pillar rust — despite the fact that some crystallised phosphates could locally be identified. These phosphate phases have not been found in this study.

More surprising are the high quantity of minor elements (Ca, Si, and S) that can be found in the ancient rust layers. The presence of this elements in the external zone of the rust can be explained, on the one hand, by an endogenous pollution coming from the cement or the stones of building walls where the samples were withdrawn, and on the other hand, by the fact that some slag inclusions of the metallic substrate could dissolve in the rust layers and cause the local increase of Si, Ca or S contents. Nevertheless, the values encountered are relatively high in the entire thickness of the rust (especially for sulfur). As evidence of this is the fact that anion species as large as sulfate SO_4^{2-} can be transported through a thick and dense rust layer. There is a need to focus on transport phenomena in the rust layer.

4. Conclusions

Eleven iron archaeological artefacts exposed to atmospheric corrosion have been studied. The metallic substrates have been characterised by metallographic observations and EPMA. The morphology, composition and structures of the rust layer have been studied by SEM, OM, EPMA and µXRD; and, in particular, the local evolution of these parameters on cross sections.

Despite the fact that ancient rust layers could be very complex and that the artefact history could influence its formation, general tendencies have been observed. The major phases of the rust layers are magnetite (Fe_3O_4), goethite (α-FeOOH) and lepidocrocite (γ-FeOOH). Lepidocrocite seems to be present more in the outer layer. Goethite seems to be the major constituent of the inner layer. No amorphous phases have been clearly identified. The older the rust layers, the higher the Protective Ability Indexes (α/γ and α^*/γ^* ratios). Finally, high quantities of minor elements (P, Si, S and Ca) have been detected in the rust layers.

These first results bring very important information for understanding the rust layer morphology and structure. These parameters are of importance for modelling the wet–dry cycles of atmospheric corrosion mechanisms, and then will be more thoroughly investigated. Characterisation has also to be continued on more samples to increase the statistical reliability of the study. Moreover, to understand the mechanisms of species transport inside the rust, some complementary experiments need to be performed on the rust scales (for example, diffusion of inert ions through the rust). Last but not least, some electrochemical experiments will be carried out on the ancient rusted samples to understand the electrochemical behaviour of atmospheric rusts that are more than a hundred years old.

References

1. M. Stratmann, *Ber. Bunsenges. Phys. Chem.*, 1990, **94**, 626–639.
2. M. Stratmann, *Metallurgia i oldenwnictwo*, 1990, **16**, 45–52.
3. M Stratmann, K Bohnenkamp and H.-J. Engell, *Corros. Sci.*, 1983, **23**, 969–985.
4. K. Kashima, S. Hara, H. Kishikawa and H. Miyuki, *Corros. Engng*, 2000, **49**, 25–37.
5. H. Katayama, M. Yamamoto and T. Kodama, *Corros. Engng*, 2000, **49**, 77–84.
6. K. Shiotami, W. Tanimoto, C. Maeda, F. Kawabata and K. Amano, *Corros. Engng*, 2000, **49**, 99–109.
7. M. Takemura, S. Fujita, K. Morita, K. Sato and J.-I. Sakai, *Corros. Engng*, 2000, **49**, 111–121.
8. M. Yamashita, T. Misawa, R. Balasubramanian and D. C. Cook, *Corros. Engng*, 2000, **49**, 133–144.
9. H. Fujiwara, T. Sugano, M. Aosawa and T. Nagatani, *Corros. Engng*, 2000, **49**, 145–158.
10. M. Yamashita, H. Miyuki, Y. Matsuda, H. Nagano and T. Misawa, *Corros. Sci.*, 1994, **36**, 283–299.
11. T. Misawa, K. Asami, K. Hashimoto and S. Shimodaira, *Corros. Sci.*, 1974, **14**, 279–289.
12. T. Misawa, T. Kyuno, W. Suetaka and S. Shimodaira, *Corros. Sci.*, 1971, **11**, 35–48.
13. R. Balasubramaniam and A. V. Ramesh Kumar, *Corros. Sci.*, 2000, **42**, 2085–2101.
14. R. Balasubramaniam and P. Dillman, This volume, p.261–279.
15. R. Balasubramaniam, *Corros. Sci.*, 2000, **42**, 2103–2129.
16. P. Dillmann, R. Balasubramaniam and G. Beranger, *Corros. Sci.*, 2002, **44**, 2231–2244.
17. A. Raman, B. Kuban and A. Razvan, *Corros. Sci.*, 1991, **32**, 1295–1306.
18. P. Dillmann, PhD, Université de Technologie de Compiègne, France, 1998.
19. T. Nishimur, H. Katayama, K. Noda and T. Kodama, *Corros. Engng*, 2000, **49**, 85–97.
20. P. Dillmann, *et al.*, *C. R. Acad. Sci. Paris*, 1997, **324**, 763–772.
21. V. F. Buchwald and H. Wivel, *Mater. Charact.*, 1998, **40**, 73–96.
22. P. Dillmann and R. Balasubramaniam, *Bull. Mater. Sci.*, 2001, **24**, 317–322.
23. D. Neff and P. Dillmann, *Nucl. Instrum. and Methods Phys. Res. B*, 2001, **181**, 675–680.
24. J. W. Stewart, J. A. Charles and E. R. Wallach, *Mater. Sci. Technol.*, 2000, **16**, 283–290.

22

Long Term Behaviour of Iron in Clay Soils: A Study of Archaeological Analogues

E. PONS, L. URAN, S. JOIRET*, A. HUGOT-LE GOFF*, C. LEMAITRE
and D. DAVID

Laboratoire Roberval (UMR CNRS 6066), Université de Technologie de Compiègne,
BP 20529, 60205 Compiègne, France
*Laboratoire de Physique des Liquides et Electrochimie (UPR CNRS 15), Université Paris 6, Tour 22,
4 place Jussieu, 75252 Paris Cedex 05, France

ABSTRACT

Iron objects from a battlefield of World War I were studied to obtain a better understanding of corrosion phenomena in clay soils. The identification of the corrosion products by Raman Spectroscopy highlighted iron oxides and oxy-hydroxides. The orange–brown external layer is composed of oxy-hydroxides, mixed with soil crystals, while the oxides are always associated with the internal layer. Electrochemical experiments (Voltammetry and Electrochemical Impedance Spectroscopy) in Evian mineral water (pure or with sodium chloride), confirmed the porosity of the external layer, and the protective role of the internal products. The influence of the burial environment on the corrosion rate was also established.

1. Introduction

The study of iron archaeological objects is part of the French national programme concerned with a nuclear waste deep repositories. The programme is conducted by ANDRA (French National Agency for Radioactive Waste Management), which is examining the possibility of the disposal of long-lived and high level wastes (HLWs) in deep geological formations. Within this framework, a collaboration with the Technological University of Compiègne was initiated by the ANDRA in 1995, to consider the corrosion of ancient artefacts.

Iron archaeological analogues make a valuable contribution to the specifying of containers for HLWs (overpacks for vitrified wastes, or canisters for spent nuclear fuel). The radioactivity of these wastes will last several thousand years and archaeological analogues provide access to this sort of time scale.

The objective of our work is to achieve a better understanding of corrosion phenomena in clay soils, by studying archaeological artefacts. The experimental issue is divided into two major parts:

• a physico-chemical characterisation of corrosion products,

• an electrochemical study of the properties of the different corrosion layers.

In the following study, we considered objects from a battlefield of World War I (Nouvron-Vingré, Aisne — France), chosen for the similar aspect of their corrosion layers, and their availability. The work will take the time period of burial into account by studying older archaeological objects, corresponding to more advanced corrosion phenomena.

2. Physico–Chemical Characterisation of Corrosion Products

Visual and microscopic observations revealed that all these objects had two corrosion layers: an external layer, which appears orange–brown, about 200 μm to 1 mm thick and, in contact with the metallic core, an internal layer, which is dark grey and of equivalent thickness to the external layer (Fig. 1). A metallographic study, coupled with Energy dispersive X-ray analysis, showed that the 1914–1918 remains were hypoeutectoid steels (low carbon content). They did not contain any alloying element which could modify their corrosion behaviour.

Energy dispersive X-ray spectroscopy analyses (EDS) under the Scanning Electron Microscope identified alumina inclusions, which were probably associated with the manufacturing process (aluminium killed steels) or contamination by the blast furnace refractories. Manganese sulfide inclusions, coming from iron ore, were also detected.

Among characterisation methods, Raman spectroscopy is particularly well-adapted, because it gives access to the constituent phases of corrosion layers. The experiments were carried out with a Jobin-Yvon Horiba Raman spectrometer (LABRAM), with an integrated laser source at the 632.8 nm wavelength (red line). The spectrometer was coupled with an air-cooled CCD detector, and an Olympus microscope (× 80 objective). With this objective, an area of about 1 μm^2 was analysed on the sample surface.

Figure 2 gives Raman spectra obtained for magnetite, goethite and lepidocrocite on World War I artefacts, compared with some bibliographical spectra. For this oxide and these two oxy-hydroxides, the wavenumbers of the major peaks observed are in good agreement with the bibliographical values of De Faria *et al.* [1], which provide

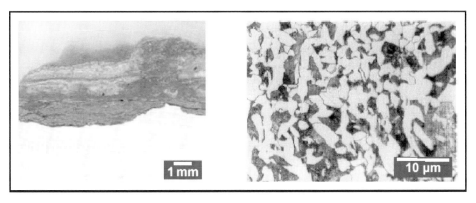

Fig. 1 *Object NV00_02 (a piece of shrapnel of a mortar shell, named 'crapouillot'), showing corrosion layers' appearance and micrography after etching with 1% nital.*

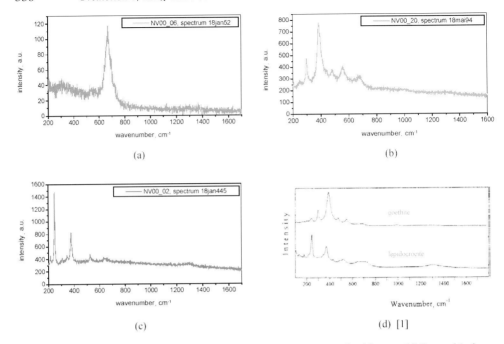

Fig. 2 *Raman spectra obtained on World War I objects, compared with some bibliographical results (a) magnetite, (b) goethite, (c) lepidocrocite, (d) bibliographical spectra of goethite and lepidocrocite.*

a quite complete study of the various iron oxides and oxy-hydroxides by Raman spectroscopy. Our experimental spectrum of magnetite (*spectrum 18jan52*, obtained on the object NV00_06) has a rather large peak centred on 670 cm^{-1}. According to the scientific literature, magnetite has a spinel structure (O_h^7) above 119K, giving rise to five Raman bands: three T_{2g}, one E_g and one A_{1g}. Only two papers quoted the same five peaks, at 300 (*298*), 320, 420, 560 (*550*), and 680 cm^{-1} (*676*) [2, (3)]. The comparison with other experimental works [1–8] highlighted some differences in the appearance and location of the minor peaks, but all authors agree on the significance of the major peak, located between 663 and 680 cm^{-1}. De Faria *et al.* showed that the Raman spectrum of magnetite underwent a significant change depending on the laser power, temperature, and surface morphology (rough and flat surfaces of magnetite samples were tested). This can explain the differences found in the literature for the magnetite spectrum. Concerning goethite, our Raman spectrum (18mai94, object NV00_20), has peaks at 246, 300, 387, 481, 558 and 680 cm^{-1}. According to De Faria *et al.*, the Raman spectrum of goethite has peaks at 243, 299, 385, 479, 550, 685 and 993 cm^{-1}. Our values are very similar, except the peak at 993 cm^{-1} which is lost in noise. But this peak has a relatively low intensity on the bibliographical spectrum (see Fig. 2). The experimental results of De Faria *et al.* tally with many other references [5–7]. The spectrum of lepidocrocite was obtained on the object NV00_02. It has peaks at 251, 377, 527 cm^{-1}, a shoulder at 642 cm^{-1} and a broader peak of low intensity around 1300 cm^{-1}. De Faria *et al.* found for this compound peaks at 245, 373, 522, 719 and 1303 cm^{-1} and shoulders at 493 and 650 cm^{-1}. The peak at

719 cm⁻¹, which does not appear in our spectrum or in other bibliographical references [5–8], seemed to be due to a small amount of maghemite formed during the synthesis of lepidocrocite.

Figure 3 shows the evolution of the Raman spectra obtained on the internal corrosion layer of the object NV00_30 (piece of iron wire, diameter around 6 mm), starting from the metallic core to the outside. The first spectrum — *sp18mai70* — (inner corrosion product), corresponds to a product rich in maghemite of low crystallinity (major band centred around 705 cm⁻¹). The greater the distance from the metallic core, the more the 705 cm⁻¹ band moves to the low wavenumbers (around 700 cm⁻¹ on the spectrum *sp18mai74* and 683 cm⁻¹ on *sp18mai75*), and the more peaks appear and grow at lower wavenumbers. These peaks are located around 250, 300, 385, 485 and 540 cm⁻¹ (this last peak seems to be divided into two peaks centred around 525 and 550–555 cm⁻¹). This evolution in the spectrum can be associated with a progressive disappearance of maghemite, which is replaced by a combination of goethite and lepidocrocite (see Fig. 2).

To take a global view on all the artefacts studied, we can say that the Raman identification of the corrosion products highlighted iron oxides (magnetite Fe_3O_4, maghemite $\gamma\text{-}Fe_2O_3$) and oxy-hydroxides (goethite $\alpha\text{-}FeOOH$, akaganeite $\beta\text{-}FeOOH$, lepidocrocite $\gamma\text{-}FeOOH$). The external layer is made up of oxy-hydroxides, mixed with soil crystals, while the internal layer seems to be more precisely defined. The presence of oxides is always associated with the internal corrosion layer. Figure 4 sums up the different phases identified and their location among corrosion products for the different objects.

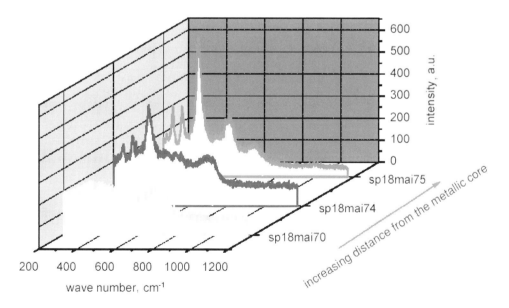

Fig. 3 *Raman identification of the products encountered in the internal corrosion layer of NV00_30, according to the distance from the metallic core.*

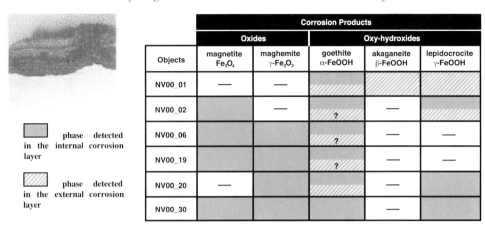

Objects	Corrosion Products				
	Oxides		Oxy-hydroxides		
	magnetite Fe₃O₄	maghemite γ-Fe₂O₃	goethite α-FeOOH	akaganeite β-FeOOH	lepidocrocite γ-FeOOH
NV00_01	—	—	ext	ext	ext
NV00_02	int	—	ext ?	—	ext
NV00_06	int	int	int ?	—	—
NV00_19	int	int	ext ?	—	—
NV00_20	—	int	ext	—	int
NV00_30	int	int	int	—	int

Legend: ■ phase detected in the internal corrosion layer; ▨ phase detected in the external corrosion layer.

Fig. 4 *Corrosion products identified by Raman spectroscopy and their location in the corrosion layers.*

3. Electrochemical Characterisation of the Properties of Corrosion Layers

3.1. Behaviour of Each Corrosion Layer

First, electrochemical impedance spectroscopy diagrams (EIS) were recorded at the free corrosion potential (E_{corr}) on raw samples, that is to say on samples covered with the two corrosion layers, as identified by physico–chemical characterisation. Then an examination was made of samples with only the internal corrosion layer, i.e. after removal of the external layer by polishing. Finally, the procedure was repeated with samples of the metallic core of the objects after removal of corrosion products (Fig. 5). For these electrochemical tests, an area of 1 cm² was isolated on the sample (working electrode) by applying varnish on the remaining surface (Lacomit Varnish, Agar Scientific LTD). The experimental apparatus was a Tacussel Z.EIT potensiostat, coupled with a Hewlett-Packard 9826 computer. A saturated calomel electrode SCE was used as the reference electrode, and a platinum counter electrode. The impedance diagrams were established in Evian mineral water, chosen for its compositional stability, in the 10^4–1.58×10^{-2} Hz frequency range, with a 12 mV signal perturbation at the free corrosion potential.

- For the sample with two corrosion layers, the diagram presents three loops: a first loop corresponding to high frequencies: points 1–5, a second loop in the middle range of frequencies, and a straight line at low frequencies, which corresponds to the diffusion process. This result relates to the number of interfaces of the system; the first is the metal–internal corrosion layer interface, the second is between the internal and the external corrosion layers, and the third is the external layer–electrolyte interface. The general behaviour of samples with two corrosion layers suggests that the external layer is porous as shown in the following part of this study.

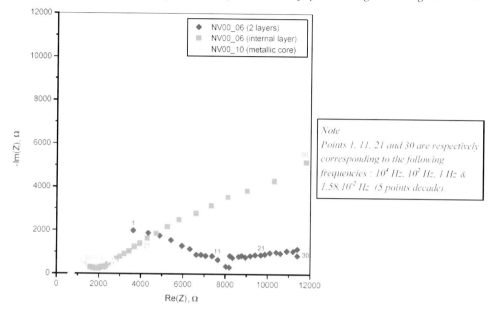

Fig. 5 *Nyquist plot according to the number of corrosion layers of the sample.*

- With only the internal layer, the sample presents a very different impedance diagram. The elimination of the external layer led to the disappearance of the first loop, whereas the low frequency part of the diagram became closer to the diffusion model.

- The metal sample of the object NV00_10 has a more common impedance diagram, which is close to a semicircle centred on the real axis. This diagram can be simulated using a parallel RC circuit in series with a resistor R_0. The values of R_0, R_1, and C_1 are deduced from the experimental diagram (Fig. 6). R_0 corresponds to the intercept between the semicircle and the real axis at high frequencies. The angular frequency corresponding to the top of the semicircle is equal to the inverse of the system time constant, according to:

$$\omega_c = \frac{1}{\tau} \text{ with } \tau = R_1 C_1$$

R_1 is easy to find: indeed, the impedance of the system is equivalent to $R_1 + R_0$ at low frequencies. R_0 represents the resistance of the electrolyte, R_1 corresponds to the transfer resistance of the system, and C_1 is the double layer capacitance.

Metal samples from other World War I artefacts were tested in the same way, and we established an equivalent $R_0 - (R_1 // C_1)$ circuit for each. Table 1 sums up the values of the parameters R_0, R_1, and C_1 obtained for the different samples.

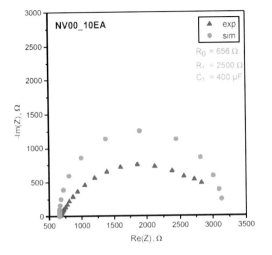

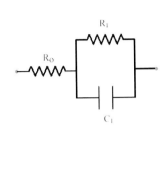

Fig. 6 Equivalent electrical circuit and experimental versus simulated impedance diagrams.

Table 1. *Equivalent circuit parameters for different metal samples*

Samples	R_0 (Ω)	R_1 (Ω)	C_1 (μF)
NV00_10, Sample A	656	2500	400
NV00_19, Sample A	1700	1080	93
NV00_20, Sample Rect	1000	1880	134
NV00_20, Sample Rond	984	1852	215
NV00_30, Sample A	1430	3760	67

3.2. Electrochemical Characterisation of the External Corrosion Layer

Electrochemical tests were carried out on samples with two corrosion layers, so as to confirm the hypothesis that the external layer was porous. We followed the evolution of the impedance diagram at the free corrosion potential (E_{corr}) with time in two different electrolytes. First, the samples were immersed in slightly conductive Evian mineral water, and, afterwards, in a more conductive solution (Evian mineral water with 25 g L^{-1} sodium chloride, according to the ASTM standard for sea water).

The immersion in Evian mineral water with sodium chloride led to a rapid modification of the diagram (Fig. 7). The capacitive component was removed in the first 24 h, and at the same time the resistive component decreased by ten times. The impedance modulus changed from values between 4 and 10 kΩ in Evian mineral water to values between 0.5 and 1.2 kΩ in salt-containing Evian. This decrease can be associated with the layer pores, which led to the impregnation of the external

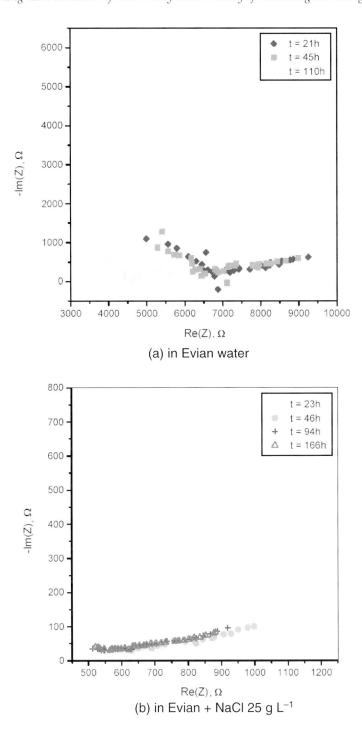

(a) in Evian water

(b) in Evian + NaCl 25 g L^{-1}

Fig. 7 *Impedance diagrams at the free corrosion potential (object NV00_19).*

layer. Voltammetry experiments (polarisation curves) also showed that the external layer did not slow down corrosion.

3.3. Electrochemical Characterisation of the Internal Corrosion Layer

In a second stage, we worked on samples with only the internal corrosion layer, after removal of the external layer by polishing, in parallel with samples of metal. The experiments consisted in obtaining polarisation curves. The anodic and cathodic polarisation curves were obtained using a potential sweep rate of 0.1 mV s^{-1}, starting from the free corrosion potential (10 or 20 mV above for the cathodic curve and below for the anodic), to + 200–250 mV from E_{corr}, and then from this value back to the initial potential (return curve). The curves we obtained had no linear portion, and so a Tafel representation could not be used. An indicative corrosion current, i_{ind}, was therefore defined to compare them (Fig. 8). By definition, i_{ind} is the intensity (per cm^2) which corresponds to the intercept between the tangent to the polarisation curve (log $|i| = f(E)$), at ± 150 mV from the free corrosion potential, and the vertical line at the free corrosion potential (straight line $E = E_{corr}$). The values of i_{ind}, obtained for the different samples tested, are summarised in Fig. 9.

For a given object, the indicative corrosion current is lower for the samples with an internal corrosion layer, than for the samples of the remaining metal. The measurements of the indicative corrosion current confirm the protective role of the internal layer against corrosion.

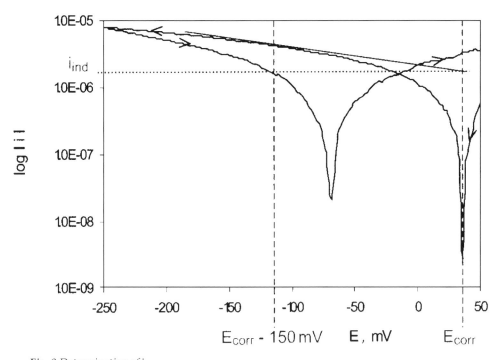

Fig. 8 Determination of i$_{ind}$.

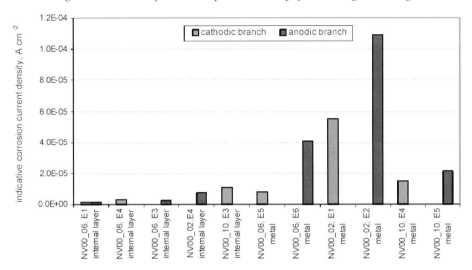

Fig. 9 *Measurements of an indicative corrosion current for World War I steel samples.*

Furthermore, on the basis of the indicative current obtained on samples with the internal corrosion layer and of Faraday's law, it is possible to obtain an indicative instantaneous corrosion rate in Evian mineral water, according to:

$$v_{ind}^{cor} = \frac{M}{\rho n_1 F} \times i_{ind}$$

where M represents the atomic mass of the metal, ρ its density, n_1 the number of electrons lost by an atom of metal, and F = Faraday constant.

If we replace i_{ind} in the former formula by the minimal and maximal numerical values obtained for the samples with the internal corrosion layer, the indicative corrosion rate falls in the following range:

$$15 \ \mu m/year \ < v_{ind}^{cor} \ < \ 130 \ \mu m/year$$

These values are ten time higher than bibliographical figures [9–11]. According to Johnson and Francis (Fig. 10), who are quoted by De Franco *et al.*, and Chapman *et al.*, average corrosion rates for iron base artefacts were almost all between 0.1 and 10 μm/year. These estimates were deduced from description, photographs, photomicrographs, drawings, radiographs, etc., assuming uniform corrosion. In Fig. 10, a considerable homogeneity of the results can be seen, despite the variety of the burial environments in which the artefacts were found. Only the first two objects have a higher corrosion rate than the average, but these were cannon balls found in sea water and this is a highly oxidising environment, very different from the conditions foreseen for the disposal of radioactive wastes in deep geological formations.

The bibliographical figures quoted above were average corrosion rates in the burial environment, whereas we performed our experiments in Evian mineral water. So,

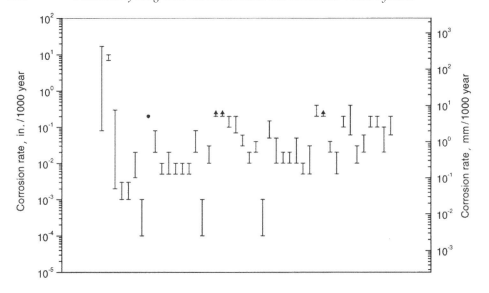

Fig. 10 *Average corrosion rate estimated for archaeological iron base objects (based on [9]).*

we can deduce that the nature of the environment, i.e. clay soils, probably plays a significant role in the decrease of corrosion rate during burial, by controlling the accessibility of oxidising species. Moreover, the definition of the indicative corrosion current itself, leads probably to an overestimation of the real corrosion current, and consequently to an overestimation of the instantaneous corrosion rate. Another hypothesis is that the equilibrium between the corrosion layers and Evian mineral water was not reached at the time the polarisation curves were obtained.

4. Conclusions

This study of World War 1 remains, using physico–chemical characterisation and electrochemical measurements, has demonstrated the protective role of corrosion layers, and particularly of the internal layer. Characterisation methods, and specially Raman spectroscopy, showed that the external layer, which appeared orange–brown, was composed of oxy-hydroxides, mixed with soil crystals, while the oxides were always associated with the dark grey internal layer. Dynamic electrochemical measurements clearly illustrated different behaviour between the two corrosion layers. Impedance spectroscopy applied to the external corrosion layer in two electrolytes of different conductivity, confirmed that it was porous. An indicative corrosion current was defined from polarisation studies of the internal corrosion layer. The measurements of this indicative current, for samples with the internal layer and for samples of metal, confirmed the role of the internal layer against corrosion. However, the comparison between the indicative corrosion rate obtained in Evian mineral water, and the average corrosion rate of archaeological artefacts, as listed in the scientific literature, seemed to highlight a significant role of the burial environment, in the slowing down of corrosion. The contribution of each of these two hypotheses should be clarified.

This work has shown the value of a combined approach of physico–chemical characterisation methods and electrochemical methods, in reaching a better understanding of the corrosion of archaeological objects in clay soils. After this first stage devoted to the study of World War 1 objects, the next stage will be concerned with Gallo–Roman iron, so as to link the development of corrosion layers with the period of burial, and to provide a comparison with the long duration that interests ANDRA for the prediction of long term corrosion behaviour in nuclear waste deep repositories.

References

1. D. L. A. De Faria, S. Venâncio Silva and M. T. De Oliveira, *J. Raman Spectrosc.*, 1997, **28**, 873–878.

2. J. L. Verble, *Phys. Rev., B*, 1974, **9**, (12), 5236–5248.

3. T. R. Hart, S. B. Adams and H. Tempkin, in *Proc. 3rd Int. Conf. on Light Scattering in Solids*, p.254–258 (M. Balkanski, R. Leite and S. Porto, eds).Flammarion, Paris, France, 1976.

4. N. Boucherit, A. Hugot-Le Goff and S. Joiret, *Corros. Sci.*, 1991, **32**, 497–507.

5. D. Thierry, D. Persson, C. Leygraf, N. Boucherit and A. Hugot-Le Goff, *Corros. Sci.*, 1991, **32**, 273–284.

6. T. Ohtsuka, K. Kubo and N. Sato, *Corrosion*, 1986, **42**, 476–481.

7. R. J. Thibeau, C. W. Brown and R. H. Heidersbach, *Appl. Spectrosc.*, 1978, **32**, (6), 532–535.

8. J. Dünnwald and A. Otto, *Corros. Sci.*, 1989, **29**, 1167–1176.

9. A. B. Johnson and B. Francis, *Durability of metals from archaeological objects, metal meteorites and native metals*, U.S. Department of Energy, 1980.

10. B. Miller and N. Chapman, *Radwaste Magazine*, 1995, **2**, (1), 32–42.

11. M. De Franco, J. M. Gras and J. P. Moncouyoux, *RGN (Revue générale Nucléaire)*, 1996, (3), 27–33.

Part 4

R & D Approaches and Results

23

A Methodology for Assessing the Integrity of Spent Nuclear Fuel in Long Term Dry Storage

W. S. WALTERS and J. HENSHAW

AEA Technology plc, Fuel & Materials Performance Department, 168, Harwell International Business Centre, Didcot, Oxfordshire OX11 0QJ, UK

ABSTRACT

One of the key concerns in the long-term storage of spent nuclear fuel is the possible degradation of the fuel whilst in storage. Degradation can be by a number of means, particularly corrosion, leading to cladding failure. It is necessary to have confidence that at the end of the storage period the fuel will be in a condition where it can be handled safely, whatever the ultimate fate of the fuel may be. If the fuel is to be reprocessed, there will be minimum standards of physical robustness before it can be handled in the plant. If it is to be consigned for repository disposal, then it may need to be capable of being re-packaged. In both cases, in addition to physical robustness issues, minimal discharges of activity are also desirable, therefore minimal degradation whilst in storage is highly desired.

In order to predict the extent of corrosion, cladding failure, etc., it is first necessary to have a thorough understanding of the storage environment and its effect on corrosion processes. Even dry stores are not fully dry, there is always a trace of moisture in the gas, especially if the fuel has been pond stored and may have partly failed. Similarly there is often a trace of air in the fill gas — in practice it is very difficult to remove every last trace. Inside the storage container the radiation emitted by the fuel interacts with the fill gas, water vapour and trace air, by radiolysis processes. These can produce corrosive species such as oxides of nitrogen, even nitric acid molecules. These may be adsorbed on the surfaces of all components, in the same way that there is always an adsorption film of water present, maybe only a few molecular layers thick. The effect of an adsorption layer containing nitric acid needs to be carefully evaluated in assessment of long term storage performance. This paper presents a flowchart and an outline method for evaluating these effects in a quantitative way. The method is in principle universally applicable to all methods of fuel dry storage.

1. Introduction

The spent nuclear fuel dry store concept has been gaining in attractiveness from both a commercial and environmental standpoint. Over the past decade or more, there has been a considerable amount of literature written in support of the concept and addressing the technical issues which arise from it. (See for example [1–3].) This paper considers one of the more complex of these technical issues; the possible enhancement of corrosion of the stored fuel cladding by the radiation field emanating from the spent fuel itself.

It will at once be appreciated that any metallic material placed in long term storage will be subject to normal 'thermal' corrosion. The materials used in the construction of reactor fuel elements are normally considered to be extremely resistant to corrosion in the context of their environment of use. However, for long term dry storage, in a different environment to that for which they were designed, the performance of the fuel element materials has to be validated. Besides the actual cladding materials, the other key factor in controlling the corrosion process is that of the environment; a dry, low temperature environment will assist in this respect, and a non-oxidising gas as the surrounding medium to the fuel will be of further benefit. In this respect some general considerations of the storage medium have been written [4]. However, it should be noted that the now international thrust towards dry store technology includes several types of nuclear reactor; assessments of storage media for one type of fuel may not necessarily be appropriate for other fuel types. A number of studies of fuel element corrosion in dry storage have been published [5–9], but mainly on LWR (Light Water Reactor) fuel and predominantly with Zircaloy cladding, otherwise usually with stainless steel cladding.

It must be understood that the main point of concern, the failure of the fuel cladding whilst in storage, is a low probability event whilst the fuel is in a well-managed facility. The storage of spent fuel has been reviewed, both in wet and dry storage [10] and guidelines have been published on the best practice for spent fuel storage [11]. In these reviews the overwhelming consensus is that fuel failure is not an issue in a well-managed facility. However, there are other fuel types (such as from prototype, research or materials testing reactors) with other cladding metals, where there is comparatively little information on cladding performance in dry storage. Further, there may be issues arising regarding fuel from facilities which have been shut down and where the standard of management may have become less than ideal. Subsequent dry storage of these fuels requires careful assessment before action is taken.

2. Assessment Method

The performance of spent nuclear fuel whilst in wet storage (immersed in water, either in ponds or specially designed containers) has been studied over a considerable time period [12]. The corrosion of the fuel cladding has been observed, although the corrosion rates have been variable [13]. The effect of radiolysis in changing the gas composition within transport or storage containers has also been reported and calculation methods have been developed [14].

Under dry storage conditions there are generally fewer data, as the fuel is less accessible for observation, and there are fewer dry storage installations. The overall approach to assessing fuel performance in dry storage is illustrated in the flow-sheet shown in Fig. 1. This Figure groups the possible features of the initial conditions, including fuel temperature, cladding condition, cover gas, dryness and residual air. The various processes which may then follow are shown, including radiolysis, pressurisation, corrosion, etc. The Figure then shows how the effects of these processes maybe quantified, by a combination of absolute calculation, mechanistic computer modelling and empirical modelling. The interaction of these effects is then shown as effects on the calculation process. The various features and processes identified in

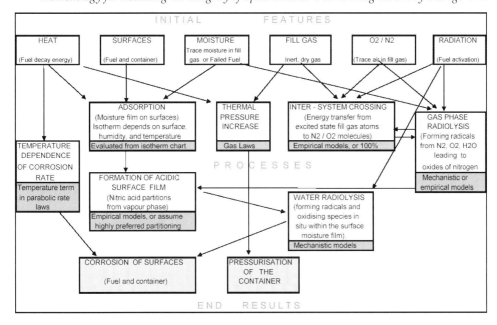

Fig. 1 *Summary of processes affecting spent fuel in storage.*

the flow sheet are now described in greater detail. Also, the method is used below (as an example) in the context of spent MTR (Materials Test Reactor) fuel.

2.1. Features and Processes

The spent fuel will normally be contained within a sealed package, which may (optionally) also be an approved transport package for movement to another location. Within the package there will be a closed system consisting of the fuel, the cover gas, and the internal surfaces of the package. The features of these components are as follows:

- The fuel may have some residual decay heat, and may contain some moisture (if the cladding has been penetrated by corrosion, during pond storage, and the fuel has become waterlogged). The fuel will also emit radiation, effectively only beta/gamma as far as it affects the other components of the system.

- The cover gas will be nominally pure and dry, however in practice it is impossible to exclude all moisture and quite difficult to exclude all traces of air. The probability is that there will be some traces of these components in the cover gas.

- The package internal surfaces will largely depend on the structural materials of the package. They may, possibly, have a surface coating (e.g. paint if the container is mild steel) or not need one if the container is stainless steel. The surfaces may also include corrodible materials such as Aluminium or Boral, if the packages require deliberate criticality control or are freely adapted from existing designs used for shipments of higher activity fuel.

The processes which may result from these features are as follows:

- *Decay heat.* It is assumed that decay heat is not sufficient to cause significant pressurisation of the cover gas, when equilibrated, or that any such rise in internal pressure is within the design base of the package approval. However, decay heat may assist in vapourising any trapped moisture and expelling it from the internals of a failed fuel pin. Also, decay heat may accelerate corrosion processes which will take place within the container during the storage period. During pond storage the fuel temperature is low, as heat is removed effectively. In dry storage this is not necessarily true (gases are always better thermal insulators than water) and fuel element temperatures will be higher.

- *Radiation.* The emitted radiation will interact with the remainder of the system and result in radiolysis of the cover gas (together with any traces of moisture or air). If these traces are present, the result of this process is inevitably oxides of nitrogen and ultimately nitric acid molecules. If the package internal surfaces are painted, the paint may degrade and volatile radiolysis products may be emitted; depending on the paint formulation, these may be corrosive species (such as chlorine or oxides of sulphur).

- *Surface adsorption.* Any traces of moisture, arising from the cover gas or from waterlogged failed fuel, will distribute through the gas volume and water molecules will adsorb on every surface in contact with the gas. The thickness of the adsorption layer depends on the inventory of water available, the temperature, and the nature of the surface. There is a relationship between the amount of water vapour adsorbed, per unit surface area and the temperature, based on the adsorption isotherm for that particular material. Note that adsorption isotherms also depend on surface treatments, such as pre-oxidation, shot peening, sand blasting, and others. An example of an adsorption isotherm is shown in Fig. 2. The adsorption layer may be only a few molecular layers thick, up to hundreds of molecular layers, and yet invisible to the eye at humidities which may reasonably be considered 'dry'.

- *Corrosion.* Ultimately, the main concern of long term storage of spent fuel is the corrosion of the fuel and its ultimate condition at the end of the storage period. The corrosion of cladding metals such as Zircaloy, stainless steel and aluminium is generally a uniform corrosion process, but can become localised if crevices are present. The corrosion rate will depend on the material, temperature, humidity, and the concentration of aggressive ions. There may well be a discontinuity in corrosion rate at low enough humidities; an example of this (for mild steel, stored in nominally dry air, under gamma irradiation) is shown in Fig. 3. The discontinuity relates to a humidity of about 8% RH at 50°C, the experimental temperature. We now consider the quantitative

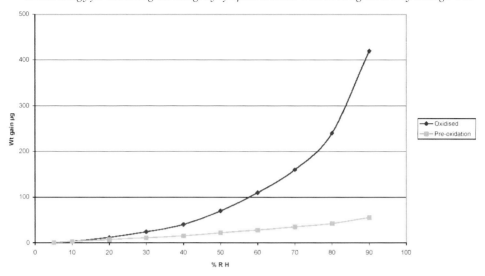

Fig. 2 *Typical adsorption isotherm for 304 stainless steel.*

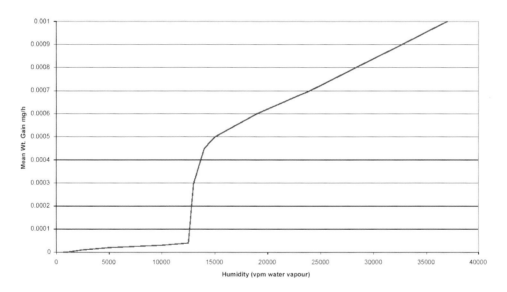

Fig. 3 *Radiation assisted oxidation as a function of humidity, for mild steel irradiated in air.*

evaluation of these processes; it should be noted how they interrelate by way of the flow chart shown in Fig. 1.

Decay heat. From original enrichment, exposure and burn-up the isotopic inventory of fission and activation products within the fuel element can be calculated (e.g. using computational codes such as FISPIN [15] or ORIGEN

[16]. The energy released per unit time and the decay curve allows a calculation of the energy release rate at the start of dry storage and throughout the storage period.

Temperature. Depending on the design geometry of the container, the rate of heat loss to the walls, through the cover gas can be calculated. The effect may be a temperature rise of the fuel and of the cover gas. Also, the external heat sink terms for the package as a whole may need to be considered. Water adsorption. From the temperature and the known surface condition of the materials, the water adsorbed on the surfaces may be evaluated from the isotherm. This may also be limited by the availability of water, which depends on the source items (purity of the gas, waterlogged fuel, etc).

Radiolysis. The formation of oxides of nitrogen depends on the availability of trace air, and moisture [17]. In reality there is always some formation of oxides of nitrogen. The radiation will usually interact with the cover gas atoms or molecules, as these are far more abundant, and produce energetically excited states of these species. On interaction with a molecule of nitrogen or oxygen (on a diffusional basis), the excess energy can be transferred by a process known as inter-system crossing [18]. This results in excited state oxygen or nitrogen molecules, and from there the process of formation of oxides of nitrogen is energetically favoured. It is a paradox that although there may be very little of the trace species (air, water vapour) present, it is the abundance of the cover gas species which virtually guarantees that the trace species will react. The process of inter-system crossing is energetically favoured for inert gases such as would be used for cover gases in dry storage, and is more effective for helium than for argon. Excited state krypton or xenon would not have the necessary energy for inter-system crossing, but their prohibitive cost precludes their use for dry storage.

The amount of nitrogen oxides, and nitric acid, can be calculated either empirically or mechanistically. On an empirical basis, it is assumed that the production is limited only by the availability of reactants (nitrogen, oxygen, water vapour) and that the inventory of nitric acid can then be calculated to give a maximum figure, assuming complete reaction. Alternatively, the formation of nitrogen oxides can be calculated mechanistically using FACSIMILE models if the necessary parameters (reactant concentration, dose rate etc.) are known. This has already been done for Swedish BWR (Boiling Water Reactor) fuel destined for ultimate disposal [19]. The mechanistic model has the advantage that it permits time resolution of the production process, and avoids any pessimism implicit in the empirical approach.

Partitioning. The nitrogen oxide (and specifically nitric acid) molecules will exist both in the vapour phase and will partition with the surface moisture layers as determined by the adsorption isotherms for the various materials present [20]. The extent of partitioning depends on the thickness of the moisture film; if the film is only a few molecular layers thick, then the water will be highly co-ordinated to the surface sites and there will be little freedom for water molecules to hydrate any incoming nitric acid molecules. In principle, nitric acid molecules prefer to hydrate (it is energetically favoured) but cannot unless there are the free water molecules within the surface moisture film. Even for a known thickness of surface moisture film, the availability of free water molecules will vary from material to material, as different numbers of molecular layers will be needed to satisfy the co-ordination requirements of the solid surface. As a general rule (and here there are some sweeping assumptions made) if there are fewer than about 10 molecular layers, then there will be no available water for hydrating any nitric acid molecules. (As an example, in Fig. 2 the discontinuity is at about 6 molecular layers.) For between 10 and 40 molecular layers, the situation is one of transition, and for more than 40 molecular layers there will be free water capable of hydrating any nitric acid molecules. It does not take much surface moisture to meet this requirement; for example a film of 40 molecular layers would only be 11 milligrams of water per square metre of surface (for stainless steel). This is represented schematically in Fig. 4. From a known adsorption isotherm, the thickness of the surface film can be calculated and the extent of partitioning can be evaluated.

Surface film chemistry. Assuming that there is sufficient water to form a viable surface moisture film, the partitioning of nitric acid then leads to a surface film with an acidic pH. Quantitatively, the pH can be calculated from the amount of nitric acid formed (calculated by either empirical or mechanistic models) and the thickness of the water layer which hydrates the nitric acid molecules. This will be time - dependent and will evolve as more nitric acid is produced, up to the inventory limit set by the availability of air and water as sources for nitric acid production.

Corrosion. The corrosion of metal surfaces is invariably dependent (to some extent) on the surface pH. For the metals used in nuclear fuel cladding (Zircaloy, stainless steel and aluminium) there is good resistance to general

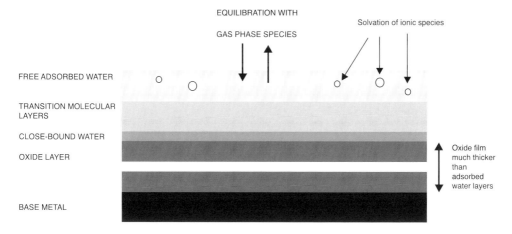

Fig. 4 Schematic diagram of water mono-layers on a metal oxide surface.

corrosion, even when there is an acidic surface film (although this is less true of aluminium than the others). However, the main cause of failure of nuclear fuel cladding is not through general corrosion, but localised corrosion such as pitting or crevice corrosion. For this form of corrosion, the surface pH becomes much more important. The propagation rate of a pit in stainless steel corrosion, for example, depends strongly on pH, corrosion potential, and the presence of aggressive species such as Cl^-, SO_4^{2-}, and the concentration of hydrogen peroxide (formed by radiolysis, which is dependent on dose rate) [21]. It is well known that the corrosion of aluminium by pitting is influenced by a number of anionic species, including nitrate [22]. Assuming that the aluminium has already experienced wet corrosion, and the surface is covered with a (thin) inner oxide layer of Boehmite and a (thicker) outer oxide layer of Bayerite, pitting depends on the breakthrough of the Boehmite layer. Corrosion propagation normally follows a parabolic rate law, similar to that for sensitised stainless steel.

It should also be noted that there is an effect of temperature on corrosion rate. This can also be accommodated in the rate expressions that describe pitting propagation.

From these considerations, the corrosion rates of pitting in the relevant materials can be quantified. It must be noted that these procedures have only recently become available, with the development of modern computing technology and the relevant mathematical models, together with the practical measurements which calibrate the models.

2.2. Calculation Route Summary

1. Characterise the conditions at the start of dry storage, evaluating the following:

- Residual moisture inventory
- Trace oxygen/nitrogen inventory
- Radiation dose rate
- Temperature.

2. Evaluate the adsorption of moisture on all surfaces within the storage container, by their relevant isotherms.

3. Evaluate the production of oxides of nitrogen/nitric acid, by empirical or mechanistic methods.

4. Allow partitioning of nitric acid to the adsorbed moisture layer and evaluate the resulting nitric acid concentration (and pH).

5. Assess the effect of the pH and nitrate ion concentrations on the corrosion rate, including any effect of temperature from the decay heat of the fuel.

3. Application to Spent MTR Fuel

Materials Test Reactor fuel of this type is normally clad with an aluminium alloy, and is irradiated in a water-moderated reactor and subsequently stored in a pond. During irradiation the reactor water chemistry is well controlled, the surface temperature is below 100°C, and the expected surface oxide is boehmite. During pond storage, the chemistry control may be less rigorous, and there are up to 32 solute species which affect the corrosion rate of aluminium. These include anions such as chloride, nitrate, sulphate. As noted above, it is expected that during pond storage the outer oxide layer of bayerite will become more significant, and the presence of aggressive ions may lead to conditions where pitting corrosion is initiated. Therefore, the fuel condition at the beginning of dry storage is variable, and the full range of possibilities from very good condition, through to failed fuel, should be expected.

The degree of hydration of bayerite, and any occluded moisture within the porous oxide layer, should also be taken in to account as a possible moisture source term. At high temperatures and over long time scales, bayerite (α-Al_2O_3.$3H_2O$) transforms to boehmite (β-Al_2O_3.H_2O); as it does so, further moisture is released.

The formation of surface moisture layers on aluminium oxide is not well characterised; reliable adsorption isotherms are generally lacking. As the oxide is already partly hydrated, it would be expected that surface moisture would adsorb more readily, and the number of monolayers to produce conditions where solvation of nitrate could occur would be reduced. Therefore, the initiation of pitting corrosion would be favoured, compared with similar conditions and a stainless steel fuel cladding.

By way of contrast, UK experience with aluminium-clad MTR fuel at the (now decommissioned) research reactors at Harwell was that under operational conditions, corrosion was not experienced. Due to the large number of possible aggressive species, the corrosion of aluminium was prevented by careful attention to the conductivity of the water in the fuel storage pond as a general means of ensuring all aggressive species were kept to a minimum. The Harwell reactor fuel was not stored dry at any stage, and was routinely reprocessed after about two years pond storage.

4. Conclusions

The long term dry storage of spent nuclear fuel requires careful assessment of the conditions (including radiolytic and other processes) inside the storage container. The possible consequences of radiolysis and moisture adsorption may affect the corrosion of the stored fuel. A method has been presented which identifies each process that contributes to the corrosion chemistry and provides a means of calculating the effect of these, thereby allowing a complete assessment of the corrosion performance of the dry storage arrangement.

References

1. Third International Spent Fuel Storage Technology Symposium/Workshop, Proceedings, Volumes 1 & 2, Pacific Northwest Laboratory, CONF-860417 (1986).

2. Thirteenth Water Reactor Safety Research Information Meeting, Volume 3 (A.J. Weiss, compiler), U.S. N.R.C. report NUREG/CP-0072 Vol. 3 (1986).

3. The 9th International Symposium on the Packaging and Transportation of Radioactive Materials, Oak Ridge National Laboratory, CONF-890631 (1989).

4. D. Wheeler, *Nucl. Eng. Int.*, 1984, 26–31.

5. R. E. Einzeiger, Tutorial Review of Spent Fuel Degradation Mechanisms under Dry Storage Conditions, U.S. Dept. of Energy report DE83 012825 (1983).

6. A. B. Johnson and E. R. Gilbert, Current status of fuel degradation studies in dry storage, Pacific Northwest Laboratory report PNL-SA-11894 (DE84 011250) (1984).

7. Workshop on spent fuel/cladding reaction during dry storage (D.W. Reisenweaver, ed.). U.S. N.R.C. report NUREG/CP-0049 (1984).

8. E. R. Gilbert, C. A. Knox and G. D. White, Behaviour of spent LWR fuel in nitrogen and in air, Pacific Northwest Laboratory report PNL-SA-13902 (DE86 011115) (1986).

9. R. W. Knoll and E. R. Gilbert, Evaluation of cover gas impurities and their effects on the dry storage of LWR spent fuel, Pacific Northwest Laboratory report PNL-6365 (DE88 003983) (1988).

10. Storage of water reactor spent fuel in water pools, IAEA Technical Reports Series No. 218, (1982) and Survey of experience with dry storage of spent nuclear fuel and update of wet storage experience, IAEA Technical Reports Series No. 290 (1988).

11. 'Guidebook on spent fuel storage', IAEA Technical Reports Series No. 240 (1984).

12. A. B. Johnson, 'Behaviour of spent reactor fuel in water pool storage', Batelle Pacific Northwest Laboratory report BNWL-2256 (1977).

13. 'Final report of a co-ordinated research programme on the behaviour of spent fuel assemblies during extended storage(1981–1986)' IAEA TECDOC-414, IAEA, Vienna (1987).

14. W. S. Walters, 'Computerised modelling of radiolysis and corrosion effects in transport containers for spent nuclear fuel', *Int. J. Radioact. Mats. Transp.* (RAMTRANS), 1994, **5**, 2–4, 253–260.

15. R. F. Burstall, FISPIN — A Computer Code for Nuclide Inventory Calculations, UKAEA report ND-R-328(R). 1979, and FISPIN: A Code for Nuclide Inventory Calculations, User Guide for Version 7A (ANSWERS/FISPIN(95)2), AEA Technology plc.

16. A. G. Croff, 'ORIGEN2: A versatile computer code for calculating the nuclide compositions and characteristics of nuclear materials', *Nucl. Technol.*, 1983, **62**, 335–352.

17. J. Henshaw and A. M. Baverstock, A comparison of experimental and theoretical radiation chemistry for the $N_2/O_2/H_2O/H_2$ gas phase system, AEA Technology report AEA-RS-2369 (1993).

18. See, for example, C. H. J. Wells, *Introduction to Molecular Photochemistry*, Chapman and Hall Chemistry Textbook Series, Chapman and Hall/John Wiley & Sons (1979).

19. J. Henshaw (AEA Technology), SKB Technical Report 94-15, Modelling of Nitric Acid Production in the Advanced Cold Process Canister due to the Irradation of Moist Air , January 1994.

20. F. C. Tompkins, *Chemisorption of Gases on Metals*. Academic Press, 1978.

21. J. Henshaw and J. C. McGurk, A model for radiation chemistry effects on crack corrosion. BNES conference, *The Water Chemistry of Nuclear Reactor Systems — 7*, Bournemouth, UK, 1996, .Vol. 2, p 513.

22. H. P. Godard, in *The Corrosion of Light Metals* (Godard, Jepson, Bothwell and Kane, eds). John Wiley & Son, New York, 1967.

24

Long Term Thermal Stability and Corrosion Behaviour of Welded and Unwelded Alloy 59 for Rad-Waste Containers

H. ALVES, O. IBAS, V. WAHL, J. KLÖWER* and C. COLLET*

KRUPP VDM GmbH, Altena, Germany

* KRUPP VDM S.A.R.L., France

ABSTRACT

Highly corrosion resistant materials are specified in the design of containers under consideration in the United States nuclear waste disposal program. Ni–Cr–Mo alloys are the main candidate materials for the corrosion resistant barrier. In this study the long-term thermal stability and corrosion resistance of the new but well established Alloy 59 (UNS N06059) is discussed and compared with that of other alloys of the Ni–Cr–Mo-family, e.g. Alloy 22 (UNS N06022) and C-276 (UNS N10276).

The results show that after long term ageing in air at 200, 300 and 427°C for 10 000 h the microstructure and the mechanical properties as well as the corrosion resistance of Alloy 59 in both welded and unwelded conditions remain unchanged. Alloy 59 is most probably the best candidate material for nuclear waste containers.

1. Introduction

Highly radioactive waste from the last 50 years of the nuclear age urgently need to be stored. One site currently under consideration for the construction of a permanent repository is an underground facility at Yucca Mountain, Nevada. The disposal containers to be used must ensure long-term isolation of the high level nuclear waste. The current design for such containers is based on a metallic multibarrier system, which includes a 20 mm thick shell of a nickel base super alloy, e.g. Alloy 22, as a main barrier to prevent corrosion penetration over the years. Nickel base alloys are under consideration because candidate alloys must have not only a low general corrosion rate but also excellent resistance to localised corrosion. This is so because the underground water contains halides which may promote localised corrosion of many other alloys such as iron base alloys.

Thermal stability is another key issue in the long-term performance of container materials. Due to the radioactive decay of the waste, it is predicted that during the first 1000 years of service lifetime temperatures as high as 250–350°C can be reached if the design of the repositories do not involve ventilation. Thus, possible ageing effects resulting from long time exposure to this range of temperatures have to be considered as well. Irrespective of the fabrication technique, some form of welding of the container materials will be involved in producing cylindrical packages of large diameters. The performance of weldments is therefore of great concern and has to be studied along with that of the base material. The selection of Alloy 22 is claimed to be based on its good resistance to localised corrosion [1]. However, another alloy of this family, the Alloy 59 (UNS N06059), has even better resistance both to uniform and localised corrosion particularly in chloride-containing environments as measured in standard ASTM laboratory tests [2]. In addition, this pure ternary alloy of the Ni–Cr–Mo family exhibits superior thermal stability due to its balanced composition without addition of other elements. Tests were carried out to obtain new data on the long-term thermal stability and corrosion resistance of Alloy 59 and its weldments. Microstructure, mechanical properties and corrosion behaviour were investigated in both solution annealed and long-term aged conditions; ageing was conducted at 200, 300 and 427°C for 10 000 h. The results are described in this paper.

2. Experimental

Plates of Alloy 59, 16 mm thick were thermally aged in air for 10 000 h at 200, 300 and 427°C. The microstructure, mechanical properties and corrosion resistance of solution annealed and air aged specimens were investigated. All specimens were studied in the as-wrought and welded conditions. GTAW-HW welding was performed using the matching filler metal. Heat tints were removed by brushing the hot weld joints. Welded samples consist of approximately 1/3 of weld and 2/3 of base material.

Standard polishing techniques were used to prepare the samples for metallographical studies.

Mechanical properties were measured at room temperature by tensile and notch impact-bending test. The tensile test was carried out following the DIN EN 10002 – Part1 standard. Dogbone-shaped specimens were machined according to DIN 50125 – B 6 x 30. At least 2 specimens were tested to obtain the values of yield strength at 0.2% plastic deformation, ultimate tensile strength and elongation to fracture. The Charpy-V-notch impact test was performed according to DIN EN 10045 – Part 1. At least three parallel samples were tested in order to evaluate the impact energy.

Corrosion resistance was evaluated by ASTM G 28 A and 'green death' solution laboratory tests. Cut edges were wet ground with 80 grit abrasive paper. Before testing samples were degreased in acetone/isopropanol. The ASTM G 28 A standard involves the immersion of specimens for 24 h in a boiling solution of 50% H_2SO_4 and 42 gL^{-1} of $Fe_2(SO_4)_3$; after immersion corrosion rates were determined. The critical pitting temperature (CPT) was measured in 'green death' solution. For this, samples were immersed in a solution containing 11.5% H_2SO_4, 1.2% HCl, 1% $FeCl_3$ and 1% $CuCl_2$ at a starting temperature of 80°C. After a 24 h exposure period a gradual temperature increase of 5°C increments was applied and a freshly prepared solution was used. This procedure was repeated until samples failed by pitting; above the boiling point an autoclave was used. To confirm the visual observation of pitting corrosion rates were measured.

3. Results

3.1. Microstructure

Optical microscopical investigation showed that the microstructure of all samples aged for 10 000 h remained unchanged. No precipitation could be observed in the base material as a result of the ageing treatment. In the same way, no particular feature was detected in the weldments or in the heat-affected zone. Figure 1 shows the region of the heat-affected zone after ageing at 427°C.

The average grain size of base material aged at 200, 300 and 427°C was observed to be 147, 160 and 170 μm, respectively. For comparison, the grain size of the as-wrought material was 140 μm. This variation is statistically not very significant if one takes into consideration the measurement uncertainties associated with the grain size scatter already observed within one sample.

3.2. Mechanical Properties

The results from the mechanical testing of wrought and welded materials are presented in Figs 2–4 as a function of the ageing temperature. Data obtained for solution annealed samples are included for direct comparison.

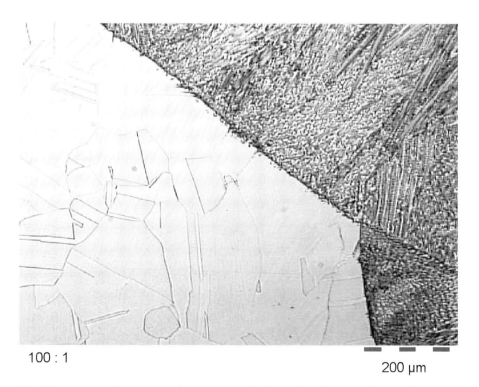

100 : 1

200 μm

Fig. 1 Cross-section showing the microstructure of the heat-affected zone of Alloy 59 after ageing at 427°C for 10 000 h.

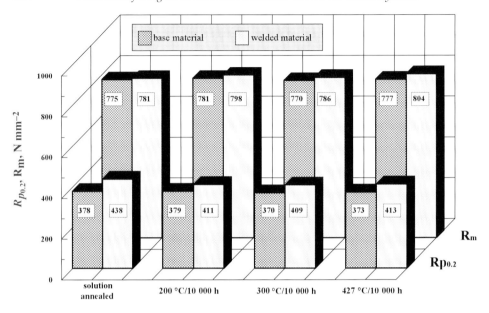

Fig. 2 *Effect of ageing for 10 000 h on the yield strength (0.2% offset) and ultimate tensile strength of Alloy 59 base and welded materials.*

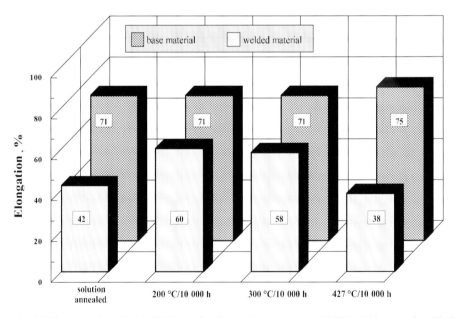

Fig. 3 *Effect of ageing for 10 000 h on the elongation to fracture of Alloy 59 base and welded materials.*

From Fig. 2 it can be concluded that the yield strength at 0.2% offset as well as the ultimate tensile strength (UTS) were not influenced by the heat treatment, even at the higher temperature of 427°C. Variations can be considered to be within the experimental scatter. In general, weldments exhibit slightly higher yield strength values.

The values of elongation to fracture are shown in Fig. 3. The elongation of the base material was almost not influenced by the ageing treatment. Weldments exhibited lower elongation values than the base material. In addition, the elongation of the welded material increased upon ageing at 200 and 300°C; at 427°C it compares with the elongation of the non-aged welded sample.

The results of the Charpy impact test are represented in Fig. 4. There was also no measurable influence of the ageing treatment on the impact energy of the base material. Concerning the welded material the impact energy showed a similar behaviour to the elongation: ageing at lower temperatures slightly increased the impact strength, while ageing at 427°C decreased it.

3.3. Corrosion Behaviour

The critical pitting temperature (CPT) is the most important parameter for characterising the resistance to localised corrosion in chloride-containing environments. Figure 5 shows that the critical pitting temperature of Alloy 59 did not change even after ageing at 427°C for 10 000 h. Aged and solution annealed unwelded samples have the same critical pitting temperature of 130°C. The pitting resistance of the welded material is also excellent and was not affected by the long-term heat treatment. All welded samples except that annealed at 300°C exhibit a critical pitting temperature of 130°C, as well. As mentioned above the starting temperature of the 'green death' test was 80°C.

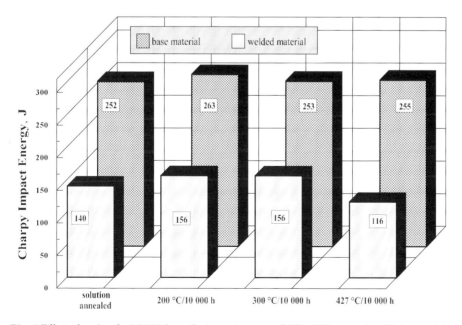

Fig. 4 *Effect of ageing for 10 000 h on the impact energy of Alloy 59 base and welded materials.*

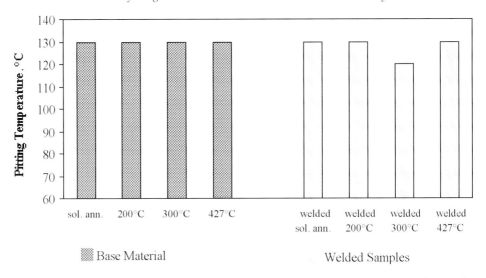

Fig. 5 *Effect of ageing for 10 000 h on the critical pitting temperature of Alloy 59 base and welded materials in 'green death' solution.*

Similar behaviour is observed when the corrosion rate determined by mass loss in the standard ASTM G 28 A test is considered. The results shown in Fig. 6 indicate that the rate of dissolution of both base and welded materials almost did not change as the ageing time increased. Ageing appears to decrease the corrosion rate very slightly, however this effect is not very significant. On average, corrosion rates amounts to approximately 0.71 and 0.75 mm/year for base and welded material, respectively. The rate of dissolution of weldments is only negligibly higher than that of the base material.

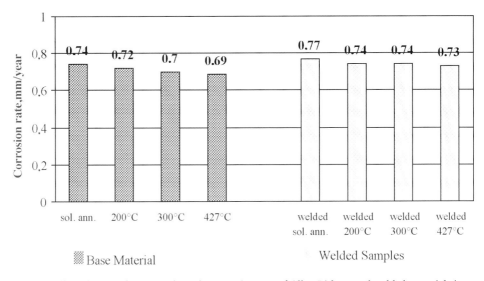

Fig. 6 *Effect of ageing for 10 000 h on the corrosion rate of Alloy 59 base and welded materials in ASTM G 28 A solution.*

4. Discussion

Table 1 gives the chemical composition of various alloys of the Ni–Cr–Mo family. The major alloying elements of these nickel base alloys are chromium and molybdenum with some alloys containing other elements like iron and tungsten. The pitting resistance equivalent number, PRE = %Cr + 3.3 (%Mo), is also included in the table.

These alloys have a very low carbon content as a result of the improved AOD (argon–oxygen–decarburisation) and VOD (vacuum–oxygen-decarburisation) melting technologies. This overcame the often serious intergranular corrosions in the heat-affected zone of weldments which characterised the first alloy of this family, the alloy C [3].

Although low in carbon and silicon, the Alloy C-276 is under certain conditions not adequately thermally stable in regard to precipitation of both carbides and intermetallic phases, thus becoming susceptible to corrosion [4]. This fact is well illustrated by the time–temperature-sensitisation (TTS) diagram shown in Fig. 7 [5,6]. Alloy C-276 has a high tendency to precipitation of the intermetallic µ and P phases along with carbides of the M_6C-type in the middle temperature range. For example, at temperatures between 700 and 1080°C precipitation along the grain boundaries can already be observed after approx. 6 min [5]. Accordingly, susceptibility to intergranular corrosion in the ASTM G 28 A test occurs after few minutes (see Fig. 7). C-276 is not a candidate alloy for radioactive waste containers. New developments in the Ni–Cr–Mo family like Alloy 2000 (UNS N06200), which contains copper, and Alloy 686 (UNS N06686), which contains tungsten, also cannot be recommended due to their proneness to sensitisation illustrated by the TTS-diagram. This phenomenon could occur when welding the thick sections of waste containers, which require multiple weld passes, leading to undesirable phase precipitation in the heat-affected zone and to lower corrosion resistance. Risk of sensitisation also would exist in the weld vicinity, if a local induction annealing of the weld were to be conducted. This heat treatment is sometimes suggested in order to mitigate the stresses resulting from welding of the closure-lids of the container and thus to avoid failure by SCC. Figure 7 further shows that susceptibility to sensitisation strongly decreases for

Table 1. Nominal chemical composition in mass% of some Ni–Cr–Mo alloys

Alloy	UNS No.	Ni	Cr	Mo	W	Fe	Others	PRE
C-276	N10276	Bal.	16	16	4	5	<2.5 Co	69
22	N06022	Bal.	22	13	3	3	<2.5 Co	65
686	N06686	Bal.	21	16	4	2	—	74
2000	N06200	Bal.	23	16	—	2	1.6 Cu	76
59	N06059	Bal.	23	16	—	<1	—	76

Alloy 22 and that the best construction material to prevent this undesirable phenomenon would be Alloy 59.

Alloy 59 is the purest form of a Ni–Cr–Mo alloy. It is one of the highest nickel-containing alloys of this family without the addition of other alloying elements, such as tungsten or copper which are known to impair thermal stability. It is this compositional balance that is responsible for its superior thermal stability. The metallographical investigation conducted in this work showed that none of the performed long-term heat treatments induced visible precipitation of carbides or intermetallic phases, for example μ and P, which could increase corrosion susceptibility.

Changes in the mechanical properties are frequently used as a measure of the thermal stability of materials against ageing processes. In this study no evidence was found that heat treatment at 200, 300 and 427°C for 10 000 h affects the mechanical properties of Alloy 59. Independently of the ageing temperature all samples kept their good mechanical performance after 10 000 h: The yield strength ranged between 370 and 380 Nmm^{-2}, while the ultimate tensile strength varied between approx. 770 and 780 Nmm^{-2} (see Fig. 2). These high values of strength guarantee the excellent mechanical behaviour characteristic of Alloy 59 and confirm its great thermal stability. The minor variation of grain size observed on ageing has no impact on the material strength. In general, welded samples exhibit slightly higher yield strength values than the base material. This is due to the characteristic microstructure of a weld, which results from a simple re-melting plus solidification process being therefore quite different from the optimised microstructure of a wrought and solution annealed base material. Both elongation and impact energy are important parameters used to characterise the ductility of a material. For better mechanical performance and

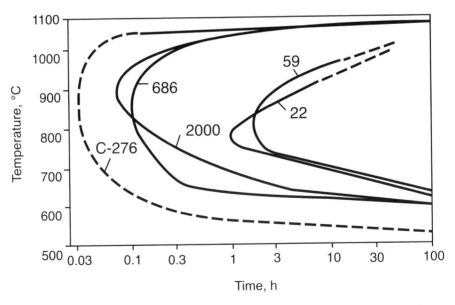

Fig. 7 *Time–temperature-sensitisation diagram of some Ni–Cr–Mo alloys tested according to ASTM G 28 A; curves define the depth of intergranular penetration > 50 μm as evaluated by metallography.*

fabricability a combination of high strength and high ductility is desired. Alloy 59 has excellent ductility, which is much larger than values requested by the pressure vessels code (e.g. in the German code VdTÜV $A_5 \geq 30\%$ and $A_v \geq 26$ J are required for base material and weldments). Neither the elongation nor the impact energy of the base material are influenced by the ageing treatment. For weldments it was observed that ageing at lower temperatures (200 and 300°C) increases these properties, while ageing at 427°C decreases them to the level of the non-aged material. Taking into consideration that processes involving motion of dislocations are thermally activated and can occur above approx. $1/3T_m$ [7], where T_m is the melting temperature in K, it is justified to assume that recovery may take place at 300°C and even at 200°C after longer times. This would account for the higher ductility after ageing at the intermediate temperatures. Studies at longer ageing time are already underway and will further explain these effects. The welding of Alloy 59 (similar to other C-alloys, as shown later in Figs 8–10) brings together with some strengthening also a certain loss of ductility, as expected from the as-cast microstructure resulting from this process. Nevertheless, the mechanical values of weldments are still very high and none of these trends has a detrimental effect in the corrosion properties as can be seen from Fig. 5 and Fig. 6. The results of this work demonstrate that ageing for 10 000 h even at 427°C does not impair the mechanical properties of Alloy 59. In Figs 8–10 Alloy 59 is compared with Alloy 22 and C-276 using data obtained from the literature [1,8].

As far as the yield strength is concerned all alloys exhibit similar performance (see Fig. 8). Maximum differences amount to 30 Nmm^{-2} and can be considered to be within the normal scatter associated with a given microstructure. Aging at 427°C for 10 000 h does not influence the yield strength. As observed for Alloy 59, the welded counterpart alloys show higher strength than the corresponding base material.

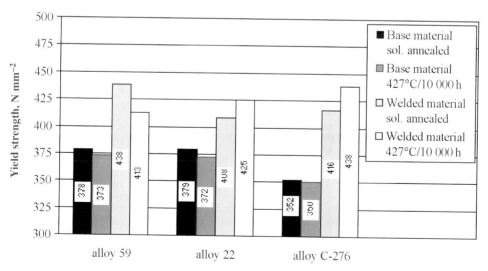

Fig. 8 *Yield strength of some Ni–Cr–Mo alloys. (Samples from left to right: base material solution annealed, base material aged at 427°C for 10 000 h, welded material solution annealed, welded material aged at 427°C for 10 000 h.)*

The ultimate tensile strength of base material is similar for all three alloys and is not affected by the ageing treatment at 427°C (see Fig. 9). However, while weldments of Alloy 59 remain stable, those of Alloy 22 and C-276 seem to undergo certain microstructural changes during the ageing treatment as reflected by the UTS variation. As shown in Fig. 10 the base material of Alloy 59 is slightly more ductile than that of the other C-alloys; after ageing at 427°C all three alloys have ductility values comparable to the non-aged state. As for Alloy 59, the weldments of Alloy 22 and C-276 exhibit a certain loss of ductility along with some strengthening in relation to the base material. The elongation to fracture of welded Alloy 22 and C-276 samples also shows stronger variation when compared to Alloy 59. Although at the first glance the increase in elongation of Alloy 22 and C-276 may appear positive, it means that the microstructure is indeed changing in an uncontrolled way during ageing and therefore the alloy performance, in particular concerning corrosion, becomes unpredictable. To avoid possible discrepancies in the experimental procedure used in this work and in that of Rebak *et al.* [1,8], e.g. plate thickness and welding parameters such as shielding, our measurements are planned for Alloy 22 and C-276.

As already mentioned, Alloy 22 as well as C-276 are expected to have lower thermal stability than Alloy 59 due to the presence of tungsten. This seems to be in accordance with the changes observed for the mechanical properties of welded Alloy 22 and C-276 on ageing at 427°C for 10 000 h. In a study involving several Ni–Cr–Mo alloys some evidence of a long range ordering after ageing at 427°C for 30 000 and 20 000 h was observed in alloys containing tungsten, i.e. Alloy 22 and C-276 [1]. At lower temperatures like those observed in a nuclear waste repository the precipitation kinetics of such solid state reactions will be sluggish. However, since the deposition of high radioactive waste involves very long periods of time (more than 10 000 years) it should be expected that precipitation may indeed occur during the lifetime planned

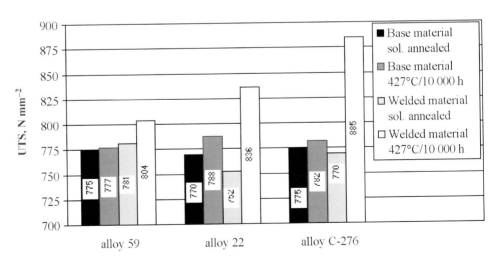

Fig. 9 Ultimate tensile strength of some Ni–Cr–Mo alloys. (Samples from left to right: base material solution annealed, base material aged at 427°C for 10 000 h, welded material solution annealed, welded material aged at 427°C for 10 000 h.)

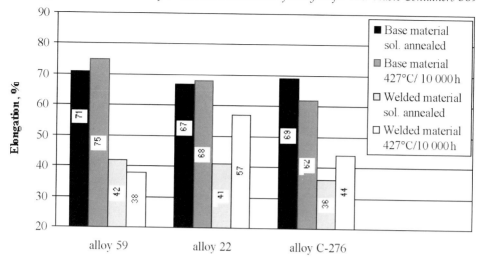

Fig. 10 *Elongation to fracture of some Ni–Cr–Mo alloys. (Samples from left to right: base material solution annealed, base material aged at 427 °C for 10 000 h, welded material solution annealed, welded material aged at 427 °C for 10 000 h.)*

for the container. For long-term disposal materials that may undergo uncontrolled changes of the microstructure are not suitable. Therefore, the more thermally stable Alloy 59 would be a better choice for the construction material.

It is well known that the resistance to localised corrosion correlates with the beneficial elements Cr and Mo according to the PRE number. The higher the PRE, the larger is the resistance to pitting and crevice corrosion. As a result of its high chromium and molybdenum content Alloy 59 exhibits a very high PRE. Accordingly, the literature data given in Table 2 make it clear that Alloy 59 has indeed higher critical pitting and crevice corrosion temperatures (CPT and CCT), and thus a superior resistance to localised corrosion than its counterpart Alloy 22 and C-276. These data refer to the solution annealed state. Alloy 22 is the currently proposed material for the container; it is a high Cr-containing alloy, however its Mo content is rather low (only 13%, while all other Ni–Cr–Mo alloys contain 16%). This strongly reduces its resistance to pitting and crevice corrosion.

Table 2. *Localised corrosion resistance in 'green death'* solution of some Ni–Cr–Mo alloys [2]*

Alloy	UNS No.	PRE	CPT (°C)	CCT (°C)
C-276	N10276	69	110	105
22	N06022	65	120	105
59	N06059	76	>120	110

*The test solution contains 11.5% H_2SO_4, 1.2% HCl, 1.0% $FeCl_3$, and 1.0% $CuCl_2$ (b.p. 103°C).

Since the resistance to localised corrosion is one of the key performance parameters used to evaluate materials for nuclear waste disposal, Alloy 59 has to be considered a strong candidate. The studies presented here demonstrate that this alloy keeps its outstanding performance after ageing. As plotted in Fig. 5 the high CPT of 130°C obtained in 'green death' solution for Alloy 59 base material remains after 10 000 h of heat treatment at 200, 300 and 427°C. Furthermore, no degradation of localised corrosion resistance is expected due to welding because the weldments show a performance similar to the base material. Only one welded coupon has failed at 120°C. Nevertheless, a CPT of 120°C is still excellent: It is higher than the CPT of C-276 base material and equal to that of Alloy 22 base material. It is well documented that Alloy 59 has excellent resistance to corrosion in a wide variety of environments [9,10]. Comparative data on the corrosion rate of Alloys 59, 22 and C-276 in various standard and non-standard boiling corrosive media can be found in Table 3 which shows that in fact, Alloy 59 exhibits practically overall the lowest corrosion rate.

In this work Alloy 59 was tested in ASTM G 28 A solution (see Fig. 6), and the results are in agreement with the data plotted in Table 3. This solution may induce intergranular corrosion in nickel base alloys and is used to detect susceptibility

Table 3. *Typical corrosion resistance rate of some Ni–Cr–Mo alloys in boiling corrosive environments (mm/year)*

Media	Alloy 59	Alloy 22	Alloy C-276
ASTM G 28 A	0.6	0.9	6.0
ASTM G 28 B	0.1	0.18	1.4
Green death	0.1	0.1	0.65
10% HNO_3	0.05	0.05	0.48
65% HNO_3	1.0	1.3	18.8
10% H_2SO_4	0.2	0.45	0.58
50% H_2SO_4	4.4	7.7	6.0
1.5% HCl	0.08	0.35	0.28
2% HCl	0.08	1.6	1.3
10% HCl	4.5	9.8	6.0
10% H_2SO_4 + 1% HCl	1.8	8.9	2.2
10% H_2SO_4 + 1% HCl (90°C)	0.08	2.3	1.0

to this form of localised corrosion as influenced by variations of processing or composition. It is a common test to evaluate mill annealed material and to evaluate the effects of subsequent heat treatments. According to this test, the corrosion rate of Alloy 59 was not affected by the long-term heat treatment; on the contrary it even seems to decrease with the annealing temperature. However, these minor variations are considered to be within the experimental error. Studies at longer annealing time (20 000 h) are now underway for a better characterisation of trends.

Figure 11 is intended to give a direct comparison between corrosion rates in ASTM G 28 A solution of Alloy 59, 22 and C-276. Data for Alloy 22 and C-276 are published elsewhere [1,8]. The corrosion rate of Alloy 59 is lower than that of C-276 by a factor of almost six and is also slightly lower than the rate of Alloy 22. Long-term ageing at 427°C does not affect the excellent resistance to corrosion shown by Alloy 59. Alloy 22 and C-276 also do not show apparent impairment of corrosion resistance after the same ageing treatment. However, this aspect has to be verified for much longer ageing time since these alloys are more prone to precipitation. As already mentioned, Alloy 59 contains little iron and silicon and does not contain tungsten in order to improve thermal stability [6]. Thus, it is expected that Alloy 22 (which contains 3% W) and C-276 (which contains 4% W) undergo microstructural changes and lose performance more easily than Alloy 59.

Based on the data obtained in this work it can be assumed that Alloy 59 will show higher long-term reliability in the harsh and unpredictable conditions expected in the nuclear waste repository. In future work the ageing time will be extended to 20 000 h and a more accurate study of the microstructure by means of transmission electron microscopy will be performed in order to further enlighten the trends discussed here.

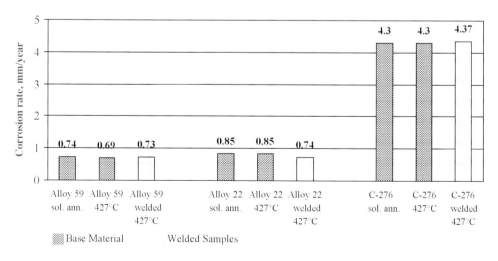

Fig. 11 *Corrosion rate of base and welded materials of some Ni–Cr–Mo alloys in ASTM G 28 A solution. Solution annealed material is compared with material aged at 427°C for 10 000 h.*

5. Conclusions

- Alloy 59 exhibits very good metallurgical stability, high strength and ductility and an excellent corrosion resistance, in particular to localised corrosion. Additionally, this alloy has excellent fabricability and weldability.

- Ageing at 200, 300 or 427°C for 10 000 h does not affects the superior performance of Alloy 59. The results presented here show that microstructure, mechanical and corrosion properties of Alloy 59 remain stable during the long-term heat treatment. In future studies the ageing time will be extended to 20 000 h.

- The properties of weldments of Alloy 59 are comparable to those of the base alloy. The ageing treatment does not deteriorate the good performance of the weldments. While the mechanical properties of welded Alloy 59 remain stable during ageing at 200, 300 and 427°C for 10 000 h, those of welded Alloy 22 and C-276 exhibit significant changes.

- Alloy 59 can be considered the best alloy of the Ni–Cr–Mo family. The comparison with the current candidate alloy for the radioactive waste disposal programme in the United States — Alloy 22 — shows great advantages for Alloy 59 with regard to thermal stability and resistance to uniform as well as localised corrosion.

- From the results presented in this work it can be concluded that Alloy 59 having superior thermal stability and the highest resistance to localised corrosion would be the best material for the nuclear waste container.

References

1. R. B. Rebak and N. E. Koon, *Corrosion '98*, Paper No.153. Published NACE International, Houston, TX, USA, 1998.
2. D. C. Agarwal, U. Brill and R. A. Corbett, *Corrosion 2001*, Paper No.1120. Published NACE International, Houston, TX, USA, 2001.
3. D. C. Agrawal and Herda, *Mater. Corros.*, 1997, **48**, 542–548.
4. M. Köhler and D. C. Agarwal, *Corrosion '98*, Paper No. 313. Published NACE International, Houston, TX, USA, 1998.
5. U. Heubner and M. Köhler, *Werkst. Korros.*, 1992, **43**, 181–190.
6. M. Köhler, *Mater. Corros.*, 1997, **48**, 528–534.
7. *Werkstoffe* (E. Hornbogen, ed.). Springer Verlag, Berlin, 1986.
8. R. B. Rebak, *Workshop on Phase Stability in Nickel Alloys for Radioactive Waste Containers*, Nickel Development Institute, San Diego, CA, USA, 19–20 March, 1998.
9. D. C. Agarwal, W. R. Herda and J. Köwer, *Corrosion 2000*, Paper No. 501, Published NACE International, Houston, TX, USA, 2000.
10. R. Kirchheiner, M. Köhler and U. Heubner, *Werkst. Korros.*, 1992, **43**, 388–395.

25

Assessment of Crevice Corrosion and Hydrogen-Induced Stress-Corrosion Cracks in Titanium–Carbon Steel Composite Overpack for Geological Disposal of High-Level Radioactive Waste

G. NAKAYAMA, N. NAKAMURA, Y.FUKAYA, M. AKASHI
and H. UEDA*

Research Laboratory, Ishikawajima-Harima Heavy Industries Co., Ltd, 1. Shin-Nakahara-cho, Isogo-Ku,
Yokohama 235-8501, Japan
*Tokyo Electric Power Company, 1-1-3 Uchisaiwaicho, Chiyoda-ku, Tokyo 100-0011, Japan

ABSTRACT

Overpacks for high-level radioactive waste (HLW) must be reliable for geological disposal for as long as 1000–10 000 years. From a study of parameters such as the critical potential for initiation of crevice corrosion, $E_{R,CREV}$ and the free corrosion potential E_{sp} in neutral aqueous environments it is concluded that composite overpacks composed of a corrosion resistant Ti alloy (Ti–0.06 Pd, or Ti–Gr.17) outer layer and a carbon steel inner layer should never be subject to crevice corrosion and hydrogen-induced stress corrosion cracking when stored deep underground environments. Hydrogen-induced stress corrosion cracking has been shown not to occur in alloys exposed to conditions of disposal based according to results based on accelerated constant current tests and constant load tests.

1. Introduction

High-level radioactive waste (HLW) discharged following the processing of spent nuclear fuels from nuclear power stations at fuel reprocessing plants [1] should be stored in canisters, cooled and then placed in ground-level intermediate storage facilities for more than fifty years. The contents are then to be transferred to overpacks for permanent storage in deep underground sites, as depicted in Fig. 1. An engineered barrier system will be used containing vitrified HLW, overpack and buffering material of bentonite, and finally a natural barrier system of host rock. HLW seepage into areas of human habitation has to be prevented for an indefinite duration. The overpack should last for as long as 1000–10 000 years [1]. Overpack materials of carbon steel and titanium alloys are presently under assessment in Japan. A titanium alloy for overpacks is examined in the present study.

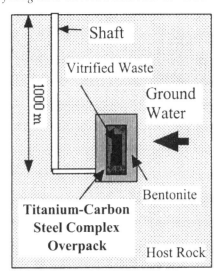

Fig. 1 *Schematic high-level nuclear waste geological disposal.*

2. Disposal Sites and Selection of Potential Alloys

2.1. Environments for Disposal Sites

Environments envisaged for geological disposal are neutral natural ground water environments free from dissolved oxygen at a depth of 1000 m at 45–55°C [1]. However, oxidation of groundwater can take place by oxygen entering via the disposal shaft. When the ground water originates from sea water, the chloride content is 3.5% [NaCl] (or 17 000 ppm [Cl$^-$]). Overpacks will be exposed to ground water of pH = 6 to 10 [2,3] when neutral water penetrates through bentonite layers (which provide a buffering action) placed between the overpacks and the host rock. Corrosion due to deep ground water may be the factor having the greatest adverse effect on overpack performance.

2.2. Selection of Overpack Material

Carbon steel [1] and titanium alloys are presently being used as overpack materials in Japan. The problems according the use of those materials are listed in Table 1.

3. Corrosion of Carbon Steel as HLW Overpack Material

A corrosion diagram for a simulated natural water environment (1 mmol L^{-1} [HCO$_3^-$] +10 ppm[Cl$^-$] at 20°C) for carbon steel is shown in Fig. 2. The transition pH from general corrosion to passivity, the depassivation pH, has been found to be pH$_d$ = 9.4 [3–5]. The pH values of natural water depends on the equilibrium between carbonate and bicarbonate salt. Carbon steel may be passivated in environments

Table 1. *Corrosion of carbon steel and titanium alloys as HLW overpack materials*

	Carbon steel	**Titanium alloy**
Assessment items	• General corrosion to passivity transition pH • Crevice corrosion, SCC under passivity domain • Corrosion propagation depth for general corrosion	• Crevice corrosion • Hydrogen induced SCC
Main factors	• pH • Role of bentonite	• Chloride concentration • Temperature • Alloys
Subordinate factors	• Chloride concentration • Temperature • Carbonate and bicarbonate salt concentration	• pH
Typical assessment index	• pH_d; depassivation pH • pH for disposal environment	• $E_{R,CREV}$; crevice corrosion repassivation potential • Hydride thickness/crack depth vs electric charge based on cathodic reactions under deaeration
Results	• Empirical E–pH diagram	• Crevice corrosion map • Hydride thickness/crack depth prediction model
Other significant factors	• MIC • Corrosion acceleration due to magnetite products	—

SCC = stress corrosion cracking; MIC = microbially induced corrosion

with pH values higher than $pH_d = 9.4$, but in this pH domain it may generate pitting corrosion and crevice corrosion.

Corrosion of carbon steel in neutral water containing dissolved oxygen involves electrochemical reactions that include anodic and cathodic reactions as follows:

$$Fe \rightarrow Fe^{2+} + 2e \tag{1}$$

$$1/2O_2 + H_2O + 2e \rightarrow 2OH^- \tag{2}$$

As an example, waste packages can be considered [6] each with dimensions 1.2 m dia. × 2.2 m length buried at a depth of 1000 m in a storage pit 3 m in diameter and

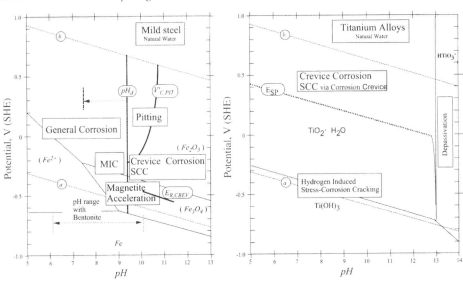

Fig. 2 *Empirical corrosion diagrams (potential–pH diagram) for carbon steel in natural water environment at 20°C, and titanium alloy.*

5 m in length per package. The steady state concentration of dissolved oxygen in ground water may be considered to be zero in such a case and corrosion by the oxygen of the package would be the only process that occurs. Based on the volume and space of any one package, a package would contain 32.8 m^3 of air. Air occupies 22.4 L/mol under standard conditions (i.e. 0°C and 1atm). The volume would consequently become 24.0 L/mol at 55°C at 1000 m underground with the oxygen concentration in atmospheric air of 21%, the volume would be 287 mol in the package. Should all the oxygen be consumed by carbon steel corrosion, with at least, 2 mol iron reacting per one mol oxygen, the steel consumed by corrosion would be 287(mol) × 2 × 55.8(g mol^{-1})/7.86(g cm^{-3}) × 10^6 = 4.07 × 10^{-3} m^3. The package surface area being 10.6 m^2, the average penetration depth, a, is computed as

$$a = 0.384 \text{ mm [6]} \tag{3}$$

Carbon steel (JIS SS400) was used as material for the bottom of a corrosion cell, with a central 100 × 100 (mm) area exposed to a 0.5 mmol L^{-1} [Cl$^-$] + 1mmol L^{-1} [HCO$_3^-$] solution of pH = 8.0. Galvanostatic tests were conducted at room temperature. The penetration depth distribution and the greatest penetration depth were determined for each specimen by ultrasonic inspection technique. Figure 3 shows corrosion penetration depth at a corrosion current of 2 A m^{-2} over a period of 2016 h. The average corrosion penetration depth was 0.573 mm, and a maximum corrosion penetration depth, d_m of nearly three times as much (1.920 mm), this being consistent with the actual observation.

Maximum corrosion depth data were examined by approximation to a Gumbel

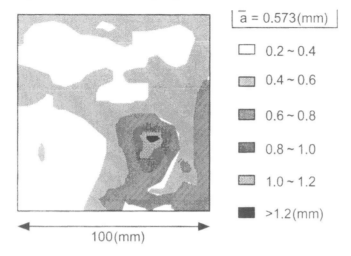

Fig. 3 *Corrosion penetration depth. Distribution (2 Am⁻², 2016 h, a = 0.573 mm, d_m =* $\overline{a} = 0.573$ (mm), $d_m =$ *1.920 mm)[6].*

Distribution as the sympototic distribution of the maximum value distribution. The cumulative distribution function, $F(x)$ may be expressed in terms of the double exponential function by the following equations:

$$F(x) = \exp\left[-\exp\{-(x-\lambda)/\alpha\}\right] \tag{4}$$

where λ is the location parameter (= distribution mode), and α is a scale parameter. Figure 4 summarises the relations between λ and average penetration depth, a, and α/λ and a as determined in this study. The empirical relation between λ and α is

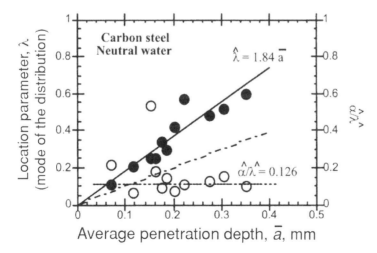

Fig. 4 *Relation of average penetration depth, a, with maximum penetration value, λ, and of a with α/λ[6].*

constant and independent of corrosion propagation. With increase in α, α becomes greater so that their ratio is essentially constant.

$$\lambda = 1.84\, a \tag{5}$$
$$\alpha/\lambda = 0.126 \tag{6}$$

An overpack surface area of 10.6 m² was found to greatly exceed the specimen area of 0.01 m². Yet, estimation of the mode of the greatest localised penetration depth distribution, λ_m could be determined here by assuming a return period*, T, to be proportional to a surface area ratio of $T = 10.6/0.01 = 1{,}060$. $\alpha = 0.126\lambda = 0.126 \times (1.84 \times 0.384) = 0.0890$. Namely,

$$\lambda_m = \lambda + \alpha \ln T = 1.33 \text{ mm} \tag{7}$$

Gumbel distribution of greatest penetration depth could be expressed as,

$$F_m(x) = \exp\left[-\exp\{-(x-1.33)/0.0890\}\right] \tag{8}$$

This assumes one disposal lot consists of 20 000 packages, with only one of them experiencing through-penetration. The requirement for corrosion allowance, d given as 99.995% in the distribution of eqn (8) is thus computed as,

$$D = \lambda_m - \alpha \ln[-\ln\{F_m(x)\}] = 2.21 \text{ mm} \tag{9}$$

A corrosion margin of 1000 years is thus 2.21 mm of general corrosion [6].

Carbon steel has been shown to experience crevice corrosion in weak alkaline solution environments of pH = 10.5 at a chloride concentration of 10 ppm [3]. With carbon steel used as overpack material for geological disposal of HLW, the pH value of the disposal environment should be in the range pH = 6~10 [2,3,5,7], assuming bentonite also to be present. But an aqueous environment on the surface of the overpacks will change continuously with time, and cannot be defined precisely. Therefore, carbon steel at a given time may not necessarily undergo general corrosion after a long period because the pH of the disposal environment will rise above pH_d, into the passivated region, where carbon steel may very well be passivated [3–5,7].

A common corrosion preventive measure in general industrial equipment is to adjust the environmental pH values into weakly alkaline to alkaline regions in which passivation can occur. When carbon steel is in the passivation region during actual applications, the corrosion may be quite small in terms of general corrosion. However, pitting corrosion and crevice corrosion may occur in industrial applications that are presumed to be in the passive region; daily maintenance and inspection schedules should detect any corrosion incident, which then could be appropriately treated.

Crevice corrosion will be quantitatively evaluated in the future for the case of carbon steel as the overpack material. The corrosion of such material during 1000 years should be considered to confirm its reliability. The corrosion margin for

---

*The return period, T, is the ratio between the area of the specimen used in this test and that for the actual structure.

overpack material is presently not adequate and present studies indicate the incidence of crevice corrosion to be quite high. Considerably more data on propagation and countermeasures should be made available to establish a useful quantitative model.

Carbon steel may be used only for cases of general corrosion, subject to confirmation of the absence of MIC (microbiologically-induced corrosion). Corrosion rates under MIC can be 0.3 mm/year [8] in special cases. Also, corrosion acceleration due to magnetite corrosion products [9,10] can occur in a completely deaerated environment. In the case of 14.58 g cm^{-2} magnetite corrosion products on the surface of specimen which is equivalent to a corrosion depth of 13.41 mm, the corrosion rate will rise up to 0.9 mm/year [9]. Thus, carbon steel is therefore shown not to be feasible as an overpack material. The present study was therefore conducted to assess whether titanium alloys can be used as adequate material.

4. Localised Corrosion of Titanium Alloys as HLW Overpack Material

The stability of titanium alloys is due to the presence of a surface film of $TiO_2 \cdot H_2O$, as is evident from Fig. 2. This stability would be compromised most by crevice

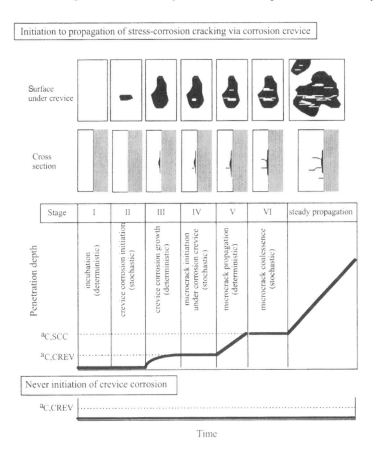

Fig. 5 *Model for stress-corrosion crack initiation inducement by crevice corrosion.*

corrosion. Stress-corrosion cracks are likely to arise from crevice corrosion in a neutral chloride environment, as demonstrated in Fig. 5. In this study, search was made to find a material, which would never permit any kind of localised corrosion, such as pitting corrosion, crevice corrosion, and stress-corrosion cracking. Pitting corrosion would not occur in a natural water environment, and this would facilitate material selection.

4.1. Methods

The test materials selected (i.e. 4 types of titanium alloys) are shown in Table 2. The crevice corrosion region was determined based on critical potential for crevice corrosion, $V_{C,CREV}$, and free-corrosion potential, E_{SP}, as shown in Fig. 6 [11]. Initiation conditions for localised corrosion of titanium alloys were evaluated based on differences between E_{SP}, and crevice corrosion repassivation potential [12], $E_{R,CREV}$,

Table 2. Chemical compositions of samples (mass%)

	C	H	O	N	Fe	Pd	Mo	Ni
Ti–Gr.1	0.007	0.0020	0.062		0.032	–	–	–
Ti–Gr.7	0.008	0.0007	0.090	0.004	0.046	0.22	–	–
Ti–Gr.12	0.012	0.0010	0.125		0.078	–	0.3	0.86
Ti–Gr.17	0.010	0.0006	0.044	0.003	0.029	0.06	–	–

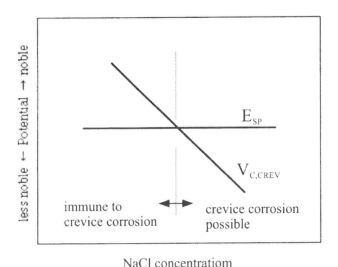

Fig. 6 Determination of immunity to crevice corrosion based on critical potential for crevice corrosion, $V_{C,CREV}$, and free-corrosion potential E_{SP} [11].

the critical potential for initiation of crevice corrosion at a given chloride concentration. If E_{SP} is more noble, i.e. of a higher potential, than $E_{R,CREV}$, crevice corrosion may occur.

The free-corrosion potential, E_{SP}, determined for free-surface specimens immersed in aerated aqueous 1 mmol L^{-1} [HCO$_3$] has been shown to become more noble with time, and attain an eventual free-corrosion potential, E_{SP}, within a few months [11]. On the other hand, the corrosion-crevice repassivation potential [12], $E_{R,CREV}$, should be determined by cyclic polarisation for the creviced specimens [12,13] as shown in Fig. 7 [13]. To study this, a specimen with a pair of polysulfone gasket crevices was immersed in the test solution, and the electrode potential was varied in the noble direction at a scanning rate of 30 mV min^{-1} so that crevice corrosion would take place under the potentiokinetic polarisation. The corrosion current representing propagation of crevice corrosion reached 50 μA, and was held at this level for 2 h to allow the crevice corrosion to progress further; the potential was then reduced by 10 mV; and potentiostatically held for some time to see if the corrosion current would increase by 5 μA, then indicating continuation of corrosion-crevice propagation; if this happened the potential was reduced further by 10 mV. This process was repeated until increase in the corrosion current was no longer recognised after 2 h holding, and then the most noble potential at this point was taken as $E_{R,CREV}$ [12,13]. A potentiostat controlled by microcomputer was used in this operation.

4.2 Determination of E_{SP}

Electrode potential vs time, and determination of free-corrosion potential, E_{SP} are presented in Fig. 8 [13], and, pH dependency of E_{SP} is indicated in Fig. 9 [14]. Electrode

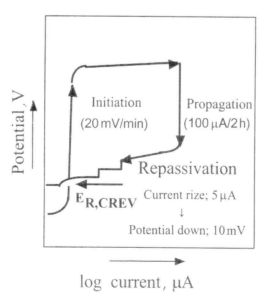

Fig. 7 Potential–current curve for cyclic polarisation in determination of E$_{R,CREV}$ *[13].*

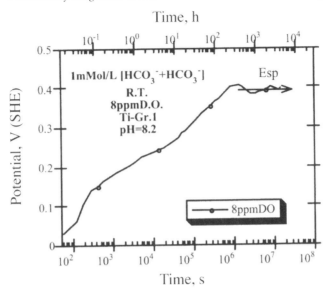

Fig. 8 *Electrode potential vs time, and determination of* E_{SP} *for Ti–Gr.1 in aqueous solution* [13]. *DO = dissolved oxygen.*

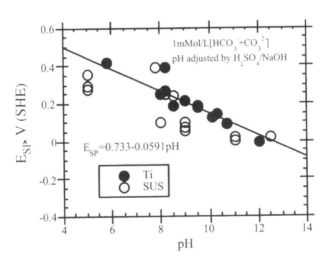

Fig. 9 *Effects of pH (5~13), temperature (20~90°C), on free corrosion potential,* E_{SP} [14].

potential was noted to increase to +0.4 V vs SHE after several hundred hours. E_{SP} in natural water may be determined solely from pH and not alloy composition or temperature from room to 90°C, as follows:

$$E_{SP} \text{ V(SHE)} = 0.733 - 0.0591 \text{ pH [11,14]} \tag{10}$$

where SHE refers to the Standard Hydrogen Electrode. E_{SP} corresponding to neutral ground water may thus be computed as $E_{SP} = +0.32$ V at pH = 7.0 from eqn (10).

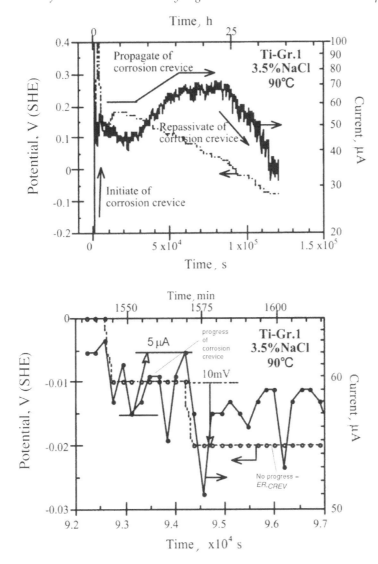

Fig. 10 *Potential and current vs time for creviced specimen of Ti–Gr.1 in 3.5% NaCl at 90°C, and determination of crevice corrosion repassivation potential,* $E_{R,CREV}$, *for the critical potential for initiation of crevice corrosion [13].*

4.3. Determination of $E_{R,CREV}$

Potential and current vs time for creviced Ti–Gr.1 in 3.5% NaCl at 90°C and determination of crevice corrosion repassivation potential, $E_{R,CREV}$, for the critical potential for initiation of crevice corrosion inducement are shown in Fig. 10 [13]. At –0.01 V (SHE), current increased by more than 5 µA to 62 µA, so the potential was lowered by 10 mV to –0.02 V and held for 2 h. There was no tendency to further increase in current during the 2 h at –0.02 V, and so $E_{R,CREV}$ was determined to be –0.02 V (SHE) [13].

Figure 11 shows the relationship between chloride concentration and $E_{R,CREV}$ for Ti–Gr.1 and Ti–Gr.12 at 90°C [11, 13–16]. $E_{R,CREV}$ clearly increased with decrease in chloride concentration. In Fig. 12 comparison is made of $E_{R,CREV}$ and crevice corrosion initiation lifetime of the creviced specimen of Ti–Gr.1, as determined in potentiostatic tests at 90°C. Crevice corrosion initiated within a few minutes or a few hours at potentials higher than $E_{R,CREV}$. Crevice corrosion initiation lifetimes were longer at potentials only just above $E_{R,CREV}$. No crevice corrosion was initiated even after immersion for 2000 h at potentials less than $E_{R,CREV}$. Thus, $E_{R,CREV}$ is found to be the critical potential at which crevice corrosion is initiated.

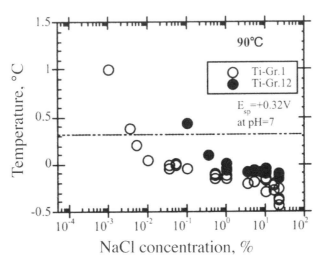

Fig. 11 *Effects of NaCl concentration on* $E_{R,CREV}$ *for Ti–Gr.1 and Ti–Gr.12 in 90°C [13].*

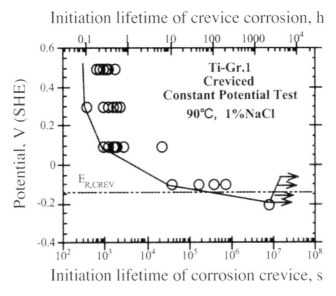

Fig. 12 *Effects of electrode potential on crevice corrosion initiation lifetime of Ti–Gr.1 in 1% NaCl solution at 90°C.*

4.4. Alloy Selection based on Degree of Crevice Corrosion and HLW Environment

The point at which E_{SP} = +0.32 V intersects with the $E_{R,CREV}$ determines the critical chloride concentration of Ti–Gr.1 to be 0.004% NaCl at 90°C. In this manner, the critical chloride concentration for an alloy in an aqueous environment may be determined. Figure 13 shows crevice corrosion initiation for Ti–Gr.1 in a neutral natural water environment at a given temperatures and chloride concentrations. Ti–Gr.1 was quite clearly shown to undergo crevice corrosion compared with the envisaged HLW disposal environment.

Alloys with high corrosion resistance should thus be used as overpack material. The dependency of $E_{R,CREV}$ of Ti–Gr.12 (Ti–Ni–Mo alloys) on chloride concentration at 90°C is also superimposed in Fig. 11. $E_{R,CREV}$ for Ti–Gr.12 exceeded that for Ti–Gr.1. A critical chloride concentration at E_{SP} = +0.32 V was noted for 0.2 % NaCl. The corrosion resistance of those alloys in this environment would thus appear insufficient. The Ti–Pd alloy (Ti–Gr.17) has never been reported to undergo crevice corrosion at 90°C. $E_{R,CREV}$ for Ti–Gr.17 at 150 °C was found to be +0.178 V.

The critical temperature for initiation of crevice corrosion for Ti–Gr.1, 7, 12, and 17 alloys (chemical compositions shown in Table 2) were plotted against chloride concentration in Fig. 14. The curves for Ti–Gr.1 and Ti–Gr.12 intersected with the HLW disposal environment region indicating susceptibility to crevice corrosion. Ti–Gr.7 and 17 underwent crevice corrosion, but were free from crevice corrosion, in that the curves did not intersect with the HLW disposal environment region. Thus, Ti–Gr.17 containing less Pd than Ti–Gr.7 is shown to be usable as an overpack material.

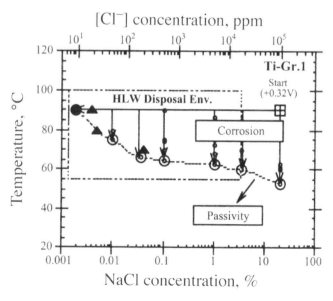

Fig. 13 *Repassivation NaCl concentration for crevice corrosion, $C_{R,CREV}$, repassivation temperature, $T_{R,CREV}$, for E = +0.32 V, $E_{R,CREV}$ = +0.32 V for Ti–Gr.1 and comparison of HLW disposal environment.*

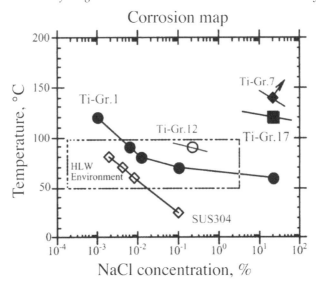

Fig. 14 *Crevice corrosion initiation temperature vs NaCl concentration for titanium alloys in HLW disposal environment.*

5. Titanium Alloy Resistance toward Hydrogen Induced Stress-Corrosion Cracking

5.1. Hydrogen Induced Stress-Corrosion Cracking

A deep ground water environment is primarily completely deaerated. Essentially the same situation is created when oxygen entrapped by overpacks is consumed by corrosion soon after disposal.

Cathodic reactions on a titanium surface reduce water under complete deaeration, and slowly generate hydrogen over prolonged periods, as:

$$2H_2O + 2e^- \rightarrow H_2 \ (2[H]^*) + 2OH^- \qquad (11)$$
$$[H^*]; \text{ atomic hydrogen}$$

The hydrogen generated is mostly gaseous, and some is absorbed into the titanium matrix, which subsequently precipitates as titanium hydride (TiH_x) [17] or TiH_2 [18,19]. The titanium alloy may become brittle through uniform precipitation of the titanium hydride phase throughout the alloy following uniform diffusion of absorbed hydrogen to the alloy [20]. Titanium hydride may also precipitate as a titanium hydride layer on the alloy surface [18,21,22]. The capability of absorbed hydrogen for initiating and propagating hydrogen-induced stress corrosion cracking in titanium should thus be evaluated. Titanium alloys with a single α-phase, such as commercially pure titanium and Ti–Pd alloys can experience hydride forming type stress corrosion cracking though to a very limited extent. Previous studies on cracking were carefully examined here so as to establish a quantitative lifetime assessment model that could be supplemented with additional considerations.

5.2. Estimation of the Amount of Hydrogen Generation
over a 1000-Year Period

The cathodic current density for maintaining the titanium alloys passive in a simulated aqueous underground environment at 90°C was reduced to 6.8×10^{-5} Am^{-2} after 10^5 s, as shown in Fig. 15 [23].

Figure 16 shows temperature and time dependences of electric charge density corresponding to hydrogen generation by the cathodic reactions, based on

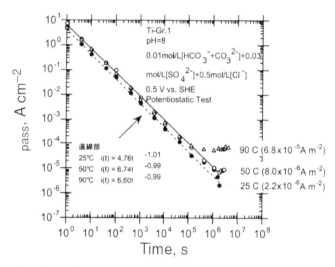

Fig. 15 *Time dependence of current density for maintaining the alloy passivity [23].*

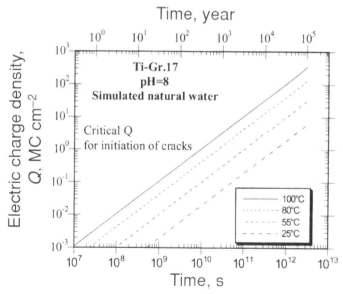

Fig. 16 *Temperature and time dependence of electric charge density corresponding to hydrogen generation based on cathodic reactions.*

temperature dependence [21,23] of current density for maintaining the passive state of the Ti–Gr.1. The electric charge density at 100°C after exposure period of one year was 3.4×10^3 Cm^{-2}, 3.4 MCm^{-2} (mega coulombs per sqaure metre) at 1000 years, and 34 MCm^{-2} at 10 000 years. The value for electric charge density is 0.3 MCm^{-2} at 55°C at 1000 years.

5.3. Experimental

The central portion of a round bar tensile specimen of 5 mm dia. and 10 mm in length was polished, and the entire surface except a test area of 10 mm in length was coated to prevent corrosion. The specimen was placed on a Uni-axle Constant Load (UCL) tensile testing machine equipped with a test solution chamber at constant applied stress of 200 MPa, as indicated in Fig. 17. The aqueous test solution, containing 3.5% NaCl at 80°C, was circulated through the test chamber while a constant current of –4 Am^{-2} was applied to the specimen.

Here, the test duration is taken as a test parameter for defining the amount of hydrogen loading on the specimen. The actual electric current density and charge density were determined after measuring the area of the specimen exposed to the environment.

The surfaces of the specimens were first examined by a microscope. Then, specimens were cut parallel to the axis, polished and etched so as to measure the thickness of the titanium hydride layer formed on the surface and the depth of cracks generated on the cross section. A reagent containing HF/HNO$_3$ was used for surface etching.

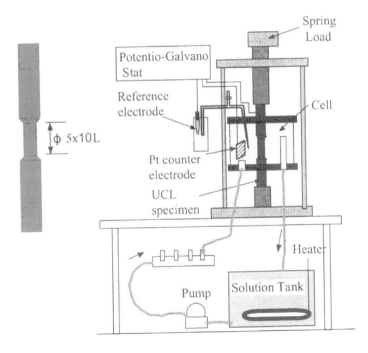

Fig. 17 *Setup for constant load, constant current test.*

5.4. Titanium Hydride Layer and Crack Formation at Constant Current and Loading

Figure 18 [21] shows microcracks on cross sections of specimens held galvanostatically in a hydrogen generation region with the electric charge density of 12.3 MCm^{-2}. The white and acicular phases in the Figure were those of titanium hydride. Cracks were perpendicular to the stress direction. Growth of the titanium hydride phase is illustrated in Fig. 19.

The mean titanium hydride layers thickness, H_d (μm), was measured. Dependence of H_d on electric charge density, Q (MCm^{-2}), is plotted in Fig. 20, in which H_d is clearly seen to increase with Q [21], for Ti–Gr.1 in accordance with the following equation [22]:

$$H_{d\,\text{Ti.Gr}-1}(\mu m) = 6 \times (Q\,(\text{MCm}^{-2}) - 1)^{0.5} \tag{12}$$

The same relationship for Ti–Gr.17 applies as follows:

$$H_{d\,\text{Ti.Gr}-17}(\mu m) = 12 \times (Q\,(\text{MCm}^{-2}) - 1)^{0.5} \tag{13}$$

which gives a titanium hydride thickness as nearly twice that of Ti–Gr.1. A plausible mechanism for the involvement of Pd in the formation of a thicker layer cannot be put forward at present.

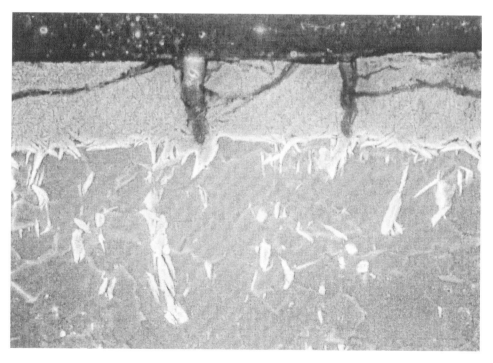

Fig. 18 SEM observation of a cross section after constant load test (Ti–Gr.1; Stress: 343 MPa; electric charge density: 12.3 MCm^{-2}) [21].

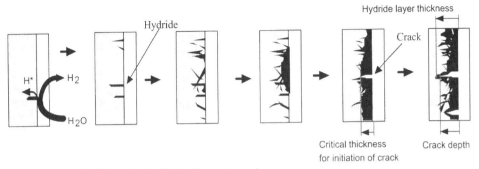

Fig. 19 Growth of titanium hydride phase and cracks.

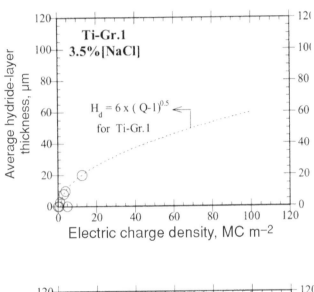

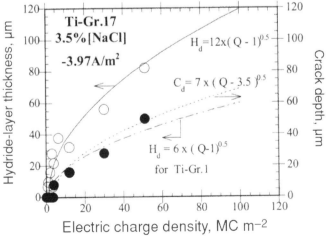

Fig. 20 Electric charge dependence of titanium hydride layer thickness and crack depth.

The dependence of Q on the greatest depth for cross sections, C_d parallel to specimen axis is shown in Fig. 20. For Ti–Gr.17 may be expressed as

$$C_{d\,EG.17}(\mu m) = 7 \times (Q\,(MCm^{-2}) - 3.5)^{0.5} \tag{14}$$

No crack could be detected below $Q_c = 3.5\,(MCm^{-2})$, and above the Q_c, crack depth was initially increased with increases in Q, as apparent from Fig. 20. Figure 21 illustrates the relationship between crack depth and titanium hydride thickness.

Figure 21 shows the smallest (critical) Q at which crack initiation can be observed, and shows that critical crack depth, C_d, corresponding to Q is always one half the mean thickness of the titanium hydride phase. A crack is thus smaller than the titanium hydride layer thickness. The minimum (critical) titanium hydride layer thickness required for crack initiation is shown to be about 20 µm (see Fig. 21). The mechanism for initiation at this thickness is concluded to be essentially the same as for SCC induced by the so-called Tarnish Rupture mechanism [24]. Progressive crack growth was noted here to continue with continued alloy exposure to corrosion promoting factors. Crack depth with exposure time of 1000–10 000 years should be less than 50 µm, a value too small to affect material performance significantly.

Possible corrosion/crack depths are specified in Table 3.

6. Conclusions

Various titanium alloys as the corrosion resistant layer of composite overpack materials were examined with respect to the effects of a deep ground water

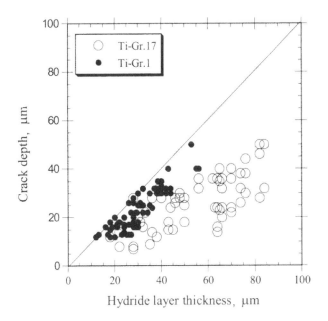

Fig. 21 *Relationship of crack depth to titanium hydride layer thickness at the site of cracks.*

Table 3. *Surmised corrosion/crack depth of titanium alloy and carbon steel overpack in 1000 years*

Material	Type of damage	Corrosion damage			Evaluation/ problems	Selection
		per year	1000 year	10 000 year		
Titanium alloy (Ti–Gr.17)	Crevice corrosion	–	None	None	Clearance	Recommendations
	SCC via crevice corrosion	–	None	None	Clearance	
	Hydride thickness	–	18.6 μm	<80 μm	Clearance	
	Hydrogen induced SCC	–	None	<50 μm	Clearance	
Carbon steel	General corrosion	–	Ave. 0.384 mm Max. 2.21 mm		Clearance	Unsatisfactory
	MIC	0.3 mm	?	–	Misgivings	
	Localised corrosion	Several mm	?	–	Misgivings	
	Acceleration by magnetite	0.9 mm	?	–	Misgivings	

environment likely to compromise material performance. Certain titanium alloys were found in the present study to qualify as such materials in place of carbon steel.

(1) At the early stage of HLW disposal, the environments are oxidising because oxygen entrapped in the disposal shaft can reach deep underground. Commercially pure titanium (Ti–Gr.1) readily undergoes crevice corrosion at this stage. A resistant titanium alloy containing 0.06% Pd (Ti–Gr.17) experienced no crevice corrosion in the present study, thus showing it to qualify for use as an overpack.

(2) With the eventual consumption of entrapped oxygen, the environment will have only a reducing effect. The titanium alloy may then as a result undergo hydrogen-induced stress corrosion cracking initiation and propagation promoted by the hydrogen produced over a long period via cathodic reactions that occur under a condition of complete deaeration. Titanium alloys incur hydrogen-induced cracking with current densities of a few MCm^{-2} through hydride layer formation. Total hydrogen generated in 1000 years would correspond to $3.41\ MCm^{-2}$. Accumulated hydride layer thickness in this period would be 9.3 μm for Ti–Gr.1 and 18.6 μm for Ti–Gr.17, both those values being

less than critical hydride layer thickness required to induce cracking. Thus, from a practical view, no cracking due to hydride layers would occur.

(3) The titanium (Ti–Gr.17)-carbon steel composite overpack should thus prove suitable for high-level nuclear waste disposal.

7. Acknowledgement

The authors gratefully acknowledge the contribution of Joint utility studies to some part of present work.

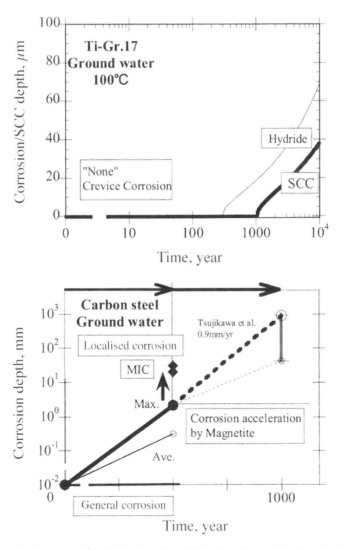

Fig. 22 *Surmised corrosion/crack depth vs time of titanium alloy and carbon steel.*

References

1. JNC report, "H12, Project to Establish the Scientific and Technical Basis for HLW Disposal in Japan", JNC, 1999.
2. H. Ishikawa, K. Amemiya, Y. Yusa and N. Sasaki, *Int. Clay Conf.*, Strasbourg, France (1989).
3. G. Nakayama and M. Akashi, *Mat. Res. Soc. Symp. Proc.*, 1992, **257**, 415, Material Research Society.
4. G. Nakayama and M. Akashi, *Mat. Res. Soc. Symp. Proc.*, 1991, **212**, 279, Material Research Society.
5. G. Nakayama and M. Akashi, *Zairyo-to-Kankyo* [in Japanese], 2000, **49**, 222, JNCE.
6. M. Akashi, T. Fukuda, and H. Yoneyama, *Mat. Res. Soc. Symp. Proc.*, 1990, **176**, 525, Material Research Society.
7. Y. Fukaya and M. Akashi, *Mat. Res. Soc. Symp. Proc.*, 1999, **556**, 871, Material Research Society.
8. F. Kajiyama, K. Okamura, Y. Koyama and K. Kasahara, ASTM Publications Code (PNC) 04-012320-27, p.266 (1994).
9. Y. Kojima, T. Yabuuchi and S. Tsujikawa, *Proc. Zairyo-to-Kankyo '98* [in Japanese], 1998, p.233, JSCE.
10. Y. Fukaya and M. Akashi, *Proc. 48th Zairyo-to-Kankyo* [in Japanese], 2001, p.329 JSCE.
11. T. Fukuda and M. Akashi, *Proc. Nuclear Waste Packaging Focus '91*, p.201, Las Vegas, Nevada, American Nuclear Society, (1991).
12. S. Tsujikawa and Y. Hisamatsu, *Corr. Eng.* [in Japanese], 1980, **29**, 37.
13. G. Nakayama, *Proc. 129th Corrosion and Protection Symp.*, "Application and problem of $E_{R,CREV}$ method for titanium alloys" [in Japanese], JSCE (2000).
14. G. Nakayama, T. Fukuda and M. Akashi, *Proc. Fusyoku-Bousyoku '93* [in Japanese],1993, p.415, JSCE.
15. T. Fukuda and M. Akashi, *Corrosion '92*, Paper No.92, NACE, Houston, TX, USA (1992).
16. M. Akashi, G. Nakayama and T. Fukuda, *Corrosion '98*, Paper No.158 NACE, Houston, TX, USA (1998).
17. G. Sanderson, *Corros. Sci.*, 1966, **6**, 541.
18. N. Nakamura, M. Akashi, Y. Fukaya, G. Nakayama and H. Ueda, *Corrosion 2000*, Paper No.195, NACE, Houston, TX, USA (2000).
19. *Binary Alloy Phase Diagrams*, 2nd edition (T.M. Massalski, ed.), ASM International, Materials Park, Ohio, USA (1990).
20. K. Shimogori, *Proc. 46th Corrosion and Protection Symp.*, p.23, JSCE (1983).
21. M. Akashi, G. Nakayama and K. Murakami, *Proc. 45th Zairyo-to-Kankyo* [in Japanese], 1998, p.301, JSCE.
22. I. I. Phillips, P. Poole and L. L. Shreir, *Corros. Sci.*, 1974, **14**, 533.
23. N. Nakamura, M. Akashi, Y. Fukaya, G. Nakayama, M. Kimoto and H. Ueda, *Proc. Zairyo-to-Kankyo 2000* [in Japanese] p.63, JSCE.
24. A. J. McEvily, Jr. and A. P. Bond, *J. Electrochem. Soc.*, 1965, **112**, 181.

26

The Performance of a Copper Canister for Geologic Disposal of Spent Nuclear Fuel in Granitic Rock

L. O. WERME and P. SELLIN

Svensk Kärnbränslehantering AB (SKB) Box 5864, SE-10240 Stockholm, Sweden

ABSTRACT

The evolution of the geochemical conditions in a repository for spent nuclear fuel in Sweden is outlined. The experimental and modelling backgrounds to lifetime predictions for these conditions for copper spent fuel disposal canisters are summarised. Much is known about the general corrosion behaviour of copper under repository conditions. In a sealed repository, the extent of general corrosion is limited by the general lack of oxidants. Because of the limited amount of available oxidant, general corrosion will not limit the canister lifetime.

For pitting corrosion, analyses based on literature pit depth data (either the pitting factor or extreme value approaches) implicitly assume that pits propagate indefinitely. Predictions of pit depth based on these approaches must be considered as conservative.

1. Introduction

Copper has been the corrosion barrier of choice for the canister in the Swedish nuclear waste disposal programmes for over 20 years. Werme [1] has recently reviewed the background to the choice of copper. The materials selection is based on the analysis of the geochemical conditions at depth in granitic rock, the interaction between the groundwater and the components in the engineered barrier system and on the analysis of the mechanical situation at the repository level. The design goal has been to isolate the spent nuclear fuel in a canister that would withstand known corrosion processes for at least 100 000 years. Based on these analyses, SKB has designed a reference waste canister, which has an outer 50 mm thick corrosion barrier of copper and an inner load-bearing nodular cast iron insert. Figure 1 shows this canister.

The plans are to bury this canister at a depth of 500–700 m in the Swedish bedrock in deposition holes, bored from the floor of access tunnels, surrounded by a 35 cm thick buffer layer of compacted sodium bentonite. After burial, the canister will be subjected to conditions that will vary after water saturation from initially oxygenated conditions to oxygen-free, reducing conditions. The corrosion behaviour of the copper canister under these conditions has been studied experimentally and by modelling for more than 20 years. Recently, a review of the current knowledge of copper corrosion under repository conditions has been published and the present paper presents, to a large extent, highlights from that more comprehensive review [2].

Fig. 1 *SKB's canister for geologic disposal of spent nuclear fuel, version for BWR fuel.*

2. Geochemical Conditions

The chemistry of the groundwater at the possible repository sites in Sweden is basically well known. Tables 1 and 2 present the predicted values for two Swedish sites at closure of the repository, after resaturation, and up to 10 000 years into the future. The data in the table were used in the SR-97 post closure safety assessment [3].

In the long term (i.e. over a glacial cycle of 150 000 years), major changes in chemical composition can be expected as a result of the climate changes. Based on the climate scenarios defined in SR 97 we might expect:

Temperate/boreal condition: During the period when the climate is gradually changing, shoreline displacement is expected to be the only process of any significant importance for the groundwater composition. This might change the chemistry by replacing the freshwater by brackish or saline water. The salinity of the sea water may vary from brackish to salt. During some periods, the Baltic may be a lake. Based on current knowledge, one must assume that for long periods of time (of the order of thousand years) the salinity at sites close to the current shoreline may correspond to that currently observed in the Atlantic.

During periods of permafrost: Groundwater turnover is expected to be lower than under current conditions. Almost impermeable areas of frozen ground are expected to force groundwater movements to greater depth. Another important process related to freezing is that solutes are frozen out. Freezing-induced salt exclusion and the reduction in groundwater turnover can contribute to a significant increase in salinity. The salinity is expected to reach slightly more than 35 g L^{-1}.

Table 1. *Predicted chemical composition of the groundwater at repository depth in the Simpevarp area in Oskarshamn (at coast). The concentrations are given in mol L^{-1}*

Constituent	At closure	After resaturation (<100 years after closure)	10 000 years into the future
pH	6–8	7.7	8–9
E_h (mV)	oxic to –400	–300	–200 to –300
Na^+	$(4.4–13.1) \cdot 10^{-2}$	$9.1 \cdot 10^{-2}$	$(0.4–8.7) \cdot 10^{-2}$
K^+	$(1.3–5.1) \cdot 10^{-4}$	$2.0 \cdot 10^{-4}$	$(0.5–2.6) \cdot 10^{-4}$
Ca^{2+}	$(2.5 – 7.5) \cdot 10^{-2}$	$4.7 \cdot 10^{-2}$	$(0.05–5.0) \cdot 10^{-2}$
Mg^{2+}	$(0.4–8.2) \cdot 10^{-3}$	$1.7 \cdot 10^{-3}$	$(0.04–1.7) \cdot 10^{-3}$
HCO_3^-	$(0.2–16.4) \cdot 10^{-3}$	$0.2 \cdot 10^{-3}$	$(0.2–0.4) \cdot 10^{-3}$
Cl^-	$(0.8–2.8) \cdot 10^{-1}$	$1.8 \cdot 10^{-1}$	$(0.06–1.4) \cdot 10^{-1}$
SO_4^{2-}	$(1.0–6.3) \cdot 10^{-3}$	$5.8 \cdot 10^{-3}$	$(0.01–4.2) \cdot 10^{-3}$
HS^-	$(0–0.3) \cdot 10^{-3}$	$0.5 \cdot 10^{-5}$	$(0–0.3) \cdot 10^{-4}$

During glaciation: The ice sheet governs the boundary conditions for groundwater flow. In the melting zone and at the ice margin, high water pressures are expected to occur locally. This will drive glacial meltwater deep down to the subsurface. The composition of the glacial melt water, when it has reached repository depth, is 5–10 mg L^{-1} of sodium, calcium, magnesium, sulfate and chloride, 20–40 mg L^{-1} of bicarbonate and pH 8–9. The water is anoxic but does not contain any reducing compounds of iron, manganese or sulfide.

The water that actually contacts the waste canister will have been conditioned through its contact with the compacted bentonite. Bentonite materials normally consist of the clay mineral montmorillonite and accessory minerals, e.g. feldspars, quartz and calcite. The desired physical properties are mainly governed by the montmorillonite, which also dominates the bentonite material in commercial products. The type and amount of accessory minerals vary quite substantially between the different commercial products depending on the mining site. Montmorillonite has a relatively low solubility, and the accessory minerals in combination with the added water solution, therefore, determine the pore-water composition. The pore water composition will also vary with time since the accessory minerals may have solubilities that are temperature dependent. Muurinen *et al.* have determined the pore water chemistry in compacted bentonite and they observed substantially increased levels of sulfate and total carbon (most likely carbonate/bicarbonate) [4]. Bruno *et al.* have modelled the interaction of bentonite and

Table 2. Predicted chemical composition of the groundwater at repository depth in the Forsmark-North Uppland area (inland). A difference of the conditions after resaturation compared with Table 1 is related to the location (at coast or inland). The concentrations are given in mol L^{-1}

Constituent	At closure	After resaturation (<100 years after closure)	10 000 years into the future
pH	6–8	7.0–7.9	7–9
E_h (mV)	oxic to –400	–200/–250	–200 to –300
Na$^+$	$(1.3–8.7)·10^{-2}$	$(1.2–7.4) · 10^{-2}$	$(0.4–4.3) · 10^{-2}$
K$^+$	$(0.5–3.3)·10^{-4}$	$(0.5–3.3) · 10^{-4}$	$(0.5–2.6) · 10^{-4}$
Ca^{2+}	$(0.4–4.1) · 10^{-2}$	$(0.4–4.1) · 10^{-2}$	$(0.05–2.5) · 10^{-2}$
Mg^{2+}	$(0.7–4.5) · 10^{-3}$	$(0.7–4.5) · 10^{-3}$	$(0.2–4.1) · 10^{-3}$
HCO$_3^-$	$(0.8–4.9) · 10^{-3}$	$(0.8–4.6) · 10^{-3}$	$(0.3–0.7) · 10^{-3}$
Cl$^-$	$(0.1–1.4) · 10^{-1}$	$(0.2–1.6) · 10^{-1}$	$(0.06–1.4) · 10^{-1}$
SO$_4^{2-}$	$(0.4–4.2) · 10^{-3}$	$(0.5–3.9) · 10^{-3}$	$(0.01–4.2) · 10^{-3}$
HS$^-$	$0–0.3 · 10^{-3}$	$0–<0.3 · 10^{-6}$	$0–0.3 · 10^{-4}$

groundwater by applying instantaneous equilibrium and a number of exchange cycles with incoming groundwater [5]. Their results showed an increase in sulfate and carbonate but that otherwise, the reacting groundwaters do not have a marked effect on the actual bentonite pore-water composition in a million year perspective. The presence and activity of microorganisms in Fennoscandian groundwater have been studied extensively. Pedersen has reviewed the state of knowledge [6,7]. The microbial activities at the repository horizon will have a large effect on the geochemical conditions. Puigdomènech *et al.* have shown that microbes in hard rock aquifers and tunnels have a dominating role in reducing oxygen [8]. The microbial activity will, therefore, be of importance in rapidly re-establishing the reducing conditions that prevailed at the repository depth before the construction work started. Potentially detrimental effects of the microbial activity can, however, also be envisaged. Under oxygen-free conditions, copper will corrode in sulfide-containing water. The worst case scenario for corrosion of the copper canister would be if sulfate-reducing bacteria formed biofilms on the canisters or grew intensively in the buffer close to the canister. The corrosion process would be controlled by the transport of sulfate to the canister, if enough hydrogen or degradable organic carbon were available for such growth. This could lead to considerably accelerated corrosion since the transport of sulfate is expected to be much faster than the transport of sulfide. Motamedi *et al.* [9] and Pedersen *et al.* [10,11] have shown, however, that the microbes will not survive in the compacted bentonite buffer that will surround the canister in the deposition hole. Wersin *et al.* [12] have calculated the evolution of oxygen in the nearfield taking into account the important processes affecting oxygen migration, diffusion, and oxidation of pyrite and dissolved Fe(II). The effect of increased temperature (60°C) and hydrostatic pressure were allowed to dissolve all the available O_2 in the porewater. The time estimated for the O_2 concentration to decrease to 1% of the initial value ranged between 7 and 290 years, with the elapsed time at which transition to anoxic conditions occurred was estimated to be within the same time range. Thus, the oxidation potential of –100 to –400 mV was estimated to prevail after 10–300 years of emplacement.

3. General Corrosion

Following saturation of the repository, the environment surrounding the canister will continue to evolve. As a consequence, the corrosion behaviour of the canister will also change with time. Eventually, however, the environment and the corrosion behaviour will attain a steady state. In general, the evolution in conditions will lead to less aggressive forms of corrosion. The corrosion behaviour of the canister will evolve from an initial period of relatively fast general corrosion accompanied by possible localised corrosion, to a long-term steady state condition of a low rate of general corrosion with no, or little, localised corrosion.

Due to the corrosion during the saturation phase, when high levels of oxygen will be present initially, the canister surface may become covered with a duplex corrosion product layer. The inner layer will be Cu_2O and the outer layer will consist of basic Cu(II) salts, most likely either malachite ($Cu_2CO_3(OH)_2$) or atacamite ($CuCl_2 \cdot 3Cu(OH)_2$), depending on the relative concentrations of carbonate and chloride in the water.

Once the section of the repository, in which the canister is placed, is closed the remaining oxygen will control the redox potential. The redox potential will drop as the oxygen is depleted from the system from an initial value of greater that +600 mV to close to the H_2/H_2O equilibrium, especially in the presence of sulfide. When this happens, the thermodynamically stable solids will change from CuO, or $CuCl_2 \cdot 3Cu(OH)_2$, depending on the pH and Cl^- activity, to first Cu_2O and then Cu, or copper sulfides if sulfides are present. Figure 2 shows a Pourbaix diagram for water at 25°C with $[Cu]_{TOT} = 10^{-6}$ mol kg^{-1} and $[CO_3{}^{2-}]_{TOT} = 2$ mmol kg^{-1} and $[Cl^-]_{TOT} = 1.7$ mol kg^{-1}. Figure 3 shows a Pourbaix diagram for water at 25°C with $[Cu]_{TOT} = 10^{-6}$ mol kg^{-1} and $[HS^-]_{TOT} = 0.2$ mmol kg^{-1}. In the case of chloride-containing water, the stable solution species will be copper chloride complexes.

In order to assess the extent of canister corrosion in the repository, the effects of oxygen, chloride and sulfide must be considered. Initially, the corrosion will be controlled by the reduction of O_2 to OH^- according to the reaction

$$O_2 + 2H_2O + 4e^- \rightarrow 4OH^-$$

In addition to this reduction process, the corrosion reaction is characterised by such factors as oxide surface films and the concentration of O_2. The rate of O_2 reduction

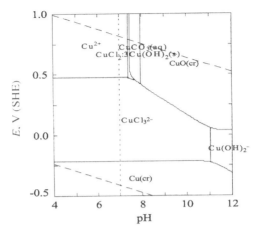

Fig. 2 *Pourbaix diagram for water at 25°C with $[Cu]_{TOT} = 10^{-6}$ mol kg^{-1} and $[CO_3^{2-}]_{TOT} = 2$ mmol kg^{-1} and $[Cl^-]_{TOT} = 1.7$ mol kg^{-1}.*

on the surface will be slower on oxide-covered surfaces than on film-free Cu. The nature of the surface film on the canister will also affect the rate of O_2 reduction. The Cu(II) salts ($Cu_2CO_3(OH)_2$) and $CuCl_2 \cdot 3Cu(OH)_2$) are electrically insulating and if a continuous layer of these species is formed, the reduction rate of O_2 will be reduced significantly [13]. Generally, the outer layer of the duplex corrosion products will be non-continuous, so that O_2 reduction can proceed on the inner semiconducting Cu_2O layer. Although the rate of reduction on this oxide layer will be slower than on a Cu surface, the rate of reduction in the repository is likely to be controlled by the supply of O_2 rather than by the rate of the surface reaction. A consequence of the presence of

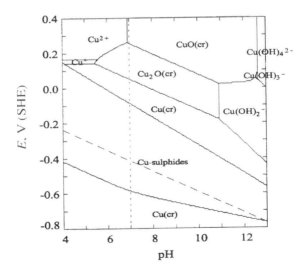

Fig. 3 *Pourbaix diagram for water at 25°C with $[Cu]_{TOT} = 10^{-6}$ mol kg^{-1} and $[HS^-]_{TOT} = 0.2$ mmol kg^{-1}.*

electrically insulating Cu(II) salts is the possible spatial separation of anodic and cathodic surface reactions, possibly leading to localised corrosion.

A great deal of evidence is available concerning the effect of Cl$^-$ on the corrosion of Cu. As can be seen in Fig. 2, chloride ions stabilise dissolved copper as complex anions, such as $CuCl_2^-$ and $CuCl_3^{2-}$ depending on the Cl$^-$ concentration. At sufficiently low pH, Cu corrosion is accompanied by the evolution of H$_2$ in Cl$^-$ solutions. In the presence of oxygen, the thermodynamic stability of $CuCl_2 \cdot 3Cu(OH)_2$ with respect to Cu$_2$O and CuO increases.

Chloride ions will also affect the properties and stability of the surface films. By incorporation of CuCl 'islands' in the surface Cu$_2$O film, defects are created, which are believed to be initiation points for pitting. Substitution of monovalent Cl$^-$ ions for divalent O^{2-} in the Cu$_2$O lattice enhances the semi-conducting properties of the surface film. The Cu$_2$O film formed in Cl$^-$ solutions may, therefore, support O$_2$ reduction and the anodic Cu dissolution and, therefore, be less protective than Cu$_2$O films formed in the absence of Cl$^-$. It should be borne in mind, however, that the extent of canister corrosion by oxygen, with or without the presence of high levels of Cl$^-$, would be limited by the availability of oxygen in the repository. The available amount of oxygen in the repository, per copper canister is, with the present design for a Swedish repository, 4.5 m^3 or 200 mol. Assuming that Cu$_2$O is formed as corrosion product this corresponds to 50 kg of oxidised copper, or a corrosion depth of 300 μm if evenly distributed over the canister surface. In reality, the corrosion will be considerably lower since transport of oxygen to the canister surface will be limited and microbial activity and reactions with minerals in the host rock and the backfill will also consume oxygen [8].

The behaviour of copper in sulfide-containing environments is important because of the potential for the corrosion of Cu to be supported by the reduction of H$_2$O. As can be seen in Fig. 3, cuprous sulfide (Cu$_2$S) is stable at potentials below the H$_2$/H$_2$O equilibrium line. Because of this thermodynamic stability and because of the presence of sulfide minerals in many types of bentonite and in deep Swedish groundwaters, Cu canisters may be subject to corrosion in the presence of sulfide under the long-term reducing conditions expected to develop in the repository.

Various workers have studied the electrochemical and corrosion behaviour of Cu alloys in sulfide environments. The majority of these studies are associated with corrosion of Cu alloys in polluted sea water. Care should be taken in applying the results of these studies to the corrosion of Cu canisters, however, because most of them involved Cu–Ni alloys (commonly used in marine heat exchangers) and because the particularly aggressive forms of corrosion observed in these applications are associated with *alternating* oxidising and reducing conditions.

Several electrochemical studies of the early stages of film formation on Cu in sulfide environments have been published. As commonly observed in other environments, Cu forms a duplex corrosion product film in sulfide solutions. It consists of an inner layer of Cu$_2$S (variously reported to be between 0.4 nm [14] and 25–50 nm thick [15] and a thicker outer layer of CuS. Non-stoichiometric sulfides, Cu$_{2-x}$S, may form during the conversion of Cu$_2$S to CuS. The rate of growth of the CuS layer is believed to be controlled by the rate of transport of sulfide to the surface, which would be a particularly slow process in the compacted bentonite to be used in a repository.

A simulation of the evolution of redox conditions in a Swedish repository was performed by King and Stroes–Gascoyne [16], although the experiment was actually performed to determine the effect of sulfate-reducing bacteria on the corrosion of a Cu canister in a Canadian repository. Copper electrodes were exposed to a 1 mol dm^{-3} NaCl solution under well-controlled mass-transport conditions, either by rotating the electrode in bulk solution or by placing a 0.1 cm thick layer of compacted bentonite (clay dry density 1.2 Mg m^{-3}) between the electrode and the bulk solution. The E_{CORR} of the electrode was measured as the purge gas was changed or as sulfide ions were added to the bulk electrolyte. The solution was initially aerated, but the purge gas was sequentially changed to 2 vol.% O_2/N_2, 0.2 vol.% O_2/N_2, and 100% Ar (nominally deaerated). In two experiments, sufficient Na_2S was added to the solution to give a bulk [HS$^-$] of either 10 µg g^{-1} or 100 µg g^{-1}.

Figure 4 shows the time dependence of E_{CORR} for three experiments, one in bulk solution and two with the compacted clay electrode. The E_{CORR} of the rotating electrode in bulk solution responded quickly to changes in the purge gas, decreasing with decreasing [O_2] (curve (a)). Under these rapid mass-transport conditions, the anodic reaction is mass-transport controlled and the cathodic reaction is limited by the rate of the interfacial reduction of O_2 at E_{CORR} [17]. Upon the addition of 10 µg g^{-1} HS$^-$, E_{CORR} drops immediately by ~500 mV to a value of *ca.* –0.90 V(SCE). The very sharp drop in E_{CORR} upon the first addition of HS$^-$ was explained in terms of a switch in the anodic reaction from dissolution as $CuCl_2^-$ to the formation of a

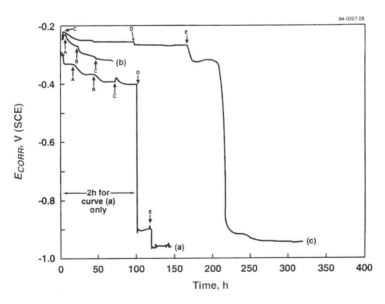

Fig. 4 *Variation of the corrosion potential of a copper/compacted clay electrode and of a copper rotating disc electrode in 1 mol dm^{-3} NaCl solution as a function of oxygen and sulfide concentration [16]. (a) Rotating disc electrode in bulk solution (the time axis for the first 2 h of this experiment has been expanded by a factor of 50 for clarity); (b) copper/compacted clay electrode in O$_2$-containing solution only; (c) copper/compacted clay electrode with various [O$_2$] and sulfide additions. The arrows represent the times at which either the overpurge atmosphere was changed or sulfide additions were made according to: (A) 2 vol.% O$_2$/N$_2$; (B) 0.2 vol.% O$_2$/N$_2$; (C) Ar; (D) 10 µg g^{-1} sulfide; (E) 100 µg g^{-1} sulfide.*

Cu_2S or CuS surface film, and a switch in the cathodic reaction from the reduction of residual O_2 to the reduction of H_2O (or HS^-).

In the presence of compacted clay, a similar decrease in E_{CORR} is observed upon the addition of HS^-, but more slowly as a consequence of the restricted mass-transport conditions. The E_{CORR} of the compacted clay electrode also decreases as the $[O_2]$ is decreased (curves (b) and (c)). The magnitude of the decrease suggests complete transport control of the anodic and cathodic reactions [17]. Upon the addition of 10 µg g^{-1} HS^- (point D, curve (c)), E_{CORR} decreases by 10–20 mV but then stabilises for the next 75 h. The addition of 100 µg g^{-1} HS^- (point E, curve (c)) causes a further decrease in E_{CORR} by 50–60 mV followed by a second plateau of about 40 h. These small decreases in E_{CORR} were thought to be due to the consumption of residual O_2 in the clay layer by reaction with HS^-. After the second plateau period, however, E_{CORR} drops sharply by ~0.6 V(SCE) to a final steady-state value virtually identical to that observed in bulk solution at that $[HS^-]$.

As mentioned above, the most serious corrosion in sulfide polluted water occurs in alternating aerated and deaerated environments. These alternating conditions may have one of two effects. First, exposure to alternating environments can result in alteration of the corrosion products and spalling of otherwise protective surface films due to volume changes associated with the growth of the altered layer. Thus, truly protective surface films are not stabilised under alternating oxidising and reducing conditions. Secondly, Cu sulfide films are more catalytic towards the reduction of O_2 than Cu_2O films, due to their more-defective structure. Thus, following the growth of a Cu sulfide film under reducing conditions, the introduction of dissolved O_2 would cause a rapid increase in corrosion rate. In either case, such effects should not occur in a repository, since the environment is expected to evolve from initially aerated to reducing in the long term, and to remain so indefinitely.

4. Modelling of General Corrosion

There are both thermodynamic and kinetic approaches to predict the long-term general corrosion behaviour of the copper canisters in a repository. Canister lifetime predictions based on a thermodynamic description of the corrosion process(es) generally involve an assumption of rapid interfacial kinetics and rate control by the rate of (diffusive) mass transport. Kinetically based lifetime-prediction models combine the finite kinetics of interfacial reactions with possible limitation by mass transport to and from the corroding surface. Thermodynamic models represent a 'worst-case' assessment because of the assumption of infinite interfacial kinetics, and produce a conservative estimate of the corrosion rate.

Werme *et al.* [18] describe a mass-transport limited model for predicting the extent of corrosion of copper canisters in a Swedish repository due to sulfide. In a previous assessment [19], it had been conservatively assumed that all of the sulfide in the bentonite surrounding the canister plus that produced by microbial activity in the deposition hole was consumed in the first 1000 years resulting in a corrosion depth of 55 µm. The total corrosion depth after 10^6 years was approximately 0.3 mm. Werme *et al.* re-considered the conservative assumption that all the sulfide in the deposition hole was consumed in 1000 years. Using a 1-dimensional sulfide consumption model assuming instantaneous consumption of sulfide on the canister surface, they estimated that the sulfide in the deposition hole would be consumed in 850 000 years. The amount of sulfide estimated to be present in the deposition hole was higher than that assumed in ref. [19], and gave a maximum depth of corrosion after 850 000 years of just over 1 mm. Until the sulfide in the deposition hole is consumed, there can be no diffusion of sulfide from the tunnel or from the groundwater, since the bentonite pore-water [HS$^-$] exceeds that in the tunnel or groundwater.

While Werme *et al.* considered the dissolution of sulfide from sulfide impurities in the bentonite and diffusional transport to the canister surface, Taxén [in ref. 2] considered the dissolution of copper as chloride complexes and diffusional transport of these species to the sulfide impurities where copper sulfide was formed. The hydrogen formed the reaction

$$Cu(s) + H^+ + n\ Cl^- \bullet {}^1/_2\ H_2(aq) + CuCl_n\ (1{-}n)$$

was assumed be removed from the system through reactions Fe(III) in the bentonite. Taxén calculated the corrosion depth to be <6 μm after 50 000 years. This is a factor of 5 lower than what is obtained using the approach of Werme *et al.*

In contrast to the thermodynamically based models used in the Swedish programme, the assessment of the rate of general corrosion in the Canadian programme has been based on both thermodynamic (mass-transport limited) and kinetic models. Two models have been used to predict canister lifetimes, each having their particular strengths and weaknesses. The models are: the Cu(II) mass-transport limited model for oxidising conditions [20, 21]) and the transient kinetic Copper Corrosion Model (CCM) [22–24]. All these models are developed for corrosion in environments with oxygen present at various levels. No kinetic modelling has yet been performed for the corrosion of copper in sulfide environments.

In the Cu(II) mass-transport limited model, which is based on corrosion experiments in compacted buffer material under oxidising conditions, the diffusion of Cu(II) away from the Cu surface limits the rate of corrosion. Implicit in the model is the assumption that the mechanism does not change over time, i.e. the corrosion rate does not become limited by the supply of O$_2$ to the canister surface. This is clearly a conservative assumption for a sealed repository and, together with the assumption of infinitely fast interfacial kinetics, this model results in an extremely conservative assessment of the canister lifetimes. Nevertheless, the minimum predicted lifetime for a 25 mm thick Cu canister (only 16 mm of which was used as the corrosion allowance) in a Canadian repository was of the order of 30 000 years. Predictions based on a more-realistic interfacial Cu concentration produced lifetimes of ~10^6 years, even for the conservative assumption of an infinite supply of O$_2$.

The most advanced general corrosion model developed in the Canadian programme, and consequently the one that has proven most useful, is the transient kinetic Cu corrosion model. The model is based both on corrosion studies of copper in oxygen-containing saline groundwater in compacted bentonite and on electrochemical studies of copper in chloride- and sulfate-containing solutions. The reaction scheme considered in the model is shown in Fig. 5 and is based on a very substantial experimental database.

A series of ten mass-balance equations is solved subject to various initial and

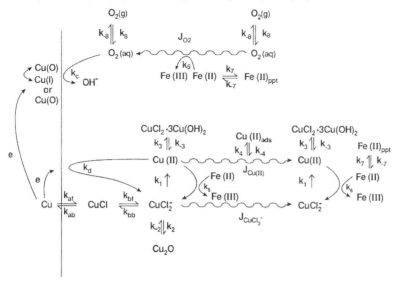

Fig. 5 *Mechanism for the general corrosion of copper in compacted buffer material with O²⁻ containing saline groundwater (from ref. [25]).*

boundary conditions. One of the useful features of the model is the wide range of output data that is produced. In addition to predicting the time dependence of the corrosion potential (E_{CORR}) and corrosion rate of the canister, the model also predicts the spatial and time dependence of the concentration of each of the ten chemical species, and of the rate and extent of each of the individual reactions. These latter data can be used, for example, to predict the time dependence of the increase in salinity in the repository, or of the consumption of O_2. Since the model does not consider sulfide environments the corrosion will stop when the initially trapped oxygen is consumed. In a simulation of the canister corrosion in the Canadian repository, the trapped O_2 in the buffer and backfill is consumed by reaction on the canister surface, by reaction with Fe(II) minerals in the backfill, and by diffusion out of the repository, where it is consumed by reaction with Fe(II) minerals in the excavation-disturbed zone (EDZ), and intact rock. The $[O_2]$ at the canister surface drops to ~1% of its initial value after 1 month, suggesting the reduction of O_2 is transport limited. Oxygen is consumed most rapidly in the EDZ and backfill, because of the abundance of Fe(II) minerals in these layers. The O_2 in the buffer layer is the last to be consumed after ~2600 years when the total corrosion depth will amount to only 11 μm. The predicted variations in the dissolved oxygen as a function of time is shown in Fig. 6 and the integrated corrosion in Fig. 7.

5. Localised Corrosion

Localised corrosion refers to a range of corrosion phenomena that result in localised, as opposed to general, corrosion and which do not fall under the category of

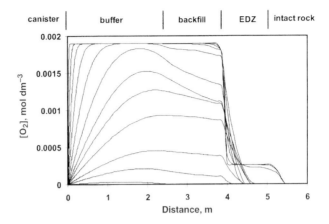

Fig. 6 *Predicted variation of the concentration of dissolved oxygen as a function of time and position in a Canadian repository. The uppermost profile is for a time of 1 month, with the [O₂] decreasing gradually with time until the final profile illustrated at a time of 1840 years. For the purpose of these calculations it was assumed that all the O₂ in the vapour-filled pores was instantaneously dissolved in the pore water.*

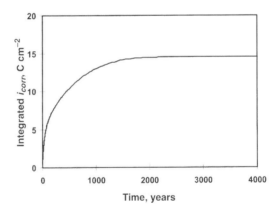

Fig. 7 *Predicted time dependence on the extent of corrosion of a copper canister in a Canadian repository. A charge density of 15 C cm⁻² is equivalent to 11 μm of corrosion.*

environmentally assisted cracking. Environmentally assisted cracking will not be discussed in the present paper. For pure Cu alloys, which do not undergo dealloying, the most important form of localised corrosion is pitting. Before considering pitting in some detail, crevice corrosion will be briefly discussed.

Crevice corrosion is of significant concern for many alloy systems, especially passive materials containing alloying elements, which strongly hydrolyse and so form locally acidified environments in occluded regions. The hydrolysis of Cu(I), especially when complexed by Cl⁻, is weak, and local acidification in crevices is unlikely to occur. Cupric species hydrolyse more strongly, but, the formation of Cu(II) requires the presence of O₂, and is unlikely to occur in occluded regions, such as crevices, where O₂ access is restricted. As a consequence, the crevice corrosion of pure Cu is uncommon, and when it is observed occurs via a differential Cu-ion concentration cell mechanism, with areas exposed to a high Cu-ion concentration acting as cathodic sites supporting the dissolution of areas in contact with a low concentration of dissolved Cu. This form of localised corrosion, however, is inherently self-limiting, because the differential [Cu] cell driving localised corrosion is destroyed by the dissolution itself. In long-term irradiated corrosion tests under simulated conditions of a Canadian repository, no crevice corrosion was observed on either creviced U-bend of creviced planar samples [26].

Pitting of Cu has been investigated in relation to a number of different applications of Cu alloys and for various environments. The most well known phenomenon is the pitting of Cu water-distribution pipes. These studies are of limited use for predicting the pitting behaviour of Cu canisters, however, because of the difference in salinity between fresh potable waters and saline pore waters. However, a number of Cu pitting studies have also been reported in Cl^- solutions, with and without the addition of HCO_3^-. In addition, a number of corrosion studies have been performed under conditions that simulate those expected in a repository and provide direct evidence for the pitting behaviour of Cu canisters.

Pitting of Cu water pipes has been studied extensively since the 1960s. At least three types of pitting have been recognised; Types I and II pitting [27] and a type of pitting apparently induced by microbial activity [28] (not discussed further here because of the absence of microbial activity in highly compacted bentonite). Types I and II pitting occur under distinctly different conditions. Type I pitting is associated with cold, hard and moderately hard waters free of naturally occurring inhibitors but containing HCO_3^-, SO_4^{2-}, Cl^- and O_2, and on Cu pipes with a residual surface carbon film from the manufacturing process. Pit initiation involves formation of a CuCl pocket in an otherwise protective Cu_2O film. Dissolution of Cu as $CuCl_2^-$ occurs at the defect produced by the CuCl pocket. The dissolved Cu(I) is oxidised to Cu(II) by O_2 and precipitates forming a crust of $CuCO_3 \cdot Cu(OH)_2$ and $CaCO_3$. The crust forms an occluded region in which localised dissolution continues. Type II pitting is associated with hot potable waters ($>60°C$) with pH <7.4 and a $[HCO_3^-]/[SO_4^{2-}]$ ratio <1, and tends to produce pits with a larger depth/width ratio than Type I pits.

Oxygen is a pre-requisite for pit propagation. In Type I pitting, O_2 either serves directly as the oxidant supporting pit growth, or oxidises Cu(I) to Cu(II), with the latter species then acting as the oxidant. The reduction of O_2 to OH^- also produces local alkalinity, which supports and maintains the crust over the pit, which in turn maintains local acidity within the pit and sustains pit growth. A decrease in $[O_2]$ would result in less Cu dissolution and an increase in the pit pH, both of which would eventually cause the pit to stop propagating. Pitting of Cu water pipes is only sustained because of the high $[O_2]$ in fresh water and because it is continually replenished by the movement of water in the pipe. This would not be the case for pits on Cu canisters, both because of the limited amount of O_2 available and because of the restricted mass-transport conditions, which will limit the supply of O_2 to the corrosion sites. Therefore, pits on Cu canisters will be far more likely to die than pits on Cu water pipes.

Localised corrosion of Cu has been reported in sea water polluted by sulfide, but only under conditions of alternating aerated and deaerated conditions. As discussed above, cyclic redox conditions can disrupt the otherwise protective surface film (either

Cu_2O or Cu_2S/CuS). In marine applications, disruption of the surface film is exacerbated by flow effects, which further destabilise the protective layer. Such alternating redox conditions are not expected in a repository.

A number of corrosion studies have been performed under conditions that simulate the canister nearfield environment. Copper coupons have been exposed to compacted buffer material wetted by (initially) aerated saline pore waters, and exposed for extended periods of time (up to 2 years) usually at elevated temperature [29–33]. These experiments simulate the likely environmental conditions soon after emplacement of the canisters and saturation of the buffer material. It is during this period in the evolution of vault conditions that localised corrosion is most likely, since the environment will be oxidising. Furthermore, the pore-water $[HCO_3^-]$ may be significant because of the dissolution of carbonate minerals and the $[Cl^-]$ may be sufficient to cause film breakdown, but not so high as to cause general dissolution of the surface. Thus, short-term lab and field tests can be used to study the period of most aggressive localised corrosion. Despite the relative aggressiveness of the conditions in such tests, no evidence for pitting has been observed on copper exposed to compacted clay over period up to 2 years. The only instance of non-uniform corrosion reported under such conditions is the so-called under-deposit corrosion reported by Litke *et al.* [33]. New experiments have been initiated at the same site as in ref. [30] and the preliminary results will be reported by Rosborg *et al.* [34].

Various modelling approaches have been developed to study the pitting of Cu under repository conditions. Taxén has described a model for the growth of pits based on mass transport and chemical equilibrium principles [35,36]. An alternative approach is to predict the probability of pit initiation based on observed film breakdown potentials (E_b). This approach requires knowledge of the variation of E_b with environmental conditions and of the time dependence of E_{CORR} for a canister. The final approach to predicting the extent of localised corrosion on Cu canisters is to make projections based on observed pit depths from the literature. This approach was taken in ref. [19] and similar data were analyses using extreme value statistics by King and LeNeveu [20].

6. Conclusions

Much is known about the general corrosion behaviour of Cu under repository conditions. Detailed thermodynamic analyses on possible corrosion reactions have been performed, particularly in the Swedish and Finnish programmes. In Canada, more emphasis has been placed on kinetic studies under well-controlled mass transport conditions. Combined, these complementary approaches provide a detailed understanding of the general corrosion behaviour of Cu canisters under the evolving conditions expected in a repository. The most important parameters controlling the rate of general corrosion are: the rates of mass transport of species to and from the canister surface, the availability of O_2, the influx of Cl^- ions from the groundwater, and the supply of sulfide ions to the canister.

In a sealed repository, the extent of general corrosion is limited by the general lack of oxidants. Trapped atmospheric O_2 will support corrosion in the initial stages of the evolution of vault conditions, but the amount of available O_2 is limited and will be partially consumed by microbes and by reaction with oxidisable minerals (principally sulfide minerals) in the repository. Under reducing conditions, corrosion will be supported by the slow supply of sulfide to the canister surface. Because of the limited amount of available oxidant, general corrosion will not limit the canister lifetime.

Although the extensive database on the pitting of Cu water pipes provides some useful mechanistic information, the results of corrosion experiments under simulated repository conditions suggest that canisters will not undergo classical pitting, but rather a form of under-deposit corrosion, in which there is no permanent separation of anodic and cathodic sites. The mechanistic Cu pitting studies indicate that an oxidant (either O_2 or Cu(II)) is a pre-requisite for pit propagation. Since the nearfield environment in the repository will evolve from initially oxidising to ultimately reducing, this implies that pitting will only be possible (if at all) in the early stages of the repository life. Thus, the environment within the repository is evolving to one in which only general corrosion will occur. In addition, the difficult problem of predicting localised corrosion is made easier by the fact that predictions only have to be made for the early oxidising period. Given the evolution in localised corrosion behaviour with repository conditions, the most suitable pitting models are those that include the possibility of pit death. Of the currently available models, only those based on a critical potential (either for pit initiation or pit propagation) are capable of predicting when pitting may cease to be an active degradation mechanism in the repository. For analyses based on literature pit depth data (either the pitting factor or extreme-value approaches) it is implicitly assumed that pits propagate indefinitely. Predictions of pit depth based on these latter two approaches must be considered conservative.

References

1. L.O. Werme, 'Design premises for canister for spent nuclear fuel', Technical Report TR-98-08, Svensk Kärnbränslehantering AB, Stockholm, Sweden (1998).
2. F. King, L. Ahonen, C. Taxén, U. Vuorinen and L. Werme, 'Copper corrosion under expected conditions in a deep geologic repository', Technical Report TR-01-23, Svensk Kärnbränslehantering AB, Stockholm, Sweden (2001).
3. SR 97 — Post-closure safety. Deep repository for spent nuclear fuel, Technical Report TR-99-06, Svensk Kärnbränslehantering AB, Stockholm, Sweden (1999).
4. A. Muurinen, J. Lehikoinen, 'Porewater chemistry in compacted bentonite', POSIVA 99-20, POSIVA Oy, Helsinki, Finland (1999).
5. J. Bruno, D. Arcos and L. Duro, 'Processes and features affecting the near field hydrochemistry. Groundwater–bentonite interaction', Technical report TR-99-29, Svensk Kärnbränslehantering AB, Stockholm, Sweden (1999).
6. K. Pedersen, 'Microbial processes in radioactive waste disposal', Technical Report TR 00-04. Svensk Kärnbränslehantering AB, Stockholm, Sweden (2000).
7. K. Pedersen, 'Diversity and activity of microorganisms in deep igneous rock aquifers of the Fennoscandian Shield', *in Subsurface Microbiology and Biogeochemistry*, (J.K. Fredrickson and M. Fletcher, eds). Wiley-Liss Inc., New York, pp. 97–139 (2001).
8. I. Puigdomenech, *et al.*, 'O_2 depletion in granitic media: the REX project', Technical Report TR-01-05 Svensk Kärnbränslehantering AB, Stockholm, Sweden (2001).
9. M. Motamedi, O. Karnland and K. Pedersen, *FEMS Microbiol. Lett.*, 1996, **141**, 83–87.
10. K. Pedersen, M. Motamedi and O. Karnland, *Engin. Geol.*, 2000, **58**, 149–161.

11. K. Pedersen, M. Motamedi, O. Karnland and T. Sandén, *J. Appl. Microbiol.*, 2000, **80**, 1038–1047.

12. P. Wersin, K. Spahiu and J. Bruno, 'Time evolution of dissolved oxygen and redox conditions in a HLW repository', Technical Report TR-94-02, Svensk Kärnbränslehantering AB, Stockholm, Sweden (1994).

13. C. Kato, B. G. Ateya, J. E. Castle and H. W. Pickering, *J. Electrochem. Soc.*, 1980, **127**, 1890–1896.

14. M. R. G. de Chialvo and A. J. Arvia, *J. Appl. Electrochem.*, 1985, **15**, 685–696.

15. D. Vasquez Moll, M. R. G. de Chialvo, R. C. Salvarezza and A. J. Arvia, *Electrochim. Acta*, 1985, **30**, 1011–1016. 16. F. King and Stroess-Gascoyne, in *Proc. 1995 Int. Conf. on Microbially Influenced Corrosion*, NACE International and American Welding Society (Houston, TX and Miami, FL), 1995, pp. 35/1–35/14.

17. F. King, C. D. Litke, M. J. Quinn and D. M. LeNeveu, *Corros. Sci.*, 1995, **37**, 833–851.

18. L. Werme, P. Sellin and N. Kjellbert, 'Copper canisters for nuclear high level waste disposal. Corrosion aspects', Technical Report TR-92-26, Svensk Kärnbränslehantering AB, Stockholm, Sweden (1994).

19. The Swedish Corrosion Institute and its reference group, 'Corrosion resistance of a copper canister for nuclear fuel', SKBF/KBS Technical Report 83-26, Svensk Kärnbränslehantering AB, Stockholm, Sweden (1983).

20. F. King and D. LeNeveu, 'Prediction of the lifetimes of copper nuclear waste containers', *in Proc. Topical Meeting on Nuclear Waste Packaging, FOCUS '91*, (American Nuclear Society, La Grange Park, IL), pp. 253–261, (1992).

21. F. King, C. Litke, S. Ryan and D. LeNeveu, 'Predicting the long-term corrosion behaviour of copper nuclear fuel waste containers', in *Life Prediction of Corrodible Structures* (R.N. Parkins, ed.). NACE International, Houston, TX, USA, pp. 497–512 (1994).

22. F. King and M. Kolář, *Corrosion '95*, Paper No. 425, NACE International, Houston, TX, 1995.

23. F. King and M. Kolář, 'A numerical model for the corrosion of copper nuclear fuel waste containers', *Mat. Res. Soc. Symp. Proc. 412*, Materials Research Society, Pittsburgh, PA, pp. 555–562 (1996).

24. F. King and M. Kolář, 'The copper container corrosion model used in AECL's second case study', Ontario Power Generation, Nuclear Waste Management Division Report 06819-REP-01200-10041-R00, Toronto, Ontario (2000).

25. F. King, 'A copper container corrosion model for the in-room emplacement of used CANDU fuel', Atomic Energy of Canada Limited Report, AECL-11552, COG-96-105 (1996).

26. S. R. Ryan, *et al.*, 'An investigation of the long-term corrosion behaviour of selected nuclear fuel waste container materials under possible disposal vault conditions', Atomic Energy of Canada Limited Technical Record, TR-489, COG-94-55 (1994), Chalk River, Ontario, Canada K0J 1J0.

27. E. Mattsson, *Brit. Corros. J.*, 1980, **15**, 6–13.

28. W. Fischer, I. Hänßel and H. H. Paradies, 'First results of microbial induced corrosion of copper pipes', in *Microbial Corrosion — 1* (C.A.C. Sequeira and A.K. Tiller, eds). Elsevier, London, UK, p. 300–327 (1988).

29. P. Aaltonen and P. Varis, 'Long term corrosion tests of OFHC-coppers in simulated repository conditions — final report', Nuclear Waste Commission of Finnish Power Companies, Report YJT-93-05 (1993).

30. O. Karnland, *et al.*, 'Long term test of buffer material. Final report on the pilot parcels', Technical Report, TR-00-22, Svensk Kärnbränslehantering AB, Stockholm, Sweden (2000).

31. F. King, C. D. Litke and S. R. Ryan, *Corros. Sci.*, 1992, **33**, 1979–1995.

32. F. King, S. R. Ryan and C. D. Litke, 'The corrosion of copper in compacted clay', Atomic Energy of Canada Limited Report, AECL-11831, COG-97-319-I (1997). Chalk River, Ontario, Canada K0J 1J0.

33. C. D. Litke, S. R. Ryan and F. King, 'A mechanistic study of the uniform corrosion of copper in compacted clay–sand soil', Atomic Energy of Canada Limited Report, AECL-10397, COG-91-304 (1992).

34. B. Rosborg, O. Karnland, G. Quirk and L. Werme, 'Measurement of copper corrosion in the LOT project at the Äspö Hard Rock laboratory, Sweden', this Volume, pp.412–423.

35. C. Taxén, in *Proc. 13th Int. Corros. Conf.*, Paper No. 141, Melbourne, Australia (1996).

36. C. Taxén, 'Pitting corrosion of copper. Equilibrium — mass transport limitations', *Mat. Res. Soc. Symp. Proc.*, 2000, **608**, 103–108.

27
Measurements of Copper Corrosion in the LOT* Project at the Äspö Hard Rock Laboratory

B. ROSBORG, O. KARNLAND[†], G. QUIRK[§] and L. WERME[¶]

Rosborg Consulting, Östra Villavägen 3, SE - 611 36 Nyköping, Sweden
[†]Clay Technology AB, IDEON Research Center, Scheelevägen 19F, SE - 223 70 Lund, Sweden
[§]InterCorr International Ltd, 2 Fodderty Way, Dingwall Business Park, Dingwall, Rosshire, IV15 9XB, UK
[¶]Svensk Kärnbränslehantering AB, Box 5864, SE - 102 40 Stockholm, Sweden

ABSTRACT

Real-time monitoring of corrosion by means of electrochemical noise and other electrochemical techniques may offer interesting possibilities to estimate the kind and degree of corrosion in a sample or component, and further visualise the corrosion resistance of pure copper in repository environments. As a pilot effort, three cylindrical copper electrodes for such measurements, each of about 100 cm^2 surface area, have been installed in a test parcel in the Äspö Hard Rock Laboratory and electrochemical measurements using InterCorr's SmartCET system were initiated in May 2001.

The first results from real-time monitoring of copper corrosion in the Äspö HRL under actual repository environment conditions by means of linear polarisation resistance, harmonic distortion analysis and electrochemical noise techniques are presented, and compared with the results obtained from one of the retrieved test parcels.

1. Introduction

1.1. A KBS-3** Repository

In Sweden the principal strategy is to enclose the spent nuclear fuel in tightly sealed copper canisters that are embedded in clay about 500 m down in the Swedish bedrock (see Fig. 1) [1]. The rock provides a stable and durable environment where changes occur very slowly. The canister isolates the fuel from the groundwater. The clay prevents groundwater flow around the canister and protects against minor movements in the rock. Also, when the canister is eventually breached, the bentonite clay will retain or retard the release of radionuclides. The rock is the last barrier in this multibarrier system between the fuel and the environment.

1.2. The Äspö Hard Rock Laboratory

The construction of the Äspö Hard Rock Laboratory (HRL) was completed in 1995. The activities at the HRL can be regarded as a dress rehearsal for the siting and

---

*Long Term Test of Buffer Material.
**The main alternative for deep disposal of spent nuclear fuel in Sweden is called the KBS-3 method.

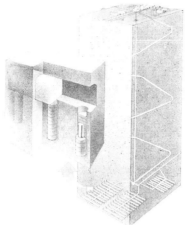

Fig. 1 *A KBS-3 repository.*

construction of the future deep repository for spent nuclear fuel. The laboratory offers a realistic environment for different experiments and tests under the conditions that will prevail in a deep repository. Table 1 gives a typical groundwater composition from the Äspö HRL [2].

1.3. The LOT Test Series

The test series 'Long Term Test of Buffer Material' (LOT) has been initiated at the Äspö HRL with conditions similar to those in a final KBS-3 repository [3]. The main purpose is to study the mineralogical stability and behaviour of the bentonite clay. However, additional testing has been included, of which the investigation of corrosion on copper coupons in bentonite blocks is one.

1.4. Real-Time Monitoring of Corrosion Attack

Real-time monitoring of corrosion by means of the electrochemical noise and other electrochemical techniques may offer interesting possibilities to estimate the kind and degree of corrosion in a sample or component, and further visualise the corrosion resistance of pure copper in repository environments. The added advantage is that by using real-time monitoring the development of the corrosion behaviour with time can be observed, which can give insights into the performance of the copper/bentonite/groundwater interactions. As a pilot effort for real-time monitoring of corrosion, cylindrical copper electrodes for such measurements have been installed in one of the test parcels in the LOT project.

Table 1. *A typical groundwater composition at Äspö (mg dm^{-3})*

Na^+	K^+	Ca^{2+}	Mg^{2+}	HCO_{3-}	Cl^-	SO_4^{2-}	HS^-	pH	E_{redox} (mV (SHE))
2100	8	1890	42	10	6410	560	0.15	7.7	−308

2. Experimental

The LOT test series comprises seven test parcels, which will be run for 1, 5 and 20 years. Each test parcel contains prefabricated bentonite blocks placed on top of each other around a central copper tube and has copper coupons in bentonite blocks 22 and 30. Wyoming bentonite with the commercial name MX-80 has been used. The test parcel A2 is shown in Fig. 2. The test parcels were placed in core- or percussion-drilled bore holes with a diameter of 300 mm and 4 m below the tunnel floor in granitic rock at a depth of 450 m below ground at the Äspö HRL. The final density of the bentonite clay, at full saturation and after swelling in the test holes, was calculated to be 2000 kg m^{-3}. The major proportion of the bentonite clay in the LOT test parcels has now been saturated by the groundwater.

The pilot test parcels S1 and A1 placed at one location in the Äspö HRL have so far been retrieved, and test parcels S2, S3, A0 (duplicate to test parcel A1), A2 and A3 placed at another location are in progress. Test parcel A2, in which the real-time corrosion monitoring takes place, will be retrieved in 2004.

2.1. Exposure of Copper Coupons

Test parcel S1 was emplaced in late 1996 and retrieved in early 1998. See Ref. [3] for experimental details of the exposure of copper coupons in the test parcel.

2.2. Real-Time Monitoring of Corrosion

Test parcel A2 including the cylindrical copper electrodes was emplaced on 29 October 1999. The three cylindrical copper electrodes, each of 98.7 cm^2 surface area, have been installed in bentonite block 36 of the test parcel. Thus, the electrodes are positioned near the top of the test parcel (see Fig. 2), where the temperature is about 24°C.

The electrode arrangement is shown in Fig. 3. There are two solid copper connections to each electrode, making a total of six connections for the monitoring equipment. The copper connections are fitted into drilled holes in the top edge of the copper cylinder which are then hammered closed to capture the electrical connection. In this manner, there are no other extraneous alloys introduced to the system, which might cause galvanic corrosion.

The real-time monitoring of the corrosion is performed with linear polarisation resistance (LPR), harmonic distortion analysis and electrochemical noise (EN) techniques using InterCorr's SmartCET system. The measurements were initiated in May 2001.

Linear polarisation resistance technique. The LPR technique polarises the electrodes by a small amount (±10 mV) using a sine-wave perturbation in the linear region of the E–I curve around the corrosion potential, and the current response is measured. It is a useful measurement of the corrosion rate when corrosion is relatively uniform, but it has some limitations. The assumption is made that the system under study has reached a steady state when the measurement is made (true in this application), and the measured polarisation resistance is a composite of the solution resistance and

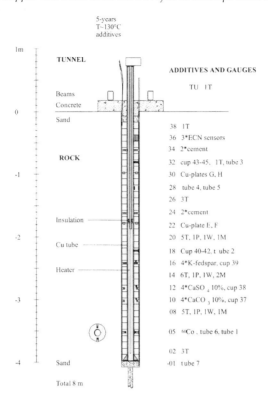

Fig. 2 *Cross-section view of test parcel A2. (Note: The first figures in column denote bentonite block number and second figures denote the number of sensors. T denotes thermocouples, P total pressure sensor, W water pressure sensor, and M moisture sensor.)*

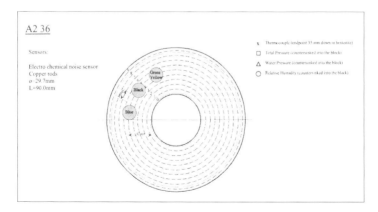

Fig. 3 *Electrode arrangement in bentonite block A236.*

charge-transfer resistance. In this application, the solution resistance is low, which provides a greater confidence in the derivation of corrosion rate from the polarisation resistance. Corrosion rates are calculated using the Stern–Geary relationship [4].

Harmonic distortion analysis. This technique relies on the non-linear nature of electrochemistry and is related to electrochemical impedance spectroscopy, in that an alternating potential perturbation is applied to one sensor element in a three element probe, measuring a resultant current response. The technique applies an AC signal typically of about 0.1 Hz to the cell and a measurement is made of the 1st, 2nd and 3rd harmonic as well as the fundamental of the current response. Using these results, the polarisation resistance and Tafel slopes are mathematically calculated [5,6].

Electrochemical noise techniques. Electrochemical noise (EN) is the generic term used to describe the low amplitude, low frequency random fluctuations of current and potential observed in many electrochemical systems, and has been used to characterise both corrosion rate and mechanism [7–13]. Electrochemical noise data are taken at a frequency of one reading per second and statistical analyses are employed to compute the EN resistance [12], which is analogous to the polarisation resistance obtained from LPR measurements. General corrosion rates are calculated from the EN resistance using the Stern-Geary method [4]. However, EN measurements also provide information on the type of corrosion. The localisation index (LI) is defined as the ratio of the standard deviation of the current noise to the root mean square of the EN current (I_{rms}), which provides an indication of the stochastic distribution of microscopic events. For uniform corrosion, the LI is typically 1×10^{-2} or less. When localised corrosion occurs, the raw data typically exhibit stochastic transients and the LI tends to approach unity [13]. The 'pitting function' is a similar index, except that the corrosion current from harmonic distortion analysis is employed instead of the I_{rms} of the EN signal.

3. Results

3.1. Exposure of Copper Coupons

The results from about one year exposure of copper coupons in test parcel S1 have already been presented [3,14]. In summary, the corrosion rate was estimated to be 3 µm per year at 50°C, and no signs of pitting were found. Several types of corrosion products such as cuprite (Cu_2O) and malachite ($Cu_2CO_3(OH)_2$) were present. The presence of cupric corrosion products indicates that at least part of the corrosion had taken place under quite oxidising conditions.

3.2. Real-Time Monitoring of Corrosion

The results from the linear polarisation resistance, the harmonic distortion analysis and the electrochemical noise measurements on the cylindrical electrodes in bentonite block 36 of test parcel A2 (bentonite block A236) are given below.

The Stern–Geary constant. A value of the Stern–Geary constant (sometimes called the *B*-value) is required to calculate the corrosion rate from the measured linear polarisation resistance data [4]. The field equipment used allows the Stern–Geary constant to be estimated by means of harmonic distortion analysis of the current response from the sine wave LPR potential polarisation [5], without the need for a reference electrode. The obtained result is shown in Fig. 4. (The first transient part of the graph should be ignored for reasons of electrical checks during the installation.) A Stern–Geary constant value of 6.5 mV has been used to calculate the corrosion rate from the LPR and EN data.

Linear polarisation resistance measurements. The corrosion rate of copper in bentonite block A236 at a temperature of about 24°C is shown in Fig. 5. The corrosion rate is about 1.7 μm per year. The data show a slight variation with time. It is not obvious from the present monitoring period if the corrosion rate is decreasing as expected or not.

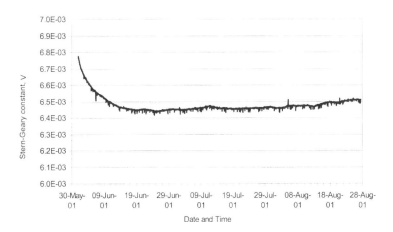

Fig. 4 *Stern–Geary constant estimate from harmonic distortion analysis.*

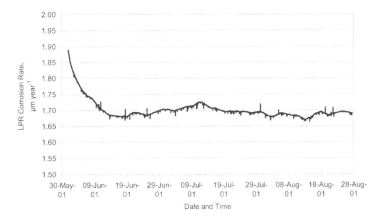

Fig. 5 *The corrosion rate obtained from linear polarisation resistance measurements (using a Stern–Geary constant of 6.5 mV).*

Harmonic distortion analysis measurements. A comparison of the measured corrosion currents from the harmonic distortion and the LPR data is shown in Fig. 6. There is good agreement between the data, which are calculated by different algorithms.

Electrochemical noise measurements. The current noise data and the calculated 'pitting function' and localisation index are shown in Figs 7–9 respectively. The 'pitting function' is defined as the standard deviation of the current noise (ECN) divided by the corrosion current, and the localisation index is defined as the ECN divided by the root mean square of the EN current.

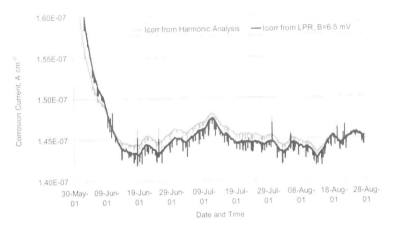

Fig. 6 *Comparison of corrosion current obtained from harmonic distortion analysis and linear polarisation resistance techniques.*

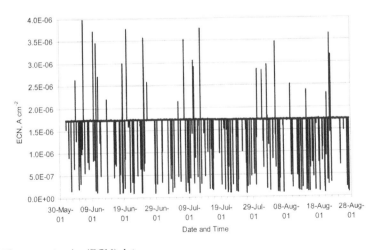

Fig. 7 *The current noise (ECN) data.*

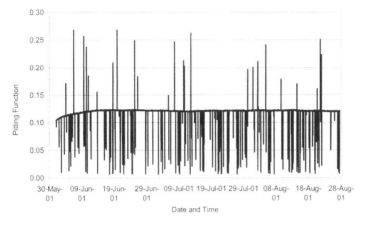

Fig. 8 *The 'pitting function' (ECN/corrosion current).*

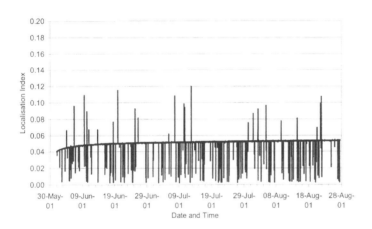

Fig. 9 *The localisation index (ECN/root mean square of the current).*

The EN data reveal a slight tendency to localised corrosion. With purely uniform corrosion the 'pitting function' tends to be in the region of 0.05 or less [13]. Here it is approximately 0.12. While this is far from values of >0.9 for obvious pitting, it is not as low as expected from purely uniform corrosion.

4. Discussion

Below the results from real-time monitoring of corrosion on the copper electrodes in bentonite block A236 are compared with the observations of corrosion found on a copper coupon in bentonite block 22 of the retrieved test parcel S1. A comparison with the copper coupons in bentonite blocks 22 and 30 in test parcel A2 might have been better. However, test parcel A2 will not be retrieved until 2004.

4.1. The Stern–Geary Constant

A reliable value for the Stern–Geary constant is required to calculate the corrosion rate from measured LPR and EN data [4]. Most commercial instrumentation employing electrochemical techniques to measure a corrosion rate have a pre-set value for the Stern–Geary 'constant' in their configuration. This can, however, result in quite erroneous results if the set value is inaccurate. The corrosion rate is directly proportional to the Stern–Geary 'constant' which can vary considerably for different material–environment systems (usually in the range 5 to 50 mV) [15]. The Stern–Geary constant should preferably be determined for the actual case.

The Stern–Geary constant can be obtained from well-controlled Tafel-slope measurements. The experimental setup requires a reference electrode. Here, the constant has been estimated from harmonic distortion analysis measurements without the need for a reference electrode. A value of 6.5 mV was measured (see Fig. 4), which can be compared with a value of 10.3 mV obtained from DC polarisation techniques for pure copper in a very similar, however not identical, deoxygenated copper/bentonite environment [16].

4.2. General Corrosion

Objections can be raised to a comparison between the gravimetric data from the copper coupon in bentonite block S122 and the real-time monitoring data from the copper electrodes in bentonite block A236. However, presently this is the only comparison that could be made as test parcel A2 will not be retrieved until 2004. With this in mind, it can be said that the data are of the same order of magnitude (3 µm per year for S122 at 50°C compared to 1.7 µm per year for A236 at 24°C). Since most of the entrapped oxygen in test parcel A2 should have been consumed by now a lower corrosion rate was in fact expected [14]. Thus, the corrosion monitoring over the next few months will be interesting to follow.

The harmonic distortion analysis measurements gave a similar estimate of the corrosion rate to the LPR measurements (see Fig. 6). However, the corrosion rate calculated from the EN data shows a higher than expected value. This is not surprising since it is known that very low corrosion rates are usually overestimated by the EN technique due to the sampling size employed in the EN calculation. Observations have been made that if the polarisation resistance (R_p) is very high (e.g. passive or coated electrodes), the EN resistance may be substantially lower than R_p, unless very long time records are taken [17–20]. This may explain why, in the low corrosion activity that is observed, the corrosion rate predicted from the EN measurements is higher than that predicted by LPR monitoring, as the time records employed are 300 s in duration. Nevertheless, the purpose of the EN technique is to investigate localised corrosion activity rather than the uniform corrosion rate.

4.3. Tendency to Localised Corrosion

The EN data depict a slight tendency to localised corrosion on the copper electrodes in bentonite block A236. Optical and scanning electron microscopy on a copper coupon exposed in bentonite block S122 did not reveal any signs of pitting after about one year exposure. However, it was observed that the corrosion products were not the same all over the specimen surface [3,14].

Even if efforts were made to avoid crevices, it cannot be ruled out that tiny crevices may exist at the lead connections to the electrodes. The presence of such crevices might be the reason for the observed tendency to localised corrosion, and thus be an artefact in the corrosion monitoring. Crevice corrosion of pure copper is, however, uncommon and in long-term corrosion tests under simulated conditions of a Canadian repository, no crevice corrosion was observed on either creviced U-bend or creviced planar samples [21]. Another (maybe more probable) explanation for the apparent slight tendency to localised attack could be the fact that for very low general corrosion activity that one might expect under these conditions, corrosion events are stochastic (random) and statistically can represent pit initiation so the localisation index and 'pitting function' could produce indications of nascent pitting. If localised corrosion is present, it is expected that the localisation index will increase further as pits grow; if the measure is an artefact of the technique it will remain constant.

4.4. Benefits of the Real-Time Monitoring

It is still too early to make any firm statements about the results obtained from the real-time corrosion monitoring. Accumulation of data over a longer test period is needed. Real-time monitoring of corrosion is of course a far better way of estimating and following corrosion than merely weighing a sample before and after exposure. However, in order to estimate corrosion rates from electrochemical data the Stern–Geary 'constant' has to be known or measured for the actual material–environment system. A benefit of the equipment employed is that an estimate of the Stern–Geary constant is obtained by means of the harmonic distortion analysis technique without any need for a reference electrode. This is of great value when a possible reference electrode leakage constitutes a risk for the experiment.

5. Summary and Conclusions

- The long-term corrosion of copper in actual repository environment conditions is being studied by different techniques.

- The gravimetrically determined average corrosion rate of copper in bentonite block S122 over a period of about one year (at a temperature of about 50°C) was estimated to 3 µm per year.

- The corrosion rate of copper in bentonite block A236 after one year exposure (at a temperature of about 24°C) shows a prevailing value of about 1.7 µm per year from the linear polarisation resistance measurements (using the Stern–Geary constant value 6.5 mV measured at Äspö).

- The harmonic distortion data have produced a measurement for the Stern–Geary 'constant' and a check on the corrosion current for the linear polarisation resistance data, which are in good agreement.

- The electrochemical noise data shows a tendency towards some low-level localised corrosion, such as nascent pitting. The electrochemical noise data gives a too high estimate of the uniform corrosion rate, which is explainable.

- The equipment employed for real-time corrosion monitoring has proved capable of measuring very low corrosion activity of the order of less than 2 μm per year, with an apparent resolution of < 0.1 μm per year. This has been assisted by the choice of installing electrodes of relatively large surface area, in this case about 100 cm^2 each.

6. Acknowledgements

The contributions of Lars Andersson and Gunnar Ramqvist of SKB, and Jason Duncan of InterCorr International, in installing the equipment are gratefully acknowledged.

References

1. SR 97 Waste, repository design and sites, Swedish Nuclear Fuel and Waste Management Co, TR-99-08, 1999. SKB, Box 5864, SE-102 40 Stockholm, Sweden.
2. Deep repository for spent nuclear fuel. SR 97 — Post closure safety, Swedish Nuclear Fuel and Waste Management Co, TR-99-06, Main Report Volume 1, 1999, p. 189. SKB, Box 5864, SE-102 40 Stockholm, Sweden.
3. O. Karnland *et al.*, Long term test of buffer material — Final report on the pilot parcels, Swedish Nuclear Fuel and Waste Management Co, TR-00-22, 2000. SKB, Box 5864, SE-102 40 Stockholm, Sweden.
4. M. Stern and A. L. Geary, *J. Electrochem. Soc.*, 1957, **104**, (1), 56–63.
5. K. Darowicki and J. Majewska, *Corros. Rev.*, 1999, **17**, (5–6), 383–399.
6. M. I. Jafar, J. L. Dawson and D. G. John, Electrochemical impedance and harmonic analysis measurements on steel in concrete, in *Electrochemical Impedance: Analysis and Interpretation*, ASTM STP 1188, p. 384–403 (J. R. Scully, D. C. Silverman and M. W. Kendig, eds). American Society for Testing and Materials, West Conshohocken, Pa, USA,1993.
7. K. Hladky and J. L. Dawson, *Corros. Sci.*, 1981, **22**, (4), 317–322.
8. K. Hladky and J. L. Dawson, *Corros. Sci.*, 1982, **23**, (3), 231–237.
9. A. N. Rothwell and D. A. Eden, *Corrosion '92*, Paper No. 223, NACE, Houston, Tx, USA, 1992.
10. A. N. Rothwell and D. A. Eden, *Corrosion '92*, Paper No. 292, NACE, Houston, Tx, USA, 1992.
11. D. A. Eden, A. N. Rothwell and J. L. Dawson, *Corrosion '91*, Paper No. 444, NACE, Houston, Tx, USA, 1991.
12. D. A. Eden, J. L Dawson, and D. G. John, UK Patent Application 8611518, May 1986.
13. D. A. Eden, *Corrosion '98*, Paper No. 386, NACE International, Houston, Tx, USA, 1998.
14. B. Rosborg, O. Karnland and L Werme, The corrosion resistance of pure copper in repository environments, *10th Int. Conf. on Environmental Degradation of Materials in Nuclear Power Systems – Water Reactors*, NACE/TMS/ANS (CD-ROM), 2001, Lake Tahoe, NV, USA.
15. R. Grauer, P. J. Moreland and G. Pini, A literature review of polarisation resistance constant (*B*) values for the measurement of corrosion rate, NACE, Houston, Tx, USA, 1982.
16. A. Brennenstuhl, Kinectrics Inc., Toronto, Canada, Private communication, May 2001.
17. A. Aballe, A. Bautista, U. Bertocci and F. Huet, *Corrosion 2000*, Paper No. 00424, NACE International, Houston, Tx, USA, 2000.

18. U. Bertocci, C. Gabrielli, F. Huet, M. Keddam and P. Rousseau, *J. Electrochem. Soc.*, 1997, **144**, (1), 37–43.

19. F. Mansfeld and C. C. Lee, *J. Electrochem. Soc.*, 1997, **144**, (6), 2068–2071.

20. A. Bautista and F. Huet, *J. Electrochem. Soc.*, 1999, **146**, (5), 1730–1736.

21. S. R. Ryan *et al.*, An investigation of the long-term corrosion behaviour of selected nuclear fuel waste container materials under possible disposal vault conditions, Atomic Energy of Canada Limited, TR-489, COG-94-55, 1994, Chalk River, Ontario, Canada, K0J 1JO.

28

Effect of Magnetite as a Corrosion Product on the Corrosion of Carbon Steel Overpack

N. TANIGUCHI

Japan Nuclear Cycle Development Institute, Tokai Works, 4-33 Muramatsu, Tokai-mura, Ibaraki, Japan

ABSTRACT

The mechanism of the acceleration of the corrosion of carbon steel due to the presence of magnetite was studied. Immersion tests of carbon steel in the presence of magnetite powder were performed using sealed glass ampoules. Hydrogen gas enclosed in the ampoule was analysed by gas chromatography, and the material balance of hydrogen gas and reduced Fe(III) in the magnetite against the weight loss of the specimens was analysed. The analysis showed that the main cathodic reaction in the presence of magnetite powder was the reduction of Fe(III). The effect of magnetite on the lifetime of carbon steel overpacks is also discussed in this paper.

1. Introduction

The duration of radionuclide containment by overpacks in high-level radioactive waste repositories is currently required to be at least 1000 years [1]. Carbon steel is one of the candidate materials for overpacks in Japan as a corrosion allowance material. The maximum corrosion depth of carbon steel overpacks was estimated to be about 32 mm for 1000 years in a vertical emplacement concept taking account of the corrosion due to oxygen and water reduction [1]. The outline of this assessment is summarised in the next section. However, Kojima *et al.* reported that the corrosion rate of carbon steel was accelerated by the addition of magnetite powder on the metal surface as a dummy corrosion product [2–4]. Their estimated average corrosion rate after 1000 years was about 100 times larger than that determined in the lifetime prediction of carbon steel overpacks [4]. If such large corrosion rates cannot be avoided for a long time, the carbon steel overpacks will lose their integrity in a short period. If large amounts of hydrogen gas are generated accompanying the corrosion and no limitation is placed on the severe corrosion, then the impacts on the buffer material and host rock will be serious enough to change any concept using iron or steel as overpack material. This problem cannot be avoided by the selection of a titanium overpack or a copper overpack because a carbon steel inner container is necessary to satisfy the mechanical strength, radiation shielding function and redox buffering function [1].

It is therefore important to investigate the mechanism of corrosion in the presence of magnetite and the possibility of severe acceleration of corrosion. There are two possible mechanisms for the corrosion acceleration by magnetite. One is a catalytic action by magnetite on the cathodic reaction of hydrogen generation and the other is oxidation by Fe(III) in the magnetite. If oxidation by Fe(III) dominates the cathodic reaction, the corrosion acceleration will be stopped with consumption of Fe(III). Various studies have been conducted to understand the mechanisms of acceleration of corrosion by magnetite. Haruna *et al.* [5] measured the hydrogen gas generation rate by gas chromatography for the immersion of carbon steel with magnetite in acidic carbonate solution. They reported that the hydrogen generation reaction was not accelerated by addition of magnetite powder. Tsuru [6] measured the pressure change by generation of hydrogen gas in the coupling of carbon steel and magnetite in deaerated 0.5M NaCl solution. They concluded that although almost all the cathodic reaction was the reduction of magnetite, the hydrogen generation reaction was also accelerated several times. Fushimi *et al.* [7] using scanning electrochemical microscopy reported that 50% of the galvanic current between a crystalline magnetite electrode and a carbon steel electrode in 0.1M Na_2SO_4 solution was due to the hydrogen generation reaction. Tsushima *et al.* [8] indicated that the corrosion of carbon steel should not be accelerated by the hydrogen generation reaction on magnetite because the hydrogen overvoltage is increased by the formation of the magnetite film. Ali *et al.* [9] on the basis of a study of the mechanism of the hydrogen generation reaction on iron and magnetite reported that corrosion of carbon steel accompanying hydrogen generation would not be accelerated by the pressure of magnetite. Thus, various studies have given attention to the hydrogen generation behaviour of carbon steel and magnetite. According to these, oxidation by Fe(III) seems to dominate the cathodic reaction in the accelerated corrosion by magnetite, although the contribution of the hydrogen generation reaction cannot be ignored. A major focus of previous studies has been to find a method to detect hydrogen accurately because only a trace of hydrogen is generated on magnetite in neutral solution. In addition, the leakage of hydrogen gas from the experimental system has made it difficult to detect hydrogen precisely. Because of these difficulties, there has been no attempt to measure the quantity of hydrogen where carbon steel is in direct contact with magnetite in neutral or weakly alkaline environment. Prior attempts [6,7] to detect hydrogen from magnetite in neutral condition have been made in galvanic coupling systems in which carbon steel and magnetite were placed separately in test solutions and connected through a zero resistance ammeter.

One of the aims of the present paper is to contribute to the understanding of the hydrogen generation behaviour where carbon steel is in direct contact with magnetite in neutral or weakly alkaline environment. In this study, immersion tests of carbon steel with magnetite were conducted in completely sealed glass ampoules. All hydrogen gas generated by the corrosion during the immersion period was retained in the ampoule. The quantity of accumulated hydrogen gas in the ampoule was then analysed by gas chromatography. The ratio of Fe(III)/ Fe(II) in the magnetite powder was also analysed both before and after immersion. The material balance of the amount of hydrogen gas and reduced Fe(III) against the weight loss of the carbon steel specimens was analysed. Based on the results of experiments, the ratio of hydrogen generation reaction to the total amount of corrosion in the presence of magnetite was evaluated. The experimental procedure and results are described in Section 3 and Section 4. Another aim of this study was to examine the impact of magnetite on the corrosion lifetime of a carbon steel overpack. In Section 5, the possibility of the formation of magnetite on a carbon steel overpack was reviewed from the viewpoint of thermodynamic stability. Finally, based on the experimental data on the hydrogen generation ratio and thermodynamic stability of magnetite, the maximum corrosion depth by magnetite was estimated in Section 6.

2. Outline of Latest Corrosion Lifetime Prediction for Carbon Steel Overpacks

Before estimating the corrosion depth, the type of corrosion of carbon steel under the expected repository condition has to be understood because the propagation behaviour of corrosion will vary with the type of corrosion. Corrosion of carbon steel is divided into two types: general corrosion and localised corrosion. Localised corrosion is initiated by the breakdown of the passive film on carbon steel. If carbon steel does not passivate, general corrosion will propagate. The passivation behaviour of carbon steel will be affected by the presence of buffer material placed around the overpack as shown in Fig. 1. We have performed anodic polarisation experiments of carbon steel in compacted bentonite simulating the situation of the overpack being in contact with buffer material [10,11]. Based on the results of the experiments, we concluded that carbon steel overpacks cannot passivate in buffer material in Japanese groundwater conditions. Therefore, the main task for the lifetime prediction of carbon steel overpack is to estimate the general corrosion depth under repository conditions. There are two kinds of oxidant affecting the carbon steel overpack: oxygen and water. The assessment of corrosion depth for each oxidant is outlined below. Oxygen will be brought from the ground by the excavation, operation, ventilation and backfilling of the disposal zone. The corrosion by oxygen was assessed assuming that all the oxygen in the buffer material and the backfill will react with carbon steel. The resulting average corrosion depth due to the oxygen reaction was estimated to be 1.8 mm in a vertical emplacement concept (Fig 1) [1] assuming production of ferrous corrosion product as shown by the following equation:

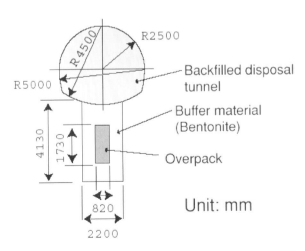

Fig. 1 *The current engineered barrier design in vertical emplacement concept [1].*

$$2Fe + O_2 + 2H_2O = 2Fe(OH)_2 \tag{1}$$

General corrosion of carbon steel is not always completely uniform. A certain degree of heterogeneous corrosion was observed in the immersion test in neutral aqueous solutions [12]. The relationship between maximum corrosion depth and the average corrosion depth was given by

$$P = Xm + 7.5Xm^{0.5} \tag{2}$$

where P is the upper limit of the maximum corrosion depth (mm) and Xm the average corrosion depth (mm) [12]. By using eqn (2), the maximum corrosion depth was assessed at about 12 mm.

After the consumption of oxygen, corrosion by water reduction will propagate on carbon steel. The corrosion depth as a result of this reaction was estimated by an extrapolation of the corrosion rate measured by the immersion test in buffer material. The result of the immersion test in compacted bentonite simulating a buffer material is shown in Fig. 2 [1]. Experimental data indicate that the average corrosion rate of carbon steel immersed in compacted bentonite reduces with time to less than 0.01 mm/year by the formation of protective corrosion products on the surface of carbon steel. Blackwood *et al.* [13] and Simpson [14] also observed a decline of the corrosion rate or hydrogen generation rate by the formation of solid corrosion products. Results from immersion tests of up to 4 years duration indicate an average corrosion rate less than 0.005 mm/year after 2 years. The average corrosion depth would be less than 5 mm in 1000 years. The average corrosion rate may decrease with time further but it was assumed to be constant at 0.01 mm/year in the lifetime prediction to make a conservative assessment allowing for uncertainty in environment conditions.

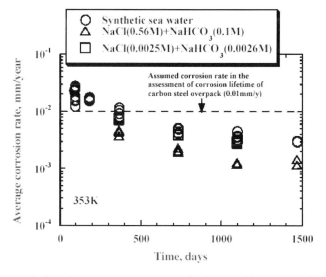

Fig. 2 Time variation of average corrosion rate of carbon steel in compacted Bentonite at 353K under anaerobic conditions [1].

The average corrosion depth was calculated to be 10 mm for 1000 years and a pitting factor of 2 was used to take account of heterogeneity of corrosion [11]. The maximum corrosion due to water was assessed at 20 mm.

The corrosion depth due to oxygen and water is summarised in Table 1. The maximum corrosion depths are added, and the total corrosion depth in 1000 years calculated to be about 32 mm. By preparing a corrosion allowance of over 32 mm, carbon steel could maintain a lifetime of over 1000 years unless the corrosion rate is accelerated by magnetite. Based on the estimation of corrosion depth, the corrosion allowance of the carbon steel overpack had been determined to be 40 mm [1].

If acceleration of corrosion by magnetite occurs on the carbon steel overpack, the corrosion may exceed this corrosion allowance. If the cause of acceleration of corrosion is the catalytic action of magnetite on hydrogen generation reaction, then the acceleration may continue for a long time. The worst scenario is that magnetite is continuously produced by the corrosion and that the corrosion rate will increase with increasing quantity of magnetite. The simulation of corrosion propagation for this scenario was performed by Kojima *et al.* [4]. According to their simulation, the corrosion rate would increase with time and the average corrosion depth would reach up to about 900 mm in 1000 years. If the simulation is correct, a carbon steel overpack with the present design will be penetrated within 100 years.

3. Experimental Procedure for Immersion Test in Sealed Glass Ampoule

A carbon steel wire having a diameter of 2 mm was cut into pieces 30 mm long. Its surface was mechanically polished with No. 800 emery paper and rinsed in acetone. The chemical composition is shown in Table 2. The test solutions were distilled water, 0.5M NaCl solution and 0.01M $NaHCO_3$ solution. The pH in distilled water and 0.5M NaCl solution was in the range of 7.6–7.8 and that in 0.01M $NaCHO_3$ solution was 8.9–9.0. The weights of magnetite powder per unit surface area of carbon steel specimens were 0.2 g cm^{-2} and 0.6 g cm^{-2}. Immersion tests using powdered glass instead of magnetite powder of 0.6 g cm^{-2} were included as a blank test to simulate the physical condition where some other material is in contact with the metal surface. The 30 pieces (test in 0.2 g cm^{-2} magnetite) or

Table 1. The latest corrosion lifetime assessment for carbon steel overpack by Japan Nuclear Cycle development Institute [1]

Oxidant	Average corrosion depth (mm)	Maximum corrosion depth (mm)
Oxygen	1.8	11.8
Water	10.0	20.0
Total	11.8	31.8

Table 2. *Chemical composition of carbon steel specimens (wt%)*

C	Si	Mn	P	S
0.02	0.01	0.26	0.012	0.021

12 pieces (test in 0.6 g cm^{-2} magnetite) of carbon steel wire were placed in the glass ampoules with magnetite powder. The ampoules were then brought into a glove box to remove the trapped air in the pores of the magnetite powder and replace this with nitrogen gas so that the oxygen concentration in the atmosphere was less than 0.1 ppm. The test solutions were also deaerated by purging nitrogen gas in the glove box for 24 h before being poured into the glass ampoules. After mixing the test solution, carbon steel specimens and magnetite powder, the stopcock of each ampoule was closed. The ampoules were brought out of the glove box and each neck of the ampoule was completely sealed by a gas burner. A schematic of a sealed ampoule is shown in Fig. 3. The sealed ampoules were put in a constant-temperature water bath and kept at 353K for 30 days and 90 days. After the immersion period, each neck of the ampoule was broken in an airtight chamber connected to a gas chromatography to analyse the total amount of hydrogen gas generated by the corrosion of carbon steel. The weight of each specimen was measured to calculate the average corrosion rate of carbon steel after cleaning in 10% diammonium hydrogen citrate solution at a cathodic current of 1 mA cm^{-2}. The ratio of Fe(III)/Fe(II) in the magnetite powder was measured by potassium dichromate titration of Fe(II) and atomic absorbtion analysis of total Fe. The schematic of the procedure for the immersion tests in the completely sealed glass ampoules is shown in Fig. 4.

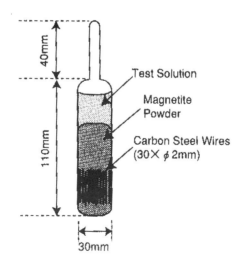

Fig. 3 *Schematic of completely sealed glass ampoule.*

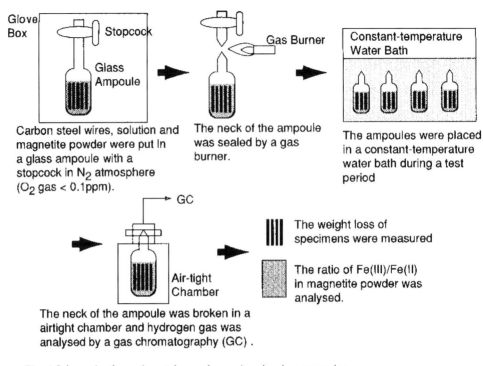

Carbon steel wires, solution and magnetite powder were put In a glass ampoule with a stopcock in N_2 atmosphere (O_2 gas < 0.1ppm).

The neck of the ampoule was sealed by a gas burner.

The ampoules were placed in a constant-temperature water bath during a test period

The neck of the ampoule was broken in a airtight chamber and hydrogen gas was analysed by a gas chromatography (GC).

The weight loss of specimens were measured

The ratio of Fe(III)/Fe(II) in magnetite powder was analysed.

Fig. 4 Schematic of experimental procedure using the glass ampoules.

4. Results of Experiments

The average corrosion rates of carbon steel observed in the immersion tests in each solution are summarised in Table 3. The quantity of hydrogen analysed by gas chromatography was converted to an average corrosion rate based on the balance of following anodic reaction (3) and cathodic reaction (4).

$$Fe = Fe^{2+} + 2e^- \tag{3}$$

$$2H_2O + 2e^- = 2OH^- + H_2 \tag{4}$$

The corrosion rates of carbon steel without magnetite powder obtained by hydrogen rates were almost in agreement with that calculated from the weight loss of the specimens except for the test of 30 days. The reason for the disagreement in this test is not known but there is a possibility that dissolved oxygen in the solution was not sufficiently removed. The weight loss and hydrogen generation rate in the presence of powdered glass applied instead of magnetite powder was almost equal to that without addition of powdered glass. Therefore, there was little effect of physical contact of the material with the surface of carbon steel in this experiment.

The relationship between the weight of magnetite powder and the average corrosion rate of carbon steel obtained by both weight loss and hydrogen generation

Table 3. Results of immersion tests in completely sealed glass ampoule

Test solution	weight of magnetite (g cm^{-2})	Test period (days)	Average corrosion rate (10^{-6} m/year) by weight loss	by hydrogen generation rate	Fe(III)/Fe(II) ratio in magnetite before immersion	after immersion
Distilled water	0	30	6.9	2.6	–	–
			6.1	2.5		
	0	90	1.3	1.2		
	0	90	2.3	1.7	–	–
	0*	90	1.7	1.7		
	0.2	30	35.3	12.3	2.72	2.51
			29.7	11.3		2.42
	0.2	90	15.4	5.2	2.72	2.47
			14.5	3.3		2.51
	0.6	30	45.3	15.2	2.72	2.58
			23.1	18.4		2.64
	0.6	90	23.1	5.3	2.72	2.42
			28.6	7.5		2.52
0.5M NaCl solution	0	30	9.1	2.8	–	–
			8.9	2.8		
	0	90	1.3	1.2		
	0	90	1.7	1.5	–	–
	0*	90	1.8	1.7		
	0.2	30	38.6	16.8	2.72	2.39
			39.8	13.1		
	0.2	90	22.4	5.5	2.72	2.35
			24.2	6.9		
	0.6	30	64.3	15.8	2.72	2.50
			74.6	17.3		
	0.6	90	30.8	8.5	2.72	2.16
			36.7	9.6		
0.1M NaHCO$_3$ solution			2.9	2.0		
			3.2	1.9	–	–
			8.8	2.7		
			8.0	3.0	3.22	2.89
			11.5	3.8	3.22	2.72
			12.6	3.3		

* = with powdered glass.

rate is shown in Figs 5–7. The average corrosion rate increased with increase in the weight of magnetite powder and decreased with time. There were distinct differences between the weight loss and hydrogen generation converted into corrosion rates.

The corrosion rates calculated from hydrogen generation rates were smaller than those determined from the weight loss of specimens. In view of the fact that the Fe(III)/Fe(II) ration in the magnetite powder after the immersion with carbon steel was smaller than that obtained before the tests, it was implied that the reduction of Fe(III) to Fe(II) would represent a considerable part of the cathodic reaction. The amount of reduced Fe(III) was converted to an average corrosion rate for each test

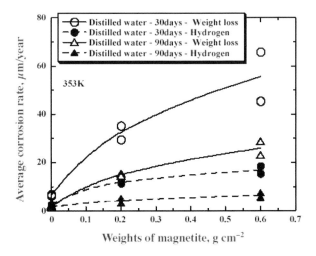

Fig. 5 *Relationship between the weights of magnetite and the average corrosion rate of carbon steel calculated from both the weight loss of the specimens and the hydrogen generation rates in distilled water.*

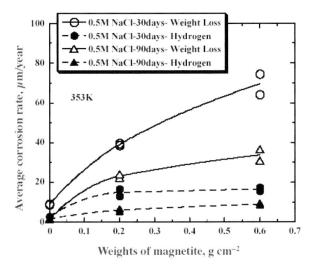

Fig. 6 *Relationship between the weights of magnetite and the average corrosion rate of carbon steel calculated from both the weight loss of the specimens and the hydrogen generation rates in 0.5M NaCl solution.*

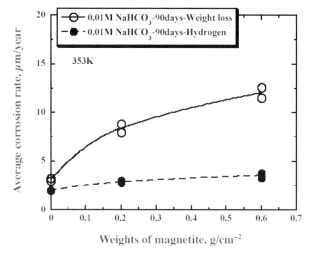

Fig. 7 *Relationship between the weights of magnetite and the average corrosion rates of carbon steel calculated from both the weight loss of the specimens and the hydrogen generation rates in 0.1M NaCHO$_3$ solution.*

and the material balance of hydrogen generation and reduction of Fe(III) against the weight loss of the specimen was analysed as shown in Fig. 8. Although fairly large dispersion was observed in the plots of solid symbols in this Figure, the sum of hydrogen generation rate and reduced Fe(III) converted to an average corrosion rate was nearly in agreement with the corrosion rate obtained by the weight loss of the specimens. The corrosion due to hydrogen generation reaction occupied only about

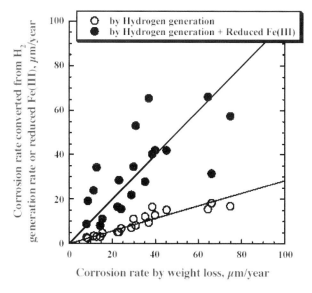

Fig. 8 *Comparison of corrosion rates obtained by the weight loss of specimens and the converted corrosion rates from hydrogen generation rates or reduced Fe(III).*

30% of the total corrosion of carbon steel. It was concluded that the main cathodic reaction in the presence of magnetite powder is the reduction of Fe(III) to Fe(II) in magnetite.

5. Possibility of Magnetite Production under Repository Conditions

Magnetite is known to be one of the representative corrosion products produced in a low oxygen concentration environment [15]. However, the corrosion products observed in the presence of compacted bentonite simulating the buffer materials were ferrous materials such as $FeCO_3$ or $Fe(OH)_2CO_3$, and magnetite has never been found by XRD analysis [16]. The condition for thermodynamic stability of Fe, Fe_3O_4 and $FeCO_3$ on the potential–pH diagram is shown in Fig. 9. In this calculation, the equilibrium eqns (5)–(7) and standard free energies of formation for $FeCO_3$, H_2O, Fe_3O_4, H_2CO_3, HCO_3^- and CO_3^{2-} by Garrels and Christ [17] were used.

$$3FeCO_3(s) + 4H_2O(l) = Fe_3O_4(s) + 3H_2CO_3\ (aq) + 2H^+(aq) + 2e- \qquad (5)$$

$$3FeCO_3(s) + 4H_2O(l) = Fe_3O_4(s) + 3HCO_3^-(aq) + 5H^+(aq) + 2e- \qquad (6)$$

$$3FeCO_3(s) + 4H_2O(l) = Fe_3O_4(s) + 3HCO_3^{2-}(aq) + 8H^+(aq) + 2e- \qquad (7)$$

The pH range of pore water in buffer material in which overpacks will be placed is estimated to be 5.9–8.4 [18]. The concentrations of bicarbonate/carbonate ions are expected to be in the range of $1.7 \times 10^{-4} – 7.8 \times 10^{-2}M$ [18]. According to Fig. 9, $FeCO_3$

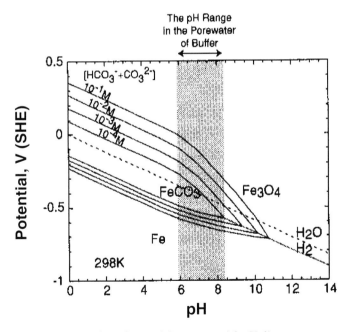

Fig. 9 *Thermodynamic stability of Fe, $FeCO_3$ on potential–pH diagram.*

is more stable than Fe_3O_4 near the H_2/H_2O equilibrium potential under the chemical conditions in the pore water of buffer materials. This thermodynamic analysis is not inconsistent with the results of XRD analysis for corrosion products generated in the presence of bentonite. Tsuru *et al.* [19] pointed out that it might be unreasonable to simulate the corrosion product generated in anaerobic conditions with magnetite, because $Fe(OH)_2$ becomes more thermodynamically stable than Fe_3O_4. Dong *et al.* [20] analysed the corrosion products in $NaHCO_3$ solution (0.01–1M) by XRD and *FT–IR* after applying various electrode potentials to carbon steel specimens between –0.75 and –0.58 V(SCE). According to the analysis, magnetite was observed only at a relatively noble potential above –0.65 and $FeCO_3$, $Fe_2(OH)_2CO_3$ and $Fe_6(OH)_{12}CO_3$ were observed at less noble potentials near the H_2/H_2O equilibrium potential.

In summary, magnetite is able to form on carbon steel overpack only under oxidising condition. The ferrous materials such as $FeCO_3$ or $Fe(OH)_2$ will be more dominant than magnetite in the corrosion products of carbon steel overpack following the consumption of oxygen.

6. Effect of Magnetite on the Corrosion Lifetime of Carbon Steel Overpacks

Even if magnetite forms on the surface of carbon steel, the film of corrosion products will probably suppress the propagation of corrosion as observed in experiments simulating the repository conditions [13,14]. However, any inhibition by a magnetite film was neglected in this study. Following the review of the thermodynamic stability of magnetite, it was assumed that magnetite is produced under oxidising conditions and will cause extra corrosion of a carbon steel overpack. The scenario for the assessment of corrosion depth due to magnetite was as follows:

(i)   The carbon steel overpack is corroded by oxygen, and all of the dissolved iron becomes ferric corrosion product.

(ii) Further corrosion occurs by the reduction of Fe(III) to Fe(II).

(iii) Corrosion by hydrogen generation also occurs coincident with the oxidation be Fe(III). The ratio of corrosion by hydrogen generation reaction to the reduction of Fe(II) was 3 to 7 based on the results in the immersion tests in completely sealed glass ampoules.

Although fully ferric corrosion products are different from magnetite, the most extreme case was assumed because the more Fe(III) present, the larger the corrosion depth. The result of the assessment of corrosion by magnetite is shown in Table 3 with a previous assessment [1] in which any acceleration by magnetite was not included. The increase in the maximum corrosion depth brought about by magnetite was only 1 mm in this case. One of the most important parameters in this assessment is this assessment is in the hydrogen generation ratio in the total corrosion reaction, which is estimated to be 30% in this study. Tsuru *et al.* observed less than 10% of the hydrogen generation reaction ratio [6] and Fushimi *et al.* reported a ratio of about

Table 3. _Calculation result of the corrosion depth by oxygen and magnetite (unit: mm)_

		No effect on magnetite [1]	With the effect of magnetite
	Corrosion by oxygen	1.79	1.20
Corrosion by magnetite	by Fe(III)	0	0.59
	by hydrogen generation	0	0.26
Total average corrosion depth, Xm		1.79	2.05
Max. corrosion depth, P ($P = Xm + 7.5Xm^{0.5}$)		11.8	12.8

50% [7]. The effect of the ratio of hydrogen generation reaction on the estimation of corrosion depth is shown in Table 4. Even if the ratio of the hydrogen generation reaction is assumed to be 50% of the maximum value at present, the increase in the maximum corrosion depth would be only 2.2 mm compared with the previous assessment in which no effect of magnetite was included. The corrosion allowance of 40 mm for the present design of carbon steel overpack [1] is enough to prevent a penetration even if the acceleration of corrosion by magnetite occurs.

Additionally, this study has examined the more severe case where magnetite is produced not only in aerobic but also in anaerobic conditions. This case would seem to contradict the account in the previous section. However, the possibility of magnetite production in anaerobic condition cannot be entirely ruled out, because there are some uncertainties in the environmental conditions. For example, Oda _et al._ indicated that the buffer material might lose its chemical buffering function after a long time [19]. They simulated the pore water of buffer material after a long time by repeating a replacement of pore water with groundwater. According to their calculation, the pH of the pore water of the buffer material rose and the carbonate/bicarbonate

Table 4. _Effect of the contribution ratio of hydrogen generation reaction on the estimation of corrosion depth (unit: mm)_

	No effect of magnetite [1]	Contribution ratio of hydrogen generation reaction in the corrosion due to magnetite		
		10%	30%	50%
Estimated maximum corrosion depth due to O_2 and magnetite	11.8	12.1	12.8	14.0
Total corrosion depth including the corrosion due to water reduction of 20 mm	31.8	32.1	32.8	34.0

concentration decreased on repeating the replacement. The time required for a full replacement of pore water depends on the geological condition and varies from less than 100 years to several ten thousand years. If buffer material loses its chemical buffering function in a short period and a high-pH environment or low-carbonate/bicarbonate conditions are achieved around the overpack, then magnetite may be produced under anaerobic condition as shown in Fig. 9. Taking into consideration the uncertainty in environmental conditions, the possibility of magnetite production in anaerobic condition cannot be ignored. Therefore, the author assumed that magnetite was produced in anaerobic conditions and that this led to extra corrosion on carbon steel. As mentioned above, the average corrosion depth in 1000 years was conservatively estimated to be 10 mm including a margin of 5 mm for uncertainty under environmental. A calculation was performed to investigate whether or not the corrosion by magnetite exceeds the margin of 5 mm. Assuming that all the iron corroded in anaerobic conditions became magnetite at a corrosion rate of 0.005 mm/year for 100 years, the amount of magnetite per unit surface of carbon steel is estimated to be 0.023 mol cm^{-2}. Therefore, the quantity of Fe(III) in magnetite can be determined as 0.069 mol cm^{-2} and extra corrosion by Fe(III) can be estimated to be 1.67 mm. Corrosion by the hydrogen generation reaction is then calculated to be 0.71 mm assuming a hydrogen generation ratio of 30%. Total increase in the average corrosion depth is, therefore, about 2.4 mm, which is smaller than the margin of 5 mm.

Throughout the simulation of corrosion by magnetite , even if overpack corrosion is accelerated by magnetite, the impact on the lifetime of overpack would be negligible. However, the ratio of hydrogen generation reactions may vary with magnetite properties such as the Fe(III)/Fe(II) ration and environmental conditions such as pH, temperature, anion concentration and so on. In order to make more reliable assessments about corrosion by magnetite, these effects will have to be understood in the future.

7. Conclusions

- The results of immersion tests of carbon steel with magnetite powder in completely sealed glass ampoule showed that the main cathodic reaction of corrosion was the reduction of Fe(III) to Fe(II) in magnetite and that the ratio of the hydrogen generation reaction to the total cathodic reaction was about 30%.

- From the viewpoint of thermodynamic stability, it is only possible to produce magnetite in oxidising conditions in the buffer material.

- Based on the experimental results, if acceleration of corrosion due to magnetite occurs, the effect on the corrosion lifetime of overpack would be negligible.

References

1. Japan Nuclear Cycle Development Institute: Second Progress Report on Research and Development for the Geological Disposal of HLW in Japan, Supporting Report 2 Repository Design and Engineering Technology, Chapter 4, JNC TN1410 200-003, 2000.

2. Y. Kojima, T. Hioki and S. Tsujikawa, Simulation of the State of Carbon Steel *n* Years after disposal with *n* Years of Corrosion Product on its Surface in a Bentonite Environment, *Mat. Res. Soc. Symp. Proc.*, 1995, 353, 711–718.

3. Y. Kojima, T. Taguchi and S. Tsujikawa, Corrosion Behaviour of Steel Material with Corrosion Product Layer in Compressed Bentonite Environment, *Proc. JSCE '97*, pp.297–300, 1997.

4. Y. Kojima, T. Yabuuchi and S. Tsujikawa, Acceleration of Corrosion Rate of Carbon Steel in Compressed Bentonite Environment by Deposited Corrosion Product, *Proc. JSCE Materials and Environment*, pp.233–236, 1998.

5. T. Haruna, M. Murakami and T. Shabita, Hydrogen Generation with Corrosion of Carbon steel in Bicarbonate Aqueous Solution, *Proc. 46th Japan Conf. on Materials and Environment*, pp.271–274, 1999.

6. T. Tsuru, Acceleration Mechanism of Corrosion of Carbon Steel by Corrosion Product — II, Modelling Study on Corrosion Mechanism for Corrosion of Candidate Materials for Overpacks, p.69, JNC TJ8400 2000-013, 2000.

7. K. Fushimi, T. Yamamuro and M. Seo, Hydrogen Generation from a Single Crystal Magnetite Coupled Galvanically with a Carbon Steel in Sulfate Solution, *Corros. Sci.*, 2002, **44**, 611–623.

8. S. Tsushima, N. Akao, N. Hara and K. Sugimoto, Effect of Minor Alloying Element on the Corrosion Resistance of Carbon Steel for a Container of High-level Radioactive Waste Disposal, *Proc. 48th Japan Conf. on Materials and Environments*, pp.325–328, 2001.

9. M. R. Ali, A. Araoka, A. Nishikata and T. Tsuru, Mechanism of Hydrogen Evolution Reaction on Iron and Magnetite in Neutral Solutions, JNC TJ8400 2001-008, 2001.

10. N. Taniguchi, A. Honda and H. Ishikawa, Experimental Investigation of Passivation Behaviour and Corrosion Rate of Carbon Steel in Compacted Bentonite, *Mat. Res. Soc. Symp. Proc.*, 1998, 506, 495–501.

11. N. Taniguchi, A. Honda, M. Kawasaki, M. Morita, M. Morimoto and M. Yui, A study for Localization of Corrosion on Carbon Steel Overpack, JNC TN8400 99-067, 1999.

12. H. Ishikawa, A. Honda, K. Tsurudome, K. Inoue, M. Obata and N. Sasaki, Selection of Candidate Materials for Overpack and Lifetime Prediction of Carbon Steel Overpack, PNC TN8410 92–139, 1992.

13. D. J. Blackwood, A. R. Hoch, C. C. Naish, A. Rance and S. M. Sharland, Research on Corrosion Aspects of Advanced Cold Process Canister, SKB Technical Report 92-12, 1994.

14. J. P. Simpson, Experiments on Container Materials for Swiss High-Level Waste Disposal Project — Part II, Nagre Technical Report NAGRA-NTB 84-01, 1984.

15. M. Pourbaix, *Atlas of Electrochemical Equilibria*. Pergamon Press, 1996.

16. N. Taniguchi, A. Honda, M. Kawasaki and T. Tsuru, The Assessment of Corrosion Type and Corrosion Rate of Carbon Steel in Compacted Bentonite, JNC TN8400 99-003, 1999.

17. R. M. Garrels and C. L. Christ, *Solutions, Minerals, and Equilibria*. Jones and Bartlett Publishers, 1990.

18. C. Oda, M. Shibata and M. Yui, Evaluation of Porewater Chemistry in the Buffer Mineral for the Second Progress Report H12, JNC TN8400 99-078, 1999.

19. T. Tsuru, H. Watanabe and A. Nishikata: Acceleration Mechanism of Corrosion of Carbon Steel by Corrosion Product, Modelling Study on Corrosion Mechanism for Corrosion of Candidate Materials for Overpacks, p.85, JNC TN8400 99-047, 1999.

20. J. Dong, T. Nishimura and T. Kodama, Corrosion Behaviour of Carbon steel in Bicarbonate (HCO_3^-) Solutions, *Proc. 48th Japan Conf. on Materials and Environments*, pp.321–324, 2001.

N.B. JNC reports can be obtained from Technical Cooperation Section, Technology Management Division, Japan Nuclear Cycle Development Institute, 4-49 Muramatsu, Tokai-mura, Nakagun, Ibaraki, 319-1184 Japan.

29

Carbon Steel Behaviour in Compacted Clay: Two Long Term Tests for Corrosion Prediction

F. PAPILLON, M. JULLIEN and C. BATAILLON*

Service Analyse et Migration des Radioléléments, Commissariat à l'Energie Atomique, Cadarache,
F-13108 Saint Paul Lèz Durance, France
*Service de la Corrosion et du Comportement des Matériaux dans leur Environnement, Commissariat à
l'Energie Atomique, Saclay, F-91191 Gif sur Yvette, France

ABSTRACT

Two long-term corrosion experiments were conducted in order to study the behaviour of a carbon steel within the framework of general studies dealing with container corrosion in deep geological disposal. The experiments were designed to reproduce the anaerobic and non-oxidising conditions that prevail once all the residual oxygen is consumed by the corrosion of the container and by the oxidation reactions with the surrounding environment. The objectives of the study were to characterise the corrosion process and to compare the experimental results with the predictions of the existing semi-empirical model for carbon steel and low alloy steel corrosion. The experiments used a corrosion cell in which five sheet samples of a carbon steel were placed in contact with a compacted clay. The clay matrix was saturated and percolated with an alkaline water (having properties similar to those of the groundwater) with a flow rate of 1 mL d^{-1} and a hydrostatic pressure of 30 bar. The experiments were performed at two different temperatures: 25°C (6 months duration) and 80°C (8 months duration). The post-test study of the samples showed general corrosion on the sheets from the 25°C experiment while general and localised corrosion were found in the 80°C experiment. The estimated corrosion rates were 4–5 μm/year for the experiment at 25°C and 7–8 μm/year for the experiment at 80°C. For the general corrosion process, these results agree well with the predictions of the Semi-empirical Comprehensive Model (developed earlier by Electricite de France) which is valid for carbon steel corrosion under anaerobic conditions.

1. Introduction

The interaction between nuclear waste containers and their geological environment (rock or engineered barrier) is the subject of various international research programmes [1–6]. One of the objectives of the corrosion experiments performed by the French Atomic Energy Commission (CEA) is to contribute to the study of carbon or low-alloy steels, which are candidates for the fabrication of nuclear waste containers. These steels can develop general and homogeneous corrosion. Instead of being an inconvenience, this property has the advantage of allowing a better prediction to be made of the long-term corrosion behaviour of the material. This paper describes the corrosion experiments performed by the CEA in collaboration with the French Agency for Nuclear Waste Management (ANDRA) and the French Electricity Company (EDF).

Our study consisted of two long duration experiments designed to reproduce the anaerobic conditions of the corrosion of a carbon steel in contact with compacted FoCa7 smectitic clay (see section 2.2) and groundwater. The experiments also reproduced the conditions found in a geological site, for example, the hydrostatic pressure (corresponding to the depth of the potential site), the physical–chemical properties of the clay (compositions, compression rate and porosity) and the presence of groundwater with a given flow rate and a given chemical composition. The objectives of the study were the determination of the corrosion rate, the corrosion type, the corrosion products and the reactivity of the clay. These parameters are of great interest as the clay could be used as a buffer material or could be the host rock in some geological disposal concepts because of its property of retention of water and radio-nuclides. The results obtained from the experiment were also compared with the predictions of the Semi-empirical Comprehensive Model (SCM) developed for carbon steel and low-alloy steel corrosion [7,8]. The purpose of this model is to determine the life expectancy of an iron container in a clay environment.

The paper is arranged as follows. Section 2 describes the experimental device, the preparation of the samples and the start-up of the experiment. Section 3 presents the results regarding the post-test visual inspection of the samples, the corrosion rate (gravimetric method), the corrosion products (X-ray diffraction analysis, optical and electron microscopy) and the properties of the interface between the steel and the clay. Results obtained in this study are compared with previous experiments and with the predictions of the Semi-empirical Comprehensive Model. In Section 4 a corrosion mechanism is proposed to explain the results of the experiments. Finally, the conclusions are presented in Section 5.

2. The Experiment

2.1. Experimental Device

The experimental device was composed of a water generator and the corrosion cell containing the steel sheets and the clay.

2.1.1. Water generator.
The water generator (WG) provides water, representative of a referential geological site, for the feed of the corrosion cell. In order to obtain the adequate physical, chemical properties, the WG was designed to percolate natural water (extracted from a granite site in France at about 630 m of depth) through columns composed of materials found in the same geological site. The water obtained from this process was alkaline (pH = 8,0) and reducing (Eh = –230 mV(NHE)) with the chemical composition reported in Table 1.

2.1.2. Corrosion cell.
The corrosion cells (CC), made of stainless steel, were designed to accommodate the high hydrostatic pressure (30 bar) and temperature conditions of the experiment (25°C and 80°C). Each CC contained five metal samples kept in place by a structure made of polymer. The sheets were put in contact with a compacted clay and exposed

Table 1. Chemical composition of the water obtained from the WG

Element	Concentration (mg L^{-1})	Element	Concentration (mg L^{-1})
Al	0.10	Cl	54
Ca	12	HCO$_3$	272
Fe	0.05	Si	9.5
K	38	SO$_4$	3.0
Li	0.33	S	2.4
Mg	3.5	F	9.9
Na	132	Mn	0.3
Zn	0.05–0.51		

to the flow of the water supplied by the WG (see Fig. 1). The flow rate was fixed by setting the pressure gradient between inlet and outlet of the CC. Two diffusers were used to obtain a laminar flow in the cell.

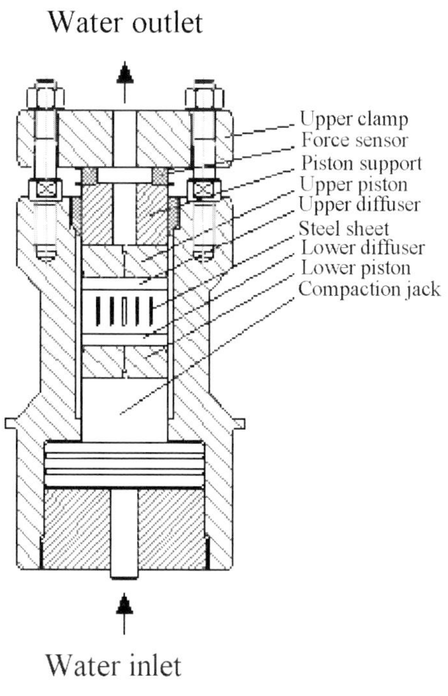

Water outlet

Upper clamp
Force sensor
Piston support
Upper piston
Upper diffuser
Steel sheet
Lower diffuser
Lower piston
Compaction jack

Water inlet

Fig. 1 *Corrosion cell diagram.*

2.2. Preparation of the Samples

2.2.1. Metallic sheets.

The metallic sheets used in our experiments were 50 mm × 20 mm × 2 mm plates made of carbon steel No.1035 (ASTM / AISI standards) with the chemical composition given in Table 2. The samples were prepared by mechanical polishing with successive grades of emery paper (1000–1200) and rinsed with alcohol and distilled water between each operation. In addition, before introducing them into the corrosion cell, the sheets were ultrasonically degreased (using a cleaning solution composed of acetone, toluene and alcohol), rinsed, dried and finally weighed.

2.2.2. Clay (FoCa7).

The clayey material chosen for the study was the French clay FoCa7. The FoCa7 is a kaolinite/smectite mixed-layer [9] with accessory minerals like quartz, goethite, calcite, gypsum and haematite as reported in Table 3.

Table 2. Chemical composition of the carbon steel

Element	mass%	Element	mass%
Al	0.005–0.012	Ni	0.090–0.110
C	0.364–0.370	P	0.010–0.014
Cr	0.120–0.220	S	0.022–0.023
Cu	0.200–0.230	Si	0.230–0.260
Mn	0.650–0.760	Sn	0.012–0.013
Mo	0.020–0.030	Ti	0.009–0.013
Nb	0.001	V	0.002–0.003

Table 3. Chemical composition of FoCa7 clay

Element	mass %	Element	mass %
SiO_2	44.08	MgO	0.91
TiO_2	1.77	CaO	2.66
Al_2O_3	24.31	Na_2O	0.32
Fe_2O_3	7.53	K_2O	0.31
MnO	0.25	Total	82.14

One important feature of the FoCa7 clay is its high cation exchange capacity (CEC) as can be seen in Table 4. This characteristic was found to be useful for retention of radio-nuclides in the event of a leak of the container integrity. On the other hand, some studies showed that the movement of exchangeable cations (like Ca^{2+}, Na^+) could contribute to the ionic conductivity of the clay [10]. This conductivity seems to be improved with the compression rate of the clay and it could be independent of saturation. Consequently, in our experiments the compacted FoCa7 could participate actively in the corrosion process of the steel.

Another feature that must be taken into account in our experiment is the fact that when the clay is saturated with water, it will exchange ions until chemical equilibrium is reached. Therefore, in order to avoid this chemical transitory, the preparation of the clay was performed by repeating the following operations five times:

1. Leaching with the natural groundwater, then centrifugation and filtration;

2. Drying in an oven at 60°C for 48 h and crushing.

2.3. Start-up of the Experiment

The start up of the experiment involved the following steps:

(i)   Determining the clay mass necessary to obtain a dry density of 1.3 g cm^{-3}.

(ii) Packing the steel plates and the clay into the corrosion cell (see Fig. 2).

(iii) Performing three cycles of argon purge and vacuum cleaning to eliminate the residual oxygen inside the cell.

(iv) Compressing the clay to achieve the necessary density by tightening the higher clamp (uni-axial compression).

(v)   Performing three cycles of argon purge and vacuum cleaning.

Table 4. Cation exchange capacity of the clay FoCa7

Cations	CEC (in meq/100 g)
Ca^{2+}	73.1
Na^+	3.6
K^+	0.8
Mg^{2+}	6.5

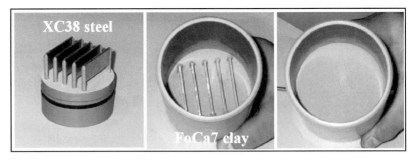

Fig. 2 Placing of carbon steel and FoCa7 clay into the corrosion cell.

(vi) Connecting the cell to the water generator and performing the saturation of the medium composed of the metal sheets and the clay. The saturation process required about 15 days because of the low permeability of the clayey matrix. This stage was performed at 25°C, 30 bar and with a flow rate of 1 mL d^{-1}.

Two experiments were conducted: one set of steel samples was corroded at 25°C for a duration of 6 months and another at 80°C for 8 months. In the latter case, the final temperature was reached by step increments using two heaters, one placed at the outlet of the WG and the other embedded in the cell (a heating coil).

3. Experimental Results

3.1. Visual Inspection of the Post-Test Samples

Once the experiments were stopped, the position of the steel sheet in the clay bulk was determined by X-radiography. Then the corrosion cells were dismantled in a glove box under anaerobic atmosphere. The visual inspection of the post-test samples indicated that there had been a reaction between steel and clay. At 25°C, the initially ochre colour of the clay changed to green in the proximity of the metal: a green homogeneous clay layer with a maximal thickness of 300 μm was observed (see Fig. 3a). In addition to this characteristic colour change of the clay, in the experiment at 80°C, an heterogeneity of the texture of the bulk clay material (see Fig. 4a) was also observed. Chemical analyses showed that these heterogeneities correspond to composition changes of the clay.

After the extraction of the metallic samples from the clay matrix, the steel sheet surface was examined. *General and homogeneous corrosion* (see Fig. 5) was found for the samples of the experiment at 25°C while *general corrosion and localised corrosion* (and more particularly pitting corrosion (see Fig. 6)) were found for the samples at 80°C. The steel sheet surface corresponding to the 25°C experiment also revealed the presence of a green adherent layer of clay resistant to water cleaning and ultrasound. In this case the rupture line between clay and steel was very homogenous and very close to the interface (see Fig. 3b). On the contrary, for the experiment at 80°C, the rupture line between steel and clay was found to occur very heterogeneously. In addition, a green smooth coat covered some sections of the steel surface while the remainder were covered by

**Green clay at the interface
between XC38 and FoCa7**

XC38 steel embedded in clay

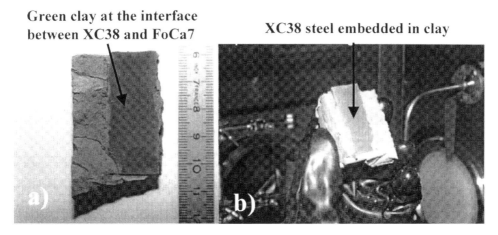

Fig. 3 *Macroscopic photographs of clay and steel of the experiment at 25 °C (6 months) after the test.*

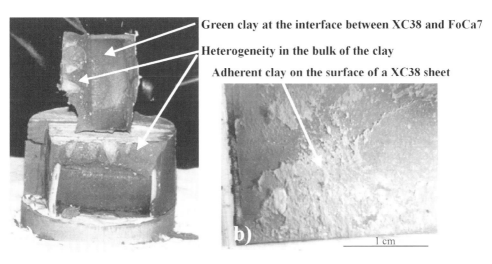

Green clay at the interface between XC38 and FoCa7

Heterogeneity in the bulk of the clay

Adherent clay on the surface of a XC38 sheet

1 cm

Fig. 4 *(a) Photograph of the clay and the steel of the experiment at 80 °C (8 months) after the test; (b) Photograph of a sheet recovered by adherent clay.*

an ochre clay layer, 500 µm thick (see Fig. 4b). This last layer was also associated with white nodules that did not exist in the initial clay.

3.2. Corrosion Rate

Estimation of the total corrosion damage was performed by a gravimetric procedure. This procedure commenced with a chemical cleaning of the samples (using a 50% vol. hydrochloric acid solution with 5 g L^{-1} of hexamethylenetetramine) followed by mass loss measurement. These mass measurements indicated a thickness reduction between 2.17 ± 0.07 µm and 2.51 ± 0.06 µm for the samples of the 25°C experiment. These results correspond to a calculated corrosion rate of *ca.* 4–5 µm/year. In the

Fig. 5 *General corrosion of carbon steel at 25 °C (6 months).*

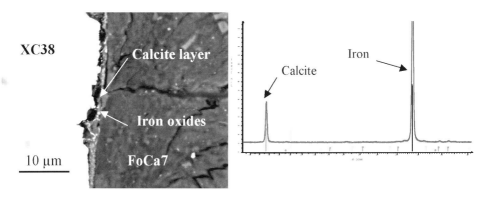

Fig. 6 *Corrosion products at the surface of carbon steel at 25 °C: (a) faces of the calcite layer adhering to the steel; (b) XRD diagram obtained for the surface of a sheet and revealing the presence of pure calcite.*

case of the experiment at 80°C, the thickness reductions were between 5.52 ± 0.28 µm and 4.96 ± 0.25 µm. In this case the equivalent corrosion rate would be 7–8 µm/year. As shown in Table 5, both corrosion rates are in good agreement with the predictions

Table 5. *Corrosion rates obtained from the experiment and from the SCM [7]*

Temperature	Experimental corrosion rate (µm/year)	SCM corrosion rate [7] (µm/year)
25°C	4–5 µm/year	2.3 µm/year
80°C	7–8 µm/year	8.3 µm/year

of the Semi-empirical Comprehensive Model for carbon steel and low-alloy steel corrosion under anaerobic conditions [7]:

$$V_{corr} = 8200 \cdot \exp\left(-\frac{2435}{T}\right)$$

(1)

where V_{corr} is the corrosion velocity (or corrosion rate) expressed in $\mu m/year$ and T is the temperature in Kelvin.

3.3. Corrosion Products

3.3.1. At the interface.
The analysis of the steel samples, using Optical Microscope and Scanning Electron Microscope SEM (Philips XL30) techniques showed, at 25°C, metallographic corrosion (under the green layer of clay) characterised by the presence of ferrite boundaries and pearlite strips. No formation of a thick passive layer was detected at 25°C in this analysis. These results agree well with those obtained in a similar parametric corrosion study of carbon steel in contact with a mixture of FoCa7 clay (50%) and water (50%), under anaerobic and reducing conditions at 50°C. However, in our samples, the iron was found in the form of nanometric iron oxides (probably magnetite) within the first 50 µm of the clay which were in contact with the steel and a pure calcite layer was detected partially adherent to the steel as can be seen in Fig. 6.

The same analyses performed on the samples corroded at 80°C showed that pits were distributed over the whole surface of the metallic sheet, except for one sheet, where the pitting was located in a limited triangular area. The pits had a spherical shape, reaching a maximum depth of 50 µm (see Fig. 7). A compacted and adherent layer of corrosion products was also observed on the bottom of the pits. The composition of these products did not correspond to the composition of the initial clay but rather to a mixture of iron oxides associated with silicon (see Fig. 8). This type of chemical compound, forming a protective layer, has already been identified in previous corrosion experiments using the same carbon steel and under anaerobic and reducing conditions. Therefore the presence of Fe-Si mixed products in the bottom of the pits in the samples at 80°C could indicate the existence of a repassivation process. However, even if pitting propagation was going to stop in this experiment, the risk of pitting corrosion in such conditions has to be underlined.

3.3.2. In the clay near the carbon steel.
No significant mineralogical transformation was observed in the clay at 25°C except in the contact region between the clay and the steel. The crystallisation of magnetite into nanometric clusters was associated with the presence of micrometric calcite crystals in the first 50 µm of clay (see Fig. 9). On the other hand, we observed in the first 100 µm of analysed clay (using Transmission Electron Microscope TEM-Jeol 2000 FX equipped with Link ISIS EDS analytical device), a preferential dissolution of only the nanometric crystals of goethite (50 nm in size) initially present in the clayey matrix (less than 5% of the initial goethite content in the clay). The chemical–transport model of Trotignon *et al.* [11] suggests that hydrogen produced during the corrosion could be the origin of the reduction of goethite, i.e.:

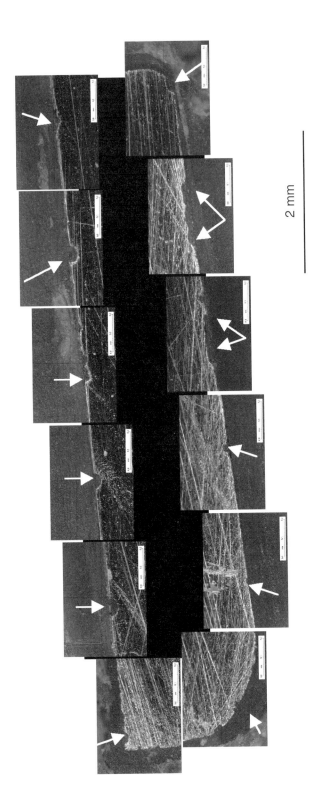

2 mm

Fig. 7 Localised corrosion of carbon steel at 80°C (8 months).

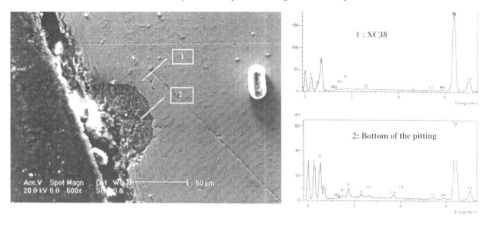

Fig. 8 *Pitting corrosion on the carbon steel at 80 °C (SEM image, SE); EDS spectra of particular zones: (1) Carbon steel and (2) bottom of the pit.*

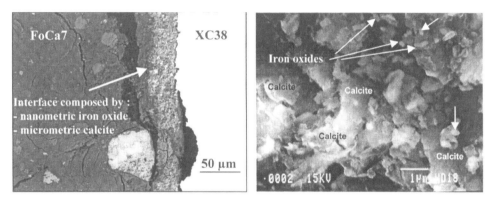

Fig. 9 *Interface between carbon steel and FoCa7 clay, SEM images: (a) global view, (b) detail of the zone composed by calcite and iron oxides.*

$$FeO(OH) + {}^1\!/_2\,H_2 + 2H^+ \Leftrightarrow Fe^{2+} + 2H_2O \tag{2}$$

The liberation of Fe^{2+} in the reaction at 25°C could also explain the formation of magnetite crystallites in the proximity of the steel and even in the first 300 μm of clay. Moreover, the fact that the magnetite is black in the mineral form and green in the hydrated form [12] could account for the observed colour change of the clay matrix.

In the experiment performed at 80°C, two types of zones characterised the clay near the steel: ochre zones and green zones. Different XRD analyses showed that the mineralogical evolution of these zones differed (see Fig. 10). Ochre zones were characterised by the presence of pure calcite while the green zones corresponded to the presence of calcite, siderite and an iron rich 0.7 nm phyllosilicate phase analogous to berthierine (see Fig. 11).

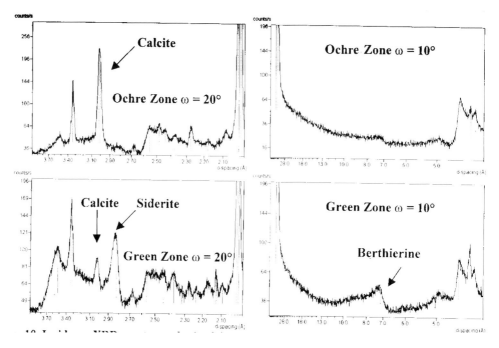

Fig. 10 *Incidence XRD spectrum obtained for the clay adhering to the surface of the carbon steel of the experiment at 80 °C: presence of pure calcite in ochre zone and calcite, siderite and berthierine in green zone.*

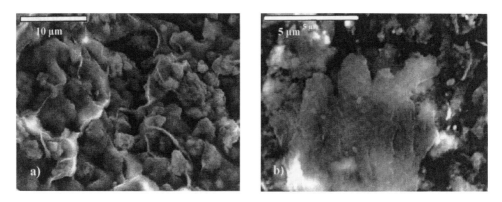

Fig. 11 *Micrography by SEM of FoCa7 clay in the experiment at 80 °C: (a) presence of calcite nodule in the ochre clay zone; (b) presence of lamellar siderite in the green clay zone.*

No correlation between pitting corrosion and presence of siderite or calcite was found. The formation of siderite has already been observed in previous experiments. In the experiments at 50°C, only traces of siderite were observed while at 90°C the siderite layer was adherent and homogeneous on the steel surface. In our experiment at 80°C, the formation of siderite seems to be close to the green clay zones.

4. Discussion

4.1. Calcite or Siderite Formation

The corrosion process of carbon steel in reducing and anaerobic condition leads to the injection of Fe^{2+} cations, H_2 and OH^- anions into the clayey matrix :

$$Fe \rightarrow Fe^{2+} + 2e^- \text{ (anode)} \tag{3}$$

$$2\,H_2O + 2e^- \rightarrow H_2 + OH^- \text{ (cathode)} \tag{4}$$

then OH^- anions react with bicarbonate:

$$HCO_3^- + OH^- \rightarrow CO_3^{2-} + H_2O \tag{5}$$

Depending on pH and on the relative concentration of Fe^{2+} and Ca^{2+}, calcite, siderite or a $Ca_{1-x}Fe_xCO_3$ solid solution can precipitate. This process depends also on the ratio between the steel corrosion rate and the ion diffusion rate in the clay (transport property). In the experiment at 25°C, the $\frac{Fe^{2+}}{Ca^{2+}}$ ratio was probably low and therefore allowed only the formation of calcite. Indeed, because of calcium excess saturation in water, calcium was the major ion at the interface. On the other hand, when the ratio was high, the major ion at the interface of the carbon steel and the FoCa7 clay was probably Fe^{2+}, thus involving precipitation of siderite. In the experiment at 80°C, these two precipitation processes coexist. The green zones were probably zones where the corrosion rate was high enough to lead to siderite formation. On the contrary, in the other zones the corrosion rate was lower, leading only to the precipitation of calcite.

As the calcite is a porous material, the diffusion of iron through the calcite layer could explain why nanometric iron oxides (probably magnetite) were formed within the first 50 µm of the clay in proximity to the steel. The corrosion was then homogeneous and generalised on the steel surface. On the other hand, in the siderite, Fe^{2+} ions form part of the layer structure and are partially immobilised near the surface of the steel. In this case, steel passivation and corrosion rate reduction could occur associated with pitting corrosion.

4.2. Localised and General Corrosion

The presence of pitting corrosion at 80°C showed that some oxidising species were injected into the system. These agents need to be available and mobile in order to reach the steel surface. The presence of oxygen in the system was unlikely because no hematite or goethite were found by XRD at the surface of the steel. These oxides should indicate an aerobic stage as shown in some parametric tests or *in situ* experiments where this condition was not controlled. In order to investigate the origin of the oxidising species, microbiological studies were carried out and revealed only the presence of anaerobic bacteria as consumers of H_2 and CO_2. As no sulfate-reducing bacteria were detected (they are usually responsible for the development of pitting corrosion), the assumption of a bacteriological origin of the oxidising species was discarded.

It seems therefore that the oxidising agent probably came from the clay, which has the structural formula:

$$\left[Si_{3.6}Al_{0.4}\right]\left(Al_{1.75}Fe^{3+}_{0.15}Mg_{0.15}\right)O_{10}\left[OH\right]_2 Ca_{0.2} \tag{6}$$

Only the Fe^{3+} cations were identified as potential strong oxidising ions (Fe^{3+}/Fe^{2+}, $E = +0.77$ V(NHE)). After consumption by corrosion processes of all the protons initially present in the clay and the destabilisation of the mineral (to preserve electro-neutrality), structural Fe^{3+} can be released in the system when the temperature increases. This alteration begins as soon as the temperature is higher than 40°C and is very significant at 80°C. The released Fe^{3+} cations can then be reduced to Fe^{2+} at the surface of steel [13,14] as it is shown in eqn (7). Under these conditions, the corrosion potential of the steel becomes anodic and thus allows pitting corrosion to develop at the nanometric to millimetric scale.

$$2Fe^{3+} + Fe^0 \rightarrow 3Fe^{2+} \tag{7}$$

One assumption put forward to explain the heterogeneous localisation of pitting corrosion, developed at the centimetric scale (triangular zone) on one of the sheets, is based on hydrodynamic mechanisms: a preferential percolation path associated with the low dry density of the initial material ($\rho d = 1.3$ g cm^{-3}) could have induced centimetric heterogeneities by increasing the quantity of Fe^{3+} near the surface of the sheet (along the percolation path).

In the case of the experiment at 80°C, the total corrosion rate could include the damage due to general corrosion by protons and the damage due to pitting corrosion by Fe^{3+}. According to the quantity of Fe^{3+} liberated from the FoCa7 and the observed alteration of about 3 mm of clay (overestimation), the estimation of the corroded thickness of the carbon steel at 80°C would be 7.7 µm. In the experiment at 25°C, no pitting corrosion was observed: the fact that liberation from the clay and mobility of Fe^{3+} are weak at this temperature could explain why no pitting corrosion occurred.

5. Conclusions

Two long-term experiments of carbon steel corrosion in contact with a compacted clay (French clay FoCa7) have been conducted. The clay matrix was saturated and percolated with alkaline water (representative of referential geological sites) with a flow rate of 1 mL d^{-1} and a hydrostatic pressure of 30 bar. The experiments were performed at two different temperatures: 25°C (6 months duration) and 80°C (8 months duration). The objectives of the study were the determination of the corrosion rate, the corrosion type, the corrosion products and the reactivity of the clay. In addition the experimental results were compared with the predictions of the Semi-empirical Comprehensive Model (SCM) relative to carbon and low-alloy steel corrosion [7,8]. The purpose of the experiment was to underline the role of the compacted clay in the corrosion process under conditions similar to those found in a potential geological site.

The post-test study of the samples (by gravimetric, SEM, TEM and XRD analysis) established two different corrosion scenarios for the carbon steel sheets. The first one, corresponding to the experiment at 25°C, is characterised by a 2 μm general corrosion of the sheets. A calcite layer and some iron oxides in the proximity of the metal (<50 μm) were also observed. However, no formation of a thick passive layer was detected. The second scenario, corresponding to the experiment at 80°C, is marked by the development of general and localised corrosion on all the tested sheets, with a maximum pit depth of 50 μm. In each case, the microscopical analysis showed that the clay in contact with the steel had experienced significant transformations. These transformations included heterogeneity in colour (redox state) and in granularity as well as the formation of new mineral phases such as berthierine and siderite.

The corrosion rates of steel sheets were estimated for each scenario: 4–5 μm/year for the experiment at 25°C and 7–8 μm/year for the experiment at 80°C. These results are in good agreement with the predictions of the Semi-empirical Comprehensive Model for carbon steel or low-alloy steel corrosion under anaerobic conditions as far as the general corrosion process is concerned. On the other hand, a possible explanation for the localised corrosion can be proposed on the basis of the temperature dependent property of the Fe^{3+} liberation by the clay. According to this assumption, this process should be sufficiently activated at a temperature of 80°C to induce localised corrosion.

6. Acknowledgements

This research was sponsored by CEA, ANDRA and EDF. The authors wish to thank Dr V. Michaud for skillful mineralogical investigations, Dr M. Libert for microbiological analyses, Dr L. Trotignon for his scientific support. Mr P. Soreau, J.C. Pétronin and P. Trabac are also gratefully acknowledged for their technical support.

References

1. B. Kursten and P. Van Iseghem, "*In situ* corrosion studies on candidate container materials for the underground disposal of High-Level radioactive waste in Boom clay", in *Corrosion '99*, Paper 473, NACE International, Houston, TX, USA, 1999.
2. B. Kursten, B. Cornelis, S. Labat and P. Van Iseghem, "Geological disposal of conditioned High-Level and long live radioactive waste — *In situ* corrosion experiments", SCK-CEN, Final report R-3121, Mol (Belgium), July (1996).
3. R. Peat, S. Brabon, P. A. H. Fennell, A. P. Rance and N. R. Smart, "Investigation of Eh, pH and corrosion of steel in anoxic groundwater", SKB, Technical report TR-01-01, Sweden, January (2001).
4. F. Wang, G. E. Gdowski, J. Estill, S. Gordon, S. Doughty, K. King and D. McCright, "Long term corrosion of waste packing candidate material for the YMP; Initial results", in *Corrosion '98*, Paper 161, NACE International, Houston, TX, USA, 1998.

5. E. Smailos, A. Martinez Esparza, B. Kursten, G. Marx and I. Azakarate, "Corrosion evaluation of metallic materials for long live HLW/Spent fuel disposal containers", Forschungszentrum Karlsruhe GmbH, Final report FZKA 6285, Germany, (1999).

6. R. Schenk "Experimente zur korrosionsbedingten Wasserstoffbildung in Endlagern für mittelaktive Abfälle", Rapport Nagra, Technischer Bericht 83–16, Suisse, (1983).

7. J. M. Gras, "Corrosion assessment of metal containers for the geological disposal of HLW; part 1: Carbon steels, low alloy steels, cast irons", Rapport EDF, HT-40/96/002/A, 87 p, 1996.

8. J. M. Gras, "Modélisation semi-empirique de la corrosion des aciers non alliés en situation de stockage - Actualisation du modèle EDF 1996", Rapport EDF, HT-40/01/004/A, 33 p, 2001. See also F. Foct and J. M. Gras, this volume, pp.91–102.

9. S. Gin, P. Jollivet, J. P. Mestre, M. Jullien and C. Pozo (1999), "French SON 68 nuclear glass alteration mechanisms on contact with clay media", *Appl. Geochem.*, 2001,**16**, 861–881.

10. C. Sladert, J. Barker, P. R. Hirst , T. K. Halstead and P. I. Reid, "Conduction and diffusion in exchanged montmorillonite clays", *Solid State Ionics*, 1987, **24**, 289–295.

11. L. Trotigon, M. H. Faure, M. Granga and H. Peycelon, " Numerical simulation of the interaction between granitic groundwater, engineered clay barrier and iron canister", in *Scientific Basis for Nuclear Waste Management XXII*, Mater. Res. Soc. Proc., 556, Boston, (1998) 599–606.

12. U. Schwertmann and R. M. Cornell, "Iron oxides in the laboratory — Preparation and characterization", VCH Editions, Weinheim, Germany.

13. E. D. Potter, *Electrochemistry: Principle and Applications*, Cleaver-Hune Press Ltd, London, UK, 1952, p.93–96.

14. I. M. Kolthoff and J. J. Lingane, *Polarography Vol. II*. Interscience Publishers, New York, 1952, p.475–480.

30

Crevice Corrosion of Passive Materials in Long Term Geological Nuclear Waste Disposal

P. COMBRADE

FRAMATOME-ANP Technical Center, 71 205 Le Creusot, France

ABSTRACT

The use of passive materials for High level Waste (HLW) containers is dependent on their resistance to crevice corrosion. Using the repassivation potential as a criterion to guarantee the resistance of passive material to crevice corrosion, susceptibility diagrams can be built up to define the safe domain of use of these alloys for a given corrosion potential. These show that, in the clay water of the French repository site, 316 L stainless steel can be used only in deaerated conditions. In oxidising conditions, Alloy C22 must be used at temperatures above 80 to 90°C, but may not be safe if severe solute concentration occurs. However, a better understanding of the meaning of the repassivation potential is still required to validate fully its use in oxidising environments.

1. Introduction

The container material for nuclear waste must resist corrosion for thousands of years. In order to use reasonable wall thickness, average corrosion rates must not exceed a few µm/year. With this requirement, passive materials are of great interest due to their very low corrosion rate.

However, passive materials, such as stainless steels, nickel base alloys, titanium alloys, etc. may suffer localised corrosion phenomena. In this respect, chloride-containing solutions are by far the most dangerous environments if they contain an oxidising species other than water, usually dissolved oxygen. In chloride environments, passive alloys may suffer pitting corrosion (Fig. 1a), crevice corrosion (Fig. 1b) and also stress corrosion cracking, SCC (Fig. 1c).

Indeed, the interstitial clay water from the French disposal location under study contains significant amounts of chloride (about 1200 mg L^{-1}, i.e. 3.4 10^{-2} mol L^{-1}, see Table 1) and the possibility of localised corrosion must be taken into account if passive materials are to be considered as candidate materials for radioactive waste containers.

Among the different forms of localised corrosion, crevice corrosion is considered as the most dangerous, because it is the easiest to initiate, if not necessarily the most critical during the propagation phase.

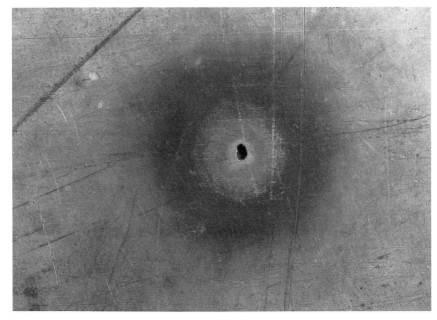

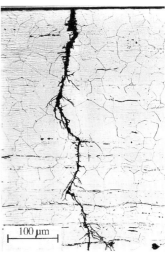

Fig. 1 Localised corrosion of 304 stainless steel in aerated chloride environments: (a) Pitting corrosion of a tank(after A. Désestret); (b) Crevice corrosion of bolts(after A. Désestret); (c) Stress corrosion cracking of a test specimen at 200°C.

Table 1. *Calculated chemical analysis (in mol L^{-1}) representative of the interstitial clay water of Meuse-Haute Marne French site*

Na	7.3 10^{-3}	Al	1.1 10^{-8}
K	2.5 10^{-3}	Cl	3.5 10^{-2}
Ca	8.6 10^{-3}	SO$_4$	1.8 10^{-3}
Mg	6.5 10^{-3}	C (as carbonates)	1.6 10^{-3}
Fe	3.0 10^{-4}		
Si	9.9 10^{-5}	pH	7.27

2. Resistance to Crevice Corrosion

For a given bulk environment, crevice initiation appears to be dependent on the corrosion potential of the free surfaces: initiation occurs when this potential exceeds a critical value E_{crev} (Fig. 2) The propagation may occur at potentials significantly lower than the potential which caused the initiation to occur, and indeed, the potential of a freely corroding material is generally lower during the propagation compared to the initiation period. However, propagation stops below a critical potential for the crevice repassivation E_{rp}. These two critical potentials may be used as criteria to avoid the crevice corrosion.

However, the *crevice potential* depends not only on the material and the environment but also on the surface condition of the material. The crevice potential may be significantly lowered by a 'bad' surface condition which can result for example

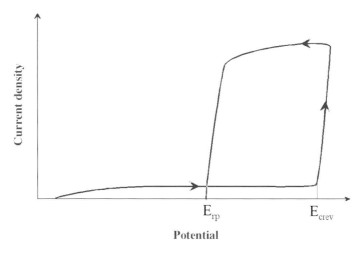

Fig. 2 *Critical potentials of crevice initiation and repassivation on a polarisation curve (forward scan and backscan).*

from modification of surface composition during heat treatments, welding or grinding, from surface pollution during manufacturing, and so on. Thus, the initiation potential cannot be a reliable criterion to avoid localised corrosion when the surface condition of the materials cannot be safely controlled during the manufacturing and handling processes. Indeed, this is the case for radioactive waste containers which will be filled, welded and handled to their final location in hostile environments.

The *repassivation potential* is a more reliable criterion because it is independent of the surface condition as soon as the corrosion has dissolved the affected metal layers. In the past, several authors claimed that this potential is dependent on crevice geometries and corrosion damage. However, more recent results tend to show that the repassivation potential reaches stable values when the corrosion damage increases (Fig. 3). We have also performed a few experiments using two very different crevice geometries with very different aspect ratios (L/h or L^2/h) and very different corrosion damages. They show that all the repassivation potentials obtained for high corrosion damage fall in the same scatter band (Fig. 4). Additionally, there has been a growing consensus on experimental techniques able to produce values of repassivation potential considered as reliable. These techniques include the production of quantified corrosion damage under potentiostatic conditions followed by a repassivation step either by a slow potential backscan, by a stepped backscan or under potentiostatic conditions.

Thus, as in many other countries (Japan, USA, Belgium, ...), the repassivation potential is being used in France as a criterion of resistance of candidate passive alloys for HLW containers: this criterion considers that a passive alloy can be safely used if its corrosion potential never exceeds its repassivation potential during the life of the container. This guarantees that no crevice corrosion will initiate on 'good' surface and that if 'bad' surface condition allows the initiation of an active corrosion, this corrosion will stop as soon as the perturbed surface layers have been consumed by the corrosion processes.

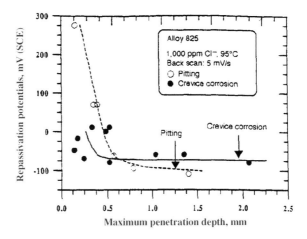

Fig. 3 Influence of the extent of corrosion damages on the repassivation potential (after [2]).

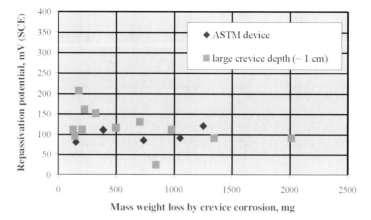

Fig. 4 *Influence of the crevice aspect ratio on the measured repassivation potential 904L Stainless steel in 1200 mg L⁻¹ Cl⁻, at 90°C [3].*

3. Parametric Effects on the Repassivation Potential

The repassivation potential of passive materials is dependent both on environment and material parameters.

3.1. Effect of the Environment

Increasing the chloride content causes a decrease of the repassivation potential. A typical effect of chloride content appears in Fig. 5. At low levels of chloride, the repassivation potential exhibits relatively high values which allow the use of the alloy in aerated environments. Increasing the chloride concentration causes a sharp

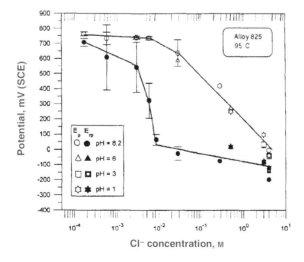

Fig. 5 *Effect of chloride content on the crevice and repassivation potentials of Alloy 825, at 95°C (after [4]).*

decrease of the repassivation potential towards low values, usually negative in the SHE scale. For high chloride concentrations, the effect of chloride on the repassivation potential is often represented by a law of the form $E_{rep} = A-B \log[Cl^-]$. This behaviour suggests the existence of a critical chloride concentration.

The temperature has a very similar effect on the repassivation potential as shown in Fig. 6. At low temperature, the repassivation potential is high. When the temperature increases, it sharply decreases to reach a range of low values. Here again, this behaviour could be described by the existence of a critical temperature for crevice corrosion.

On the other hand, pH does not seem to change significantly the repassivation potential.

To take into account these effects of the environment, susceptibility diagrams have been introduced (Fig. 7) which summarise the effect of temperature and chloride content. On these diagrams, two domains are separated by a single line: in the lower left corner which corresponds to low temperatures and low chloride contents, the alloys is resistant to crevice corrosion. In the upper right corner, crevice corrosion is possible if not necessary likely.

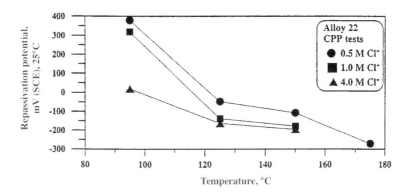

Fig. 6 Effect of temperature on the repassivation potential of Alloy C22 in solutions of different chloride content (after [5]).

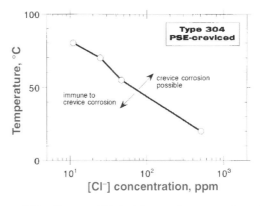

Fig. 7 Example of susceptibility diagram: 304 Stainless steel at a corrosion potential of +320 mV (SHE) (after [6]).

3.2. Material

Considering only the Fe–Ni–Cr–Mo alloys, the beneficial effect of Cr and Mo on the resistance to crevice corrosion is well established. In the offshore and oil industries, this effect is taken into account by the use of a so-called pitting index PREN:

$$PREN = wt\% \; Cr + 3.3. \; wt\% \; Mo + (16 \; to \; 30) \; wt\% \; N$$

This index is more relevant to initiation than to repassivation.

Okayama *et al.* [7] proposed the following expression to account for the effect of the main alloying elements in austenitic stainless steels on the repassivation potential:

$$E_{rp} \; (mV \; (SCE)) = 14.4 \; Cr - 96.9 \; log \; (Ni) + 29.6 \; Mo - 287 \; Cu - 492$$

According to this expression, the molybdenum has less beneficial effect on repassivation than on initiation, and both nickel and copper have an adverse effect on repassivation.

As a consequence of these effects, stainless steels may be used in low temperature, low chloride environments. As the aggressiveness of the environment increases, 'superaustenitic' (high Cr, high Mo and N additions, such as 904 L SS and higher grades) and then Ni–Cr–Mo alloys must be used.

4. Application to Candidate Materials for HLW Containers

In the site under study, located in the east of France, the ground interstitial clay water (see analysis in Table 1) has a chloride content which is around 1200 mg L^{-1}. Three different conditions of redox potential of this water must be considered during the life of the containers in contact with this water [8]:

- deaerated conditions with a corrosion potential < –250 mV (SHE),

- natural aeration which leads to moderately oxidising conditions and a corrosion potential around 200 mV (SHE), and

- highly oxidising conditions due to water radiolysis leading to a corrosion potential estimated at 400 mV (SHE).

The susceptibility diagrams of 304 and 316L stainless steels, and Ni-base alloys 625 and C22 are reported in Figs 8–10, respectively, for the three relevant environment conditions for temperature up to 100°C. They can be summarised as follows:

- in deaerated conditions, 316L SS may be used even if some chloride concentration may occur,

- in moderately oxidising (aerated) conditions, no stainless steel can be used safely. Alloy 625 is safe up to 100° C in water containing 1200 mg L^{-1} of chloride,

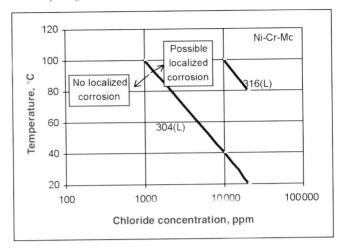

Fig. 8 *Susceptibility diagram of 304 L, 316L and Ni–Cr–Mo alloys for a corrosion potential of –250 mV (SHE) (after [8]).*

but Alloy C22 must be used if solute concentration can lead to higher values of chloride concentration at the contract of the containers, and

- in highly oxidising conditions, the only possible material is the Alloy C22, but the occurrence of crevice corrosion cannot be excluded if solute concentration may occur when the temperature exceeds 80–90°C.

5. Discussion

Although there is presently a quite large consensus about the use of this criterion, there is still some concern about the actual meaning and the intrinsic character of the repassivation potential.

In a paper elsewhere in this volume (p.225), Pourbaix describes the main mechanisms of environment evolution in a crevice and presents the concept of *protection potential* as a potential based on a 'pseudo-thermodynamic' basis (Fig. 11). This potential can be defined as the potential which stops the development of the most developed localised corrosion by

(i) suppressing the potential gradient between the crevice and the bulk solution, i.e. by suppressing the driving force for maintaining a local acidic environment and/or

(ii) by suppressing the local overpotential for dissolution in the crevice, i.e. by dissolution 'deactivation'.

This potential is slightly higher than the potential in the active crevice and it usually lies in the range of 50 to –300 mV (SHE). Thus, as indicated by Pourbaix, the protection potential is not (or is very poorly) dependent on the pH and chloride content of the

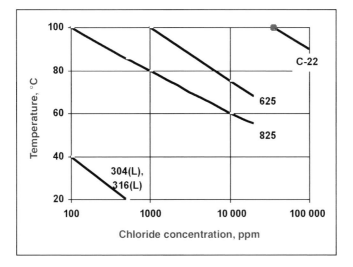

Fig. 9 *Susceptibility diagram of 304 L, 316L, Alloy 625 and Alloy C22 for a corrosion potential of +200 mV (SHE) (after [8]).*

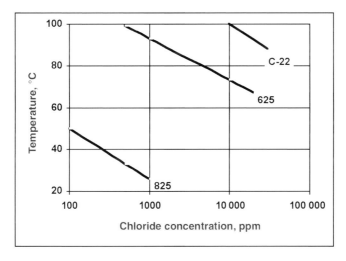

Fig. 10 *Susceptibility diagram of 304 L, 316L, Alloy 625 and Alloy C22 for a corrosion potential of +400 mV (SHE) (after [8]).*

solution outside the crevice. Indeed, Figs 5 and 12 show that the repassivation potentials measured in high chloride content solutions or at high temperature are not very dependent on the chloride content and are typical of the range expected for the protection potential.

But these figures also raise questions about the meaning of the repassivation potentials measured in low chloride content solutions which are much higher than the protection potential. The following assumptions may account for this behaviour.

It has been established that in an active crevice the local environment may become

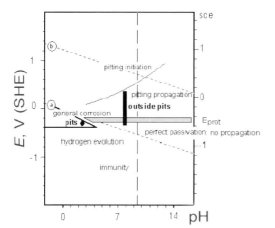

Fig. 11 Pit environment and protection potential (after Pourbaix [9]).

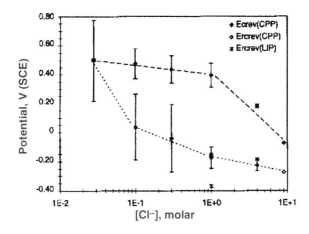

Fig. 12 Critical potentials of Alloy 625 as a function of the chloride content at 95°C (data from [10]). Note that the lower bound of the scatter band of E_{rep} is almost constant for chloride contents $\geq 3 \times 10^{-2}$ M L^{-1}.

saturated in metallic chlorides, but that the saturation is not a prerequisite for crevice initiation and propagation. In high chloride content solutions, when the repassivation potential is low, the evolution of the environment on the crevice may not be limited by the transport processes and the local environment may become easily saturated. The measured repassivation potential exhibits little or no dependence on the bulk chloride content. This appears quite clearly in Fig. 12 by looking at the lower bound of the scatter band of the repassivation potentials. This potential range can be considered as close of the protection potential.

On the contrary, in low chloride solutions, the evolution of the solution in the active crevice may be limited by transport processes such as the migration of chloride towards the crevice and the diffusion of chloride out of the crevice. In this case, the local environment cannot reach saturation and its limiting composition should be dependent on the crevice geometry. One assumption to account for the high level of repassivation potential could be the existence of a unique and reversible relationship between the local steady conditions in the crevice and the external potential. This would be consistent with the fact that initiation and repassivation potentials exhibit a common limit for long duration tests (E_u in Fig. 13).

Thus, the use of the repassivation potential as a criterion for a safe use of passive alloys in HLW containers raises two problems:

- the 'true' protection potential which is probably intrinsic of an alloy/ environment couple and can ensure a completely safe use of passive alloys, is lower than the corrosion potential in (moderately) aerated environments. Thus the use of this criterion for material selection will forbid the use of all passive materials for waste containers in oxidising environments,

- the repassivation potential measured in laboratory experiments could be a more appropriate criterion as shown in Section 3 above, but it is very likely to be dependent on the crevice geometry. The available experimental data tend

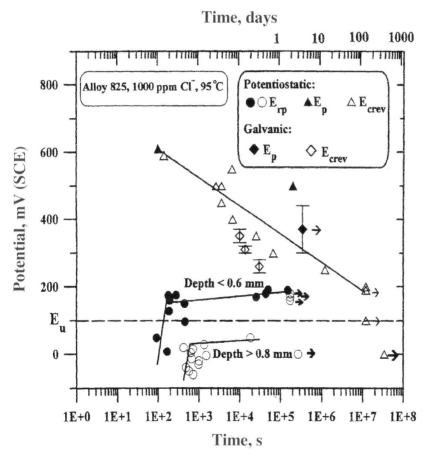

Fig. 13 *Initiation and repassivation potentials for crevice corrosion tends to the same value for very long test time — Alloy 825 in 1000 mg L^{-1} chloride at 95°C (after [2]).*

to suggest that there is a practical lower limit for this potential but the existence of such a limit must be demonstrated through a better understanding of the relationship between the local steady environment in an active crevice and the transport phenomena and crevice geometry. This is the main goal of work undertaken recently in France.

6. Conclusions

- The use of passive materials for HLW containers may guarantee a low level of general corrosion consistent with the very long expected lifetime. However, the selection of passive material must involve a criterion of resistance to localised corrosion and particularly to crevice corrosion due to the presence of chloride in the clay water which may be in contact with the containers.

- Since the surface condition of the container cannot be reliably specified, a criterion based on the repassivation potential has been selected. There is now some agreement on the techniques and the conditions of measurement of the repassivation potential in laboratories, and the effect of the chloride content and the temperature can be described by using susceptibility diagrams which define the safe conditions of use of the passive alloys for a given corrosion potential.

- The available data show that, in the clay water typical of the French repository site which contains about 1200 mg L^{-1} of chloride ions, a 316 L stainless steel may be used in deaerated conditions, but an Alloy C22 is required for moderately and highly oxidising conditions at temperature in excess of 90 to 100°C. If some process of solute concentration may occur, even Alloy C22 may not be safe if the temperature exceeds 80–90°C. Thus, the selection of a passive material must be based on a reliable estimation of the risk of solute concentration during the life of the containers.

- However, the use of the repassivation potential as a criterion for the selection of a passive material for HLW containers still requires a better understanding of the exact meaning of the measured values, particularly when these values are high, i.e. when they are consistent with the use of these materials in oxidising environments. This is the goal of a work undertaken in France to model the relationship between the potential of the free surfaces and the local chemistry inside the crevice.

7. Acknowledgements

This work is being carried out within the frame of a large programme of material selection for HLW containers driven by ANDRA and performed under the scientific control of a group of corrosion experts appointed by ANDRA. It is a pleasure to thank the members of this group for their support and particularly Messrs Crusset (ANDRA), Santarini (CEA) and Gras (EDF) who reviewed this paper.

The experimental work is being performed in several laboratories (FRAMATOME Technical Center, CORREX, Ecole des Mines de Saint Etienne and CEA).

References

1. ANDRA internal report.
2. D. S. Dunn, G. A. Cragnolino and N. Sridhar, *Corrosion*, 1996, **52**, 115–124.
3. P. Combrade, C. Bosch and C. Duret, unpublished results.
4. N. Sridhar, G. A. Cragnolino and D. S. Dunn, *Corrosion '93*, Paper No. 197, NACE, Houston, TX, USA, 1993.
5. D. S. Dunn, G. A. Cragnolino and N. Sridhar, *Mat. Res. Soc. Symp. Proc.*, Vol. 556.
6. M. Akashi, G. Nakayama and T. Fukuda, *Corrosion '98*, Paper No. 158, NACE International, Houston, TX, USA, 1998.
7. S. Okayama, Y. Uesugi and S.Tsujikawa, *Corrosion Engng*, 1987, **38**, 157–168.
8. J. M. Gras, EDF report HT-40/98/015/A, 2000.
9. A. Pourbaix, this Volume, pp.225–237.
10. K. A. Gruss, G. A. Cragnolino, D. S. Dunn and N. Sridhar, *Corrosion '98*, Paper No. 149, NACE International, Houston, TX, USA, 1998.

31

Control of Environmental Conditions during Storage of ILW Waste Packages

P. WOOD and M. WISE*

United Kingdom Nirex Ltd, Curie Avenue, Harwell, OX11 0RH, UK
*United Kingdom Atomic Energy Authority, Downs Way, Harwell, OX11 0RA, UK

ABSTRACT

The paper describes how the choice of materials, manufacturing controls and correct storage conditions are used to manage the integrity of waste packages in the UK, by (i) summarising knowledge of atmospheric localised corrosion mechanisms; (ii) identifying environmental conditions which are reported as capable of avoiding deleterious localised corrosion; (iii) discussing how this knowledge is being reflected in the designs of some UKAEA intermediate level waste (ILW) stores, together with the issues waste packagers need to consider to prevent the initiation of corrosion; and (iv) presenting information from a survey of environmental parameters and contaminants in a non-active storage building, and relating this to corrosion monitoring results from non-active waste packages.

1. Introduction

The role of Nirex is to provide the United Kingdom with safe environmentally sound and publicly acceptable options for the long-term management of radioactive materials. This includes all intermediate level waste (ILW) and some low level waste (LLW). One of the key objectives of Nirex has been to ensure that when waste is packaged by waste producers, it is in a form which is suitable for its future safe long-term management including storage, transport, handling and potential disposal.

One of the key organisations that is currently packaging ILW to Nirex standards and specifications is the UK Atomic Energy Authority (UKAEA). UKAEA has built and operated a wide range of nuclear facilities since the 1940s for the development of all aspects of atomic energy including reactor systems, fuel and reprocessing technology. UKAEA has been tasked with managing the liabilities from these facilities and has built and is operating a number of waste retrieval, packaging and storage facilities.

Nirex in conjunction with the nuclear industry has developed guidance on environmental conditions that should be applied for interim storage of ILW waste packages, which in the UK are primarily fabricated in stainless steel. This paper describes how the choice of materials, manufacturing controls and correct storage conditions are used to manage the integrity of waste packages in the UK. This paper is part of an ongoing programme of research conducted by Nirex and its contractors. It is a component of the research into one of a number of options for the long term management of radioactive waste in the UK.

2. Background

In the UK radioactive wastes are categorised according to radioactivity content and heat generating capacity [1]:

- Low-level waste (LLW) is waste with a radioactive content that does not exceed 4 GBq/t of alpha or 12 GBq/t of beta/gamma activity. LLW consists of general rubbish and other lightly contaminated plant items and equipment, as well a some materials that have been irradiated, arising predominantly from the operation and decommissioning of nuclear facilities. LLW may require containment to protect the operator but does not normally require radiation shielding.

- Wastes are classified as Intermediate-level waste (ILW) if their radioactive content exceeds the upper limits for LLW, but their radioactivity and heat output are lower than that of HLW. ILW consists principally of materials that have been irradiated in a nuclear reactor and equipment and materials that have been used in the processing of radioactive materials. ILW requires radiation shielding and/or containment to protect the operator.

- High-level waste (HLW) has the greatest concentration of radioactivity, and generates substantial quantities of heat. It is produced from the reprocessing of irradiated nuclear fuel. HLW processing and storage facilities require radiation shielding and containment to protect the operator, as well as measures to dissipate the high heat output.

The majority of all radioactive wastes produced are LLW. Currently, most LLW is routinely consigned to the Drigg facility in Cumbria, UK (operated by BNFL) for conditioning and disposal.

The UK has a wide range of ILW, over 600 waste streams from research, fuel manufacture, reprocessing, electricity generation and military activity. In order to facilitate early packaging of intermediate level wastes into passively safe forms, from its studies into phased deep geological disposal, Nirex has defined a limited number of standard waste packages suitable for the range of wastes arising in the UK. Key features and performance requirements of waste packages are defined within a Waste Package Specification [2]. These require packages to provide standardised arrangements for handling and an effective containment barrier to the dispersion of radioactive and hazardous materials. To meet these requirements stainless steel has been almost universally adopted for the manufacture of containers in which the intermediate level waste is packaged. The most commonly used stainless steels are the standard austenitic grades 316 and 316L, with more limited application of 304 and 304L (in the European material numbering system these are denoted 1.4401, 1.4404, 1.4301 and 1.4307, respectively).

Packages of conditioned waste are currently held in interim storage in surface stores. The UK Nuclear Installation Inspectorate anticipate [3] that the period of interim storage will be at least 50 years, prior to the availability of a disposal facility, followed by a second period of 50 years, during which the facility is operational, and then by a third period of 50 years during which the facility remains open and waste packages can be retrieved if required. This is also in accord with Nirex expectations. Therefore, Nirex in conjunction with the nuclear industry has developed guidance on environmental conditions that should be applied for interim storage of ILW waste packages to minimise deterioration during storage and facilitate future retrievability and onward management [4].

3. Atmospheric Corrosion Performance

The resistance to corrosion of stainless steel is well known, from anecdotes of its development, public familiarity with consumer products, and increasing use of austenitic grades in architectural, building and construction applications, where it is recognised that rainwater is beneficial to the cleaning of stainless steel [5]. However, corrosion of stainless steel by atmospheric corrosion does occur and is a potentially important degradation mechanism in the interim storage of ILW waste packages. This section presents a summary of a review of the literature on atmospheric corrosion of stainless steel, and of best practice to avoid localised corrosion. This information has been used to prepare prudent, precautionary guidance on environmental conditions that should be applied for the interim storage of ILW waste packages.

3.1. Categorisation of Corrosivity of Atmospheres

There is a history and tradition of describing the corrosiveness of atmospheres using the terms rural–marine–industrial. These tend to be qualitative terms and in some cases may be confused with geographical rather than environmental considerations. For example, it is widely assumed that chloride is absent in rural atmospheres, and only present in marine atmospheres, whereas there are many cases (as reported in this paper) where this is not the case.

ISO 9223 [6] establishes 'corrosivity category' as a general term suitable for engineering purposes, which describes the corrosion properties of atmospheres based on current knowledge of atmospheric corrosion. The five corrosivity categories are defined by the corrosion effects on standard specimens as specified in ISO 9226 [7]. The corrosivity categories may also be assessed in terms of the most significant atmospheric factors influencing the corrosion of metals and alloys, i.e. time of wetness and pollution level as specified in ISO 9225 [8]. The standards ISO 9223-6 may be used to consider the extent of atmospheric corrosion of reactive metals (carbon steel, weathering steel, zinc, copper, aluminium) given knowledge or assumptions of environmental conditions (temperature, time of wetness, airborne salinity and sulphur dioxide levels).

The corrosion of materials in *indoor* atmospheres is being considered in an ISO standard which is currently being drafted [9] based on the approach proposed by Knotkova [10]. The approach identifies five corrosivity categories, IC1-5, termed very low, low, medium, high and very high indoor corrosivity. For each of these categories there is an associated range of the rate of corrosion of *reactive* metals; an appendix considers the corrosion behaviour of basic metal representatives including nickel, lead, tin, gold and stainless steel for which it is highlighted that *"deposits of particles with ionic components (chlorides) can stimulate localised attack on surfaces in environments with higher relative humidity".*

3.2. Atmospheric Corrosion of Austenitic Stainless Steel

A review [11] of the atmospheric corrosion of stainless steel waste containers during storage included the following four points.

- Typical corrosion rates for austenitic stainless steels are in the range 0.02–3 μm year [12–16]. In rural environments, stainless steels are virtually uncorroded; in industrial atmospheres moderate staining may occur, but the depth of attack is negligible.

- Atmospheres containing high concentrations of chloride and sulphur compounds can cause surface pitting which leads to staining, but this is largely cosmetic [17,18].

- There is a possibility of atmospheric stress corrosion cracking occurring under chloride-containing deposits in certain ranges of relative humidity [19–23]. Most of the failures observed have occurred in 304 or 304L stainless steels but instances have also been seen in 301, 302, 303 and 316L.

- There is a potential for developing nitric acid films on the surface of containers as a result of radiolysis of the surrounding air [24,25]. The accumulation of nitrates on the surface of stainless steel might represent a transition to an industrial atmosphere; the impact of NO_2 as a gaseous pollutant is included in the draft ISO standard on indoor corrosion [9].

3.3. Avoidance and Minimisation of Localised corrosion

The Welding Institute have published [26] the following advice on design and structural integrity:

When designing against corrosion, one of two general philosophies may be adopted:

- *If attack proceeds fairly slowly and at a reproducible and predictable rate, a 'corrosion allowance' may be built into the design, so that a safe material thickness remains at all points after the anticipated lifetime of the structure. To assist this approach, the environment may be modified to reduce corrosion rates, where practicable.*

- *If attack may become localised and propagate rapidly once initiated, e.g. pitting and crevice corrosion of corrosion resistant alloys, it is normal to design so that attack can never initiate. This requires appropriate material selection. The simplest approach to materials selection is to rely on previous experience, although it is essential to understand the operating environment, so that changes between past and future service regimes can be accounted for ... A similar approach may be taken when designing against stress corrosion cracking.*

In the UK nuclear industry there is an example of a mixed approach where control of the environment has been used to avoid particular localised corrosion mechanisms.

- Taylor [27] carried out an extensive examination of the SCC and pitting susceptibility of 304 stainless steel over a range of surface chloride concentrations, relative humidity, temperature and exposure periods. The results of the work suggest that corrosion is sensitive to relative humidity. It is probable that the range of conditions for immunity would be wider for 316 stainless steel, whose molybdenum content confers greater corrosion resistance.

- The localised corrosion susceptibility of stainless steel is affected by the concentration of chloride on the surface. The data reported in reference [27] suggests that at 25°C ambient temperature and a relative humidity of 50–70%, a target surface chloride concentration of less than 10 μg cm^{-2} should be adopted. If this target were exceeded there would be a risk of pitting corrosion or rust staining, although the risk would be less for 316 stainless steel and by normal engineering standards the depth of corrosion is likely to be minimal [17].

The objective of this work was to define maximum allowable limits of salt contamination, depending on substrate and subsequent use, during construction and operation of power plant. If there is a risk of developing a significant concentration of chloride-containing deposits on the surface, the relative humidity will affect the risk of atmospheric stress corrosion cracking, ASCC; the data in [23] show that it would be prudent to operate at a relative humidity greater than 50%.

Nirex, in conjunction with the UK nuclear industry, has developed guidance on environmental conditions that should be applied for interim storage of ILW waste packages, drawing on the work of Taylor [27] and Shoji [23]. These laboratory studies, on the lower grade 304L stainless steel, have been combined with the experience and evidence gained from work, described in Section 5 below, on condition monitoring of a prototype 4 metre Box waste package manufactured from 304L, inspection of containers manufactured from 316L and on environmental survey work. The guidance includes:

- waste packagers should apply best practice to minimise chloride deposition on waste containers;

- to minimise surface contamination with chloride prior to use, new waste containers should be protected during transport to the processing plant;

- prior to use, stainless steel containers should be kept dry and, where possible, surfaces should be protected from contamination by dirt and dust;

- hand contact should be minimised, e.g. by use of gloves, to avoid transfer of chloride.

- store designers should consider the design of store buildings, filtration and ventilation systems and the effect these may have on internal chloride levels;

- measurements on the levels and chemical form of chloride should form part of a monitoring programme within waste package stores.

To ensure maintenance of package integrity, control of environmental conditions must be complemented by manufacturing and handling procedures; examples of these are discussed in Section 5 below.

4. UKAEA ILW Stores

UKAEA are designing packaging and storage plants, and developing packaging concepts, for waste encapsulation now and in the near future. In the absence of a finalised repository site, design and Conditions for Acceptance, UKAEA ILW packaging strategies have been developed in consultation with Nirex to ensure that the waste is packaged and stored in a manner that is consistent with Nirex plans for future waste management. Nirex has developed a strategy which facilitates packaging of ILW by providing guidance through Waste Package Specifications, supported by the formal assessment of specific packaging proposals on a case by case basis. Two documents in the Packaging Specifications are being developed to provide guidance on the storage of wastes, one on the control of store environmental conditions and one on the monitoring of packages during storage. UKAEA have used these documents in the development of store designs and in the operation of existing stores to ensure that packages will maintain their integrity for a period in excess of 150 years of storage.

UKAEA operate a number of conditioned waste stores on 4 separate sites. A number of stores are also in the design or construction phase. These stores are detailed in Table 1.

All of the stores have some degree of temperature control, largely to prevent condensation occurring on the surface of the stainless steel containers. All the stores also monitor the relative humidity of the air in the store although there is no means of providing control over the humidity. The air to all the stores is either heated or on a recirculating system to ensure the temperature within the store is controlled. All systems have inlet filtration to prevent the ingress of particulates.

Systems are being put in place to continually monitor the corrosion performance of the stainless steel packages during storage in all UKAEA stores. UKAEA are currently investigating the use of:

- Coupons — for monitoring corrosion performance of stainless steel in the specific store environment, these coupons can potentially be destructively tested at the time of transport of the packages to a future repository.

Table 1

UKAEA Site	Stores in operation or being commissioned	Stores in design or under construction
Dounreay	Dounreay Cementation Plant Store	Dounreay Cementation Plant Store Extension Dounreay Conditioned Waste Store
Harwell	Vault Store	
Winfrith	Treated Radwaste Store	
Windscale	Windscale AGR Box Store (used to store concrete boxes)	Windscale Pile 1 Box Store

- Dummy packages — again considering the performance of actual containers in the storage environment, also dummy containers filled with dummy wastes and grout can be used to consider the effect of internal corrosion on the performance of the packages. Again these can be used for destructive and non-destructive testing.

- Monitoring regimes — UKAEA plan to inspect a number of containers each year from each of their stores. UKAEA are investigating the number of containers to be inspected each year (this can depend on the total number of packages and the potential mode of failure of these containers). UKAEA are also considering what form this inspection should take, i.e. visual examination, swabbing, measurement of container dimensions, etc.

So far all efforts have concentrated on the control of the environment in which the containers are stored at the back end of the process once the containers have been filled with waste. UKAEA strategy has changed in the last twelve months as it has become aware that consideration must be given to the holistic life of the packages from the point of manufacture until final disposal. Corrosion of the container can be initiated before the container has been filled with waste, once pitting or ASCC has started it will be difficult to prevent it continuing even in well controlled environmental conditions. Some of the areas of concern that UKAEA are considering are detailed in Table 2.

UKAEA are not advocating that all the measures detailed in Table 2 are implemented for each waste packaging plant. A balance is required between the quality control arrangements that are in place for the manufacture and handling of the container, and the quality checking arrangements that have to be put in place for the encapsulated package during an extended period of storage. The waste producer must identify the most cost effective solution for their waste management strategy which ensures that the waste packages will be suitable for transport and disposal at the point when they are retrieved from the store.

Table 2

Process phases	Area of concern	Control measure
Manufacture of the containers	• Cross-contamination of stainless steel with mild steel during the manufacturing process. • Storage of new containers at the container manufacturer's site. • Contamination of the surface of the stainless steel with chemical from sticky tape or notelets.	• Segregation and control of the manufacturing process. • Cleaning of containers with acid or bead blasting following manufacture. • Control of container store conditions. • Shrink wrapping of containers in polyethylene.
Transport to site	• Cross-contamination of stainless steel with mild steel from fork lift truck used to handle the containers. • Chloride contamination of the containers.	• Containers moved on special pallet. • The containers will only be handled by operators wearing rubber gloves.
Storage prior to use	• Cross-contamination of stainless steel with mild steel fork lift truck used to handle the containers. • Storage of containers prior to use. • Chloride contamination of the containers.	• Containers moved on special pallet. • UKAEA are considering constructing a purpose built facility for the storage of new containers. The environmental conditions in this store will be controlled. • Containers will be cleaned with acid or a chloride free detergent and water prior to use.
Use of containers	• Cross-contamination of mild steel with stainless steel from the system on which the container is moved in the packaging plant.	• Movement system is manufactured from stainless steel.
Storage of containers	• Corrosion of the containers.	• Environmental conditions in the store are controlled. • Packages monitored during storage.

5. Monitoring of Indoor Environmental Conditions and Corrosion

During the course of research and development work, Nirex have produced a number of inactive waste packages for fire and impact testing. These 500 litre drum and 3m³ Box test pieces have been manufactured using grade 316L stainless steel rolled sheet with grade 304L stainless steel flanges and lifting plates; they have been stored for up to 10 years in building B2 at the UKAEA Culham site in Oxfordshire. In addition, the building is used to store a prototype 4 metre Box, manufactured in grade 304L stainless steel, which is being used to test the suitability of condition- and corrosion-monitoring methods for use in a long-term condition monitoring programme. The environment in the building has been monitored since October 1998 and inspection of the stored test pieces offers the opportunity to assess the performance of stainless steel waste packages in a storage environment, and to check the validity of information that has been used to formulate the guidance summarised in Section 3.3.

5.1. Culham Storage Environment

Building B2 at Culham is a typical mid-20th Century workshop/store building. The thin cladding and roofing material means that the thermal inertia of the building is low; it therefore resembles more closely a building for the storage of alpha-emitting or low activity radioactive wastes rather than gamma-emitting radioactive wastes. The storage environment has been monitored since October 1998 by measurement of the following items, as part of the 4 metre Box condition monitoring programme:

- Box surface temperature (hourly, at 12 locations: on each internal and external face using platinum resistance thermometers).

- Air temperature (hourly, at 2 locations: inside and outside box).

- Relative humidity (hourly, at 2 locations: inside and outside box).

These measurements have shown diurnal variations in temperature; when seasonal variations are considered the temperature range is 7–31°C. The relative humidity outside the box fluctuates and is higher in early summer, generally at 60–70% with a peak at 89%. During the first winter, heating of the building resulted in a relative humidity generally in the range 20–50% with a minimum of 12% (Fig. 1), but over the second and third winters the range was generally 50–80% and the minimum 38% as illustrated in Fig. 2. During the first three months of 2001, cold spells followed by warmer humid conditions led to at least three condensation events, in which water droplets were present on the walls of the Box and accumulated on a horizontal ledge above the bottom side rail.

The environment of the 4 metre Box is monitored further by measurement of:

- Box surface time of wetness (hourly on internal and external faces; in susceptible areas — corner fittings, lid crevice and base channels using gold microband electrodes).

- Box surface chloride concentration (quarterly, at 6 locations: 2 positions on two sides and 2 positions on the lid by swabbing).

Outside the condensation events, the time-of-wetness probes have shown no periods when the box surface was wet inside the storage building. However, the 4 metre Box was exposed to rain during February 1999. This was towards the end of

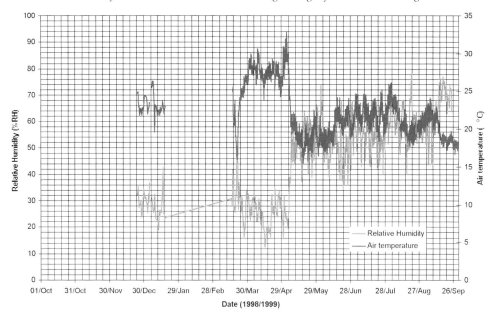

Fig. 1 *Temperature and relative humidity measured in heated building, October 1998–September 1999.*

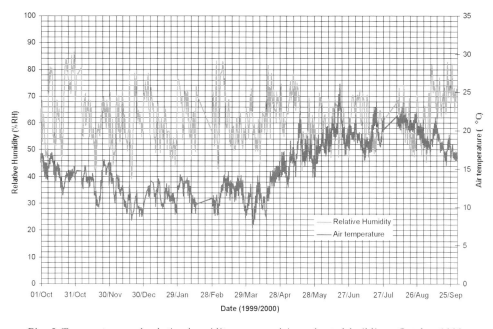

Fig. 2 *Temperature and relative humidity measured in unheated building, October 1999– September 2000.*

an operation in which the 4 metre Box was covered and transported over 1000 miles during which additional measurements of chloride surface concentration were made.

Prior to transport, the maximum concentration of chloride was 1.5 µg cm^{-2}, on a horizontal surface; following transport during which the Box was covered, higher levels in the range 1–4 µg cm^{-2} were measured on horizontal surfaces; following uncovered exposure to rain, the concentration on the lid was reduced to near zero but the level on a ledge (bottom side rail) rose to 5.6 µg cm^{-2}. Following return to storage, and cleaning of two faces of the box, chloride deposition has continued (Fig. 3). This is greatest on horizontal surfaces, where a rate of ~1 µg cm^{-2}/year has been recorded at two separate locations.

In March 2001, a study was begun to characterise the environment more widely within the storage building. The levels of contamination on a number of 500 litre drums and 3m^3 boxes was determined, an air sampler was run and coupons installed to measure the rate of accumulation. The levels of chloride contaminant are given in Table 3.

The maximum level of chloride on a surface was 13.1 µg cm^{-2}. The air sampling shows levels of chloride which are little different from those measured in the atmosphere 16 km to the south [29]. This suggests that the suspension of chloride in the atmosphere is stable and is able to penetrate the thinly clad building.

The location of the Culham site may be describes as 'rural', the nearest coastline being approximately 100 km to the south. However, chloride is detected and generated in the surrounding area:

- Nearby at Chilton, 12 km to the south-west of Culham [28], the atmospheric particulate, including levels of chloride, and rainfall composition is regularly measured. Annual mean chloride levels were in the range 1568–1685 ng.m^{-3} during 1996–98, with January–March levels of 1824–2091 ng m^{-3}. Chloride represents 9% of total suspended particulate; approximately 50% is sulfate, nitrate and ammonium ions.

- Rainwater composition, including levels of chloride, is regularly measured at Compton, 16 km to the south of Culham [29]. The annual mean chloride level

Table 3. levels of chloride contaminants measured on surfaces and in air at B2 Culham

	Chloride (µg.cm^{-2})
Drum base (vertical)	0.5
Drum wall (horizontal)	6.2
Box lid (mild steel)	3.0
Drum wall (horizontal)	13.1
Drum neck (horizontal)	11.9
	Chloride (ng.m^{-3})
Air sample (20 h)	1756

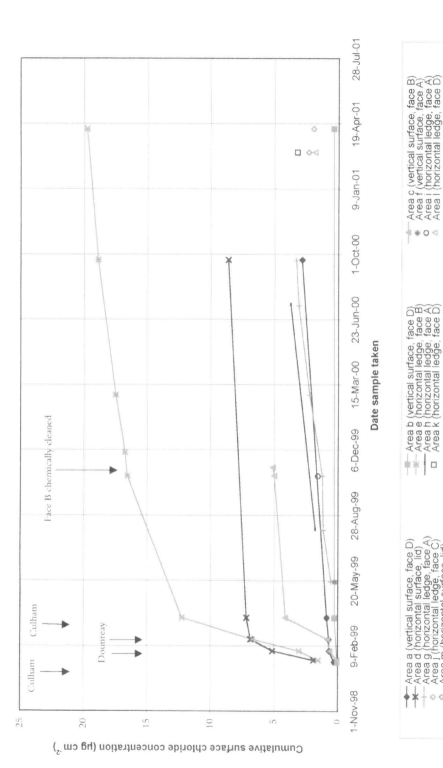

Fig. 3 accumulation of chloride on surfaces of the 4 m box.

in 1999 was 2.3 mg L^{-1}, in a rainfall of 644 mm; this corresponds to a chloride precipitation rate of 147 µg cm^{-2} /year.

- An additional, local source of chloride, downwind from the Chilton measurement site and half-way to the Culham site, is the coal-fired Didcot A power station, which emitted 2,300 t of chloride in 1999 [30]. Nearby, relatively high levels of ozone have been recorded during anti-cyclonic conditions.

5.2. Corrosion Performance at Culham

Since the cessation of winter heating, the air temperature in B2 Culham has averaged 15°C and the relative humidity has averaged 64%. Combined with surface contaminants containing up to 13 µg cm^{-2} these conditions would, based on the work of Taylor (NNC) [27], be conducive to pitting corrosion or rust staining of grade 304L stainless steel.

The prototype 4 metre Box is fabricated in grade 304L stainless steel. Regular examination during 3 years of condition monitoring [31] has revealed no evidence of pitting or cracking, although the highest levels of chloride are restricted to a small horizontal area. These observations suggest that conclusions drawn from Taylor's [27] findings are conservative. However, the Box is not devoid of corrosion interest. Instances of carbon steel pickup on the surface of the box have corroded, particularly during the transport operation of February 1999. At the same time, a streaky patina developed on the surface of the box, for which no surface finish procedure had been specified; the staining is considered to be related to inefficient removal of pickle paste residues[31].

Approximately twenty 500 litre drums are stored in B2 Culham, and about half of them are painted. Visual examination has not revealed any evidence of corrosion during storage periods of up to 10 years, apart from corrosion of some carbon steel pickup transferred from impact targets. An unpainted section of drum neck shows visible contamination, in which the chloride level is 11.9 µg cm^{-2} (Table 3), but removal of contamination reveals an unchanged surface.

Of the four 3 m^3 boxes being stored, two are fabricated in painted, grade 316L stainless steel and show no evidence of corrosion. Two carbon steel boxes are evidently corroding, with much of their paint undergoing flaking and spalling.

Overall, detailed examination has shown that the non-active stainless steel packages have not undergone detectable corrosion during storage for periods of up to ten years in their store environment.

6. Summary and Conclusions

- The resistance to corrosion of stainless steel is well known, but deposits of particles with ionic components, in particular chlorides, can stimulate attack on surfaces which may become localised and propagate rapidly once initiated, e.g. pitting and crevice corrosion. The paper has summarised a review of the literature on atmospheric corrosion of stainless steel, and of best practice to avoid localised corrosion. This information has been used to prepare prudent, precautionary guidance on environmental conditions that should be applied for the interim storage of intermadiate level waste packages.

- Examples have been given of how this knowledge is being reflected in the designs and operation of some UKAEA ILW stores, together with the issues waste packagers need to consider to prevent the initiation of corrosion.

- Monitoring and inspection of inactive waste packages, stored following Nirex studies, offers the opportunity to assess the performance of stainless steel waste packages in a storage environment, and to check the validity of information that has been used to formulate guidance on control of environmental conditions. This has shown that the chloride present in the atmosphere at the inland site (100 km from coastline) can enter the building and deposit on surfaces. However, inspection of the inactive waste packages has not revealed any instances of corrosion of the stainless steel surfaces for those packages that have been fabricated and handled in accordance with procedures for standard ILW packages. A waste package fabricated in grade 304L stainless steel has shown some staining together with corrosion of carbon steel contaminant. So far, no evidence of corrosion has been detected on packages fabricated in grade 316L stainless steel and stored for a period of 10 years.

- Because of the long timescales over which waste packages will be stored, application of best practice as in standards and in the measures adopted by UKAEA is considered prudent. Observations to date indicate that these measures are robust. Monitoring of inactive waste packages will continue, and will be supplemented by information from active stores, e.g. from coupons. This should enable guidance to be refined in the future and may allow it to be updated.

- Thus, the paper has shown how the choice of materials, manufacturing controls and correct storage conditions are used to manage the integrity of waste packages in the UK.

References

1. *The 1998 United Kingdom Radioactive Waste Inventory*, DETR/RAS/99.009, Nirex Report N3/99/01, July 1999.
2. S. V. Barlow, N. A. Carr and S. J. Wisbey, *Waste Package Specification for Intermediate Level Waste*, Nirex Report N/007, May 2000.
3. *Guidance for Inspectors on the Management of Radioactive Materials and Radioactive Waste on Nuclear Licensed Sites*, Health and Safety Executive, Nuclear Safety Directorate, March 2001.
4. W. Seddon, Engineering Considerations Associated with Plants Used for the Storage of Intermediate-level Waste — a Regulator's View, *Nucl. Energy*, 2001, **40**(1), 59–64.
5. *Guide to Stainless Steel Finishes*, Euro Inox Building Series, Volume 1, January 2000.

6. ISO 9223:1992 *Corrosion of metals and alloys — Corrosivity of atmospheres — Classification*.

7. ISO 9226:1992 *Corrosion of metals and alloys — Corrosivity of atmospheres — Determination of corrosion rate of standard specimens for the evaluation of corrosivity*.

8. ISO 9225:1992 *Corrosion of metals and alloys — Corrosivity of atmospheres — Measurement of Pollution*.

9. ISO/CD 11844 *Corrosion of metals and alloys — Classification of corrosivity of indoor atmospheres — Determination and estimation of indoor corrosivity*.

10. D. Knotkova, Corrosivity Classification in Indoor Environments, in *EUROCORR '94*, Bournemouth, UK, 1994, Publ. Institute of Materials.

11. N. R. Smart, *Atmospheric Corrosion Of Stainless Steel Waste Containers During Storage*, AEA-TPD-262, Nirex Report T/REP/326679/E/04, February 2000.

12. *Corrosion Resistance of the Austenitic Chromium–Nickel Stainless Steels in Atmospheric Environments*, NiDI Publication No. 318, 1963.

13. E. A. Baker and T. S. Lee, Long-Term Atmospheric Corrosion Behaviour of Various Grades of Stainless Steel, pg. 52 in *Degradation of Metals in the Atmosphere*, ASTM STP 965 (S.W. Dean and T.S. Lee, eds.), 1988.

14. A. L. Alexander, C. R. Southwell and B. W. Forgeson, Corrosion of metals in Tropical Environments, Part 5 — Stainless Steel, *Corrosion*, 1961, **17**(7), 345t.

15. J. R. Kearns, M. J. Johnson and P. J. Pavlik, The Corrosion of Stainless Steels In The Atmosphere, in *Degradation of Metals in the Atmosphere*, ASTM STP 965 (W. Dean and T.S. Lee, eds.), 1988, p.35.

16. *DECHEMA Corrosion Handbook: Corrosive Agents and Their Interaction with Materials*, Vol. 7 — Atmosphere (G. Kreysa, ed.), 1990.

17. N. G. Needham, P. F. Freeman, J. Wilkinson and J. Chapman, The Atmospheric Corrosion Resistance of Stainless Steels, in *Stainless Steels '87*, York, UK, 1987.

18. T. Shibata, H. Kurahashi, S. Kaneko, T. Ohya and K. Motohashi, An Effect of Airborne Salt Particles on the Atmospheric Corrosion Resistance of Stainless Steels Sheltered with the House Eaves in Marine District, *Corros. Engng*, 1995, **44**, 335.

19. R. M. Kain, Marine Atmospheric Stress Corrosion Cracking of Austenitic Stainless Steels, *Mater. Perform.*, 1990, **??** (December), 60–62.

20. J. W. Oldfield and B. Todd, Room Temperature Stress Corrosion Cracking of Stainless Steels in Indoor Swimming Pool Atmospheres, *Brit. Corros. J.*, 1991, **26**(3), 173.

21. B. K. Shah, P. K. Rastogi, A. K. Sinha and P. G. Kulkarni, Failures of Austenitic Stainless Steel Components During Storage: Case Studies, *Mater. Perform.*, 1993 (April), 65–67.

22. J. B. Gnanamoorthy, Stress Corrosion Cracking of Unsensitized Stainless Steels in Ambient-Temperature Coastal Atmosphere, *Mater. Perform.*, 1990 (December), 63–65.

23. S. Shoji and N. Ohnaka, Effects of Relative Humidity and Chloride Type on Stainless Steel Room Temperature Atmospheric Stress Corrosion Cracking, Corros. Engng, 1989, 38, 111–119.

24. J. K. Linacre and W. R. Marsh, *The Radiation Chemistry of Heterogeeous and Homogeneous Nitrogen and Water Systems*, UKAEA Report AERE R10027, June 1981.

25. G. P. Quirk, M. H. Mirzai, P. E. Doherty and W. W. Bek, Corrosion Surveillance for Reactor Materials in the Calandria Vault of Pickering NGS A Unit 1, in *UK Corrosion '93*. Publ. Inst. of Corrosion, Leighton Buzzard, UK.

26. *Design and Structural Integrity — Corrosion*. The Welding Institute, August 2000, JoinIT technical information: knowledge summaries. http://joinit.grantapark.com/

27. M. F. Taylor, The Significance of Salt Contamination on Steel Surfaces, Its Measurement and Removal, in *EUROCORR '94*, 1994. Publ. The Institute of Materials, London, UK.

28. S. J. Baker, Trace and Major Elements in the Atmosphere at Rural Locations: Summary of Data Obtained for the period 1996–1998, AEAT-4371/20174403. Issue 1, June 1999.

29. UK National Air Quality Information Archive, acid deposition networks: rainfall

composition. http://ariadne.aeat.co.uk/netcen/airqual/data/nonauto/rain.html#Data
30. *What's in Your Backyard?* Environment Agency. http://216.31.193.171/asp/introduction.asp
31. P. Wood, Monitoring Corrosion Performance of ILW Waste Packages During Storage, in *EUROCORR 2000*. Publ. Institute of Materials, London, UK.

32

Effect of Irradiation on Long Term Alteration of Oxides and Metals in Aqueous Solutions

C. CORBEL, D. FÉRON*, M. ROY*, F. MAUREL†, V. WASSELIN-TRUPIN
and B. HICKEL

Laboratoire CEA de Radiolyse, CEA/Saclay, 91191 Gif sur Yvette cedex, France
*Laboratoire d'Etude de la Corrosion Aqueuse, CEA/Saclay, 91191 Gif sur Yvette cedex, France
†Centre d'Etudes et de Recherches par Irradiation, CNRS, 3A rue de la Férollerie,
45071 Orléans cedex 2, France

ABSTRACT

The production of radiolytic species in aqueous solutions under irradiation is dependent on the nature and energy of the irradiating particles. A study is reported of the effects of gamma (^{60}Co) and light ion (H$^+$) radiolysis on corrosion. The ^{60}Co source irradiates platinum, carbon steel and copper electrodes immersed in clay water (synthetic).The H$^+$ ion beam delivered by a cyclotron irradiates a platinum/sodium sulfate interface. The open circuit potentials of the electrodes are measured as a function of irradiation time. The concentrations of radiolytic hydrogen peroxide (H_2O_2) and hydroxonium (H_3O^+) ions are measured in the solutions after irradiation. The mass loss rates of the carbon steel and copper electrodes are also measured. The results show that an oxidation process accelerates copper dissolution when the gamma source irradiates the clay water at a dose rate of 0.8±0.2 kGy/h. Carbon steel corrosion is totally unaffected in the same conditions. The Pt open circuit potential reaches a steady-value that is about 0.2–0.3 V higher for the sodium sulfate solution under proton irradiation than for the clay water under ^{60}Co gamma irradiation. A difference is observed although both type of irradiations produce comparable amounts of H_2O_2, (1–2) 10^{-4} M in the solutions. Furthermore, the open circuit potentials respond rapidly, in a time scale of the order of minutes, to the switching on and off of both types of irradiations. These properties suggest that, in addition to stable species, short-living radiolytic species, as radicals, control the potentials under irradiation. Modelling of the solution radiolysis as a function of irradiation time shows that the concentrations of the radiolytic species reach steady-state values. This behaviour is consistent with the experimental evolution of the open circuit potentials of the electrodes.

1. Introduction

Nuclear wastes contain gamma-, beta- or alpha-emitting radioelements which, depending on the packaging conditions, can irradiate their environment and generate radiolytic species in water. The production of oxidant (HO$^•$, O$^{•-}$, HO$_2^-$, O$_2^{•-}$, H$_3O^+$, H_2O_2, O$_2$) and reductant (e$_{aq}^{•-}$, H$^•$, H$_2$, H_2O_2) radiolytic species[1–3] can affect on-going corrosion processes or induce new ones. These processes need to be taken into account in the models which predict the long term evolution of nuclear waste packages. Irradiation effects may vary considerably depending on the materials, site water composition and dominant types of radiation. The influence of gamma radiation on the corrosion of candidate materials for the fabrication of nuclear waste containers has been recently reviewed for various aqueous environments [4]. A section concerns the performance that can be expected for the nuclear waste package designs proposed for the Yucca Mountain waste repository (NV, USA).

The effects of gamma irradiation on various candidate container materials have been investigated in a wide range of groundwater and solution compositions [4]. However, few studies have focused on clay water although a clay geological environment corresponds to some of the disposal sites under consideration in Europe. Furthermore, few studies concern nuclear spent fuel itself and the corrosion processes directly induced by its radioactivity. Various properties of solid/water interfaces under irradiation have been investigated for metallic [5–7] or semi-conducting solids [8]. Electrochemical techniques have been used to investigate *in situ* the irradiation effects on the corrosion of metallic compounds or semi-conducting oxides, such as uranium oxides. The effects of gamma [9–26] and alpha [27–31]. irradiations were investigated by irradiating water with radioactive sources As concerns beta irradiations, several studies were also performed with electron beams of variable energy delivered by accelerators [32,33]. However, few studies have clearly identified the basic reactions which control corrosion processes under irradiation.

For a given aqueous solution, the concentration of radiolytic species produced under irradiation is strongly dependent on the nature and energy of the irradiating particles — gamma, electrons or ions [1,34]. The response of a solid/water interface can thus vary depending on the amount of energy deposited at the interface by the irradiating particles.

To investigate such effects, a convenient method is to measure the open circuit potential of an electrode as a function of time for various irradiation conditions. We have applied this method to irradiations performed either with a gamma radioactive source or a proton beam delivered by a cyclotron. Various electrode/water interfaces need to be irradiated in order to select appropriate ones for systematic studies. Preliminary results are reported here. They concern Pt, carbon steel and Cu electrodes and two types of conducting aqueous solutions synthetically prepared in our laboratory — sodium sulfate and clay water. These solutions differ in the nature of their ionic composition and radiolytic properties. Neither Na^+ nor SO_4^{2-} ions react with the radiolytic products but carbonate ions in clay water do react. The whole electrode surface in contact with the solutions is irradiated in both types of irradiation. For the proton irradiation, the energy and geometry are such that the electrode/water interface is irradiated only by protons which go trough the electrode and penetrate into water.

2. Experimental

Two different types of electrochemical cells are used for the gamma and the proton irradiation. In both cases, the open circuit potential of the irradiated electrodes, E_{VCS}, is measured against the reference potential of a calomel saturated electrode. The electrodes are placed in the electrochemical cell several hours or days before each type of irradiation to monitor the potential evolution and irradiation is started when a quasi-steady state is reached. Then, the cells are mounted in the irradiation chamber. To check that the potential for the cell placed inside the irradiation chamber is the same as that measured for the cell placed outside, the potentials are also measured before irradiation when the cell is inside the irradiation chamber. Once a quasi-steady state is reached, irradiation is switched on. The potentials are then recorded as a function of irradiation time. At the end of irradiation, the irradiated solutions are left in the cells and the potentials are then recorded as a function of ageing. All the irradiations are carried out at room temperature.

Two reference glass electrochemical cells, without electrodes, and two others where the electrodes are fully immersed in the solution, are used for gamma irradiation. The 1.17 and 1.33 MeV gamma irradiation is performed with a ^{60}Co source during 17 h at a dose rate which varies from 0.5 to 1 kG/h depending on the electrode position in the glass cell. The reference cells without electrodes are filled with one litre of either deionised aerated water (cell 1) or synthetic aerated clay aqueous solution (cell 2) with a ionic composition mimicking those of groundwaters found in clay environments. The cells with electrodes contain either platinum (Pt3) and carbon steel (XC38) electrodes (cell 3) or platinum (Pt5) and copper (Cu) electrodes (cell 5). In both cells, the electrodes are immersed in one litre of the synthetic aerated clay water. Table 1 shows that the number of different ion species in the aerated clay solution is four for the cations, Na^+, K^+, Ca^{2+}, Mg^{2+}, and three for the anions, Cl^-, HCO_3^-, SO_4^{2-} with a total concentration of ions in the solutions of 2.272 g L^{-1}. As shown in Table 2, the main impurities in the XC38 steel electrode are C and Mn. The Pt electrodes are wires. The XC38 and Cu electrodes are cylindrical rods. The ratio of the electrode surface to the solution volume, S/V, varies from 5.2 $\cdot 10^{-3}$ cm^{-1} for Cu to 7.39 $\cdot$ 10^{-3} cm^{-1} for XC38. Before being placed in the water, the electrodes were cleaned for 30 minutes in an organic solution of ethanol (1/3), acetone (1/3) and toluene (1/3).

For proton irradiation, the design of the electrochemical cell is such that there is only one planar surface of a thin disc electrode leached by the solution. The proton beam delivered by the cyclotron at the Laboratory CERI (CNRS-Orléans) has an

Table 1. *Chemical composition of clay water (mg L^{-1})*

Salts	Concentration (mg L^{-1})
CaSO$_4$	245
CaCl$_2$	755
MgCl$_2$	619
NaCl	333
KCl	186
NaHCO$_3$	134

Table 2. *Chemical composition of XC38 carbon steel*

XC38 – X (mass%)														
C	Mn	Si	P	S	Cr	Ni	Mo	Cu	Sn	As	Al	Ti	V	Fe
0.37	0.65	0.23	0.014	0.022	0.22	0.09	0.03	0.23	0.013	0.021	0.005	0.009	0.003	Bal.

initial energy of 12 MeV. After passing through thin Ti windows and the 125 μm platinum Pt1 disc electrode mounted on the electrochemical cell, the ion beam emerges from the Pt1 disc into the solution with an energy of about 4.6 MeV. For this energy, the stopping power in water at the interface and the range in water calculated using the programme TRIM [35] are about 8.7 keV/μm and 301 μm, respectively. The ion beam is collimated so that the Pt1 disc is irradiated on a small section of 6 mm diameter (0.283 cm^2). Only the irradiated surface of the disc is leached by a 16 mL volume of an aerated 0.1M Na$_2$SO$_4$ solution and the ratio S/V is 0.0157. The same Pt1 electrode is used for several cycles of irradiation performed at different dates. Between the irradiation cycles, the electrochemical cell is empty. Before each cycle, it is refilled with either the 0.1M Na$_2$SO$_4$ aerated solution used in the previous cycle or a fresh one. The cycle of irradiation consists of a set of sequential proton irradiations where, for the results reported here, each irradiation lasts one hour at a flux of 6.6×10^{11} H$^+$ cm^{-2}s^{-1} corresponding to a dose rate of 97.6 kGy/h. The values measured for the open circuit potentials are independent of the earthing of the electrodes.

The concentrations of the radiolytic species measured in solutions after irradiation are those of the hydroxonium ions and hydrogen peroxide. The H$_2$O$_2$ concentrations are measured by absorption spectrophotometry using the Ghormley method. The corrosion of the XC38 and Cu electrodes in aerated clay water is determined by mass measurements performed before and after immersion. The measurements are performed for two immersion cycles lasting the same time but with different irradiation conditions. The first cycle is conducted without irradiation to determine the corrosion in aerated clay water. The second cycle includes irradiation and the post-irradiation ageing in the irradiated solutions.

3. Electrodes in Aerated Clay Water under Gamma Irradiation

The evolution of the open circuit potentials and corrosion is described in detail for cell 3. It is then summarised for cell 5.

3.1. Electrochemical Potentials of XC38 C Steel and Pt3 Electrodes in Cell 3

3.1.1. Before gamma irradiation.
The open circuit potentials E_{VSC}(Pt3) and E_{VSC}(XC38) of the Pt3 and XC38 C steel electrodes drift during several days after the electrode immersion in cell 3 containing synthetic aerated clay water. The evolution of E_{VSC}(Pt3) before gamma irradiation

has four distinct stages as a function of time as shown in Fig. 1(a). In the first, there is a fast increase of E_{VSC}(Pt3) of 0.04 V, from ≈0.050 to 0.090 V, in about 4 h 30 min. The increase rate is ten times lower in the second stage: i.e. 0.035 V, from 0.090 to 0.125 V, in 42 h (1.75 days). Then, in the third stage, E_{VSC}(Pt3) remains constant. After 76 h (3.17 days), E_{VSC}(Pt3) starts to decrease slowly. It finally tends to reach a quasi-steady state value of 0.106 V after 160 h (6.6 days).

The potential evolution of E_{VSC}(XC38) before gamma irradiation shows three distinct stages as a function of time, see Fig. 1(b). After a fast decrease of 0.074 V, from –0.547 to –0.621 V, in about 4 h 30 min, E_{VSC}(XC38) decreases slowly to reach the value of –0.686 V in 70 h (2.91 days). It then remains constant during the next 90 h (3.75 days). it can be seen that the XC38 electrode reaches a quasi-steady state faster than the Pt3 electrode taking only 2.9 days for XC38 but 6.6 days for Pt3. The values of the stable potentials measured for each electrode are reproducible. In two independent experiments performed in cell 3, the values are found to be 0.105±0.001 V for E_{VSC}(Pt3) and –0.681±0.005 V for E_{VSC}(XC38). After two weeks of immersion, there is clear evidence of an oxide layer on the XC38 C steel electrode. The electrode mass has increased by 0.23% (Table 3). The oxide layer can be removed easily and the mass loss (Table 3) is comparable to the mass gained during immersion. These values can be used to estimate the thickness of the corrosion layer. Its growth corresponds to a corrosion rate of about 164 μm/year.

3.1.2. Under gamma irradiation.

As soon as gamma irradiation is switched on, the potentials of both electrodes start to evolve differently. E_{VSC} (Pt3) of the Pt3 electrode increases abruptly by 30 mV (Fig. 2a) in about 5 min. Then, it slowly increases and is still evolving after 17 h of irradiation

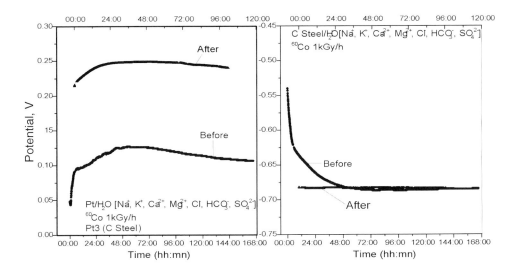

Fig. 1 *Platinum (1a) and carbon steel (1b) open circuit potentials as a function of time in aerated clay solution (Table 1) before and after 17 h ^{60}Co gamma irradiation at a dose rate of 0.8±0.2 kGy/h.*

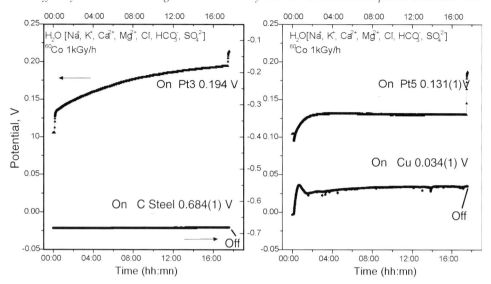

Fig. 2 *Carbon steel and platinum (2a), copper and platinum (2b) open circuit potentials as a function of irradiation time in aerated clay solution (Table 1) during 17 h ^{60}Co gamma irradiation at a dose rate of 0.8±0.2 kGy/h.*

at a dose rate of 0.8±0.2 kGy/h. The value of E_{VSC} (Pt3) is then 0. 194 V. The gamma irradiation has no effect on the potential E_{VSC} (XC38) of the C steel electrode which remains stable at the value — 0.685±0.001 V (Fig. 2a) over the 17 h irradiation at a dose rate of 0.8±0.2 kGy/h.

3.1.3. After gamma irradiation.
When the gamma irradiation is switched off, the electrodes react differently. Figure 2 (a) shows that E_{VSC} (Pt3) increases abruptly to values about 0.239 V. Then, as shown in Fig. 1(a), it evolves slowly through a broad maximum at 0.250 V between 30 and 70 h and reaches a value of 0.240±0.001 V after 120 h (5 days) of immersion in the irradiated clay solution. As seen in Figs 1(b) and 2(a), the potential E_{VSC} (XC38) of the C steel electrode remains stable when gamma irradiation is switched off. After 120 h in the irradiated clay solution, Fig. 1(b) shows that its value – 0.685±0.001 V is the same as before and under irradiation.

At the end of the off–on–off cycle of measurements, the oxide layer, which has grown on the C steel electrode, results from the oxidation occurring during the three phases. The electrode mass has increased by 0.27% (Table 3). After removal of the oxide layer, the mass loss (Table 3) is comparable to the mass gained during immersion without irradiation. The layer growth corresponds to a corrosion rate of about 164 μm/year.

3.2. Electrochemical potentials of Cu and Pt5 electrodes in cell 5

3.2.1 Before gamma irradiation.
The Pt5 and Cu electrodes immersed together in cell5 evolve differently as a function of immersion time. The time that is necessary to reach a quasi-steady state is

Table 3. Mass variations of the XC38 and Cu electrodes in aerated clay water during off–off–off and off–on–(gamma)–off immersion cycles. The values* of the oxide growth rate are obtained from the mass difference before and after removal of the oxide layer

Gamma irradiation (aerated clay water)	Electrode immersion cycle	Rate (kGy.h^{-1})	Initial mass (g)	Final mass (g)	Mass loss (g)	After oxidised layer removal (g)	Oxide growth (μm/year)
Off	XC38_1	0	13.58764	13.61875		13.55583	−0.03181* 163 μm/year
On	XC38_2	0.8(2)	13.59695	13.63400		13.56593	−0.03102* 164μm/year
Off	Cu_1	0	12.25176	12.25142	$34 . 10^{-5}$		2 μm/year
On	Cu_2	0.8(2)	12.38614	12.37613	10^{-2}		59 μm/year

dependent on the electrode. It takes only one day for the open circuit potential E_{VSC} (Pt5) of the Pt5 electrode in Fig. 3(a) to reach a stable value of 0.105±0.001 V in the synthetic aerated clay water. The potential E_{VSC} (Cu) of the Cu electrode increases continuously during 5 days before it reaches a stable value of −0.003V. After two weeks of immersion, there is evidence of a weak dissolution of the Cu electrode. The electrode mass has decreased by about $3 \cdot 10^{-3}$% (Table 3) corresponding to a layer removal rate of 2μm/year.

3.2.2. Under gamma irradiation.
As soon as gamma irradiation is switched on, the electrode open circuit potentials (Fig. 2b) start to change. Figure 2(b) shows that the potential E_{VSC} (Pt5) of the Pt5 electrode decreases from 0.104 to 0.095 V in 1 min. Then, it starts to increase and reaches a steady value of 0.131±0.001 V after only 2 h of irradiation at a dose rate of 0.8±2 kGy/h. Figure 2 (b) shows that the potential E_{VSC} (Cu) of the Cu electrode goes through a weak maximum of 0.036 V after only 30 min of irradiation. Then, after 1 h 30 min, the potential starts again to slowly and continuously increase from 0.025 to 0.034 V during the next 16 h of irradiation.

3.2.3. After gamma irradiation.
When gamma irradiation is switched off, the electrode behaviours differ markedly as seen in comparing Figs 3(a) and 3(b). Figures 2(b) and 3(a) show that E_{VSC} (Pt5) increases abruptly from 0.131 to 0.192 V within 9 min in the irradiated clay solution. Then, it goes trough a broad maximum at 0.192–0.188 V during the next 12 h. After 12h, as shown in Fig. 3 (a), E_{VSC} (Pt5) start to slowly decrease quasi-linearly from 0.188 to 0.133 V during the next 2 days at a rate of 0.026 V/day. During the next 1.5 days, E_{VSC} (Pt5) goes on decreasing quasi–linearly from 0.133 to 0.120 V but at a much lower rate 0.007 V/day. Figure 3(b) shows that the potential E_{VSC} (Cu) of the

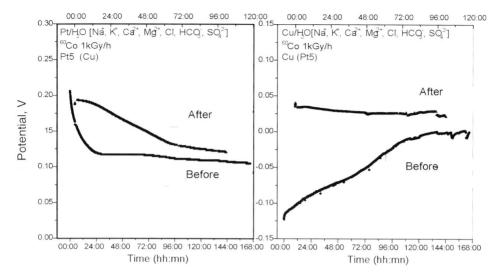

Fig. 3 *Platinum (3a) and copper (3b) open circuit potentials as a function of ageing in aerated clay solution (Table 1) before and after 17 h ^{60}Co gamma irradiation at a dose rate of 0.8±0.2 kGy/h.*

Cu electrode remains stable when gamma irradiation is switched off. During the next 2 days in the irradiated clay solution, E_{VSC} (Cu) in Fig. 3(b) decreases quasi-linearly at a slow rate of 0.006 V/day from 0.035(2) to 0.029 V during the first day and then at a lower rate of 0.003 V/day from 0.029 to 0.026 V during the second day. Then, it hardly changes. Table 3 shows that Cu mass loss is higher for the cycle of immersion under irradiation than for the cycle of immersion without irradiation. Irradiation enhances the dissolution of Cu by at least an order of magnitude.

3.3. Production of Radiolytic Species

The chemical composition of the irradiated solutions changes as a result of the production of radiolytic species. The concentration of hydrogen peroxide produced under irradiation, either in absence of electrodes in the deionised or clay reference solutions or in presence of electrodes in the clay solutions, is measured after irradiation as a function of the time of irradiation. The concentrations of hydrogen peroxide in Table 4 depend on the nature of the electrodes present in the irradiated cells. Their evolution with time of the irradiation (Table 4) is also dependent on the nature of the electrodes present in the irradiated solutions. In the reference–irradiated solutions (cell 1 and cell 2), the concentrations remain constant as a function of ageing time and the values are in both solutions $1.4 \cdot 10^{-4}$M (Table 4). In cell 3 with Pt3 and XC38 C steel electrodes, the concentrations after 15 min or 3 h irradiation (Table 4) are the same and have a high value of $1.5 \cdot 10^{-4}$M. After 70 h, the value has decreased to $1.3 \cdot 10^{-4}$M. In cell 5 with Pt3 and Cu electrodes, there is a strong decrease of the concentration as a function of irradiation time. The concentration drops from $5 \cdot 10^{-4}$M after 15 min to $3 \cdot 10^{-6}$M after 70 h.

The aerated clay water is initially slightly basic with a pH value of 8.1. At the end

Table 4. Radiolytic production of hydrogen peroxide due to gamma irradiation at a dose rate of about 1 kGy/h. Post-irradiation measurements at room temperature: influence of the time of irradiation of solutions. H_2O_2 in 10^{-4}M

Irradiation time (h)	Deionised water	Clay water	Clay water Pt3_C Steel	Clay water Pt5_Cu
0.25	1.4	1.4	1.5	0.5
3	1.4	1.4	1.5	0.4
70	1.4	1.4	1.3	0.03

of the irradiation, the pH values are dependent on the solutions. The decreases are within one unit: 7.1 and 7.8 in absence of electrodes in the irradiated reference deionised and aerated clay solutions, respectively, 7.6 in presence of the Pt3 and C steel electrodes and 7.9 in presence of the Pt5 and Cu electrodes in the irradiated aerated clay solutions.

4. Gamma Irradiation Induced Effects

4.1. Behaviour of the Electrodes under Gamma Irradiation

4.1.1. Platinum electrodes.
As clearly indicated by the changes of the free potential of Pt3 and Pt5 in Figs 2(a,b), irradiation induces new redox reactions in the cell 3 and cell 5 solutions. For both Pt electrodes, the open circuit potential increases indicating that oxidants are created under irradiation and/or that reductants disappear. One of the oxidant species produced by irradiation is H_2O_2. Its concentration at the end of irradiation reaches about $1.5 \cdot 10^{-4}$M in cell 3 and $0.5 \cdot 10^{-4}$M in cell 5. The Pt5 electrode immersed in the cell 5 with the Cu electrode has a less oxidising potential than the Pt3 electrode immersed in the cell 3 with the C steel electrode. The H_2O_2 concentration in cell 5 is much lower than in cell 3 by a factor 3. Such a reduction may explain the lower potential in cell 5. However, this needs to be checked by conducting experiments in which the potentials are measured as a function of the H_2O_2 concentration added in both types of solutions before irradiation (see below for the cell 5 with the Cu electrode).

The quasi-steady values of the Pt3 and Pt5 potentials before irradiation differ indicating that the redox potentials differ in the cell 3 and cell 5 solutions. Although the solution compositions are initially the same, the surface or the metallic impurities released by the working electrode affects the redox reactions.

Another point worth making is the rapid evolution of the redox potential of the solution when the gamma irradiation is switched off. The Pt3 and Pt5 potential exhibit a fast increase of about 30–50 mV at the switch-off. This tendency indicates that irradiation induced oxidant species have a much lower rate of disappearance than reductant species. Several processes can contribute to this effect:

(i) The species which limit the value of the redox–potential of the solution under irradiation are short-living reductant species ($e^{\bullet-}_{aq}$, $H^{\bullet}$).

(ii) A catalytic decomposition of hydrogen takes place at the Pt electrodes.

(iii) The rates at which radiolytically produced oxygen or hydrogen escape from the solutions differ. Further work is needed to evaluate the contributions of each process. The sharp evolution of the Pt potentials suggests that transient species play a role.

4.1.2. Carbon steel electrode.

One striking feature of the above results is the absence of any irradiation effect for the C steel electrode. However, as clearly shown by the increase of the potential of the Pt3 electrode under irradiation, the solutions become more oxidising both under and after, rather than before irradiation. The lack of reactivity of the C steel electrode under irradiation is in contrast with the strong corrosion of 0.16 µm/year that it undergoes when immersed in aerated clay solutions. The oxidation layer which has grown before irradiation seems to protect the electrode from the radiolytic species produced in the solutions. To our knowledge, there is no earlier work mentioning such behaviour of C steel in presence of gamma irradiated aerated clay water.

4.1.3. Copper electrode.

Irradiation has a clear effect on the Cu electrode as evidenced by the difference between the time evolution of the Cu and Pt5 electrodes in cell 5. The Cu electrode dissolves in the solution faster under irradiation than before. This increase in mass loss corresponds to an increase of the Cu electrode potential. Such a correlation suggests that an oxidation process induces a higher dissolution of the Cu electrode under irradiation. The rate of oxidation is higher under irradiation than the rate in absence of irradiation, 50–60 instead of 2 µm/year. The irradiation switch-off has no direct effect on the Cu electrode, which remains at the same potential. This lack of response suggests that Cu reacts with long-life rather than with short-life species.

4.2. Gamma Irradiation Induced Changes of the Solution Composition

Among the various species that gamma irradiation can produce in water, only two have been determined in the present measurements: H_2O_2 and H_3O^+. The gamma-irradiation of the aerated clay water results in H_2O_2 production and acidity that is dependent on the ionic composition of the solution.

A literature review of the reactions radiolytically induced in water shows that the clay water contains ions able to react with the radiolytic species [36–38]. The main species that are concerned for the present solution at pH $\approx$7.1–7.8 (Table 3) are however restricted to the carbonate species H_2CO_3, HCO_3^-, CO_3^{2-} [39]. The HCO_3^- ions can react with the $e^{\bullet-}_{aq}$, $OH^{\bullet}$ and $O_2^{\bullet-}$ radicals whereas the CO_3^{2-} ions can react with the $e^{\bullet-}_{aq}$, $OH^{\bullet}$ and $O_2^{\bullet-}$ radicals [40]. The system of reactions that describe the evolution of the clay water under gamma irradiation can then be obtained by adding these reactions to the system of reactions describing the evolution of deionised aerated water under irradiation [40]. Assuming that the spatial distribution of species is homogeneous in the system, the code CHEMSIMUL [41] can be used to calculate the evolution of the various species. The results are illustrated here for the production of H_2O_2 in aerated deionised water and carbonate solutions (Table 3) corresponding to a pH value 7.27. The time evolution has been calculated for various dose rates. The dose rates in Figs 4 and 5 vary over four order of magnitudes in the range $(10^{-4}-1)$ 3.8 Gy/s, i. e. $1.4 \cdot 10^{-3}$–14 kGy/h. In this range, the noticeable feature is that the dose dependence of the H_2O_2 concentrations is remarkably independent of the dose rates. The concentration increases as a function of dose up to 10 Gy and then reaches a quasi-steady state. The plateau value calculated in aerated deionised water (Fig. 4) is independent of the dose rates, about $1.12 \cdot 10^{-4}$M. The quasi-plateau value calculated as a function of dose in aerated carbonated water (Fig. 5) slightly depends on the dose rate. In the range $3.8 \cdot 10^{-4}$ to 13.8 Gy/s, the concentration value at 10 Gy increases by 7% from $(1.53$ to $1.64)$ 10^{-4}M and at 100 Gy by 6% from $(1.67$ to $1.68)$ 10^{-4}M. The range of dose rates used in the calculations include the range 0.5–1 kGy/s corresponding to the variation of the experimental dose rates within a cell under irradiation. On the basis of these calculations, it follows that the H_2O_2 values measured after the gamma dose of 17 Gy correspond to a quasi-steady value. It should be noted that, during the evolution towards the steady state, the local production rate of H_2O_2 varies within the cell depending on the local gamma dose rate. It is clearly

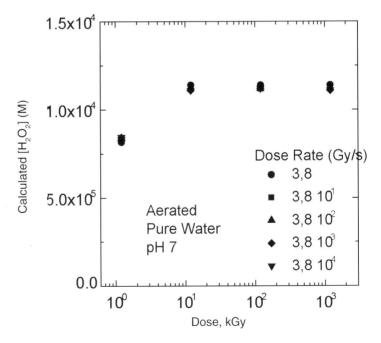

Fig. 4 *Calculated H_2O_2 concentration as a function of ^{60}Co gamma dose for various dose rates in aerated deionised water.*

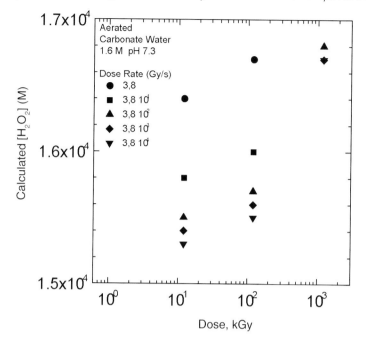

Fig. 5 *Calculated H_2O_2 concentration as a function of ^{60}Co gamma dose for various dose rates in aerated carbonated water .*

seen in Fig. 6 where the time evolution of the H_2O_2 concentration is calculated in deionised aerated water for various dose rates.

The value $1.4 \cdot 10^{-4}$ M measured for the H_2O_2 concentration in the cell 2 containing only the solution of clay water is in agreement with those found in earlier work [37,38]. The value $1.4 \cdot 10^{-4}$ M that we measure in aerated deionised water (cell 1) is higher than the value $1.2 \cdot 10^{-4}$ M that we measure in a run where a unique cell containing the deionised water is irradiated. This may be due to atmospheric contamination (CO_2) of the aerated deionised water in the run where the cells are irradiated together. The comparison between the experimental and calculated H_2O_2 concentration values shows that the calculations predict a stronger effect of the carbonates than that observed here for the clay water. The presence of carbonates in aerated water at concentrations about $1.6 \cdot 10^{-4}$ M (pH ≈ 7.3) increases the production of H_2O_2 by 16% according the measurements and 30% according to the calculations.

The presence of the Pt3 and C steel electrodes in cell 3 has little effect on the production of hydrogen peroxide under irradiation. It differs by 7% (Table 4) from the value $1.4 \cdot 10^{-4}$ M measured in the cell 2 containing only a solution of the clay water. The presence of the Cu electrode in the cell 5 has a strong effect on the production of hydrogen peroxide. The concentration is reduced by a factor of 3. Furthermore, during post-irradiation ageing, the H_2O_2 stability in the cell 5 is low. After 70 h, only 6% of the produced H_2O_2 survives in cell 5. This low stability is in strong contrast with the good stability in cell 3 where, after 70 h, 87% of the produced H_2O_2 survives. Metallic elements such as Fe, Mn, Co, Ni, Cu, Zn, Nb, Mo, Pd, Ag,

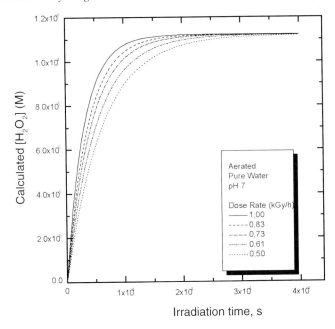

Fig. 6 [60]*Co gamma dose rate effect on* H_2O_2 *time evolution calculated in aerated deionised water.*

Cd, W, Ir, Pt, Au, Hg, Pb, Bi [42].are known to induce a H_2O_2 decomposition by catalysing the dismutation reaction

$$2\ H_2O_2 \rightarrow 2\ H_2O + O_2$$

The excellent stability both in cell 1 and cell 2 where the totality of the H_2O_2, respectively produced in deionised and clay water, survives is likely due to the lack of metallic elements able to decompose H_2O_2.

Due to the irradiation-induced dissolution of the Cu electrode, the solution in cell 5 under irradiation contains Cu^{n+} ($n = 1,2$) ions. A literature review shows that these ions react with the radicals produced by irradiation [40]. Their influence on the production of H_2O_2 under irradiation depends on their concentration. The Cu^{n+} ($n = 1,2$) concentration measured by atomic absorption photometry in cell 5 at the end of the post-irradiation ageing is about $2\ 10^{-5}$M. It is about 3 times lower than the value $6 \cdot 10^{-5}$M that can be estimated from the 0.01 g mass loss of the Cu electrode. This difference suggests that the concentration of Cu^{n+} ($n = 1,2$) ions released by the Cu electrode in the solution can affect the production of H_2O_2 under irradiation. The strong reduction in the production of H_2O_2 in cell 5 under irradiation can also result from its dismutation reaction on the Cu surface.

To check the stability of H_2O_2 in presence of the Cu electrode, an experiment was conducted in cell 5 where a high concentration of H_2O_2, $1.1 \cdot 10^{-2}$M, was added to clay water instead of being produced by irradiation. As for irradiation, the addition of H_2O_2 took place after several days once the potentials of the Pt5 and Cu electrodes had reached stable values. Figure 7(a) shows that, after its addition, the H_2O_2

concentration decreases quickly. After three hours, 20% of the added H_2O_2 concentration has already disappeared and, after one day, none is left. The continuous loss of H_2O_2 as a function of time corresponds to a continuous decrease of both Pt5 and Cu potentials by about 24 mV for Pt5 and 61 mV for Cu.

The decreases of the Pt5 and Cu potentials in cell 5 are higher during post-irradiation than post-H_2O_2 addition ageing, 72 and 12 mV instead of 24 and 61 mV, respectively. In addition the disappearance rates differ by three orders of magnitude. The rate is much lower after irradiation than addition, $6.7 \cdot 10^{-7}$ M/h (Table 4) instead of $4.6 \cdot 10^{-4}$ M/h (Fig. 7). The concentration introduced by addition is however two orders of magnitude higher, $1.1 \cdot 10^{-2}$M (Fig. 7), than that produced by irradiation,

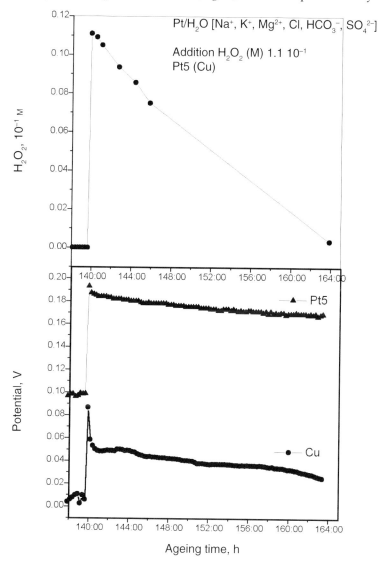

Fig. 7 Copper effect on the stability of H_2O_2. Cu, Pt5 electrode potentials and H_2O_2 concentration as a function of irradiation time in aerated clay solution (Table 1) after addition of 0.011M H_2O_2.

$0.5 \cdot 10^{-4}$M (Table 4) . It follows that the stability of H_2O_2 is clearly different in both types of solutions. It seems to be much higher after irradiation. This property suggests that the Cu surface after irradiation is much less effective with respect to the H_2O_2 dismutation. This evolution results likely from the Cu surface alteration during irradiation. Furthermore, the difference in the potential variations suggests that other species than H_2O_2 contribute to the redox potentials during post-irradiation ageing.

5. Electrochemical Potentials Before, Under and After Proton Irradiation

The results reported and discussed here correspond to cycles of proton irradiations where the proton energy at the Pt/H_2O interface, flux and fluence are the same for each irradiation: 4.6 MeV, 6.6×10^{11} $H^+cm^{-2}s^{-1}$ and 2.4×10^{15} H^+cm^{-2}, respectively. The dose rate normalised to the total volume (18 mL) of solution in the cell is then 97.6 kGy/h. When the fluence cumulated over the cycles by the Pt1 electrode is less than $1.6 \ 10^{16}$ H^+cm^{-2}, the results are found to show good reproducibility from one cycle of irradiations to the other. The total fluence cumulated by the Pt1 electrode is then too low to induce ageing effects affecting its potential. We use one irradiation giving typical results to illustrate the effects of proton irradiation on the Pt1 electrode.

5.1. Pt Electrode Potentials and Solution Composition: Measurements

Before proton irradiation. The potential of the Pt1 electrode in presence of the 10^{-1}M Na_2SO_4 solution evolves slowly before irradiation. The time to reach a stable value varies form one irradiation cycle to another one. It takes about a few hours to reach a quasi-steady equilibrium. The potential then reaches a value which varies randomly over 40% from one cycle to the other and falls within the range 0.19±0.04 V. In Fig. 8(a), the value is 0.176±0.001 V.

Under proton irradiation. As soon as proton irradiation is switched on, the potential E_{VSC}(Pt1) of the Pt1 electrode increases abruptly by about 40 mV (Fig. 8b) in about 5 min. Then, as shown in Fig. 8(b), it reaches quickly a quasi-steady value in about 15 min. This behaviour is well reproduced from one cycle to the other. The quasi-steady potential is independent of the cycle and reaches a value of 0.45±0.02 V reproducible within 5%.

After proton irradiation. When proton irradiation is switched off, E_{VSC}(Pt1) decreases abruptly and reproducibly by about 0.44 V (Fig. 8a). Then, it progressively increases to reach a quasi-steady value which varies over 20% depending on the irradiation cycle and falls in the range 0.20±0.04 V.

Production of radiolytic species. The chemical composition of the irradiated solution has changed due to the production of radiolytic species. The concentrations of hydrogen peroxide produced in each proton irradiation of 1 h at the dose rate of 97.6 kGy/h are about $(2.0\pm0.02) \ 10^{-4}$M when measured in the irradiated solution after an ageing of about one week at 4°C.

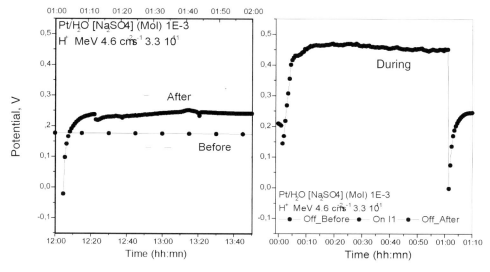

Fig. 8 *Platinum open circuit potentials as a function of irradiation time in 0.1M sodium sulfate solution before, during and after 1 h proton irradiation at a dose rate of 97.6 kGy/h.*

5.2. Proton Induced Radiolysis at Platinum Electrodes in Sodium Sulfate Solution

Before irradiation, the Pt open circuit potential in the initial solution can fluctuate depending on the residual impurities present either in the sodium sulfate solution or at the Pt1 electrode surface.

Proton irradiation has a strong effect on the Pt potential. Tests in progress show that the potential variations are induced by the solution irradiation. When proton irradiation is switched on, the sharp increase of the potential indicates the production of strong oxidant species in the solution. The solution evolves quickly towards a quasi-equilibrium suggesting that the redox species have high rates of production and disappearance. Due to the short range of 4.6 MeV protons in water (300 μm) and irradiation geometry, the production of the species in the solution is restricted to the vicinity of the Pt1 electrode in the narrow layer of about 300 μm corresponding to the proton range. The fast evolution of the potential towards quasi-equilibrium, in less than 14 min, suggests that long range diffusion processes play little role in imposing the potential. A fast species such as H_3O^+ ($D(H_3O^+)$ = $9.46 \cdot 10^{-9}$ m²s⁻¹) covers less than 2 mm in 5 min and less than 3 mm in 14 min. Consequently, it seems that either convection phenomena can play a role or that the redox species have little chance to escape from the narrow layer where they are produced.

After proton irradiation, the Pt1 electrode leached by the irradiated solution has a potential which evolves progressively towards a quasi-steady value. This post-irradiation value is higher than before irradiation. This result is consistent with the presence of hydrogen peroxide in the irradiated solution. However, for all irradiation cycles, the evolution after irradiation towards a quasi-steady state is always preceded by a sharp transient where the potential reaches a minimum value much lower than its value under irradiation, −440 mV or its long term post-irradiation value, −200 mV. When proton irradiation is switched off, this transitory minimum suggests that short-life reductant species survives longer than short-life oxidant species . These reductant species control the potential by temporarily overcoming the stable oxidant species. Their fast disappearance in about 3 min show that they are strongly reactive. The comparison of the transients at the switching on and off shows a dissymmetry in that there is an overproduction of oxidant species at the beginning and an overproduction of reductant species at the end of irradiation.

6. Conclusions

Gamma irradiation of clay water in presence of metallic electrodes results in H_2O_2 production and acidity which depend on the nature of the electrodes. This dependence reflects the influence of the electrodes on the chemical reactions involving the radiolytic species. Two processes can be involved. Radiolytic species react at the electrode surface or react with the metallic elements released by the dissolution of the electrodes. Copper reduces considerably the H_2O_2 production probably by enhancing its decomposition.

The effect of irradiation on the electrode reactivity depends strongly on its composition. The oxidation and dissolution of copper is strongly enhanced in clay water under ^{60}Co gamma irradiation at a dose rate of 0.8±0.2 kGy/h. Carbon steel remains totally inert under similar irradiations. In this case, the reactivity of the electrodes is totally different in absence or presence of irradiation. In absence of irradiation, carbon steel is strongly oxidised in clay water whereas copper is only weakly dissolved.

The electrochemical potentials of Pt electrodes in aerated clay and sodium sulfate solutions vary strongly under gamma or proton flux. The comparison of the Pt response in different cells under gamma irradiation show that the time needed to reach a steady-state depends strongly on the nature of the other electrode immersed in the cell solution. Under ^{60}Co irradiation at a rate of 0.8±0.2 kGy/h, the Pt potential becomes stable already after 2h in presence of a copper electrode whereas it is still evolving after 17 h in presence of a carbon steel electrode. Calculations in reference solutions shows that, for gamma irradiation, dose rates affect the time needed to reach the steady state but hardly affect the steady concentrations of hydrogen peroxide. This quantity is more dependent on the production of primary radiolytic species.

Platinum potentials are about 0.2–0.3 V higher under proton irradiation than under gamma irradiation. Proton irradiation creates much stronger oxidising conditions in sodium sulfate solution in the vicinity of the Pt electrode — where they stopped at a depth of about 300 μm — than gamma irradiation in clay solution. Further work is under progress to demonstrate that this property is due to the differences in the primary radiolytic yields due to proton and gamma rather than to difference in the chemical composition of solutions and received dose rates. The dose rates for proton irradiation are about two orders of magnitude higher than for gamma irradiation and the scale for reaching the steady-state is within minutes instead of a few hours.

References

1. A. O. Allen, *The Radiation Chemistry of Water and Aqueous Solutions,*1961, New York.

2. Z. D. Draganic and I. G. Draganic, *The Radiation Chemistry of Water*, 1971, New York.

3. K. Pikaev, *Pulse Radiolysis of Water and Aqueous Solutions*. Indiania University Press, Bloomington & London.

4. D. W. Shoesmith and F. King, Atomic Energy of Canada Limited Report, AECL-11999 (1999).

5. V. Byalobzheskii, Jerusalem: Israel Program for Scientific Translations Ltd. 1970 188.

6. J. J. Stobbs and A. J. Swallow, *Metall. Rev.*, 1962, **7**, 95–131.

7. R. S. Glass, Sandia National Laboratory Report,1981, SAND-81-1677.

8. V. Barou and Th. Wolkenstein, Effet de l'Irradiation sur les Propriétés des Semi-Conducteurs 1982. Editions MIR de Moscou.

9. W. E. Clark, *J. Electrochem. Soc.*, 1958, **105** (8), 483–485.

10. F. S. Feates, *Trans. Faraday Soc.*, 1960, **56**, 1671–1679.

11. P. L. Airey,. *J. Chem. Soc., Faraday Trans.*, 1972, **68** (Pt. 7), 1299–1311.

12. P. L. Airey, *Aust. J. Chem.*, 1973, **26** (7), 1443–52.

13. S. Uchida, E. Ibe and R. Katsura, *Radiat. Phys. Chem.*, 1983, **22** (3), 515–526.

14. R. S. Glass *et al.*, *Corros. Sci.*, 1986, **26** (8), 577–590.

15. G. P. Marsh *et al.*, *Corros. Sci.*, 1986, **26** (11), 971–982.

16. Y. J. Kim and R. A. Oriani,. *Corrosion*, 1987, **43** (2), 85–91.

17. Y. J. Kim and R. A. Oriani, *Corrosion*, 1987, **43** (2), 92–97.

18. S. Sunder, *et al.*, *Mater. Res. Soc. Symp. Proc.*, 1989, 127 (Sci. Basis Nucl. Waste Manage. 12), 317–24.

19. N. Saito, *et al.*, *Corrosion*, 1990. **46** (7), 531–536.

20. S. Sunder *et al.*, *Res. Soc. Symp. Proc.*, 1990, 176 (Sci. Basis Nucl. Waste Manage. 13), 457–64.

21. D. F. Taylor, *Corrosion*, 1991, **47** (2), 115–122.

22. K. Osada and S. Muraoka, *Mater. Res. Soc. Symp. Proc.*, 1993, 294(Scientific Basis for Nuclear Waste Management XVI), 317–22.

23. M. J. Danielson, *Corrosion*, 1995, **51** (6),450–55.

24. K. Nakata, *et al.*, *Nucl. Eng. Des.*, 1994, **153** (1), 35–41.

25. W. E. Wyllie, D. J. Duquette and D. Steiner, Rensselaer Polytechnic Institute, New York, 1994, 1–85.

26. T. Nagai, *et al.*, *J. Nucl. Sci. Technol.*, 1998, **35** (7), 502–507.

27. M. G. Bailey, L. H. Johnson and D. W. Shoesmith, *Corros. Sci.*, 1985, **25** (4), 233–8.

28. S. Sunder, *et al.*, *Proc. Int. Conf. Radioact. Waste Manage.*, 2nd, 1986, 625–31.

29. D. Gorse, B. Rondot and M. Da Cunha Belo, *Corros. Sci.*, 1990, **30** (1), 23–36.

30. S. Sunder, D. W. Shoesmith and N. H. Miller, *J. Nucl. Mater.*, 1997, **244** (1), 66–74.

31. J. C. Rouchaud, *et al.*, *J. Chim. Phys.*, 1997, **94**, 428–43.

32. D. J. Duquette, *et al.*, *J. Nucl. Mater.*, 1992, **191–194** (Pt. B): 992–6.

33. P. L. Airey, *J. Chem. Soc., Faraday Trans.*, 1, 1972, **68** (Pt. 7), 1299–311.

34. V. Trupin-Wasselin, 2000 Ph.D, University Paris XI, Orsay.

35. J. F. Ziegler, J. P. Biersack and U. *Littmark, The Stopping and Range of Ions in Solids*, Pergamon Press, New York, 1985.

36. T. E. Eriksen, P. Ndalamba, H. Christensen and E. Bjerbakke, *J. Radioanal. and Nucl. Chem.*, 1989, **132**, 19–35.

37. M. Fattahi, A. Eglem, C. Houée-Levin, C. Ferradini and P. Jacquier, *J. Chim. Phys.*, 1993, **90**, 767–775

38. M. Fattahi, C. Houée-Levin, C. Ferradini and P. Jacquier, *Radiat. Phys. Chem.*, 1992, **40**, 167–173.

39. C. Corbel, D. Féron, F. Maurel, M. Roy and V. Wasselin-Trupin, CEA Internal Report CEA/DSM/DECAM/SCM 12; CEA/DEN/DPC-SCCM/RT-SSSME 578.
40. A. B. Ross, *et al.*, NDRL-NIST Solution Kinetics Database: version 2.0, NIST, 1994, Gaithersburg.
41. P. Kirkegaard and E. Bjerbakke, RISO-R-395, 1999, Roskilde.
42. W. C. Schumb, C.N. Satterfield and R. L. Wentworth, *ACS.* Reinhold, New York, 1955, ch. 8, pp.447–514.

33

An Approach Towards Assessing the Effects of Microbially-Influenced Corrosion in Nuclear Waste Systems

A. M. PRITCHARD

Corrosion & Fouling Consultancy, 33 Laburnum Road, Oxford OX2 9EL, UK

ABSTRACT

Microorganisms are inevitably present in nuclear waste systems. There are no models to predict their influence on corrosion rates, since the mechanisms are poorly understood. A database of MIC corrosion rates is being assembled to allow some prediction of such rates, and thus the associated risk, for defined situations. No long-term predictions are possible, since such data are lacking, and most published short-term data lack important detail. However, both qualitative and quantitative data indicate conditions that are often associated with MIC, and which should therefore be avoided by design, operation and management of waste repositories.

1. Introduction

Containment of nuclear waste, whether of spent, unreprocessed fuel, or high-, intermediate- or low-level waste is a vital concern for the nuclear industry. The volumes of spent fuel and high-level waste are relatively small, but much larger amounts of intermediate- and low-level wastes are produced. Mechanical considerations mean that most containment systems rely on the use of metal containers, and the long-term risk of their failure is most likely to be connected with corrosion.

The choice of metals or alloys for these containers is therefore closely related to their corrosion resistance. The current concept of long-term disposal of waste in repositories, without retrieval, means that the waste needs to be contained for periods of hundreds of years or longer, until the radioactivity has decayed to a negligible level. The primary containment provided by a metal receptacle plays a very important role in this, especially during the first period when the level of radioactivity is at its highest, and a failure of containment would have the most serious consequences.

Selection of container materials and environments needs to be based on known corrosion behaviour. However, corrosion data normally relate to relatively short periods under controlled laboratory conditions. Fewer data are available for longer periods; these often refer to industrial or environmental exposures under variable conditions which are not well documented. There is therefore a need to make considerable extrapolations from data gathered over a relatively short time period. For such extrapolations to be valid, the corrosion mechanism needs to be well established. Furthermore, it is essential that the mechanism does not change with time of exposure. This cannot always be guaranteed from short-term tests, as was found for the corrosion of carbon steels and some low-alloy steels used in carbon-dioxide cooled reactors in the late 1960's. Short term tests had shown protective corrosion film formation, but at longer exposures breakaway corrosion set in, with a much higher rate, due to carbon deposition in the originally protective corrosion product layer [1]. The problem had to be controlled by reducing operating temperatures and improving control of moisture content [2].

The relatively low exposure temperatures for waste containers, as nuclear decay heat declines, mean that they may be exposed to liquid water, and thus to corrosion mechanisms that require the presence of an electrolyte. Such conditions may favour localised corrosion through the setting up of corrosion cells, causing leakage after a relatively small amount of corrosion rather than having to wait for more extensive general corrosion to cause perforation or mechanical failure of the container.

One of the corrosion mechanisms that could give rise to such localised failure is microbially influenced corrosion (MIC). Although recognised by von Wolzogen Kuhr and van der Vlugt in 1934 [3], its importance has only recently been recognised in a range of situations, including corrosion of service water systems in the nuclear industry [4,5]. Appreciation of the phenomenon requires some familiarity with materials science, engineering and microbiology.

2. Microbially Influenced Corrosion (MIC)

2.1. A Definition of MIC

MIC has been defined recently by a European network (BRITE-EURAM Network on Microbiologically Influenced Corrosion of Industrial Materials BRRT-CT98-5084) as

> "the influence of microorganisms on the kinetics of corrosion processes of metals, mineral and synthetic materials, caused by microorganisms adhering to the corroding interface (usually called 'biofilms'). A prerequisite for MIC is the presence of microorganisms. If MIC is based on their activity, (i) water, (ii) energy source, (iii) carbon source, (iv) electron donator, and (v) electron acceptor are required".

The essential feature of MIC that differentiates it from other types of corrosion is that some microorganisms are able to generate changes in the chemical nature of their environment, so that its corrosivity changes, either locally, or more generally if the products of microbial activity are released to the environment. Since different microorganisms have different metabolic requirements, and produce different metabolites, MIC refers to a group of phenomena rather than to a single one.

2.2. Basic Requirements for MIC

For MIC to be possible, three basic requirements have to be fulfilled: there must be a source of microorganisms, nutrients for their growth and metabolism, and the presence of liquid water.

The first of these is almost always fulfilled in the natural environment. Bacteria have been found to exist and reproduce in virtually every natural environment, however extreme [6]. Bacteria are also extremely resistant to nuclear radiation, as shown by their development and reproduction in water in the damaged reactor at Three-Mile Island [7]. The presence of microorganisms in nuclear waste repositories can thus be taken for granted. Additional species may be introduced through the human interventions to build the repository, as well as in the waste itself.

Microorganisms need a range of nutrients, in particular a source of carbon, which can be carbon dioxide for some species. Smaller amounts of other elements such as nitrogen, phosphorus and sulfur are required. In the natural environment nutrient levels may be low, or poorly balanced, so that the microorganisms are in a stressed state. In flowing systems such as rivers the concentrations of nutrients required may be very low, less than one part in 10^6 [8], the necessary flux being provided by turbulent mass transport from the flowing water. If nutrients are not available at a particular time it is possible for some bacteria to change their physical form and remain dormant until nutrients again become available.

In laboratory studies balanced growth media are often used. These favour the development of particular species, whose behaviour under these conditions may be very different from that in the natural environment, so that where more than one species is involved, the ratios of numbers are likely to change. For some known bacteria there are as yet no established growth media, and for others their complete nutritional requirements are not known.

Predicting relationships between the concentrations of nutrients present and the chemical nature of environments produced by the metabolism of bacteria is made very difficult by the lack of established quantitative relations between the amount of corrosive species produced and the flux of nutrients. However, for one organism, the SRB *Desulfovibrio desulfuricans*, Lee and Characklis [9] were able to show that the consumption of its nutrient lactate, when inoculated into a mixed population of SRB, was related to the amount of sulfide produced by the equation:

$$2\ CH_3CHOHCOO^- + SO_4^{2-} = 2\ CH_3COO^- + 2\ CO_2 + S^{2-} + 2\ H_2O$$

In a typical industrial situation, not only are the metabolic reactions of many of the bacteria present unknown, but the bacteria themselves may also be uncharacterised.

The presence of liquid water is not always guaranteed. However, in subsurface repositories some water is always likely to be present. Although decay heat from fission products may maintain the surface of a waste container dry for an initial period, in due course it will be inadequate to prevent the formation of a water film on its surface. The use of dry fuel stores should be an effective way of preventing MIC, provided that the ingress of moisture in any form can be controlled.

2.3. Microorganisms Involved in MIC

2.3.1. General

The three main classes of microorganisms involved in MIC are fungi, algae and bacteria. Fungi may metabolise cellulose in waste to produce organic acids. Algae have the important property of being able to convert carbon dioxide into organic compounds, but require light to do this. The main risk to corrosion usually comes from bacteria. Four main groups are involved in MIC, as described below.

2.3.2. *Slime-forming bacteria, e.g.* Pseudomonas aeruginosa

These are mainly aerobes, and are one of the first types to colonise a surface. They require a plentiful supply of metabolisable carbon to produce extracellular polymers that are an important part of biofilms, as described below. The carbon source may be produced by other organisms that can fix carbon dioxide, if other suitable sources are not available. Their chief role in MIC is the formation of a gelatinous film that restricts the diffusion of oxygen to the surface of a metal, producing attractive environments for other bacteria.

2.3.3. *Sulfate-reducing bacteria (SRB), e.g.*Desulfovibrio vulgaris

These bacteria require anaerobic conditions to metabolise and reproduce, although they can survive in oxygenated conditions. They use sulfate, thiosulfate or nitrate as an electron acceptor, and produce hydrogen sulfide. Different species can use different organic acids as nutrients, and lithotrophic varieties have been described that can obtain carbon from carbon dioxide. They are widely associated with the formation of localised corrosion cells. The sulfide produced reacts with ferrous ions produced by corrosion to form iron sulfide, often in a finely divided form, that can act as a very efficient cathode to drive the anodic pitting reaction. The presence of dissolved sulfate in many groundwaters, and the much larger concentrations in seawater means that they are the type of bacteria most frequently involved in MIC. They are often associated with the pitting of stainless steel.

2.3.4. *Sulfur-oxidising bacteria, e.g.*Thiobacillus thiooxidans

These oxidise sulfides, elemental sulfur and other sulfur-containing compounds to sulfuric acid, and can survive at pH 2 or lower. In some circumstances they may form a consortium with SRB, oxidising the sulfide they produce. They are often associated with acid corrosion of concrete in sewers, where sulfide is produced from proteins by other microorganisms. Another potential source of sulfur is blast furnace slag used in concrete. Many thiobacilli are autotrophic and can fix carbon dioxide.

2.3.5. *Iron-oxidising bacteria, e.g.*Gallionella

These oxidise soluble iron (II) to insoluble iron (III) compounds, which may form a physical barrier to the diffusion of oxygen, producing an anaerobic environment in which SRB can grow. Oxidation of Fe(II) to Fe(III) requires an ingress of anions to maintain electrical neutrality, providing a mechanism for concentrating corrosive species such as chloride. Iron oxidisers are believed to play an important role in initiating the corrosion of stainless steel at welds, where regions of higher availability of ferrous iron due to the formation of surface oxides or local reductions in the concentrations of protective alloying elements during the welding process provide favourable sites for settlement.

2.3.6. *Consortia*

It is rare for only one type of microorganism to be found on its own. As illustrated above, conditions created by the respiration and growth of one organism may provide a habitat or nutrients for another. Sulfate-reducers may be accompanied by sulfur oxidisers. Aerobic and anaerobic bacteria are often found together; where there is illumination algae may fix carbon from carbon dioxide and produce organic nutrients for bacteria. Although relatively little work has been carried out under controlled conditions with defined consortia, there is evidence that MIC rates are higher with consortia than with one species alone [10].

2.4. Microorganisms and Biofilms

A very important property of many bacteria is their ability to produce biofilms. These occur very widely in the environment, wherever conditions are favourable, and are responsible, for example, for the slimy feel of surfaces beneath the water in rivers and the sea. A number of processes are involved in the formation of a biofilm [11]. The first involves transport of biopolymers such as glycoproteins to a surface. These change its chemical and electrical properties. They may also be nutrients for bacteria. Individual bacteria then are transported to the surface. Initially held there by weak physical forces, they then develop stronger bonds. Nutrients diffuse to the surface and the bacteria respire and multiply. They also generate extracellular polymers. These produce a gel-like structure, within which gradients of chemical potential can develop, thus producing niche environments that may be favourable for other species. Changes from a fully oxygenated environment to a completely deoxygenated one have been shown, using microelectrodes, to take place over thicknesses of a few tens of micrometres [12].

As well as providing niche environments for different microbial species, biofilms may also help to deactivate biocides, by chemical or physical adsorption, or reducing the rate at which they diffuse to the organisms. Their gelatinous nature means that they capture particulate material, which can provide extra mechanical stability, and possibly nutrients as well. Recent research has shown that the structure of biofilms is very complex biologically, physically and chemically, making it very difficult to develop a realistic model [13]. In particular there is currently no detailed understanding of the relationship between the structure of a biofilm and the electrochemical interactions that may take place between different regions within it and of the metal surfaces to which it is attached. While such interactions might reasonably be expected to be the precursor of localised corrosion, MIC is notable by its absence on most metal surfaces covered by biofilms, suggesting that in general biofouling and biofilms do not enhance corrosion rates, and may provide some protection [14,15].

3. Risk Analysis

3.1. Qualitative Assessments from Field Experience

Examination of the literature on MIC in industrial environments shows that it is particularly common in a relatively small number of situations [16]. Identifying common features of these situations, and then acting to avoid their development is likely to reduce the risk of MIC occurring. Further guidance may be obtained from those situations where the highest corrosion rates are observed.

One of the commonest reasons for high localised corrosion rates is the presence of a small anodic area close to a much larger cathodic area. Such a situation could arise where coverage of a surface by a biofilm is incomplete; where coverage is complete severe MIC is not observed, despite the likely development of anodic and cathodic areas beneath the biofilm. High metabolic activity by the microorganisms, which in turn requires a plentiful supply of nutrients, facilitates the cathodic corrosion reaction, thus driving the anodic reaction. Failure of many laboratory experiments to reproduce the high corrosion rates sometimes observed in the field may be due to the ample supply of nutrients often provided leading to rapid coverage of surfaces by the biofilm, thus destroying the imbalance between anodic and cathodic areas that gives rapid localised corrosion. The possibility that the unnaturally high concentrations of added nutrients may inhibit corrosion by a non-biological mechanism also needs to be considered.

3.1.1. Metals buried in or in contact with soils

Soils contain a wide range of microorganisms, and often of nutrients as well. Groundwater often contains sulfates, allowing corrosion by SRB to occur where there is no corrosion protection and poor drainage. This is of particular concern for pipelines, which are usually made of carbon steel, and protected by coatings and cathodic protection systems. Jack [17] has shown that the highest rates of corrosion occur where coatings are damaged, or have lost their adhesion, and the uncovered metal surface of the pipe is subjected to alternate anaerobic and aerobic conditions. Under anaerobic conditions significant volumes of iron sulfide corrosion product can be produced from hydrogen sulfide generated by SRB. On changing to aerobic conditions this iron sulfide provides an efficient, large-area cathode for the reduction of oxygen, which drives the corrosion reaction. The oxygen may also oxidise sulfide to elemental sulfur, which is extremely corrosive. Corrosion rates were found to increase from 0.11 mm/year under anaerobic conditions to >1 mm/year on admission of air [17].

3.1.2. After pressure testing plant

Pressure tests of plant are too often carried out using water of low quality. If this water is allowed to remain in the system, microorganisms and nutritious impurities in it may form biofilms over part of the surface, setting up an electrochemical corrosion cell. Localised corrosion can then occur, possibly leading to perforation of the wall. This appears to be particularly common in stainless steel piping systems, with perforation often occurring at or near welds, where the corrosion resistance is locally smaller. The situation can be avoided if the water is treated with a biocide beforehand, and the system is then dried if it is not to be put into service at once. The development of stagnant conditions can also be avoided by careful design and construction of the system, and continuously circulating the fluid. A rather similar situation could arise where changes in operating conditions, such an increase in flow velocity and thus surface shear stress, remove parts of a biofilm that is otherwise having little effect on corrosion, or even providing some protection. The exposed part of the surface can then become anodic to the main area still covered by the biofilm, producing an electrochemical corrosion cell.

3.1.3. Metals in contact with sea water

Sea water contains around 2700 mg L^{-1} of sulfate, and a rich variety of nutrients at concentrations that increase near land and estuaries. This provides ideal conditions for the development of SRB once anaerobic conditions have been established. Stagnant conditions are therefore to be avoided, unless microbial activity can be controlled by the use of biocides.

3.1.4. Metals in contact with water and fuels or lubricants

In these situations the water provides the electrolyte for the electrochemical cell, and for the microorganisms. Many fuels and lubricants contain a rich source of carbon compounds that can be metabolised by microorganisms. In addition to these, chemical compounds containing sulfur, phosphorus and sometimes nitrogen are added to improve the performance of lubricants, in particular. Complete elimination of water or the addition of an appropriate biocide is essential to prevent rapid growth of microorganisms. Corrosion rates equivalent to 25 mm/year have been reported for ships' plate exposed to bilge water containing oils and sea water [18], where the motion of the ship probably plays an important part in avoiding local stagnation.

3.2. A More Quantitative Approach to Risk Assessment

A more formally based approach to risk assessment can be made by systematising features associated with reported cases of MIC, quantifying them wherever possible, and then examining the statistical relationships between those features and the incidence of MIC. The risk of MIC in a new situation can then be estimated using the established relationships and the numerical values of the features chosen in the new situation. The extent of linkages between these features or Aspects and the extent and type of corrosion can be used to reduce the number of Aspects for which data should ideally be provided.

3.2.1. Aspects and influences

The risk in a given situation can be analysed by identifying Aspects of that situation that affect the risk, and the individual Influences of those Aspects. The total Risk can then be calculated as

$$\text{Risk} = \Sigma \, (\text{Aspect} \times \text{Influence})$$

The individual Aspects and their Influences have to be identified using a body of established data. It is important that the individual Aspects should be independent of each other. Once the Aspects have been identified and the Influences evaluated from these data, values of the chosen Aspects for a new situation can be combined with the Influences measured previously to give an assessment of the Risk for this new situation.

This approach can be described as phenomenological, rather than fundamental. This means that the Aspects chosen can be taken from the data that are available, rather than the more fundamental parameters that would be more desirable scientifically. In view of the relatively few details available in most documents reporting measured corrosion rates attributed to MIC, this is a very useful property.

It is of course desirable to choose, if possible, Aspects that show a high correlation with the Risk to be assessed. The correlations can be established statistically through processing the body of established data.

In setting up a Risk Assessment System for MIC, a BRITE-EURAM Network including both academic and industrial organisations decided to divide the Aspects into three main groups: General, Microbiological, and Metal-related. A large number of features associated with each Aspect were identified as relevant or characteristic of some or all types of microbially assisted corrosion.

An Excel spreadsheet was set up to record these features from data and literature references already available. The choice of the spreadsheet had a number of advantages:

(a) Sets of data could be chosen from within the spreadsheet with characteristics closely matching the situation for which a risk assessment was required.

(b) Statistical analysis could be carried out on the chosen data to provide a quantitative evaluation of the links between different items within the data, thus helping to produce numerical values for the Influences of the chosen Aspects.

(c) Up to 32 fields could be chosen for statistical analysis. Although this was less than the number of items originally requested, it still allowed inclusion of a considerable amount of detail, in fields indicated below alphabetically from A to AE. The performance requirements for stainless steels and carbon and low-alloy steels are different, and this led to the decision to set up separate databases for the two groups of materials. For the former, any risk of pitting from MIC is sufficient for the material to be rejected, whereas for the latter group it is the rates of pitting or general corrosion that are likely to be most important. A number of the fields differ in the two databases; those outlined below refer to carbon and low-alloy steels.

(d) The Data⎮Form facility in Excel allows data to be entered in a convenient form format.

(e) Additional data can be added as they become available.

(f) Data could conveniently be exchanged by e-mail, allowing checking of data classification by different contributors to the project.

Data is only accepted for entry if quantitative or semi-quantitative measures of corrosion are provided. This was found to exclude a large majority of MIC references, which were concerned mainly with mechanisms and gave no corrosion rate data. Case histories, such as where perforation of pipes after a given time of service is recorded, allow minimum rates of corrosion to be calculated if the wall thickness is known, and such data are included. Partly because of the provision of a large number of fields of data, a majority of fields contain no data for most entries. This lack of data probably also reflects a relatively low level of understanding of MIC. The development of this system is continuing.

3.2.2. General aspects

Each entry is provided with a six-character reference: the first two numbers are the last two figures of the date of publication, followed by the first three letters of the surname of the first author, and a small-case alphabetical letter to distinguish individual entries (field A: e.g.86JACa). Three fields categorise the material involved (field B: e.g. carbon steel), the type of facility (field C: e.g. heat exchanger), and the general type of aqueous environment (field D: e.g. real sea water). These three fields would principally be used to select the most suitable data entries to use as reference data for the risk assessment. The list of general entries is completed by the temperature (field E: in degrees C) and waterflow (field F: laminar, turbulent, stagnant, intermittent).

3.2.3. Microbiological aspects

The first two fields concern data on the existence and types (field G: Yes/No/type e.g.SO4) and concentrations (field H: in mg/litre) of nutrients reported. Usually the most significant nutrient is sulfate; in laboratory conditions a range of nutrients may be provided, as for example in Postgate's media for SRB growth. A further field gives the reported details of a biofilm and its thickness (field I: Y/N/details, which may be as dry or wet weight per square centimetre, or as a thickness).

Data about microorganisms are recorded in fields J through R. These comprise the provision of data and whether they concern sessile or planktonic numbers, or both (field J: Y/N, se/pl), the species (field K: specific or general – aerobic or anaerobic – types), whether the species were identified (field L: Y/N), the numbers (field M: cfu/ml), the technique used for enumeration (field N: e.g.MPN), types of inorganic metabolite (field O: e.g.sulfide, FeS), their concentration (field P: mg/L), types of organic metabolite (field Q: includes pH) and concentration (field R: mg/L or pH value).

Aspects of microbiological control are recorded in fields S through W. The type of control is recorded (field S: Y/N/Cl(for chlorine)/pigging/mechanical), the biocide, if used (field T: Cl2/non-oxidising/specific type), and details of the treatment: period (field U: hours), interval between periods (field V: hours), and effect on corrosion (field W: Y/N/other).

3.2.4. Corrosion aspects

The corrosion data provided fill the remaining fields: the geometry (field X: inside tube/crevice/plate/buried), details of the surface preparation (field Y: includes welds), the corrosion type (field Z: crevice corrosion/pitting/ general corrosion/ corrosion product clogging), whether micrographs are provided (field AA: Y/N, for reference), corrosion product details (field AB: colour, compounds identified), the corrosion rate (field AC: in mm/year), type (field AD: mean/pitting) and method of measurement (field AE: weight loss/corrosion current/linear polarisation resistance/ leak/rust streak/clogging).

4. Application to Nuclear Waste Systems

4.1. General

In applying the risk assessment system to MIC of nuclear waste the main problems are the lack of MIC data for such systems and the need to extrapolate any results by several orders of magnitude. The latter is, of course, common to all types of corrosion. The involvement of microorganisms in MIC means that additional data are needed beyond those already required for prediction of corrosion rates in purely abiotic systems.

4.2. Dry Spent Fuel Storage Systems

One of the main advantages of dry spent fuel storage systems is that, in principle, the spent fuel and containers should remain dry. Radioactive material should therefore not be released into the store by electrochemical corrosion processes, though there may be a possibility of dry radioactive corrosion products or deposited material becoming detached from the surface of the fuel. Spent fuel is likely to have been held adjacent to the reactor for a period to allow initial decay of fission products to occur, and this may lead to a decay heat flux that is not high enough to ensure evaporation of any moisture that may be introduced into the dry store.

Decay heat is removed from the dry spent fuel store systems by air flowing through the store, preferably by natural circulation. The airflow system has to be designed to reduce the risk of moisture ingress. Larger drops of rain, spray and snow can be intercepted by screens before they reach the interior of the store, but not those in mist or fog. Such screens are likely to gather material of biological origin as well as water, so that MIC is possible. There are few MIC data for the alternating wet and dry periods that would characterise this sort of environment, and most atmospheric corrosion data take little account of the effects of biological material. The practical solution would appear to be to ensure that the design of the screens allows easy inspection and replacement as necessary. Irrespective of the provision of screens, changes in air temperature and humidity with the time of day and the season could lead to condensation on cold surfaces. If the decay heat can maintain the fuel container surfaces free of moisture, at least during the initial period of holding, MIC should not be possible.

A number of general design features of the building need to be considered. Clearly roofs and other coverings must be adequate to divert the severest rainfall or snowfall from the interior. Since the building is designed for access for spent fuel transfer, the consequences of spillages inside it of cooling water from transfer flasks have to be considered. Any radioactive water from such spillages and their removal will need to be held on site in suitable vessels, for which the design and choice of materials need to take into account the possibility of MIC occurring in pipework and the storage vessels themselves. Data from the database relating to service water systems with intermittent flow should give guidance on the risks of MIC associated with the use of different materials. Similar considerations will apply for the water services installed and used during human access in the normal operation of the store: the choice of materials and design of the plumbing needs to ensure that waste water and water from unexpected leaks are contained and drained away, without at the same time introducing a risk of flooding through the drainage system from external sources. It may be desirable to use non-metallic materials in the drainage system to avoid MIC.

Apart from the introduction of water in small drops, fog particles may contain pollutants such as sulfur and nitrogen oxides and solid particles that can serve as nutrients for bacteria. Bacteria and other microorganisms and their spores are also too small to be filtered out, so that an inoculum is always likely to be present for any liquid water that forms. Once a biofilm has formed, it provides a means of retaining water locally at relative humidities at which complete evaporation would normally take place.

The environment in a dry store is also likely to be considered favourable by a variety of animals and insects, which could introduce a variety of waste and other materials nutritious for microorganisms. Small mammals may be able to enter through a drainage system, as well as through other openings below, at, or above ground level, unless they are permanently screened. Birds and larger insects have to be kept out by similar screens on the air intake and exhaust system. However, the entry of some insects, such as spiders, is probably unavoidable. These screens need to be strong enough to resist attack by birds and animals.

4.3. Wet Spent Fuel Storage Systems

The water chemistry of spent fuel ponds is normally well controlled. Data from pond water analyses could be compared with those in the database to give a measure of the risk from MIC in that environment, and to choose appropriate treatment to avoid or minimise it. Although radiation levels are high, they are not usually high enough to prevent microbial development under the right conditions. Levels of hydrogen peroxide generated by radiolysis are also likely to be inadequate to affect microbial survival and growth.

4.4. Repositories

Repositories are designed for long-term storage (typically >100 years), and are usually sited to be above the prevailing water table. However, the inleakage of groundwater is expected. Numerous investigations have shown bacteria to be present in the virgin rock from which most repositories are excavated, and the process of construction inevitably introduces more. The survival and growth of microorganisms involved in MIC is therefore likely to depend principally on the nutrients available.

When considering MIC of the primary containment, three main sources of nutrients need to be considered. The first of these are sources contained within intermediate or low-level waste, such as cellulose in paper, ion-exchange resins and some oil-derived wastes. Apart from the carbon sources, sulfate and amine-related compounds from resins could provide substantial sources of sulfur and nitrogen. Depending on the temperature within the primary containment, and the amount of water available, microbial growth could be rapid. The use of cementitious material to fill waste drums is designed to provide a protective internal environment for the interior, since pore water in cements and concrete is at pH 12. This pH value is inimical to the development of most microorganisms, but it has been shown that some Thiobacilli are able to survive this initial pH and cause release of radionuclides from cementitious wasteforms [19] by acidifying their immediate environment.

Prediction of MIC rates from these sources is difficult, since a range of conditions can be envisaged. If such conditions can be specified, it may be possible to relate them to conditions for which data are already available in the database. An important feature of the conditions is that they are stagnant, so that mass transfer of nutrients, corrodents and corrosion products to the container surface will be diffusion-limited. The total amount of nutrients within the container will also be limited, curtailing growth and respiration unless a source of energy is present, such as from the corrosion of the container material, or possibly of other waste metals within the container. Once the energy source has been used up, it would seem reasonable to expect that no further MIC will take place, provided that there are no changes in the local conditions. It is assumed that radiolytic decomposition of the waste is insignificant and unable to provide a continuing source of nutrients. The use of relatively short-term corrosion data from the database, provided these are relevant to the local conditions expected, should give some useful idea of the risk during the relatively short timescale involved.

The second source of nutrients is from the groundwater in the rock. The existence of microorganisms in the rock shows that such water contains a residual level of nutrients. Since permeation of such water will continue, this represents a potential continuous source of nutrients. Changes in its composition, either reflecting seasonal and other changes in the source, or possibly local changes arising from radiolytic processes, could provide an energy source for bacteria. Comparison of water chemistries with those in the MIC rate database could possibly provide some idea of likely corrosion rates, though these would be expected to be small. Changes may also occur in the numbers and types of microorganisms in the water. These could lead to breakdown and regrowth of biofilms, setting up new electrochemical cells. No MIC data have been found related to such changes.

The main concern is most likely to be from nutrient materials introduced into the repository during, or as a result of, its construction. Substantial quantities of oil-based products and some of their oxidation products are likely to be introduced during the tunnelling and construction period, thus providing sources of hydrocarbons, carbon dioxide, nitrogen and sulfur oxides. The need for illumination during construction is likely to encourage the growth of algae, fixing some of the carbon dioxide and providing nutrients for bacteria. The need for construction workers to operate under reasonable conditions will ensure that temperatures are kept within the range preferred by many bacteria. Inevitably some waste carbon-based materials, whether of biological or non-biological origin, will also be introduced during the construction phase. This may continue, probably to a lesser extent, while the repository is filled. If the environment can be defined, the MIC corrosion rate database could be used to provide a measure of the likely corrosion rate.

5. Conclusions

Corrosion of nuclear waste materials may be affected by the activities of microorganisms, principally bacteria. Microbially Influenced Corrosion requires the presence of liquid water, suitable nutrients, temperatures, and a source of the microorganisms. All of these are likely to be present in spent fuel ponds and repositories, while in a dry store MIC should be avoided if the ingress and retention of liquid water can be prevented.

Although some details of the mechanisms by which some bacteria influence corrosion are known, no quantitative models are available to predict corrosion rates, and many bacteria encountered in nuclear waste systems are poorly characterised.

A database system is described in which reported corrosion rates and behaviour are linked to general, microbiological, chemical and metallurgical aspects of case histories and laboratory data. Provided that the environments likely to arise in nuclear waste repositories and fuel ponds can be defined, similar case histories can be selected from the database, and the statistical relationships between the conditions and corrosion rates used to suggest possible corrosion rates in the chosen environments.

The results of such calculations should provide guidance on corrosion rates due to MIC in the short term, and highlight engineering, sanitary and management measures required to avoid the provision of nutrients and changes in nutrient levels that are likely to lead to high MIC rates.

The absence of long-term MIC corrosion rate data means that it is very difficult to make predictions for the long periods required for nuclear repositories. A better understanding of the conditions under which a biofilm produces MIC, rather than having the protective behaviour widely observed in many service environments, would increase confidence in making corrosion rate predictions. Such work would need to include electrochemical measurements of the types normally associated with corrosion studies, particularly of pitting corrosion. Three areas where there is a need for an improved understanding of the electrochemical processes beneath biofilms, and which could be involved in the initiation and maintenance of corrosion are suggested:

(1) the effects of local changes in biofilm coverage of a surface as a result of growth, removal by mechanical action, or intermittent dryout,

(2) changes in biofilm activity associated with changes in nutrient levels, and

(3) changes in the numbers and types of microorganisms in a biofilm as a result of ecological changes.

References

1. A. M. Pritchard *et al.*, *Oxid. Met.*, 1975, **9**(2), 181–214.
2. D. Goodison, The Oxidation of Steels in Magnox and AGR Stations. Inspection and Licensing Aspects, in Corrosion of Steels in CO_2, *Proc. British Nuclear Energy Soc. Int. Conf.*, Reading University, 23–24 September 1974, pp.130–140.
3. G. A. H. von Wolzogen Kuhr and van der Vlugt, *Water* (den Haag), 1934, **18**,147–165.
4. G. J. Licina, Sourcebook for Microbiologically Influenced Corrosion in Nuclear Power Plants, EPRI, Palo Alto, CA, USA, 1988.
5. G. J. Licina, Detection and Control of Microbiologically Influenced Corrosion, EPRI, Palo Alto, CA USA, 1990.
6. A. Rosevear, Review of National Research Programmes on the Microbiology of Radioactive Waste Disposal. Report NSS/R263, 1991, United Kingdom Atomic Energy Authority, Harwell, UK.
7. J. H. Wolfram and W. J. Dirk, Biofilm Development and the Survival of Microorganisms in Water Systems of Nuclear Reactors and Spent Fuel Ponds, in *Microbial Degradation Processes in Radioactive Waste Repository and in Nunclear Fuel Storage Areas*, pp.139–147, (J.H. Wolfram, R.D. Rogers and L.G. Gazso, eds). Kluwer Academic Publishers, Dordrecht, The Netherlands, 1997.

8. J. E. Duddridge and A. M. Pritchard, unpublished work.

9. W. Lee and W. G. Characklis, *Corrosion*, 1993, **49** (3), 186–199.

10. R. F. Jack, D. B. Ringelberg and D. C. White, *Corros. Sci.*, 1992, **33**(12), 1843–1853.

11. R. E. Baier, Early events of micro-biofouling of all heat transfer equipment, in *Fouling of Heat Transfer Equipment*, (E.F.C. Somerscales and J.G. Knudsen, eds). Hemisphere Publishing, Washington, D.C., USA, 1981, pp.293–304.

12. Z. Lewandowski *et al.*, *Corrosion*, 1989, **45**, 92–98.

13. P. Stoodley, D. de Beer and Z. Lewandowski, *Appl. Environ. Microbiol.*, 1994, **60**(8), 2711.

14. J. Morley, *Civ. Eng.*, 1984, **54**(3), 14.

15. A. M. Pritchard, "Biofilms: Beneficent or Corrosive", Keynote lecture, COST 520 project meeting, Sion, Switzerland, 2–5 June, 1999.

16. A. M. Pritchard, 'A simple guide to microbially influenced corrosion', AEA Technology, March 1995.

17. T. R. Jack *et al.*, *Corrosion*, 1998, **54**(3), 246–252.

18. E. C. Hill and G. C. Hill, *Trans. Inst. Mar. Eng.*, 1993, **105**(4), 175–182.

19. R. D. Rodgers *et al.*, Microbial Degradation of Low-Level Radioactive Waste, Annual Report for FY 1994, NUREG/CR-6188, Vol.2, U.S. Nuclear Regulatory Commission, Washington, D.C., USA.

List of Abbreviations

The following abbreviations occur in the text and in the Index of contents.

ADM	Accumulated Damage Model		CORALUS	Corrosion of Active Glass in Underground Storage Conditions
ANDRA	Agence Nationale pour la Gestion des Déchets Radioactifs (French National Radioactive Waste Management Agency)		CT	Compact Tension
AFM	Atomic Force Microscopy		DCB	Double Cantilever Beam
			DFA	Damage Function Analysis
CDF	Cumulative Distribution Function		DOE	Department of Energy (US)
			DS	Drip Shield
CEA	Commissariat á l'Energie Atomique (French Atomic Energy Commission)		DVGW GW9	German Gas and Water Works Engineers Association Standard
CERBERUS	Control Experiment with Radiation in the Belgian Repository for Underground Storage		EBS	Engineered Barrier System
CFR	Code of Federal Regulations (US)		EBSFAIL	a module of the (US) TPA code
CLAB	an intermediate storage facility (Sweden)		EBSPAC	Engineered Barrier System Performance Assessment Code
CLST	Container Life and Source Term		ECP	Electrochemical Corrosion Potential
CNRS	Centre National de la Recherche Scientifique (National Scientific Research Centre (France))		EDF	Electricité de France
			EIS	Electrochemical Impedance Spectroscopy
CNWRA	Center for Nuclear Waste Regulatory Analyses		FIE	Frumkin Institute of Electrochemistry
COCON	Corrosion of Container Programme		FFRDC	Federal Funded Research and Development Center
COGEMA	Compagnie Générale des Matières Nucléaires			

GCM	General Corrosion Model		PA	Performance Assessment
			PC	Pitting Corrosion
HFM	High Field Model		PDM	Point Defect Model
HLRW	High Level Radioactive Waste			
			RBS	Rutherford Back-Scattering Spectroscopy
HRW	Hard Rock Laboratory			
HTOM	High Temperature Operating Mode		RH	Relative Humidity
			SCC	Stress Corrosion Cracking
ILW	Intermediate Level Waste			
			SCK·CEN	Studiecentrum voor Kernenergie/Centre d'Etude Nucléaire (Belgium)
J-13	near-neutral water simulating that of the Yucca mountain depository			
JNC	Japan Nuclear Cycle Development Institute		SR	Site Recommendation
			SNF	Spent Nuclear Fuel
KTI	Key Technical Issue (US)		SS	Stainless Steel
LLNL	Lawrence Livermore National Laboratory		temp.	Temperature
			TPA	Total Performance Assessment
LTOM	Low Temperature Operating Mode		TSPA	Total System Performance Assessment
LLW	Low Level Waste			
LOT	Long Term Test of Buffer Material		UKAEA	United Kingdom Atomic Energy Authority
LTCTF	Long Term Corrosion Test Facility			
			WAPDEG	Waste Package Degradation
MIC	Microbially Influenced Corrosion			
			WP	Waste Package
MTR	Materials Test Reactor			
			XRD	X-ray diffraction
NAGRA	Nationale Genossenschaft für die Lagerung Radioactiver Abfälle (Switzerland)		XPS	X-ray photoelectron spectroscopy
NRC	Nuclear Regulatory Commission (US)			

Index

Abbreviations are to be found in the Abbreviations list on p.517